Inter-Union Commission on Geodynamics
Scientific Report No. 8

Approaches to Taphrogenesis

Proceedings of an International Rift Symposium
held in Karlsruhe April, 13—15, 1972.
sponsored by the Deutsche Forschungsgemeinschaft

Editors

J. H. ILLIES and K. FUCHS
Geologisches Institut   Geophysikalisches Institut
Universität Karlsruhe, Germany

with 224 figures and 13 tables in the text and on 12 folders

E. Schweizerbart'sche Verlagsbuchhandlung
(Nägele u. Obermiller)   Stuttgart 1974

All rights reserved, including translations into foreign languages. This book, or parts thereof, may not be reproduced in any form without permission from the publishers.

© E. Schweizerbart'sche Verlagsbuchhandlung (Nägele u. Obermiller)

Stuttgart 1974

ISBN 3 510 65062 X

Printed in Germany

Printer: Gebrüder Ranz, Dietenheim

Design of the cover: Wolfgang M. Karrasch

Contents

Preface. H. BERCKHEMER X

Taphrogenesis: Rhinegraben, continental rift systems, and Martian rifts

Taphrogenesis, Introductory Remarks. J. H. ILLIES 1
The Oberrhein Graben in its European and Global Setting.
G. RICHTER-BERNBURG 13
Taphrogenic Lineaments and Plate Tectonics. R. B. MCCONNELL 43
Martian Tithonius Lacus Canyon and Rhinegraben — a Comparative Morphotectonic Analysis. K. SCHÄFER 52

The Rhinegraben: geologic history and neotectonic activity

Perm, Trias und älterer Jura im Bereich der südlichen Mittelmeer—Mjösen-Zone und des Rheingrabens. H. BOIGK & H. SCHÖNEICH 60
An Isobath Map of the Tertiary Base in the Rhinegraben.
F. DOEBL & W. OLBRECHT 71
Mikrofloristische Untersuchungen zur Altersstellung der jungtertiären Ablagerungen im mittleren und nördlichen Oberrheingraben. G. VON DER BRELIE . 72
Die Mächtigkeit des Quartärs im Oberrheingraben. J. BARTZ 78
Nivellements und vertikale Krustenbewegungen im Bereich des südlichen Oberrheingrabens. H. MÄLZER & A. STROBEL 88
Geodätische Messungen im Rheingraben. GH. BOZORGZADEH & E. KUNTZ . . . 95
Microearthquake-Activity observed by a Seismic Network in the Rhinegraben Region. K.-P. BONJER & K. FUCHS 99
Herdmechanismen von Erdbeben im Oberrhein-Graben und in seinen Randgebirgen. L. AHORNER & G. SCHNEIDER 104
In situ Stress Measurements in Southwest Germany — First Results. G. GREINER 118

The Rhinegraben rift system: deep structure and tectonic features

The 1972 Seismic Refraction Experiment in the Rhinegraben — First Results.
The Rhinegraben Research Group for Explosion Seismology 122
Beiträge der Reflexionsseismik zur Frage der Schwarzwaldrandstörung.
L. ERLINGHAGEN & G. DOHR 138
Structure and Development of the Southern Part of the Rhine Graben according to Geological and Geophysical Observations. F. BREYER 145
Magnetic Anomalies, Thermal Metamorphism and Tectonics in the Rhinegraben Area. J. B. EDEL & J. P. LAUER 155
Die Tektonik des nördlichen Schwarzwaldes und ihre Beziehung zum Oberrheingraben. D. ORTLAM 160
Le rôle des décrochements dans le socle vosgien et en bordure du Fossé rhénan.
M. RUHLAND 167

Block-faulting at a Rhinegraben scarp and a comparable recent landslide structure at the Korinthian graben

Le dispositif structural du champ de fractures de Ribeauvillé (Haut-Rhin, France). Un exemple de fragmentation tectonique (block-faulting) en bordure occidentale du Fossé rhénan. G. HIRLEMANN 172
A Landslide on the Rim of a Graben. L. MÜLLER & G. LÖGTERS 177

Geothermal anomalies and consequences for diagenesis and thermal waters

Eine geothermische Karte des Rheingrabenuntergrundes. D. WERNER & F. DOEBL   182
Diagenesis of Tertiary Clayey Sediments and Included Dispersed Organic Matter in Relationship to Geothermics in the Upper Rhine Graben. F. DOEBL, D. HELING, W. HOMANN, J. KARWEIL, M. TEICHMÜLLER & D. WELTE   192
Die Wärme-Anomalie der mittleren Schwäbischen Alb (Baden-Württemberg). W. CARLÉ   207

Volcanism of the Rhinegraben: potassium-argon ages, local setting, petrology, and gravity anomalies

Apparent Potassium-Argon Ages of Lower Tertiary Rhine Graben Volcanics. H. J. LIPPOLT, W. TODT & P. HORN   213
Der paleozäne Basalt-Vulkanismus im Raum des Unteren Neckars. O. MÄUSSNEST mit einem Beitrag von E. BECKSMANN   222
The Subvolcanic Breccias of the Kaiserstuhl Volcano (SW-Germany). I. BARANYI   226
Bouguer-Anomalienkarte für den Kaiserstuhl. P. THIELE   231

The northern end of the Rhinegraben: mantle rise, structure, and recent tectonic activity

The Northern End of the Rhinegraben due to some Geophysical Measurements. R. MEISSNER & U. VETTER   236
Block Tectonic Interrelations between Northern Upper Rhine Graben and Southern Taunus Mountains. H.-J. ANDERLE   243
Zum Verlauf der westlichen Randverwerfung des zentralen Oberrheingrabens zwischen dem Rhein südlich Nackenheim und dem Main bei Rüsselsheim sowie über eine Grundwasserkaskade und Bauschäden im Bereich dieser Störungszone. O. SCHMITT   254
Levelling Results at the Northern End of the Rhinegraben. E. SCHWARZ . . .   261

Adjacent fault troughs north of the Rhinegraben

Die tektonische Entwicklung der Niederrheinischen Bucht. R. TEICHMÜLLER . .   269
Die Fortsetzung des Rheingrabens durch Hessen. Ein Beitrag zur tektonischen Analyse der Riftsysteme. E. SCHENK   286
Das Schollenmosaik des nördlichen Michelstädter Grabens. E. BACKHAUS, A. RAWANPUR & M. ZIRNGAST   303

Southern extensions of the Rhinegraben

Relations entre le Fossé rhénan et le Fossé de la Saône. Tectonique des régions sous-vosgiennes et préjurassiennes. D. CONTINI & N. THÉOBALD   310
Das Erdbeben von Jeurre vom 21. Juni 1971 und seine Beziehungen zur Tektonik des Faltenjura. N. PAVONI & E. PETERSCHMITT   322
Deep Seismic Sounding in the Limagne Graben. A. HIRN & G. PERRIER   329

Towards a model of crust and mantle structure of the Rhinegraben rift system

Petrologische Gedanken zum Untergrund des Rheingrabens. E. ALTHAUS . . .   341
Ein Krusten-Mantel-Modell für das Riftsystem um den Rheingraben, abgeleitet aus der Dispersion von Rayleigh-Wellen. H. REICHENBACH & ST. MUELLER .   348
Hypothese zur Erklärung des Fehlens von P_n-Einsätzen im Bereich von Grabenzonen, abgeleitet aus zweidimensionalen modellseismischen Untersuchungen. J. STROWALD   355
Models for the Resistivity Distribution from Magneto-Telluric Soundings. I. SCHEELKE   362

The Distribution of Electrical Resistivity in the Rhinegraben Area as Determined by Telluric and Magnetotelluric Methods. V. HAAK & G. REITMAYR   366
A Model for the Resistivity Distribution from Geomagnetic Depth Soundings. R. WINTER   369
The Electrical Conductivity of Minerals and the Temperature Distribution in the Upper Mantle. A. SCHULT   376

Dynamic models for taphrogenesis

Crustal Dynamics in the Central Part of the Rhinegraben. ST. MUELLER & L. RYBACH   379
Model of Crustal Stresses in the Region of the Rhinegraben and Southwestern Germany. K. STROBACH   389
Thin-skinned Graben, Plastic Wedges, and Deformable-plate Tectonics. B. VOIGHT   395
Geophysical Contributions to Taphrogenesis. K. FUCHS   420
Taphrogenesis and Plate Tectonics. J. H. ILLIES   433

List of Contributors

L. Ahorner, Erdbebenstation der Universität, Bensberg bei Köln, Germany

E. Althaus, Mineralogisches Institut, Universität Karlsruhe, Germany

H.-J. Anderle, Hessisches Landesamt für Bodenforschung, Wiesbaden, Germany

J. Ansorge, Institut für Geophysik, Eidgenössische Technische Hochschule Zürich, Switzerland

E. Backhaus, Geologisch-Paläontologisches Institut, Technische Hochschule Darmstadt, Germany

I. Baranyi, Laboratorium für Geochronologie, Universität Heidelberg, Germany

J. Bartz, Geologisches Landesamt Baden-Württemberg, Freiburg i. Br., Germany

E. Becksmann, Talstr. 60, Freiburg i. Br., Germany

H. Berckhemer, Institut für Meteorologie und Geophysik, Universität Frankfurt/Main, Germany

H. Boigk, Niedersächsisches Landesamt für Bodenforschung, Hannover, Germany

K. P. Bonjer, Geophysikalisches Institut, Universität Karlsruhe, Germany

G. Bozorgzadeh, Geodätisches Institut, Universität Karlsruhe, Germany

G. von der Brelie, Geologisches Landesamt Nordrhein-Westfalen, Krefeld, Germany

F. Breyer, Kettlerweg 5, Hannover, Germany

W. Carlé, Rebmannstr. 2, Korntal, Germany

D. Contini, Laboratoire de Géologie historique et Paléontologie, Université Besançon, France

F. Doebl, Wintershall AG, Landau, Germany

G. Dohr, Preussag AG, Hannover, Germany

B. Edel, Institut de Physique du Globe, Université Strasbourg, France

D. Emter, Institut für Geophysik, Universität Stuttgart, Germany

L. Erlinghagen, Prakla-Seismos GmbH., Hannover, Germany

K. Fuchs, Geophysikalisches Institut, Universität Karlsruhe, Germany

C. Gelbke, Geophysikalisches Institut, Universität Karlsruhe, Germany

G. Greiner, Geologisches Institut, Universität Karlsruhe, Germany

V. Haak, Institut für Angewandte Geophysik, Universität München, Germany

D. Heling, Laboratorium für Sedimentforschung, Heidelberg, Germany

G. Hirlemann, Laboratoire de Géologie et de Paléontologie, Strasbourg, France

A. Hirn, Institut de Physique du Globe, Université Paris, France

W. Homann, Geologisches Landesamt Nordrhein-Westfalen, Krefeld, Germany

P. Horn, Max-Planck-Institut für Kernphysik, Heidelberg, Germany

J. H. Illies, Geologisches Institut, Universität Karlsruhe, Germany

J. Karweil, Bergwerksverband, Essen, Germany

E. Kuntz, Geodätisches Institut, Universität Karlsruhe, Germany

J. P. Lauer, Institut de Physique du Globe, Université Strasbourg, France

H. J. Lippolt, Laboratorium für Geochronologie, Universität Heidelberg, Germany

G. LÖGTERS, Sonderforschungsbereich 77 „Felsmechanik", Universität Karlsruhe, Germany

H. MÄLZER, Geodätisches Institut, Universität Karlsruhe, Germany

O. MÄUSSNEST, Tuttlinger Str. 9, Stuttgart, Germany

R. B. McCONNELL, Streatwick, Streat near Hassocks, Sussex, United Kingdom

R. MEISSNER, Institut für Geophysik, Universität Kiel, Germany

L. MÜLLER, Institut für Bodenmechanik und Felsmechanik, Universität Karlsruhe, Germany

ST. MÜLLER, Institut für Geophysik, Eidgenössische Technische Hochschule Zürich, Switzerland

W. OLBRECHT, Ulrich-von-Hutten-Str. 3, Landau, Germany

D. ORTLAM, Niedersächsisches Landesamt für Bodenforschung, Hannover, Germany

N. PAVONI, Institut für Geophysik, Eidgenössische Technische Hochschule Zürich, Switzerland

G. PERRIER, Institut de Physique du Globe, Université Paris, France

E. PETERSCHMITT, Institut de Physique du Globe, Université Strasbourg, France

C. PRODEHL, Geophysikalisches Institut, Universität Karlsruhe, Germany

A. RAWANPUR, Geologisch-Paläontologisches Institut, Technische Hochschule Darmstadt, Germany

H. REICHENBACH, Institut für Angewandte Geophysik, Universität München, Germany

G. REITMAYR, Institut für Angewandte Geophysik, Universität München, Germany

G. RICHTER-BERNBURG, Haarstr. 8, Hannover, Germany

M. RUHLAND, Institut de Géologie, Université Strasbourg, France

L. RYBACH, Institut für Kristallographie und Petrographie, Eidgenössische Technische Hochschule Zürich, Switzerland

K. SCHÄFER, Geologisches Institut, Universität Karlsruhe, Germany

I. SCHEELKE, Institut für Geophysik und Meteorologie, Technische Universität Braunschweig, Germany

E. SCHENK, Professorenweg 6, Gießen, Germany

O. SCHMITT, Hessisches Landesamt für Bodenforschung, Wiesbaden, Germany

G. SCHNEIDER, Institut für Geophysik, Universität Stuttgart, Germany

H. SCHÖNEICH, Niedersächsisches Landesamt für Bodenforschung, Hannover, Germany

A. SCHULT, Institut für Angewandte Geophysik, Universität München, Germany

E. SCHWARZ, Hessisches Landesvermessungsamt, Wiesbaden, Germany

L. STEINMETZ, Institut de Physique du Globe, Université Paris, France

P. STREICHER, Institut de Physique du Globe, Université Strasbourg, France

K. STROBACH, Institut für Geophysik, Universität Stuttgart, Germany

A. STROBEL, Landesvermessungsamt Baden-Württemberg, Stuttgart, Germany

J. STROWALD, Institut für die Physik des Erdkörpers, Universität Hamburg, Germany

M. TEICHMÜLLER, Geologisches Landesamt Nordrhein-Westfalen, Krefeld, Germany

R. TEICHMÜLLER, Geologisches Landesamt Nordrhein-Westfalen, Krefeld, Germany

N. THÉOBALD, Laboratoire de Géologie historique et Paléontologie, Université Besançon, France

P. THIELE, Geodätisches Institut, Universität Karlsruhe, Germany

W. TODT, Laboratorium für Geochronologie, Universität Heidelberg, Germany

U. VETTER, Institut für Geophysik, Universität Kiel, Germany

B. VOIGHT, Dept. of Geosciences, Pennsylvania State University, University Park, Pa. 16802, U.S.A.

D. H. WELTE, Lehrstuhl für Geologie, Geochemie und Lagerstätten des Erdöls und der Kohle, Technische Hochschule Aachen, Germany

D. WERNER, Institut für Geophysik, Eidgenössische Technische Hochschule Zürich, Switzerland

R. WINTER, Geophysikalisches Institut, Universität Göttingen, Germany

M. ZIRNGAST, Geologisch-Paläontologisches Institut, Technische Hochschule Darmstadt, Germany

The Geodynamics Project is an international program of research on the dynamics and dynamic history of the earth with emphasis on deep-seated foundations of geological phenomena. This includes investigations related to movements and deformations, past and present, of the lithosphere, and all relevant properties of the earth's interior and especially any evidence for motions at depth. The program is an interdisciplinary one, coordinated by the Inter-Union Commission on Geodynamics (I.C.G.) established by I.C.S.U. at the request of I.U.G.G., and I.U.G.S., with rules providing for the active participation of all interested I.C.S.U. Unions and Committees.

Preface

In the transition between the International Upper Mantle Project and the Geodynamics Project the Rhinegraben Research Group met to its 5. Symposium on Continental Rift Systems at Karlsruhe on April 13—15, 1972. Results obtained in more than five years of research during the Upper Mantle Project were presented and discussed and are the content of this book. Most of the papers deal with certain aspects of the Rhinegraben rift system in its broader sense but a few also with other continental rift structures. This might be justified not only because of the location where the meeting took place but mainly by the intensity with which the Rhinegraben had been investigated in past and present time in international and interdisciplinary cooperation. When in 1970 the book "Graben Problems", edited by J. H. ILLIES and ST. MÜLLER appeared it found quickly a wide distribution and appreciation. It was, therefore, felt by the Rhinegraben Research Group that the latest results at the end of the Upper Mantle Project should also be made available to a broader public. The title "Approaches to Taphrogenesis" expresses the aim to look at large scale tension features in continental blocks from the viewpoint of the evolution, the appearance at the surface, and the deep seated processes and forces.

Thanks have to be expressed to the Deutsche Forschungsgemeinschaft which sponsored much of this work and supported the symposium.

H. BERCKHEMER

Taphrogenesis:
Rhinegraben, continental rift systems, and Martian rifts

Taphrogenesis, Introductory Remarks

by

J. H. Illies

With 6 textfigures

E. KRENKEL (1922) in his classical monograph on the East African rift valleys, after discussing the morphotectonic development and seismicity of these fundamental structures, introduced in the concluding chapter of his work the term *taphrogenesis*. Derived from the Greek word τάφρος for trench and from γένεσις for formation KRENKEL understood taphrogenesis as a process of block-faulting on a regional scale. He compared the East African rift structures with the Rhinegraben and defined taphrogenesis as the formation of grabens, induced by tensional forces.

Graben means ditch or trench in German. PFANNENSTIEL (1969) reviewed the application of this word in the geological nomenclature. He pointed out that already the old miners in Thuringia used this term for downthrown segments and that J. L. JORDAN (1803) was the first to mention grabens in the geological literature. The Rhinegraben in Central Europe, a fault trough which extends from Basel to Francfort through which flows the Upper Rhine river, is considered to be the type structure for all grabens. Here, in 1841, the famous French geologist ELIE DE BEAUMONT had recognized that Vosges and Black Forest, forming together a wide-spanned vault, are interrupted along the crest of the vault by the downthrown segment of the Upper Rhine plain. In a sketchy cross section he had outlined the trough as a wedge-block which subsided along two sets of converging step faults. E. SUESS in his well-known manual "Das Antlitz der Erde" 1883 (t. 1, p. 166 ff) used the term graben for other fault troughs. From there Anglo-Saxon authors adopted the word and since the beginning of this century graben became an international term in the geological nomenclature. But it was a long and hard course from ELIE DE BEAUMONT to modern concepts of graben tectonics. One of the most striking milestones along this road is marked by the publication of HANS CLOOS "Hebung — Spaltung — Vulkanismus" in 1939. By means of synthesizing field observations as well as experimental data CLOOS established an up-to-date hypothesis on taphrogenesis of the Rhinegraben and of continental rifting in general.

The Rhinegraben is neither the largest nor the most active nor the most typical example for continental grabens. But otherwise there are some particularities in this graben that splits apart the very geographical center of Central Europe. Geoscience departments of eight universities like those of Francfort, Freiburg, Heidelberg and Strasburg and offices of five geological surveys belonging to three different nations are established in the Upper Rhine area. Geologists of many generations have mapped this area and year after year the students arrive to study the famous quarries *mente et malleo* (sometimes only *malleo*). Thousands of bore-holes have penetrated the sedimentary fill of the fault trough. Innumerable geophysical campaigns provided much information of the crust and of the upper mantle beneath the graben segment. Hence the Rhinegraben, its sedimentary fill, its internal and external structures, its root (or anti-root) became as transparent as a glass model for the geologist's eyes.

The International Upper Mantle Project and now the Inter-Union Commission on Geodynamics have stirred up the geologists and geophysicists along the Upper Rhine to see if their graben might be a key for a better understanding of the fundamentals of rifting. French, German and Swiss colleagues joined for studying the legion of facts under the scope of modern ideas. New geophysical, geodetic, geological and petrological investigations were started, most generously supported by the *Deutsche Forschungsgemeinschaft*. The Rhinegraben Research Group had just presented two symposia: "The Rhinegraben Progress Report" (1967), and "Graben Problems" (1970). This submitted third part of the Rhinegraben trilogy has been focussed under modern geodynamic concepts like plate tectonics. Primarily, it comprises new data concerning the Western European rift systems, adds new contributions on the mechanics of rifting as derived from observations in that rift belt and also comprises additional studies as required for a better understanding of the continental rifting in general.

The significant physiography of continental grabens is marked by a parallelism of the framing fault scarps. The variation in width ranges from some ten meters of smaller graben splinters to the several hundred kilometers width of the Red Sea rift system. An individual, average width between both master faults seems to be a specific attribute to nearly all graben segments. This regularity is emphasized by sharp curves of the main trend. Thus, a change in the direction of one escarpment is closely followed by a corresponding change on the opposite side, causing the woodcut-like contours of graben systems. The width of the Rhinegraben is 36 km, its length 300 km. The marginal main faults are normal faults. Their dip, measured in numerous outcrops and bore-holes, ranges between 55° and 80°. Near the surface, the most frequent values are 60°—65° (Fig. 1). By means of geophysical observations, a slight, listric configuration with a concave curve downward may be suggested (see ERLINGHAGEN & DOHR, p. 143).

The subsidence of a graben block of the described configuration implies a corresponding lateral dilatation, a yielding of the framing abutments. Such a

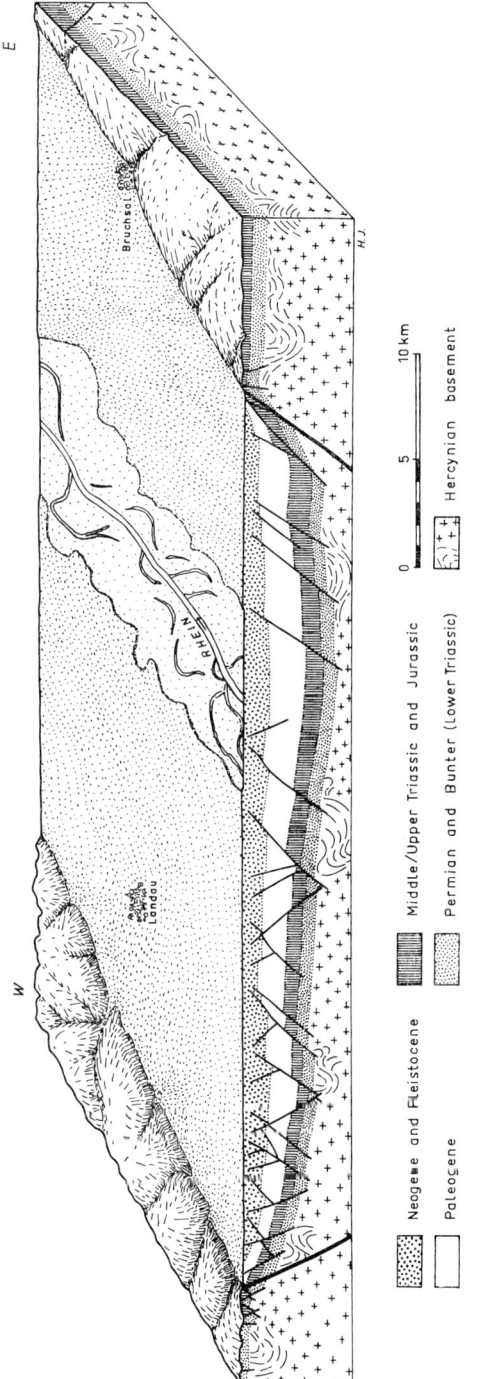

Fig. 1. Block diagram of the Rhinegraben segment immediately north of Karlsruhe. Two converging master faults are bordering the 36 km wide graben area. The graben segment itself is splintered into numerous special blocks, primarily of antithetic character. The graben is framed by elevated shoulders, the Pfälzerwald mountains on the left side and the Kraichgau hills on the right side.

movement requires a slide plane or a layer of reduced sliding friction to take over horizontal gliding movements on the flanking crustal plates. These layers might be sediments such as clays or evaporites. The friction could also be reduced by moisture, or a magmatic or viscoplastic state of the material. In any case, the original, triangular shape of the graben body implies an interrelationship between the depth of the slide plane and the surficial width of the graben segment.

However, in most cases the actual extension of the graben floor is larger than the theoretical width of the primordial wedge-block. Bundles of normal faults generally have splitten the graben segment, forming a mosaic of antithetic or synthetic narrow, tilted blocks (Fig. 2). By returning the whole graben segment and its internal, rotated blocks to its original position of departure, a lateral gap remains open, corresponding to the amount of crustal thinning normal to the graben axis. In the Rhinegraben, this tension factor has been calculated to be of about 4.8 km. Many of the internal dip-slip faults have been synsedimentarily activated by a lateral removal of the framing crustal plates during the graben formation. Active subsidence of the order of 0.5 mm per year has been ascertained by precise levellings and geological observations. Hence, there is evidence for the continuity of dilative movements up to recent times. However, in spite of the tensional character of graben tectonics, the graben body as a whole is subjected to a relative pressure produced by the wedge configuration of the subsiding graben segment.

In the East African rift system, especially in the Gregory rift valley and in the Afar depression, crustal thinning has reached a further stage. In these segments, the primary wedge-block is split apart completely and the recent seismo-tectonic and volcanic activity has migrated from the marginal master faults to the inner graben. There, gaping fissures are accompanied to magmatic dyke injections, steam jets, and rapid vertical and horizontal tectonic movements. The Red Sea rift system, with its active inner graben and smoothed outer escarpments of bilateral, mirror-inverted arrangement, demonstrates how the graben-in-graben evolution proceeds.

Tensional ruptures, as well as dip-slip movements along fault planes and closure of lateral gaps by antithetic block rotations, are the most conspicuous symptoms of taphrogenesis. The tensional forces are considered to be a component of a regional stress field. They have been recorded by various investigations, e. g. by means of microtectonic observations, especially complementary strike-slip movements, by horizontal stylolites and, in case of neotectonic activity, by means of in situ stress measurements or fault-plane solutions of earthquakes. Taking in mind the regional stress field as related to the structural inventory of graben tectonics, several problems remain open.

McConnell (1972) has pointed out that the majority of East African rift segments followed pre-existing zones of weakness within the Precambrian basement. He concluded that continental rifting might be understood as a rejuvenation of lineaments which intersect the entire continental crust from the surface to the upper mantle. Illies (1972) has shown that the Rhinegraben rift system is accom-

Fig. 2. The Rhinegraben escarpment near Baden-Baden, view to the north. In the front is the clay pit of the Hourdis brick field with Pechelbronn beds of Lower Oligocene age. The clay serie which belongs to the basal sequence of the graben fill, is down-warped grabenward and slashed by a bundle of 2nd order faults splintering off the master fault which follows 50 m right of the clay pit (not seen on the picture). The range right in the background marks the elevated graben shoulder and is formed by the Bunter (Lower Triassic). The cross-section along the wall of the clay pit corresponds almost exactly to the vibroseis profile fig. 3 published in the article of ERLINGHAGEN & DOHR (p. 143). Phot. Nov. 1972.

panied by old shear zones in the Hercynian basement which had facilitated and locally attracted the Cenozoic graben faulting. In some parts the rifting carefully followed the tracks of crustal instability. The functional interdependence between old lineaments and its revived opening under the changed realm of neotectonics is evident. This is predominantly the case when pre-existent zones of weakness are congruent to the potential fault components of the taphrogenic stress field. But the question, whether or not the lineaments have caused the rift process, cannot yet be answered. We still have to consider the grossly different tectonic realm under which the renewed tectonism was initiated.

The Rhinegraben is the central segment of a meridional rift belt which traverses the Western European continent. In addition, this system is part of an extended rift system that intersects the whole Old World from the North Sea to Southeast Africa. The total length of this intercontinental rift belt amounts to more than 9000 km. In spite of the different lithological composition, the different geological pre-history, the different tectonic framework of the crust slashed through by rifting, all graben segments in Western Europe, Near East and East Africa appear very similar in their physiographic features, structural patterns, geophysical properties, seismic activities, and volcanic extrusions. Taphrogenesis has been governed by the same construction formula throughout the entire rift belt. Even the predominant strike of the rift segments and transform elements followed a superordinated trend, and a phasic consonance or contemporaneity in geologic time of subsidence activities seems evident. A global control and an interdependent tectonic realm must be assumed. This realm used the old fracture pattern within the continental plates and revived the weakness zones to new sub-plate boundaries.

The question, whether action or reaction initiated the formation of grabens, is based upon the explanation of the origin of the "basaltic swells" or "cushions" of mantle-derived material underneath some continental graben segments. In several chapters of this book new seismic refraction data concerning the rift "cushion", which accompanies the Rhinegraben and the Limagne graben are presented. It has been suggested that a swell of upsurging substratum below the crust split the overlying sialic shell, uplifted the graben shoulders and induced the graben volcanism. The Cenozoic Rhinegraben volcanism is predominantly an olivine-nephelinitic realm, locally of melilite-ankaratritic character and derives primarily from a depth of about 80—100 km. Thus, it seems conclusive that upwelling material of the upper mantle has induced the taphrogenic cycle. On the other hand, it had been discussed that the old weakness zones have induced the ascent of the magmatic substratum. No doubt, an interplay between the beaten tracks of lineaments within the rigid lithosphere and the ascending tendencies within a plastic layer of the upper mantle, the asthenosphere, have influenced this kind of rifting. Potassium-argon ages of volcanic rocks of the Rhinegraben area, published by LIPPOLT et al. (p. 213), reveal evidence that the mantle-derived magmatic activity started considerably earlier than the surficial graben tectonism. Under this aspect, the taphrogenic action started from the upper mantle, but it was related to inequalities and anisotropies of the crustal rigidity.

The development of a graben tectogene within a crustal plate overlying an upwelling magma cushion may be explained by dilatation as a result of arching along the crest of the rising crustal dome. This single mechanism cannot explain a horizontal displacement of the graben's lateral abutments of about 5 km. To explain the observed dilatation, an additional criterion must be required. According to the interpretation of the anomalous heat flow values in the Rhinegraben area and the P-wave velocities in the lower crust, it is suggested that basal crustal layers of a non-rigid state overlie the swell of mantle material. A mechanical decoupling of the crust from the mantle by a separation along a glide-surface can be concluded. This glide plane follows the "cushion" surface and dips with a gradient increasing to 10° normal to the graben axis. Seismic reflection data has revealed a zone of laminated texture in the lower crust indicating a state of laminar plastic flow. From these observations, one may conclude that the crustal plates upwarped at both marginal faults of the Rhinegraben drifted apart by gravity sliding. Hence, it follows that the tectonic style of block faulting along both graben rims and the internal dip-slip faults reveal remarkable similarities to downslope creep structures in large slump areas, both with the typical, listric (shovel-like) fault planes. Two examples presented in this book obviously illustrate the convergency of step faulting induced by rifting along the Rhinegraben escarpment (HIRLEMANN, p. 174, fig. 2) and recent landslide phenomena as having occurred along the slope of the Korinthian graben (MÜLLER & LÖGTERS, p. 178, fig. 1).

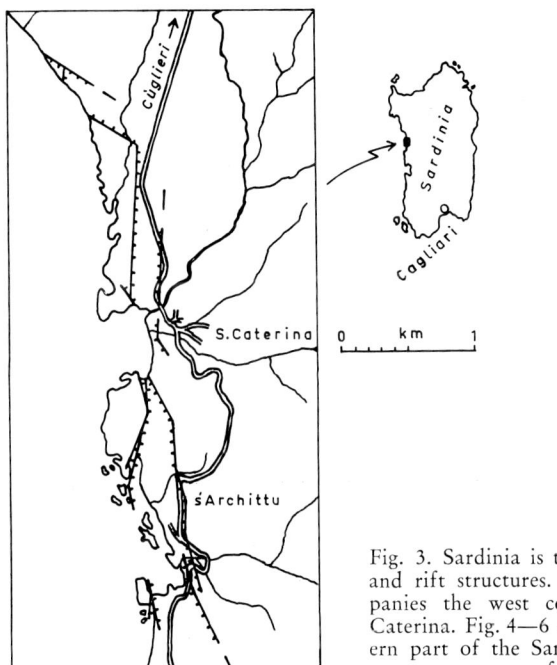

Fig. 3. Sardinia is traversed by a system of grabens and rift structures. A smaller graben splay accompanies the west coast of the island near Santa Caterina. Fig. 4—6 refer to the section at the southern part of the Santa Caterina bay and are taken from the north.

Graben formation, under conditions described above, seems to be independent of time and scale. Large grabens demonstrate equivalent tectonic patterns like micro-grabens (Fig. 3—4). Grabens, formed during millions of years are typified by the same structural inventory as shown by man-made clay models formed in a few hours. Many grabens formed under natural conditions as well as in experiments, show asymmetries in their cross-section (Fig. 5—6). An unilateral rotation of the whole graben segment with mostly an additional one-sided arrangement of the internal tilted blocks has been frequently observed (E. CLOOS 1968). The same asymmetric features characterize the graben cracks in the area of Anchorage which happened during the Alaska earthquake event in 1964 (HANSEN 1971). Asymmetry will be favoured if one framing plate is yielding or gliding more rapidly than the other. The east-west asymmetry of the Rhinegraben has been interpreted by ILLIES (1972) as a preferred westward creep of the bordering western plate.

Gravity sliding and creeping plate movements are not the only mechanisms

Fig. 4. Total view of the Santa Caterina micro-graben. The undisturbed series on both flanks consist of marine fossiliferous sandy marlstones and chalks, intercalated of reef complexes of oysters, algae and bryozoa. The inner graben which is filled by marine Pliocene clayey and glauconitic silts. This series is downthrown along step faults toward the gravels and a latite-andesitic lava flow of Lower Quaternary age (for the petrographical analysis I am indebted to Dr. R. Metz). The total vertical throw between the western (right) flank and the inner graben is about 50 m, that of the overlain Quaternary deposits only 4.5 m. The graben formation has been favoured by a slight seaward downbending of the whole coastal rock complex in this area. The preferred anti-clockwise rotation of the graben as a whole might be related to a preferred yielding of the western frame (the right one in the photo). Phot. Sept. 1972.

Fig. 5. Step faulting at the western rim of the Santa Caterina graben. The inclination of individual dip-slip faults range between 45 and 60°. The vertical throw along the single faults illustrated by the offset of an oyster reef is only of a few meters. Two systems of joints are cleaving the single step blocks, but the shoulder west (right) of the marginal master fault is almost free of jointing.

Fig. 6. Typical fault structures at the eastern margin of the Santa Caterina graben. Left: A pseudo-reverse fault as formed by an antithetic block rotation incorporating a primordial normal fault. Thus an upthrow of about 1.8 m of the hanging wall is visible along the 68° eastward dipping fault plane. But this pseudo-thrust is accompanied by typical tensional jointing with open fissures and clefts.

required for the formation of grabens under continental conditions. There is, for example, the Lower Rhine Embayment, a northern prolongation of the Rhinegraben rift system (R. TEICHMÜLLER, p. 269). Up to now no subcrustal "cushion" under this graben segment has been ascertained. Consequently, some typical symptoms of continental rift systems are lacking, there are especially no elevated graben shoulders, no parallelity of the bordering escarpments and no constant width of the fault trough. On the other hand, the tensional features and internal tilted block tectonics as well as the seismotectonic activity west and north of Cologne reveal a typical peculiarity of a graben. Thus the Lower Rhine Embayment is doubtless a real rift segment. As ascertained by fault plane solutions of earthquakes, the direction of minimum stress appears normal to the graben axis (AHORNER et al. 1972). Tensional forces have torn apart that part of the European plate and are still active without an interacting mantle swell underneath.

The strike of the axis of the Lower Rhine Embayment is parallel to the maximum stress direction. The maximum compressive stress, which is presumably effective in the northwest-southeast direction, is not consistent with the Rhinegraben axis. According to a fault pattern analysis by micro-tectonic observation and fault plane solutions of seismic events the Rhinegraben axis follows the s_2 shear component of the regional stress field. Left-lateral strike-slip movements parallel to the graben have been confirmed by slickensides at many sites. The total amount of strike-slip movements along the graben axis is not more than a few hundred meters. The Rhinegraben represents a transitional stage compared to grabens such as the Dead Sea rift valley where strike-slip movements control the entire process of taphrogenesis. Graben-like features may also be the result of other deformational reactions of a stressed rigid crust. Feather joints are en échelon arranged fissure systems which are often associated with shear zones (HANCOCK 1972). This microtectonic structure in some areas meets with its large-scale equivalent, the feather grabens. The offset zone between the Rhinegraben and the Bresse graben is traversed by a number of small grabens, mostly arranged in a staggered and sigmoidal course. According to a detailed investigation by CONTINI & THÉOBALD (p. 319) these graben systems are related to strike-slip or shear movements in the sense of feather jointing.

There are many postulations in favour of affinities or against similarities between continental and mid-oceanic rift systems. Different stages of transformation from continental grabens to oceanic rift systems are considered only in reference to major rift belts. In the Gregory rift valley of Kenya, especially in the Lake Hannington—Lake Baringo district, taphrogenesis has attained an advanced stage compared to the Rhinegraben. The upsurging mantle swell came considerably closer to the surface and magmatic dyke injections already pierced the thinned crust within the central graben area. In the Ethiopian Danakil depression, the

◄───────

Right: Dip-slip movements at the eastern rim of the inner graben. The orientation of the fault-planes is synthetically inclined with an angle of 65° towards the inner graben. But it is traversed by antithetic joints and fissures which have absorbed an additional amount of lateral dilatation.

magmatic "cushion" has pushed apart the overlying Precambrian basement. Only some sialic remainders are left. Thus, a quasi-oceanization governs the graben floor. Along the Red Sea rift system, the original cover of continental crust has been completely torn apart and the floor of the inner graben is underlain by a newly-formed oceanic crust. Sea-floor spreading now substitutes for crustal spreading without a replacement of the tectonic realm by a new driving mechanism.

Notwithstanding this spatial juxtaposition and temporal succession of different stages of an uninterrupted sequence from continental to oceanic rifting, the tectonic style and physiographic feature of both kinds of rifting are different. This difference is a result of the particular structure and composition of the continental crust. Several orogenic cycles have engraved their traces like palimpsests into the continental basement and formed its structural framework by many generations of folds, faults and joints. This anisotropy forced the continental rifting to follow invisible rails, directed and turned aside its course, molded its internal and external faces. Typical rift structures of the oceanic realm such as the transform faults appear veiled by reactivated paleo-structures. Also the low density of the predominantly granitic composition of the continental crust plays an important role since it produces specific isostatic reactions to the upsurging mantle swell below. Recent taphrogeosynclines, such as the abyssal Lake Tanganjika and Lake Baikal appear nearly isostatically compensated, and thus are the uplifted shoulder-like mountain ranges, e. g. Black Forest, Vosges and the Mount Ruwenzori. Taphrogenesis is additionally influenced by erosional processes along the elevated graben shoulders and the consequent transport of debris to the graben depressions. Several thousand meters of sedimentary fill deposited in the fault troughs correspond to proportional denudations of the elevated shoulders. Numerous creeks and rivers transport their debris load like conveyor belts down to the grabens, feeding additional potential energy for the progress of rifting.

Taphrogenic movements, in any case, imply the lateral space necessary for tensional dislocations and crustal thinning. Regularly, crustal plates appear fixed within a frame of the surrounding lithosphere. There is no freedom for distortion required for lateral side-steppings of the affected plate areas. Lateral dilatation along the rift zones must be compensated by equivalent condensation of crustal material along subduction zones. The Rhinegraben taphrogenesis demonstrates local interferences and temporal interactions with the formation of the nearby Alpine orogene. Orogenesis and taphrogenesis interacted within the same continuous crustal plate like antagonistic partners. The counterpart to the taphrogenesis in the foreland north of the Alps may be seen south of the Alpine-Himalaya fold belt in the Red Sea — East African rift system. The intercontinental rift belt which traverses the Old World from the North Sea down to Mozambique appears as an autonomous belt of the World Rift System. When the rift system crosses and interferres with the fold and subduction zone of the Tethys area a reciprocal interaction between the downward movements of the asthenosphere in the Alpidic subduction zones and the upsurging mantle swells along the rift belt seems obvious.

References

AHORNER, L. et al., 1972: Seismotektonische Traverse von der Nordsee bis zum Apennin. — Geol. Rdsch., **61**, 915—942.

CLOOS, E., 1968: Experimental Analysis of Gulf Coast Fracture Patterns. — Amer. Ass. Petrol. Geol. Bull., **52**, 420—444.

CLOOS, H., 1939: Hebung — Spaltung — Vulkanismus. — Geol. Rdsch., **30**, 401—527 u. 637—640.

HANCOCK, P. L., 1972: The analysis of en-échelon veins. — Geol. Mag., **109** (3), 269—276.

HANSEN, W. R., 1971: Effects at Anchorage. — In: The Great Alaska Earthquake of 1964, Geology, 289—356, National Academy of Sciences, Washington.

ILLIES, J. H., 1972: The Rhine Graben Rift System — Plate Tectonics and Transform Faulting. — Geophys. Surv., **1**, 27—60.

ILLIES, J. H. & MUELLER, ST. (eds.), 1970: Graben Problems. International Upper Mantle Project, Scient. Rep. 27. — 316 p., Schweizerbart, Stuttgart.

KRENKEL, E., 1922: Die Bruchzonen Ostafrikas. — 184 S., Borntraeger, Berlin.

MCCONNELL, R. B., 1972: Geological Development of the Rift System of Eastern Africa. — Geol. Soc. Amer. Bull., **83**, 2549—2572.

PFANNENSTIEL, M., 1969: Die Entstehung einiger tektonischer Grundbegriffe. Ein Beitrag zur Geschichte der Geologie. — Geol. Rdsch., **59**, 1—36.

ROTHÉ, J. P. & SAUER, K. (eds.), 1967: The Rhinegraben Progress Report 1967. International Upper Mantle Project, Scient. Rep. 13. — 146 p., Abh. geol. Landesamt Baden-Württ., **6**.

The Oberrhein Graben in its European and Global Setting

by

G. Richter-Bernburg

With 11 figures in the text

Zusammenfassung in deutsch S. 40.

1. Problems in discussion

In a Geological Map of Europe the Oberrhein-Graben (ORG) looks like a perfect surgeon's scalpel cut through the epidermis of mesozoic formations in the flesh of a great muscle — of a muscle composed of fibres the direction of which we could not guess before. The skin doesn't give us any information about the structure of the underlaying organs — what about the mesozoic cover with regard to the base? Does it also veil all essential features, or is it more transparent than an epidermis giving us some indications about the structure and the movements of the (varistian or older) basement? This is one of our questions.

The cut of the ORG has been studied during the last time from different points of view and with various spot-lights. This serious activity involved many new facts completing the knowledge of the object up to truly surprising details (see Rothe & Sauer 1967). However, we should not forget that this extremely well-studied ORG — which is very spectacular and seems to be extraordinary — has continuations towards South and North and is only one member in the Graben-Chain of Stille's "Mittelmeer-Mjösen-Zone". Apart from the ORG, we find many features which are able to complete our conceptions about the fabrication of the 4000 km long West-European Rift Zone. We observe a lot of elements other than those in rhenish direction. The Rhenish Lineament has to be regarded within its very complex environmental joint systems. In measurements of the strike of all kinds of ruptures — as faults, joints, fissures, dikes, veins etc. — in Western Europe, maxima and minima will be observed. The most common directions are named in Germany as follows:

North
    N 10—35° E — rhenish
      (Minimum at 40°)
        N 50—70° E — erzgebirgish
          (Min. 90—100°)
            N 110—125° E — hercynian
              (Min. 130°)
          N 135—145° E —franconian
        (Min. 150°)
      N 160—175° E — eggish
    (Minimum bei 180°)

This seems to be the g r o u n d p a t t e r n or the general design out of which this or that direction will be mobilized or activated — or neglected. This varies in space and time. What may be the reason for this behaviour and which specific connection exists between these other directions and the renish faults in the great NNE-lineament itself as well as in the general design of ruptures?

Furthermore, the Westeuropean Graben-Lineament and other tectonic structures of a similar kind do not cut a homogeneous crust, they run through a very complex building formed by orogenetic revolutions of different (Algomian — Caledonian — Varistian — Alpidic) age. Folding and faulting events compete with one another. Which has the primacy and which is influencing the other? Thus, we have to ask for the significance of these different types of crust deformations.

Finally, apart from the European S—N-Lineament there are other Rift Systems like e. g. the East African, Red Sea-Jordan etc. Is it allowed to compare these destruction-elements of the continental crust with the bursts in the ocean floors? Do we have to expect the same consequences for our continent? And which is the role the ORG plays in these events?

2. Varistian-rhenish relationship

The ORG itself is a very clear cut in which both of the main faults — one on the western (Vosges-) side and the other on the eastern (Schwarzwald-) side —

run straight through in rhenish direction. Additional parallel ruptures do not exist but very close to the main faults and in the downward sloped central stripe of the Graben (see ILLIES a. o.). We also miss faults in SW—NE (erzgebirgish) direction although this is the strike of the varistan folds in the basement. Some few faults of this course have only been stated in the "Süddeutsche Großscholle" (CARLÉ 1955). Our Fig. 1 shows that there is a very poor bursting of the Mesozoic cover besides of the ORG. The only exception is the Eifel-zone which is an about 120 km long parallel element in a western distance of 100 km from the ORG.

We therefore conclude that the tension of d i l a t a t i o n in about West— East could be completely released by the two Rhenish Principal Faults who probably met each other in form of an Y at a depth of 40 to 60 km, i. e. at (or deeper than) the MOHO-discontinuity (see also later).

At the northern edge of the ORG near Frankfurt, the straight main faults seem to hesitate. They are s p l i t up in one principal lineament which is running farther, and in a western branch which runs away towards NNW. It is extremely essential that the main ruptures of the eastern branch do not make use of the chance to deviate into the — rather comfortable as only a little more eastern — erzgebirgish direction. On the contrary, they follow strongly the clear rhenish NNE-course, cutting the varistian structures with an acute angle.

The separation in the two branches — NNW and NNE — happens at this particular place where, in the varistian basement, the graben course leaves the crystalline rocks (Odenwald, Spessart) of the "Mitteldeutsche Hauptschwelle") and enters the semimetamorphic southern boundary of the sedimentary and less stabile rocks of the "Rhenoherzynian Zone". It is along the rhenish striking main fault that the metamorphic zone in the basement seems to be shifted from a more southern position, as in the Taunus, up to a northernmore location in the Werra-Eichsfeld-Kyffhäuser strike (see signature 5 in Fig. 1). The quartzite-stripe Kellerwald—Acker-Bruchberg shows a parallel trend. Or: The same elements in the Varistian Orogen appear moved towards North compared with their place on the western side of the rhenish lineament (or vice versa respectively). Moreover, the width of the rhenoherzynian Zone — between the metamorphic zone in the S and the Subvaristian in the N — is less than 100 km at the eastern border, i. e. in the Hercyn-Bohemian Bloc, though it is wider than 150 km in the Rheinisches Schiefergebirge, West of the main lineament.

These facts speak for the p r e - v a r i s t i a n e x i s t e n c e of a rhenish striking element which was so d o m i n a n t that the v a r i s t i a n f o l d - u n i t s were d i s p l a c e d b y s h e a r i n g along this older lineament.

Returning to the Mesozoic cover we miss the nice clearness of the ORG when we enter the Hessische Senke — Niedersachsen, coming from the Spessart and the Vogelsberg Volcano. Here, the mesozoic seems to be split in thousands of small chips by innumerable faults. Before we investigate whether this is due to the changed mechanism because of the different character of the Varistian, we will take into consideration the frequency and the distribution of these faults.

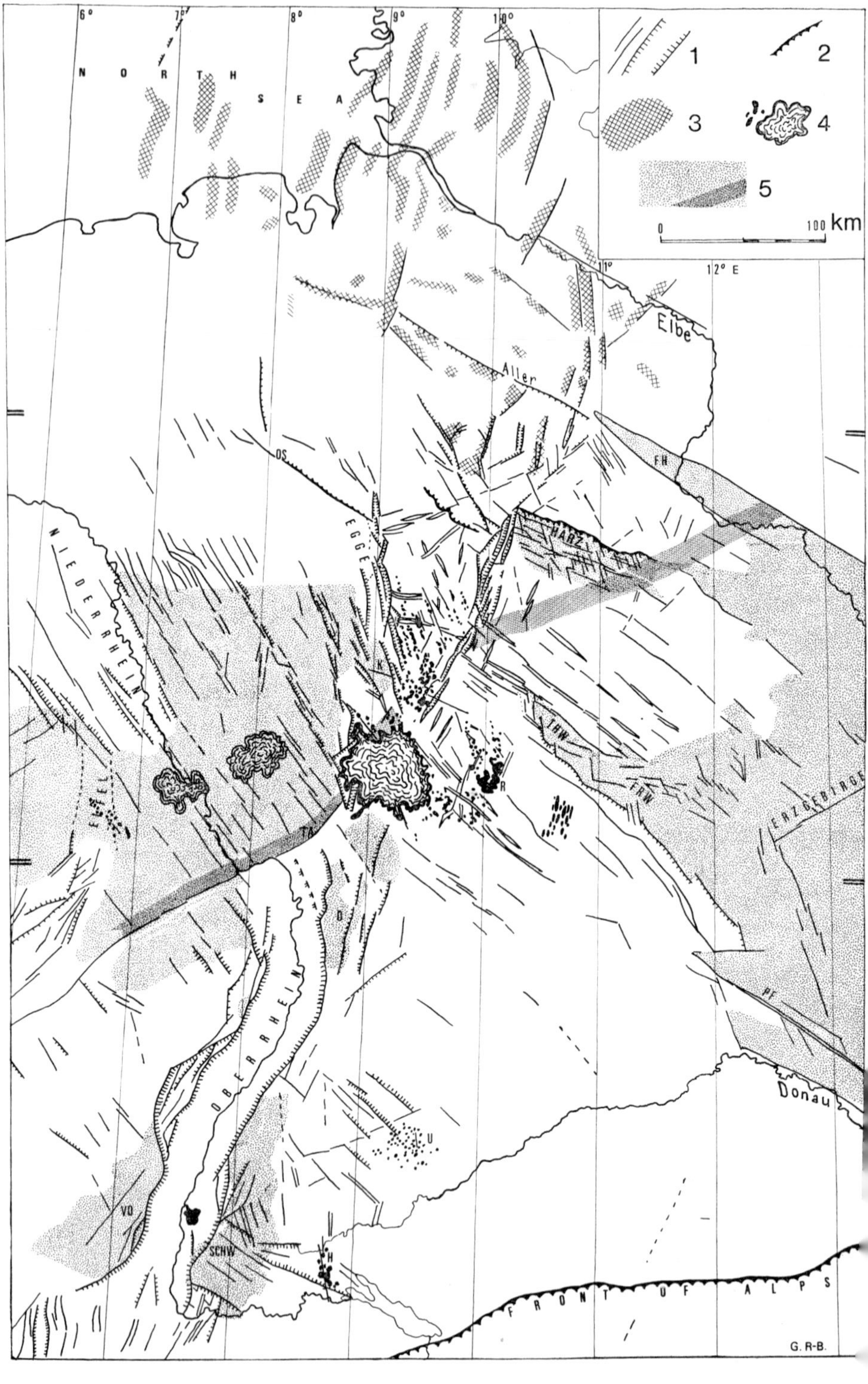
6°
7°
8°
9°
10°
11°
12° E
N O R T H S E A
1
2
3
4
5
0
100 km
Elbe
Aller
OS
EGGE
HARZ
FH
NIEDERRHEIN
K
THW
FRW
R
ERZGEBIRGE
TA
D
OBERRHEIN
U
PF
Donau
VO
SCHW
H
FRONT OF ALPS
G. R-B.

In the about 60 000 km² large area between 8° and 12° E and approx. 50° and 52° N (the two lateral marks in Fig. 1) altogether 9500 cumulative kilometers of faults have been interpreted[1]. 5700 km of these faults are distributed over 47 000 km² of mesozoic rocks, about 3800 km cut the 14 000 km² of outcropping basement (see later).

Among the faults in the mesozoic cover belong:

39 % to the hercynian direction	N 110—130° E
7 % to the franconian direction	135—145°
23 % to the eggish direction	160—175°
24 % to the rhenish direction	10— 30°
6 % to the erzgebirgish direction	45— 65°

Not the rhenish but the hercynian direction shows the highest frequency. We also see that numerous faults run in SSE—NNW, i. e. e g g i s h , l i k e t h e E g g e - H i l l s. These faults and narrow grabens split away from the rhenish direction, not quite exclusively but mainly towards "left" or West from the straight rhenish orientation, in this manner forming a V. This is not restricted on our area where it is very obvious in Fig. 1, but it occurs also in another scale as near Hannover, or far away in Sinai peninsula (Fig. 2). We have to come back to this significant relation on page 20—22.

H e r c y n i a n d i r e c t i o n. — The most frequent direction of the ruptures is the hercynian — in part because numerous long faults and even some great lineaments follow this strike. The dominant hercynian element in Europe is the southwestern border of the Russian Plate, the important "Pontus — Greenland-Lineament" (STILLE). Some significant parallel elements as the Sudeten-Fault a. o., play the role of shearing-planes in the NW-pushing Varistian Orogen. Moreover, the late-varistian granite intrusions in Sachsen (Eibenstock — Kirchberg), Frankenwald (Henneberg), Harz (Brocken, Ramberg) are connected with primary ruptures of hercynian strike (RICHTER-BERNBURG 1968). The most outstanding hercynian elements are the famous Pb—Zn-ore veins in the central area of the Harz and its — name giving — northern overthrust. The very long faults in the Mesozoic cover (in Franken and Thüringen) may come through from their basement. However, if even the Varistian has to obey the orders from below these faults are not so much to see as influences from the Paleozoic but as an inheritance of a m u c h o l d e r s t r u c t u r e.

---

[1] These basic dates have been taken from the Geological Maps 1:300 000 of Hessen and of Nordwest-Deutschland, edited by the Geological Surveys, and 1:100 000 and 1:500 000 of Thüringen.

Fig. 1. Ruptures and faults in Germany.

1 — Ruptures, faults, grabenfaults, 2 — overthrust, 3 — salt dome, 4 — kenozoic volcanoes, 5 — pre-mesozoic, mainly varistian basement (also metamorphic southern boundary of Rhenohercynian zone). OS — Osning, F H — Flechtinger Höhenzug, TH W — Thüringer Wald, FRW — Frankenwald, PF — Pfahl, K — Kellerwald, R — Rhön, TA — Taunus, O — Odenwald, VO — Vosges, SCHW — Schwarzwald, H — Hegau, U — Urach.

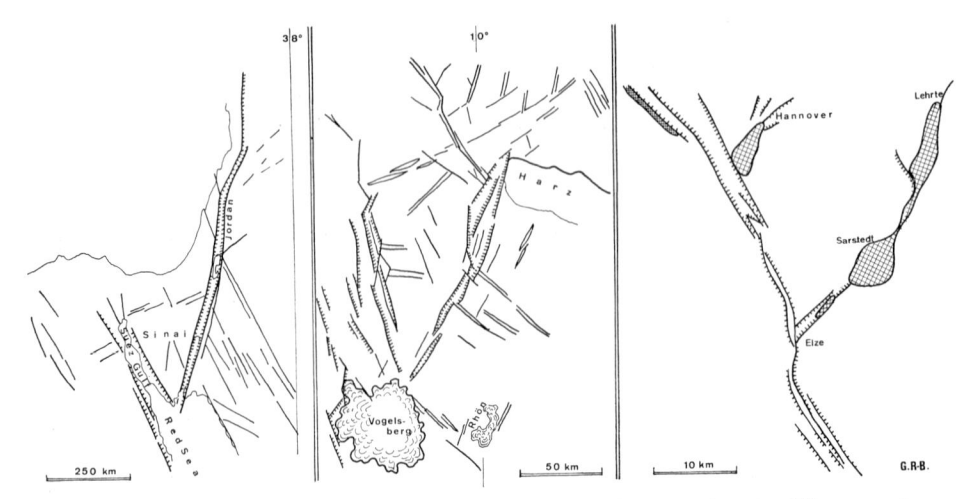

Fig. 2. The relation of Eggish and Rhenish directions. The typical V in different scale. L — Near East, M — Hessische Senke, faults and volcanos, R — Part of Niedersachsen, faults and salt diapirs.

The Fig. 1 shows that the frequency of the hercynian faults decreases rapidly from East to West. On the left side of the Rhenish Lineament, there is practically no more than the Aller-line and the Osning-overthrust — in an obvious contrast to the whole eastern side where from the lower Elbe in the North up to the Bonndorf-Zone between Schwarzwald and Bodensee (CARLÉ 1955) in the South hercynian ruptures are by far dominant in the Mesozoic cover as well as in the basement. All the more is it remarkable that the Rheinisches Schiefergebirge shows no hercynian lines, but innumerable faults in the "franconian" course. This is the Q-direction, vertical to the varistian fold axes, whereas in the Harz and Thüringerwald-Frankenwald hercynian and eggish fissures compose a joint system in diagonal direction (according to MOHR, see Fig. 3 a).

We may perhaps suppose that this is any special reaction by the thick sediments of the deeper part of the Rheno-Hercynian Geosyncline. However, we see the same difference in the course of faults between Vosges and Schwarzwald where quite another varistian facies is developed (see Fig. 3 b).

The research of all faults (in the area mentioned above) as far as the outcropping basement is concerned has been extended over about 14 000 km² where 3800 km cumulative faults have been registered and interpreted:

Rheinisches Schiefergebirge (7500 km²)	Harz — Frankenwald (6500 km²)	direction:	
3 %	58 %	hercynian	N 110—130° E
89	4	franconian	135—140°
7	38	eggish	160—175°
1	—	rhenish	10— 30°

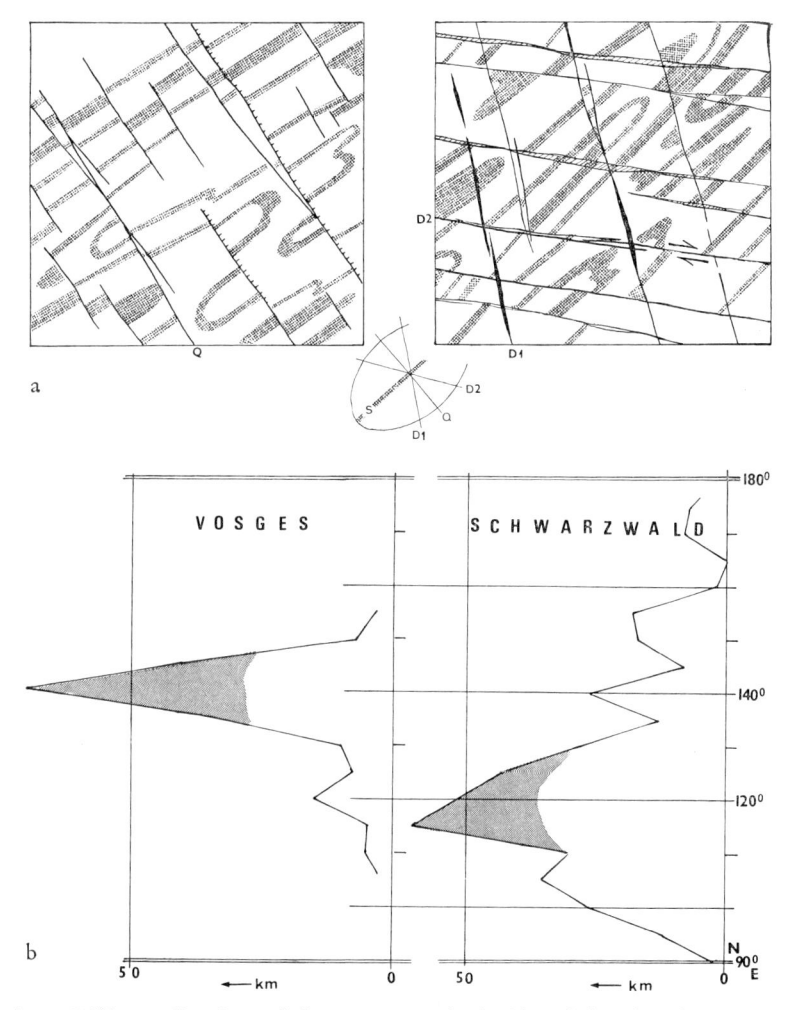

Fig. 3. Different directions of the ruptures on both sides of the Rhenish Lineament.
a) Folds and joints in the Varistian (schematic): left — Rheinisches Schiefergebirge where the Q joints are dominant, right — Thüringerwald und Harz, where hercynian and eggish directions form a Mohr system, the latter with acid volcanic rocks, the first with Pb-Zn ore veins.
b) Ruptures in Vosges and Schwarzwald (taken from Cloos 1936, p. 408) in hercynian direction: left 19%, right 48%, in franconian direction: left 73%, right 12% (rest in other directions).

Our attention may be called to two particular facts.

First: In the basement rocks, we miss the rhenish direction in spite of the immediate neighbourhood of the great lineament which proves its actuality by significant faults in the Mesozoic cover. But, it would be a heavy mistake to conclude that the rhenish faults are only younger. For, second: We observe a great preponderance of the diagonal joint system hercynian/eggish on the eastern side in contrast to the nearly exclusive Q-joints on the western side, though the striking of the varistian folds is the same in both areas. This speaks for the significance of a deep-sitting break, separating two different types of deformation within the basement — one argument more for the pre-varistian age of the great Rhenish-Lineament.

Egge-direction. — In spite of the high frequency of the hercynian elements, the mechanical comparison between the rhenish lineament and the Egge-direction is highly informative. In Thüringerwald and mainly in the Harz, quite a number of dikes and faults run SSE—NNW. The porphyry dikes and other extrusions in the middle part of the Harz are well known. Indeed, these eggish ruptures form there the right of the both diagonals, while the correspondent left one is given by the hercynian joints. But, it would be a mistake to understand them only like a varistian Mohr-System. We have already mentioned the ruptures North of Vogelsberg, bordering the bloc of Rheinisches Schiefergebirge from Giessen over Kellerwald up to the Egge-Hills near Detmold. Even in the ORG itself, from Heidelberg northwards, the tendency of the main faults is to turn over into the NNW-course. Also the Graben of Niederrhein in the whole follows the Egge-direction though the ruptures themselves often use or imitate the "franconian" Q of the basement-folds.

The Egge-direction is of particular interest because it shows a very characteristic connection with magmatic activities. Besides of the porphyries in the Harz, mentioned above, numerous basaltic extrusions in northern Hessen follow this course. The same do the Hegau Volcanoes, those in the Eifelzone and those along the middle Rhein. However, a group of basaltic dikes near Heldberg (S of Thüringerwald) strikes clearly NNE.

The supposition that the function of the eggish ruptures would be of only minor importance, since they are nothing more than secondary "en échelon" faults besides of the more significant rhenish elements, is not acceptable. It is true that the Egge-direction lies in the position of echelon faults when we believe that, along the Rhenish Lineament, the West bloc is drifting towards South in relation to the East. But, the regional distribution and the magmatophil character emphasize the mechanical independence of the Egge-element, as we will also see later.

Conclusions. — The Rhenish Lineament is an archaeo-rupture within the deep pre-Paleozoic base of the Varistian Orogen. It has not at all been influenced in its direction by the varistian orogenesis. In contrary, the Rhenish Lineament preserved its straight primary position, in which it took over a significant role, as an important shearing plane, in the varistian deformation act — without to be twisted or turned into another direction ("verschwenkt" Knetsch 1969, Abb. 2). We explicitly deny any influence of the varistian movements on the course of the rhenish main element. The visible break near Frankfurt is not real but is due to the increasing role of the ruptures in Egge-direction. This modifies the exterior shape of the rift zone without to change its r h e n i s h   c o u r s e. Though it seems that the "smoother" material of the "second floor" (Varistian) could be the reason for the fabrication of the split faults, there are also contradictory facts we find e. g. in the Massif Central.

3. Massif Central

Following the ORG towards South we enter another piece of the "Mittelmeer-Mjösen-Zone", what we may call the "Rhône-Saône-Graben" (Fig. 4), though it does not deserve this name but with reserve. The whole element, 500 km long in N—S and about 50 km wide, is a subsided stripe with — or without — a filling of Tertiary, between the Massif Central in the West and the Alps in the East. The northern end of the unit lies approximately 150 km west of the southern end of the ORG. The northern part of our object, between Port-sur-Saône (— Dijon — Châlons s. S.) and Macon, is indeed a graben with rhenish principal faults on both sides. Farther, over Lyon and Valence, the western border only is marked by a chain of faults which are running once in NNE, once in NNW, while S of Valence the western rim of the "graben" turns to NE—SW. The eastern border is not formed by normal faults but by the overthrusted folds of the Jura Suisse, the Alps and the Chaînes Vocontiennes (RICHTER-BERNBURG 1938). The southern end is given by the W—E striking Chaînes Provençales which belong to the Pyrenees-System.

In contrast to this relatively quiet zone of subsidence, the western neighbourhood is extremely broken. In the Massif Central we learn once more that the t w o m e r i d i o n a l d i r e c t i o n s a r e c l o s e l y r e l a t e d to each other. An analysis[2] of the greater faults (in the area of Fig. 4, excluded the ORG and the alpidic folded zones) involved about 10 000 km total length and admits their subdivision into:

11 %	faults of hercynian direction	N 110—130° E
6 %	faults of franconian direction	135—150°
27 %	faults of eggish direction	160—175°
43 %	faults of rhenish direction	10— 30°
13 %	faults of erzgebirgish direction	45— 60°.

An interior part of the Massif Central is, on the western side, bordered by a significant r h e n i s h l i n e a m e n t which runs through about 450 km from Najac (N Toulouse) via Figeac — Mauriac — Bort-les-Orgues — Bour Lastic — Montaigu — Noyans up to Nevers. However, within the 200 km wide interior part we observe a clear d o m i n a n c e of the E g g e - d i r e c t i o n. "Horsts" of crystalline rocks and deep "grabens", filled with Oligo-Miocene, alternate with each other. With two long legs the Limagne-Basin reaches from St. Pierrele Moutier 150 km towards SSE. The base of the Oligocene has been subsided more than 1500 m. It is not only the greatest number of the post-Oligocene faults which

---

[2] As the Carte géologique de la France 1:1 000 000 did not suffice for this interpretation the basic dates have been taken from all sheets of the Carte géologique detaillée 1:80 000 which concern the involved area.

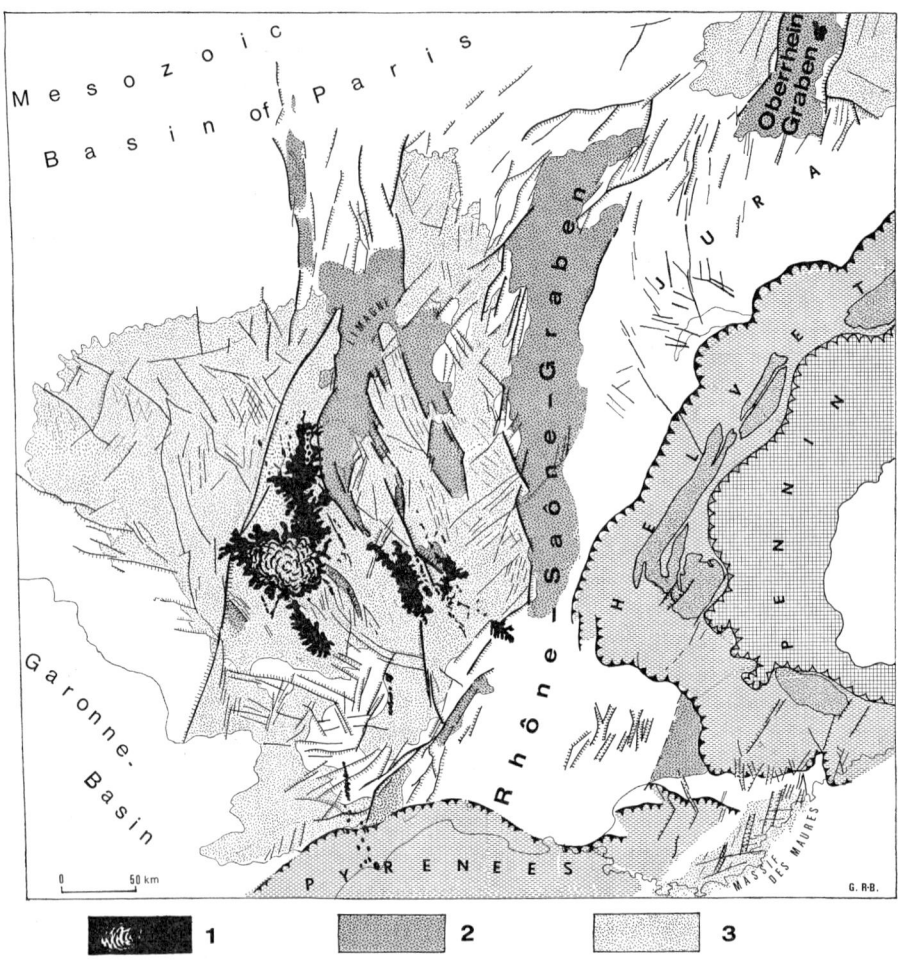

Fig. 4. Rhône—Saône-Graben and Massif Central. Faults in Eggish and Rhenish course preponderant.
1 — Kenozoic Volcanoes, 2 — subsided Tertiary Basins, 3 — pre-mesozoic basement.

follows the Egge-direction. Numerous pre-Carboniferous microgranitic and basic dikes in the metamorphic basement N of Limoges, E of Aubusson, and between Clermont-Ferrand and Lyon as well as whole the intensive volcanisme, from the Neogene until now, run NNW—SSE up to Narbonne or Privas. The faults reach northwards, with some smaller elements, nearly to the centre of the Basin of Paris. In the South, they touch the border of the Rhône-depression near Alès. Particularly here, the V-connection (mentioned above) of the directions eggish and rhenish, NNW and NNE, is very obvious.

Hercynian faults occur only West of the Causses and at the northern edge of the Garonne Basin. The erzgebirgish **direction** ENE is fairly frequent from South to North. Even if we admit that this is more or less the striking of the Varistan Orogen the coming through of this direction from the ground is unusual, mainly when we take in consideration that the NW-striking Armorican branche of the Varistan Orogen has another course. (We find perhaps an explanation; see p. 39.) Along the eastern border of the Cevennes (Flaviac — Privas — Aubenas — Alès) these faults in erzgebirgish orientation show a primarily low inclination towards East. In a later phase, these fault-planes have been deformed by tectonic stress and are overturned and bent into overthrusts (RICHTER-BERNBURG 1938, p. 115 ff.).

4. Rhenish direction and alpidic orogenesis

Western Alps. — Without leaving this area we enter the question about the relation between the rhenish — or the connected eggish — elements and the orogenic fold belts.

In SE-France, East of the Rhône-Saône-Graben s. str., we observe a swarm of faults which undoubtedly belong to our system: In the Vaucluse-Plateau (Banon-Forcalquier) a graben-family with NNW- and NNE-direction cuts the massive reef-limestones of Lower Cretaceous. Near Manosque, the Durance river runs about 50 km in rhenish direction. Further East, in the plateau of mainly Jurassic limestones (Mtge. de Barjaude etc.), in the area of Comps — La Bastide — Grasse — Nice, numerous meridional ruptures occur. Whole this wide stripe belongs to the stabile ridge, which is part of the foreland of the Alps in the North and the Chaînes Provençales in the South, and which enters the Massif des Maures and the Esterel in the East ("Vaucluse-Esterel-Schwelle" RICHTER-BERNBURG 1938). In the latter area, some graben-faults in rhenish course show the Stefanian, broken down between cristalline rocks already earlier than Autunien. Therefrom we may learn that the meridional ruptures in this region are an old heritage.

The occurrence of faults in the stabile foreland of folded belts is not strange — although the directions NNW and NNE may be remarkable in this place where neither any reasonable mechanical relation with the varistian basement nor an immediate neighbourhood with the main N—S-element seem to exist. Moreover, most of these faults run continuously against North through the alpine folds (as S of Castellane and near Nice). Some other rhenish ruptures cut the Chaînes Provençales (near St. Maximin — Barjols, Cotignac a. o. places). All these show a modification or deformation by the alpine or pyrenaic folding.

When we see that the Durance element finds its straight northern continuation to Briançon, sharply bordering the SE-side of the Pelvoux, we have in same time to remark that the Helvetic Crystalline Massives of Belledonne, Aiguilles Rouges and Mont Blanc have a straight rhenish orientation which does not correspond with the harmonic curve of the Pennin-Zone in the Western Alps (see Fig. 4). Also, we may not ignore that these massives are located in direct southern prolongation of the ORG and that the ruptures Pelvoux-

Durance and others in Provence and Maures could be regarded as g r o u n d -
f l o o r c o n t i n u a t i o n of those.

We may even suppose that the front of the alpine folds at Corsica follows an older-established NNW element which has some parallel faults in the Central-Sardinian Graben. In the southern and eastern part of Corsica also the rhenish direction is present in several faults of more than 50 km length (see Fig. 7).

Occasionally, if there were older elements, they may also have been integrated and we may only divine their primary existence.

E a s t e r n A l p s. — We get some other facts from the Eastern Alps (see Fig. 5). By measuring all significant ruptures[3] — evidently neglecting the alpine overthrusts and related planes — we got as interpretation of totally 9400 km fault-lines:

in the Alps (4800 km)	in the Hercyn-Bohemian Bloc (4600 km)	
10 %	49 %	in hercynian (+ franconian) direction
39 %	16 %	in eggish direction
48 %	30 %	in rhenish direction
3 %	5 %	in erzgebirgish direction

If the alpine orogenesis found a varistian or older pattern of ruptures like that in the Bohemian Bloc, h e r c y n i a n and franconian elements can hardly be expected for, these would surely be a s s i m i l a t e d because of their striking ± parallel to the young folds. Perhaps, the Karnish Alps — Karawanken-Lineament (Gail valley) could be one of such old elements. From Villach, the line Drau—Möll-Valley continues up to the Sonnblick (see EXNER 1964) in typical hercynian direction over 100 km. (Here, the problem rises how this may coïncide with the theory of the extreme nappism because this lineament cuts the "Tauern-Fenster" as well as the overthrusted covering units.) Parallel elements also exist, e. g. near Drauburg a. s. o.

E a s t from here, up to the Basin of Wien, the cristalline rocks and the Paleozoic show a great number of faults in the E g g e - d i r e c t i o n. The most outstanding one is the (about 120 km long) graben and fault chain of the L a - v a n t - Valley (between Sau-Alpe and Kor-Alpe) running from Rottenmann — Hohentauern — Judenburg — Reichenfels — Lavamünd — Unterdrauburg — Straže where it closes up with the Karawanken-Lineament (SCHAFFER 1943). Also at the westside, the Sau-Alpe is bordered by a more than 60 km long fault in same NNW-course. And the "end" of the East-Alps is marked by a (50 km long) eggish fault Wiener Neustadt—Köszeg.

Two facts are obvious. First: North of the tectonic border of the Alps, within the H e r c y n - B o h e m i a n B l o c, the E g g e - direction, as well as the

---

[3] What we did here is not perfect at all, for we used not so much more than the Geological Maps 1:1 Mill. of Austria and Czechoslovakia. However, for the Garda Lake area the Carta Geologica d'Italia 1 : 100 000 with all involved sheets has been searched.

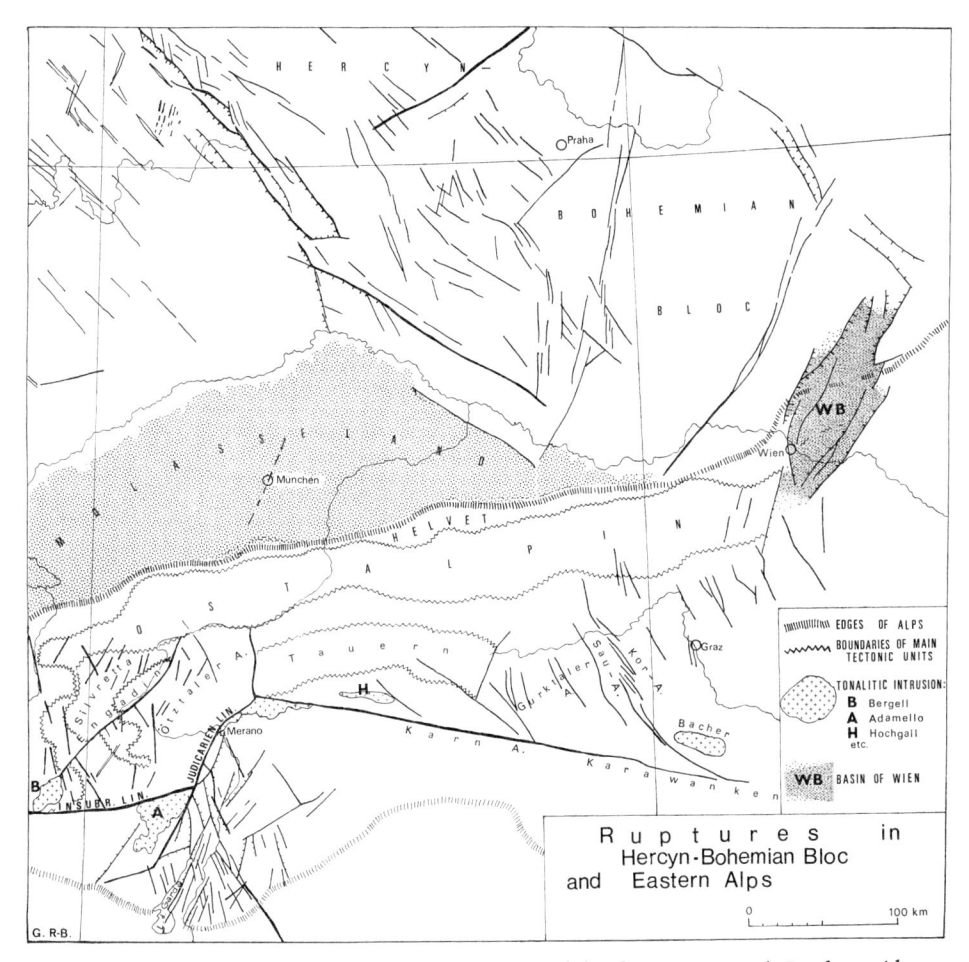

Fig. 5. Ruptures in the Hercyn-Bohemian Bloc and in the Eastern and Southern Alps.

hercynian evidently, is common again. We mainly call the attention on the continous zone of Budweis (Ceske Budejovice) — Pisek — Pilsen (Plzen) N, where it aims at the great young volcano of Doupovske Hory and reminds us of the basalts upon the Egge-faults in Hessen. We could mention other ruptures in NNW-direction. Second fact is that no comparable faults cut the nappe system of the mesozoic "Ostalpin" and "Helvetikum". It may be concluded that the varistian — or pre-varistian — j o i n t - p a t t e r n of the B o h e m i a n B l o c c o n t i n u e s in the g r o u n d - f l o o r into the old cores of the A l p i d i c O r o g e n — while the mesozoic nappes, in a much higher tectonic floor, did not get any notable infection, if they were thick enough.

Moreover, we see an interesting difference between the NNW- and the NNE-elements. Some very important r h e n i s h striking lineaments run through the

Bohemian Bloc in the whole. That is first the about 300 km long straight cut from Linz/Donau to East of Praha — Görlitz/Neisse, and the chain of ruptures between Melk and Krems/Donau up to Mohelnice a. s. o. By the involvement of Tertiary, these faults show their relation with the most significant element in this matter, the B a s i n o f W i e n. This tectonic unit has to be considered as a large g r a b e n - z o n e which got its recent form only after the alpidic overthrusting movements took place. All the more it may be remarked that — in contrast to the widespreaded Egge-elements — the presence of rhenish faults in the Eastern Alps, besides of the Basin of Wien, is extremely rare (see also p. 30 ff.).

S o u t h e r n a n d C e n t r a l A l p s. — We have to move about 300 km more West for entering another area with well marked meridional dislocations. Between Meran and the Lake Garda the "J u d i c a r i e n - L i n e a m e n t" offers a very significant r h e n i s h direction (see Fig. 5). The northern prolongation, aiming at Innsbruck, is, in same time, the western border of the "Tauern-Fenster". In the South, many parallel faults cut the area arround the Lake Garda. Some of these ruptures are transformed into steep overthrusts against SE (DAL PIAZ et al.) without loosing the indications for their primary character of normal stepfaults. The whole reminds of the border of the Cevennes, all the more as here and there the erzgebirgish direction also occurs with notable faults.

Besides of these, the NNW-direction plays a significant role too. G r a b e n - f a u l t s and other long breaks run in E g g e - direction. Here again, the narrow relation of this course with the m a g m a is documented by the volcanoes near Schio and in the Colli Eugánei which are sitting on the southern-most eggish rupture. Whole the area seems to be principally m a g m a t o p h i l, as we have b a s a l t s in the Eocene, rhyolites and dacites in the Ladinian as well as the enormous p o r p h y r y - volcanism in the Lower Permian of Bolzano etc., the latter an equivalent to the porphyries of the eastern Harz, extruding also from eggish faults. However, the most famous magmatic activity in this area is represented by the t o n a l i t i c i n t r u s i o n s which are known from Bergell, Adamello, M. Croce/Merano, Hochgall and Bacher. They thread ± parallel to the alpine folds, following the "Insubrian" or Tonale-Lineament and the Karawanken-Lineament. In the middle, the "Judicarien-Lineament" strikes in rhenish direction. The junctions of the ruptures at Merano, Adamello, Bergell are preferred by the tonalites, because there the connection with the magma were obviously the best. And as in the paragneises and micaschists of the Val Venosta, W Merano, the dikes of granodiorite and granite are already orientated parallel to the J u d i - c a r i e n - Line, we may believe that this r h e n i s h element is of p r e - a l p i - d i c a g e and comes through again from the v e r y d e e p b a s e m e n t.

We get more arguments for this consideration from the area little more West. The E n g a d i n - L i n e is another significant element almost parallel with the Judicarien-course (see KRAUS 1936). On the way from St. Moritz over Zuoz to Zernez, all tectonic units of Silvretta-, Vadret-, Aëla-, Err-nappes in the West are displaced, along the upper valley of river Inn, towards NE and find their pro-

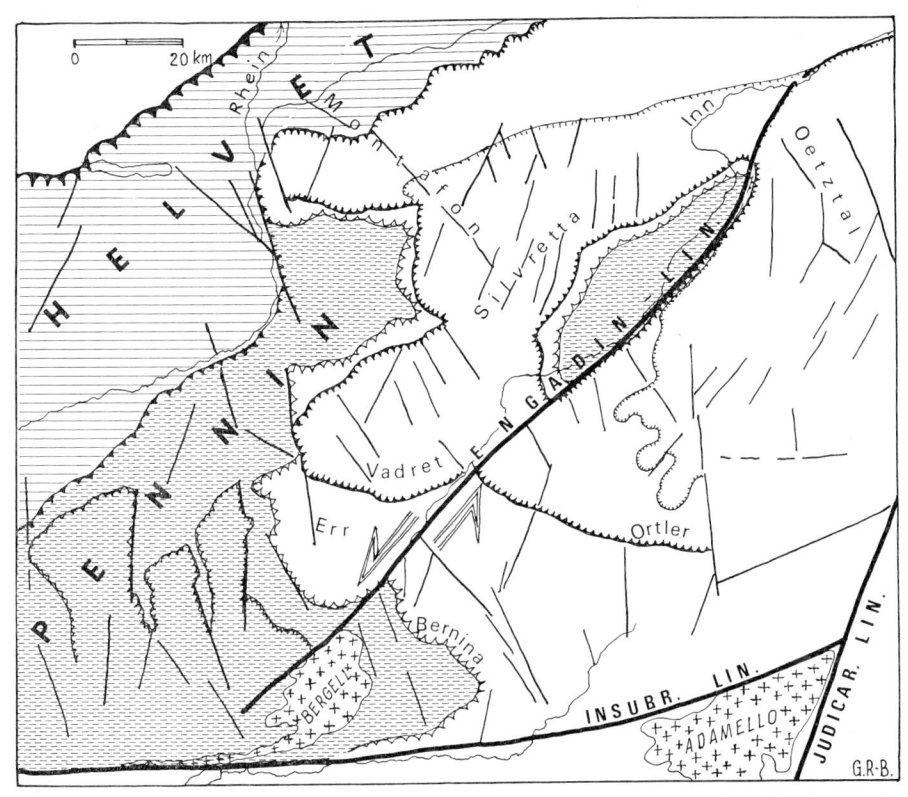

Fig. 6. The rhenish Engadin Lineament and eggish ruptures, supposed as fundamental boundary West Alps/East Alps.

longations on the other side in Bernina-, Quatervals-, Ortler- and Oetztal-nappes (see Fig. 6). Numerous parallel ruptures of NNE-course — in total more than 300 km — are to observe on both sides of the main lineament. It is remarkable that also a swarm of important ruptures — in total about 750 km — in the Egge-direction cut just this particular area which is most significant as the tectonic boundary between West-Alps and East-Alps (RICHTER-BERNBURG 1951). The principal element of these NNW dislocations runs from Tirano in the Adda-Valley over Poschiano — Pontresina and continues over Chur — Ragaz — Vaduz towards the North. We suppose that there is a groundfloor connection with the Hegau volcanoes on this northern prolongation and — some interruptions admitted — with the Colli Euganei in the South.

Conclusions. — Even in the body of the Alpidic Orogen, the old pattern in hercynian, rhenish and Egge-direction is still permanent. Some of these elements are heavily influencing the alpidic structures though, in the upper mesozoic and metamorphic nappes, the denudation-level lies much higher than in the Varistian Orogen. In the Eastern Alps, the joint design of the Paleozoic is almost no different from this in the Bohemian Bloc.

The Judicarien-Line (Innsbruck—Garda) is an old and important rhenish lineament, and the parallel element of Engadin is not only a significant shearing plane but probably the ground-floor located boundary of West-/East-Alps. In the southern prolongation of the ORG other basal ruptures modify the structures of the western Alps (Helvetic Cristalline Massifs) and reappear as rhenish faults in the Provence where the varistian age of some of them is proved. However, we admit that also some ground lineaments, here mainly the hercynian because of their similar strike, may have been integrated and overwhelmed by the younger folds, though we have indications for their untouched persistence (f. i. Villach—Sonnblick-Lineament). And the very deep e g g i s h l i n e a m e n t H e g a u — E u g á n e i with hypogene volcanoes on both sides, seems to remain latent in some intervals.

5. The northern part of the Rhenish Lineament

The northern piece of the "Mittelmeer-Mjösen-Zone" (M-M-Z) appears desintegrated. In the area of the "Saxonian Tectonics" STILLE's the Rhenish Lineament is northwards more and more choked under halokinesis and the cretaceous and neogene sedimentation. The e l o n g a t e d s a l t d o m e s i n r h e n i s h d i r e c t i o n are initiated by long faults the distribution of which is bordered by an eastern line Wolfsburg — Lübeck — Eutin (Fig. 1). There is no clear western boundary for salt domes or salt walls with NNE course. They are spread over NW-Germany and through the southern North Sea. More or less in the median line of the sea — between Norwich and the coast of the Netherlands — there is the 50 to 80 km wide "N o r t h S e a G r a b e n", altogether about 300 km long, running in the southern part approx. in rhenish, more North in Eggedirection (BOIGK 1972). Its shape seems to be similar and parallel to the structures in E n g l a n d itself where rhenish striking faults E of Wales from Bristol over Birmingham and from Shrewsbury over Stoke-on-Trent, and Egge-ruptures farther North from Liverpool to Whitehaven and in the graben of Carlisle are dominant. The Great Glen Fault in Scotland, with a length of more than 200 km, has to be mentioned as a very spectacular and most rectilinear rhenish element. The North Sea Graben doesn't touch the Norwegian Channel. Between them lies the Jutland — Fuenen-High, a NW—SE striking uplift without any meridional structures, known as remarkable.

All the more interesting is the Scandinavian part of the M-M-Z (see Fig. 8). The R e n i s h L i n e a m e n t its represented first by the famous O s l o - G r a b e n, bordered at the W side by a very clear, 175 km long fault between Nordagutu in SSW and Gjøvik/Mjøsa Lake in NNE. The E-side where the fault planes are beautiful exposed on the shoreline of the Oslo-Fjord, is built "en echelon", and NNE- and NNW-running pieces produce a nearly northern striking border — very similar to the W side of the Rhône Graben near Lyon (see above). The northern continuation runs also NNW and NNE, respectively. There are three main elements arranged upon a NNE striking line with NNW running members: one E of Glomma valley up to Tynset, another following the Trysil Elv and Femund up to Meråker, the third W of the Blåfjell granite and W of the Tromsfjell granite, East of Namsos. These are the main rupture elements. But, when we follow exactly the renish direction from the Mjøsa Lake we reach the NNE striking eastern front of the Caledonian Orogen. STILLE (1947) already emphasized that the M-M-Z has to be regarded as a "Mittelmeer—Lappland-Lineament" the northernmost piece of which plays the role of the border of the Caledonian Geosyncline or Orogen, respectively. (Whether or not this is acceptable, see later.)

Besides of the spectacular rhenish elements the E g g e - d i r e c t i o n is by far dominant. The most significant ruptures are the deep Norwegian Channel and the nearly 1000 km long K a t t e g a t t - L i n e a m e n t which runs NNW from Bornholm Island (v. BUBNOFF 1953) over Schonen, the eastern Kattegatt coast through Langesund and Ålesund/Atlantic Ocean. It is of interest that, in the Dalslandian of Telemarken, enormous a c i d e x t r u s i o n s are connected with this great rupture. Within the Oslo-Graben itself, parallel joints gave the way for h y p o g e n e magmatic material (BRØGGER) as the permian essexites, kjelsaasites etc. Ultrabasic rocks are particularly arranged on lineaments not to

Fig. 7. Map of the Faeroe Islands, showing the dominance of Egge direction in the North Atlantic.

see otherwise than by the row of these old volcanoes (RICHTER-BERNBURG 1968, fig. 11). This tradition of basic volcanism upon Egge lines is continued by the Schonen basalts as well as in the Faeroe Islands (RASMUSSEN & NOE-NYGAARD 1970) where volcanic rocks of Eocene age are evidently cut into NNW striking pieces (Fig. 7).

The meridional strike is very old in Scandinavia. In the Svekofennian structures, the NNW course is general and morphologically very obvious (see HOBBS 1911, fig. 1 and 19; SEDERHOLM 1913 a. o.). However, in southern Sweden, basic dikes and ruptures run also NNE, and Oeland and Kalmarsund are oriented in rhenish direction, too. Particularly there where we are convinced to see one of the o l d e s t p a r t s o f t h e E a r t h ' s f a c e the direction of the wrinkles is preponderantly e g g i s h and r h e n i s h.

6. View over the West European ruptures

The significant Rhenish Lineament (Fig. 7), in its run from Marocco through central Europe up to northern Norway, follows a more or less s t r a i g h t l i n e SSW—NNE. It doesn't show an uniform shape but it looks like a chain with fairly different links. The most important character of all single pieces is the d i l a t a t i o n the effect of which is pointed out at several places as around 5 km in W—E (LOTZE, RICHTER-BERNBURG, ILLIES). The structure is almost everywhere the form of a g r a b e n the border faults of which incline normally towards the subsided bloc.

The pieces of Rhône-, Rhein- and Leine-Graben are arranged in e c h e l o n r o w. That speaks for a h o r i z o n t a l s l i d i n g movement of the western bloc relatively towards South, or the eastern towards N, respectively (RICHTER-BERNBURG 1934, 1968). With local and regional arguments, the same trend has been referred for the Oslo Graben (BEDERKE 1965), the Glen Fault in Scotland, and for the ORG itself (v. BUBNOFF 1953, ILLIES 1971, p. 27). It may also be postulated because of the arrangement of the echelon faults in the Massif Central, and it is obvious along the Engadin and the Judicarien Lineaments.

The width of the "Rhenish Lineament" is very different. We have to note that, in the larger area around the North Sea, the rhenish elements are spread over a W—E distance, from the Irish Sea to Lübeck, of nearly 1000 km!

The Rhône-Saône-Graben itself is only one part of whole the width of the unit what we measure with 250 km. But, therefrom East, the B a s i n o f W i e n is a respectable graben in NNE direction, too. We risk to suppose that the bend of the Carpates, shearing northwards around the Bohemian Bloc in the North, as well as the break in the Apennines in the South, have their reason in a movement along a significant r h e n i s h g r o u n d f l o o r l i n e a m e n t Roma — Graz — Wien — Ostrava. In the mean space, the central part of the Alps, the Judicarien and the Engadin lines, both in rhenish direction, are very important lineaments. We see in these three parallel rhenish elements an essential influence on the alpidic structures: the rhodanian ruptures are the western border

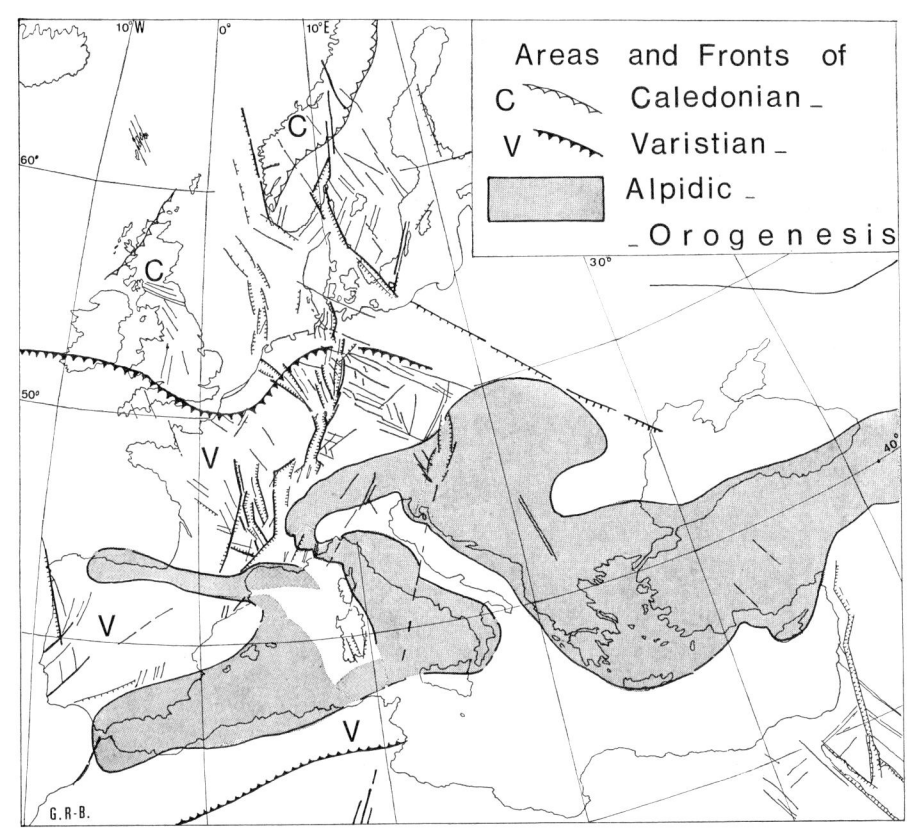

Fig. 8. The principal faults in western Europe. — To notice: the splitting of the "Rhenish Megalineament" as well as the zone of exceptional tectonic calmness — besides of ORG — between the Channel and the East Bavarian Alps. (The Glen Fault in Scotland is missing by mistake.)

of the Alps at all, the Engadin Line is the western border of the "Ostalpin", the lineament through Wien is the boundary of "Alps" and "Carpates". The question may be whether the latter is a part of or a competition for the main — rhodanian — element which lies about 800 km farther West.

The only place where parallel — or related — ruptures do n o t diffuse the Rhenish Lineament is the ORG itself. From the Channel in the West through the Basin of Paris, on the left side, and over Süddeutsche Großscholle up to Salzburg in the East, we state an about 300 km wide stripe of s i g n i f i c a n t t e c t o n i c c a l m n e s s. We suppose that this fact is the main reason for the very clear and straight shape of the ORG and its spectacular textbook figure where only two cardinal faults, in an average distance of 50 km, take over the whole amount of tension. It is like a canyon through a rocky barrier. There is no motive for this calm bridge between the eastern Alps and Middle England, for ruptures are there extremely rare also in other directions, though, at least in southern Germany, the Mesozoic cover is thin enough for being transparent for ground floor impulses.

In a distinct contrast to these facts, N of Frankfurt as well as in the Massif Central, the splitting of the uniform rhenish main faults happens nearly explosively. It is not only caused by parallel elements but also by subparallel ruptures. The brother element of the rhenish strike is the E g g e - d i r e c t i o n. All over where the Lineament appears split, the NNW faults play the most important role. They also occur in other places i n s t e a d of the NNE elements. Thus, we find in Europe extremely significant eggish lineaments in Portugal, in Corsica-Sardegna, in the Massif Central, in the Hessische Senke (Egge etc.), in the Niederrhein-Graben, in England, in the North Sea, in West Norway, in the Kattegatt-Lineament and in thousands of other ruptures in Scandinavia. Two greater NNW striking breaks, though partially hidden, cut also the Alps: the one from Hegau to the Colli Euganéi, the other runs Pilsen — Lavant etc. We believe that the poor appreciation of the NNW ruptures in the geological literature does not coincide with their omnipresence and their significance.

Moreover, we observe a marked p r e f e r e n c e   o f   t h e   m a g m a t i c   a c t i v i t y for the eggish fractures. In the older geological times, the magmatism along these lines was principally of acid character: the rhyolites in the Dalsradian, the enormous porphyries in the Harz, in Sachsen, in the southern Alps (Bolsano) during the Permian, and the microgranites in the Massif Central. Later only, from the Eocene until now, deep crustal and mafic material has been produced via NNW fractures: beginning with the essexites near Oslo, the basalts in Schonen and the Faeroe Islands, in Hessen, near Karlsbad, on the Hegau — Euganei lineament and, most effectively, in the Massif Central. Compared with this magmatic abundance on the NNW faults the volcanism is extremely poor on the rhenish elements. Concerning the Egge ruptures this is an indication for a r e a c h   i n   a   g r e a t e r   d e p t h.

There is another difference more. In general, the NNW faults dip s t e e p e r than the faults in rhenish course. The d i l a t a t i o n is therefore l e s s   m a r k e d. Although we do not see significant transcurrent displacements[4] we suppose that the E g g e elements in the whole, with more or less vertical faults or ruptures, represent mainly s h e a r i n g planes, while the r h e n i s h elements are in general d i l a t a t i o n fabrics. Both of these most significant rupture directions occur, everywhere in Europe, together with each other.

Nevertheless, it does not seem that they are a couple of a diagonal (MOHR-) system because the angle between both appears too small. It may be that each of them constitutes a partnership with the hercynian and/or the erzgebirgish direction (in the sense of STILLE or KNETSCH). However, the question rises whether the hercynian and the erzgebirgish elements which are "non-existent" in some areas, did really never exist there, or whether they remained — or still remain — latent,

---

[4] However, there is a mylonitic Egge lineament in the Massif Central, the 150 km long zone of Argentas (Bourgauneuf — Figeac).

waiting for an opportunity to get in action. For instance, although numerous h e r c y n i a n elements have been reactivated within the Hercyn-Bohemian Bloc in post-Mesozoic time, these long faults which look like satellites of the great Schonen-Pontus-Lineament on the southwestern border of the large Russian Table, seem to be d i e d o u t West of the cardinal Rhenish Lineament, excepted some sporadic and regionally unimportant faults.

Evidently the r h e n i s h and the e g g i s h f r a c t u r e s were the m o s t e f f e c t i v e e l e m e n t s — from the Precambrian until now — to d i s - c h a r g e t h e c r u s t a l t e n s i o n s in Europe. —

7. The meridional fractures of the Earth

Regarding the globe we state that no other direction in the Earth's crust is more common than the two submeridional lines (Fig. 10).

In E a s t A f r i c a , the vicarious occurrence of NNE and NNW elements belongs to the character of this famous Rift Zone. The most interesting for our

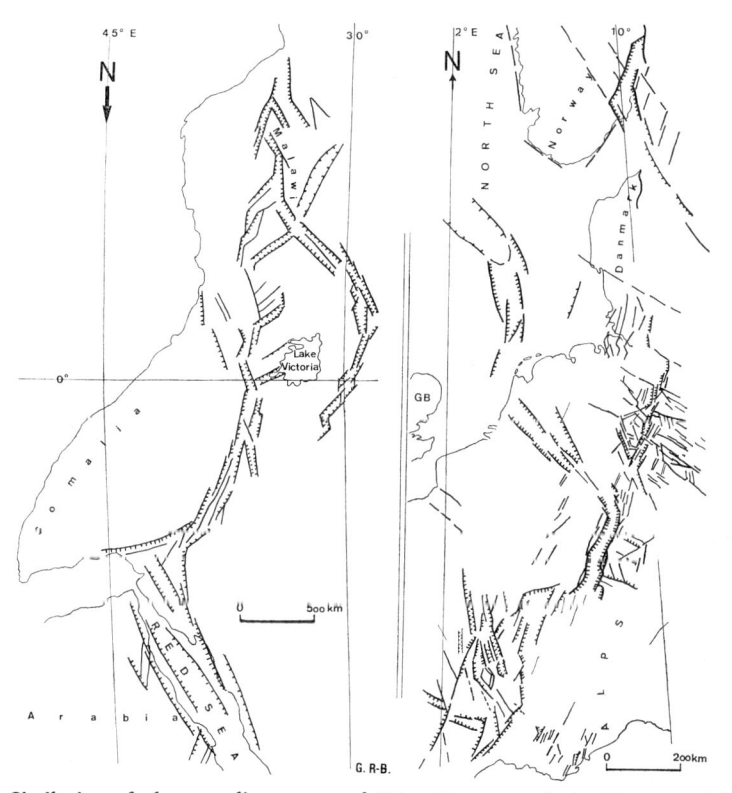

Fig. 9. Similarity of the megalineaments of West Europe and the Eastern Africa (the latter South above).

problem is the junction Red Sea/Jordan. The great V of the Sinai peninsula divides the Gulf of Suez from the Gulf of Aqaba. The latter continues over Wadi Araba — Dead Sea — Jordan — into the Bekaa Graben. This J o r d a n L i n e a m e n t strikes in exemplary r h e n i s h direction very rectilinear over 1000 km. Although this is more than three times the length of the ORG the width is only 20 km in average. The S u e z (= E g g e) d i r e c t i o n, obviously represented by the R e a d S e a, seems to be much more important because of the straight length of 2300 km and the depth of nearly 3000 m, as far as it is not filled by magmatic material. Here also the relation of the two meridional directions is evident and may stress their equivalence.

Another important fact is that the Egge direction has everywhere an excellent c o n n e c t i o n w i t h t h e U p p e r M a n t l e. In the Read Sea, in the Erythrean and in the farther E a s t A f r i c a n R i f t, the highest activity of hypogenic volcanism can be observed — when we exclude the oceans. North of the V of Sinai, faults in Suez (= Egge) direction show large basaltic lava flows, too. Contrarily, on the rhenish striking Aqaba — Jordan faults, we miss any magmatic events except sporadic ones.

The similarity of the East African Rift with the European N—S zone is evident (Fig. 9). The Jordan Lineament is regarded as a transcurrent fault with a sinistral shifting, West towards South, like in western Europe. The referred displacing amount of 90 km (GIRDLER 1965) — denied by others (PICARD 1965) — may not be equalized with the dilatation effect of the Read Sea because this is max. 50 km (KNOTT 1965) according to the median rift which is filled with magma. However, the — not only relative but also real — northward movement of the eastern bloc is made probable by the existence of the compression folds of the Syrian Arc. Of course, the dilatation effect of the Jordan Lineament may not be neglected. Altogether the drifting vector of the Arabian Bloc is NE (n o t NNE acc. GIRDLER).

In A f r i c a, we should not ignore a significant group of meridional fractures in the central western part of the continent. There, the Ahaggar Mountains are crossed by important NNE fractures of several hundreds km length each. The whole lineament Accra—Tunis runs in rhenish direction — perhaps with a continuation into the Guinea ridge below the Atlantic. Parallel ruptures occur between 0° and 20° E, coming from the mouth of the Niger and reaching the Gran Syrte in the North. In the Tjad Lake area, NNW ruptures show a course toward the Ahaggar, too. No specific knowledge exists about the mechanism of the structures (displacing etc.).

The r h e n i s h T o g o — T u n i s - L i n e a m e n t lies in the exact continuation of the Wien—Roma-Lineament. It demands our interest that the Alps seem to be translated towards S in respect to the Carpatic Arc, and we could consider the bow of the northern Apennines as an overwhelming of the south-shifting foreland by the orogenetic folding process. We therefore suppose that the whole M e g a l i n e a m e n t T o g o — W i e n is also a transcurrent fault with a sinistral displacement like the meridional megalineaments in western Europe

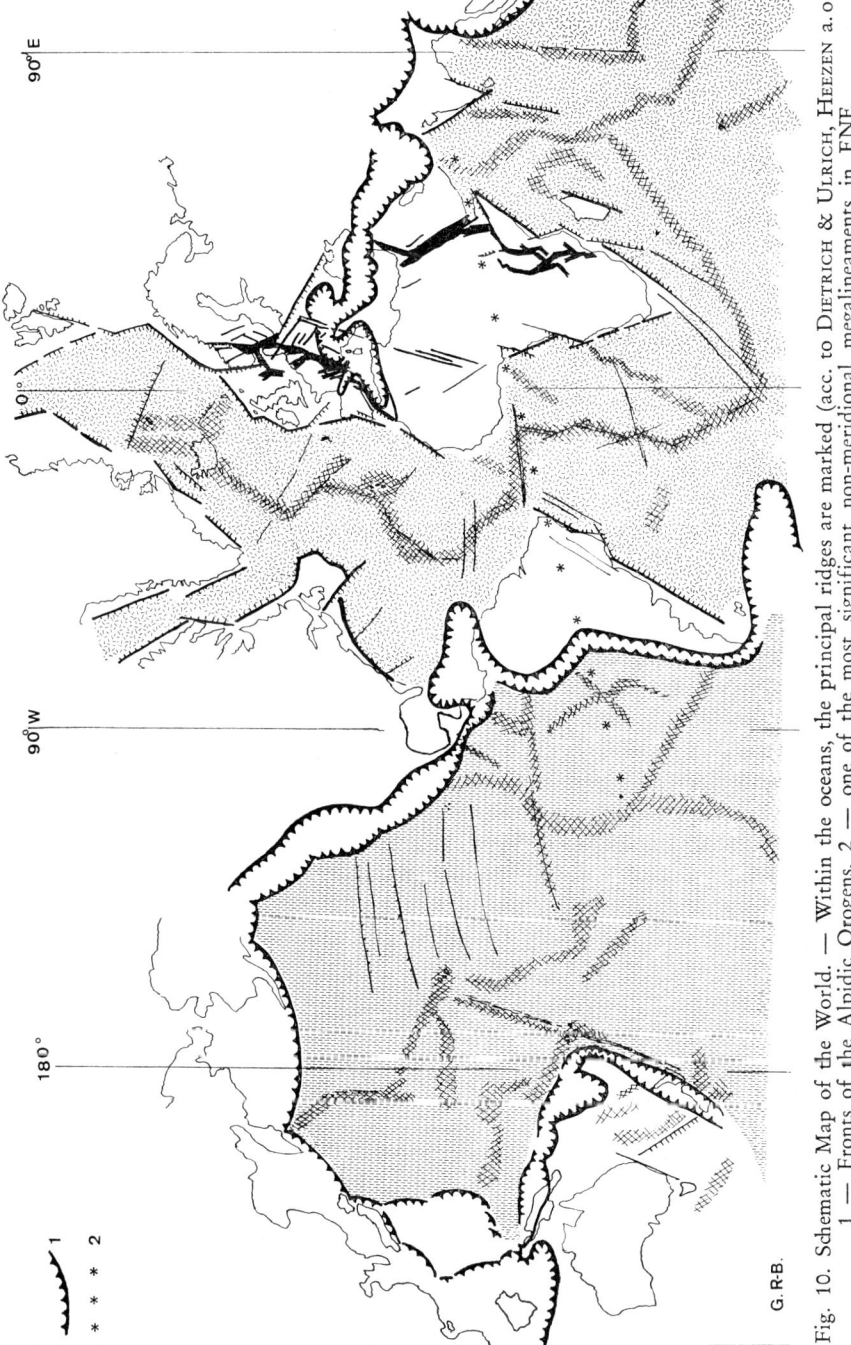

Fig. 10. Schematic Map of the World. — Within the oceans, the principal ridges are marked (acc. to Dietrich & Ulrich, Heezen a.o.). 1 — Fronts of the Alpidic Orogens, 2 — one of the most significant non-meridional megalineaments in ENE.

and along East Africa. Thus, from Scotland through Europe up to Somalia the southern trend of the West seems to be approved. Nevertheless, the dilatation in W — E is certainly the principal effect of all these meridional fractures.

Regarding the Mid Atlantic Ridge in its southern part in an only morphologic map which is not geologically interpreted yet (DIETRICH & ULRICH 1968) we observe pieces in Egge direction as well as in rhenish course, like the Walfish ridge and the Guinea ridge. They are apparently in correlation with the configuration of the African West coast — not so good with the South American East coast. Admitted that the Mid Atlantic Ridge is built up in a spreading process, the bursting of the ocean floor already followed the two meridional joint directions documented upon the continental bloc Europe-Africa. When we put together South America and Africa, according to WEGENER, the rate of turning is about 40°. The coast of South America, today in nearly classic "rhenish" direction, rotated back by 40°, in the position of touching the mid oceanic ridge, shows also elements of vicariously rhenish and eggish course.

We know that on the one hand in the Euro-African Bloc the recent and Tertiary bursts do nothing else but to follow the old lay out of Paleozoic or pre-Paleozoic archaeoruptures what we could approve. On the other hand, if it is correct that the southern Atlantic Ocean is only less than 200 million years old (FISCHER et al.) the rhenish and the eggish directions, we observe in Europe, are not a consequence but they can be seen as an originating event for the sea floor spreading. It seems to be obvious that the continental drifting does not start from the ocean but that the bursting of the continental crust which happens hundreds of million years earlier gives the impulses for the location and the shape of the origin of the oceans, by following the elements of a very old pattern.

In the whole separating process, Africa had no motive to rotate even if it was drifting East. Just as much we have no arguments for a turning movement in relation to Europe or of Europe itself. For, in the far North, we find confirmed that the meridional elements do not lose their straight direction. Following the European megalineament we reach the Scandic Ocean. There, the British Norwegian shelf is bordered on its western side by a steep slope running NNE nearly up to the North Cape. Then, the same slope turns into the NNW course. Spitzbergen, with an excellent V border, has sharp contures in rhenish and eggish directions but it remains on the East side of the Scandic deepsea. Greenland, on the other hand, is contoured by mainly NNE ruptures. North of about 73° lat. a deep and fairly narrow channel separates the Greenland shelf from the Nansen ridge on the eastern side. In rectilinear NNW course, it joints the Eurasia Basin. The latter, the Lomonossow Ridge, the Middle Ocean Ridge and the Fram Basin are running parallel to each other in meridional direction straight over the North pole, about 2500 km, nearly approaching the New Sibirian Islands (see Fig. 11).

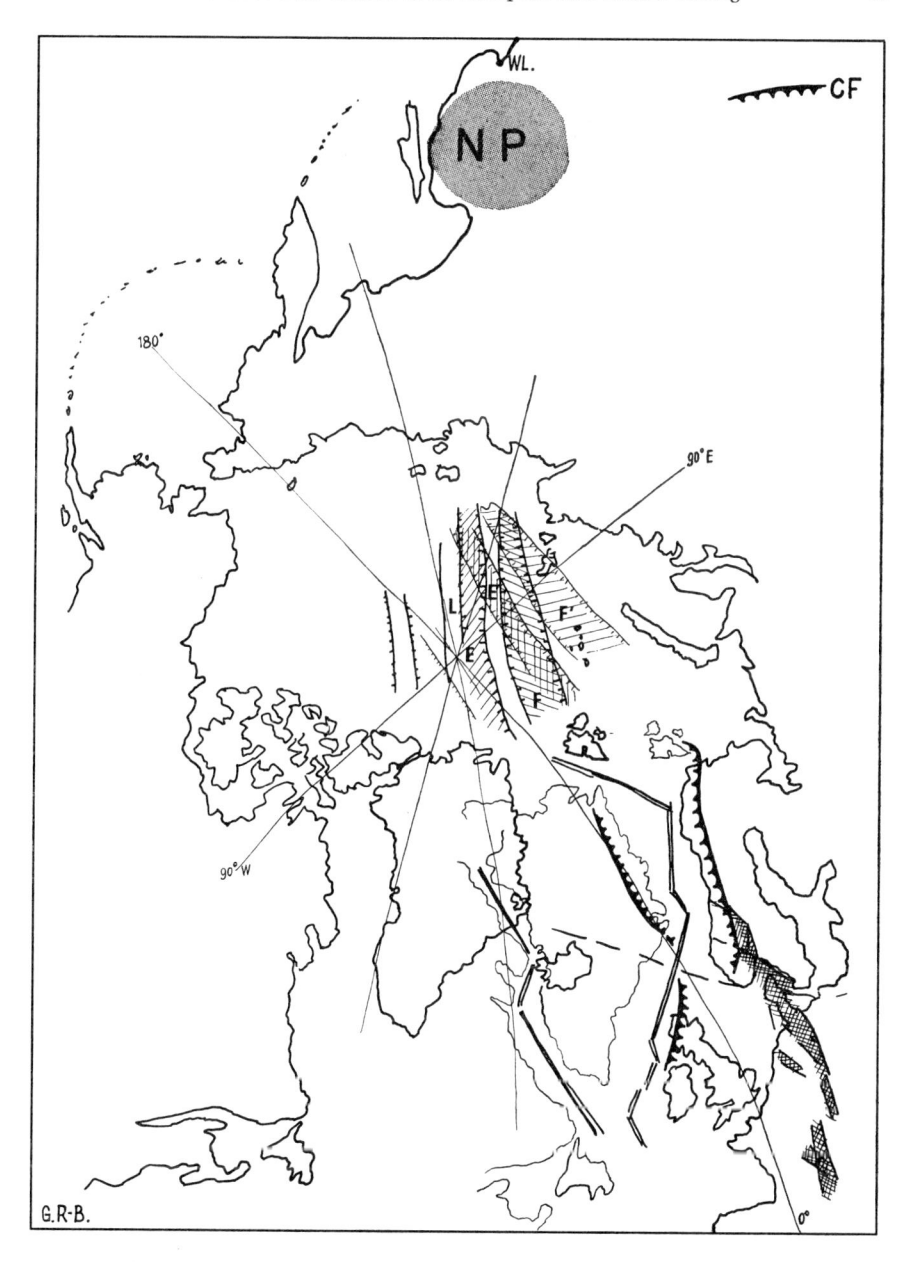

Fig. 11. The Rhenish Megalineament (RH), continuing through the Scandic Ocean. — There, the Fram Basin (F), the Middle Oceanic Ridge, the Eurasia Basin (E), and the Lomonossow Ridge (L). Meridian lines of today, N P — late-paleozoic North Pole (WL — Wladiwostok). The position of North America and Greenland etc., also Fram Basin (F, F') etc. today and before the West drift. C F — Caledonian Fronts.

The paleomagnetism gives us some further indications. The measurements show that, in late Paleozoic times, the North pole was at long. 120/130° E and lat. 40/50° N on the base of North American rocks, but at long. 165/180° E and same lat. on the base of European rocks (RUNCORN 1962). The latter only are significant for our problem. Placing North America 40° towards East, so that it touches the West European shelf according to the continental drift, the determinations get the same — European — result. Then, the North pole has to be localized in NE Asia (near Wladiwostok). Respecting this paleogeography, our Rhenish Megalineament continues from Europe with NNE and NNW elements over 50 Paleozoic degrees of latitude more than it appears in our recent geographic view.

We therefore assume that the geographical lay out of the m e r i d i o n a l d i r e c t i o n s — which are essential parts of the oldest joint pattern we have — was more or less c o n s t a n t   i n   r e l a t i o n   t o   t h e   p o l e s, at least from the Paleozoic until today. It may be that the NNE or the NNW element, respectively, was the first and that the other one has been formed somewhat later, depending on the momentary position of the pole.

Altogether, the two related directions of l o n g i t u d i n a l fractures with W—E d i s t e n s i o n exist through the geological time. They run ± perpendicular to the l a t i t u d i n a l   c o m p r e s s i o n and with that to the preponderant strike of the orogens in the non-pazific part of the Earth.

Two principal e x c e p t i o n s may be mentioned: the C a l e d o n i a n O r o g e n and the U r a l. We accept that the old Laurentic Mass, with North America and Greenland, drifted westwards from the opening Scandic Ocean while, earlier, it was close together with NE Scotland, and the nappes of Moin were the immediate continuation of the East Greenland overthrust. In that time, the Caledonian Orogen has been folded with a rhenish strike. From the cross of the "Mittelmeer — Lappland — Lineament" with the "Greenland — Island — Pontus-Lineament", STILLE (1947, fig. 5) deduced a diagonal system of the cardinal directions rhenish and hercynian. And he postulated that the different mechanism — compression or distension — may produce different deformation structures, once folds, once faults. In principle, I agree (also with KNETSCH 1969), but with some differentiation. The t e n s i o n which is the reason of all deformation acts is in part g l o b a l, in part largely r e g i o n a l, and in part also l o c a l. (The latter does not interest now. We come back to the global tensions soon.) The regional tensions may be induced from below and may be caused by currents in the Upper Mantle (VENING-MEINESZ 1964). The large Russian Table seems to be a stabil plate at the border of which the Caledonian Geosyncline has been squeezed out by a NW drift of Russia against Laurentia. In varistian time, the Russian Plate moved towards SE, folding the Ural Mountains on the frontside and shearing along the hercynian "Island — Schonen — Pontus Lineament" (see Fig. 11). The latter what we get to know as pre-varistian, was probably the gliding plane for both of these movements and was also important with a lot of parallel faults though it was losing its significance towards S and W and during the Cretaceous. All these events seem to be typical for the s u b c r u s t a l c u r r e n t e f f e c t (see also ILLIES 1965).

The g l o b a l t e n s i o n s — excepted the Pacific area — mainly produce l a t i t u d i n a l o r o g e n s and l o n g i t u d i n a l d i l a t a t i o n f r a c t u r e s. This may be completed by a view over other areas of the globe. The N o r t h A t l a n t i c, in higher latitude than 53° N, shows structures in NNE course, not only the Reykjavik ridge. Greenland is separated from the North American continent by the large graben of Labrador and Baffin Sea, bordered by faults in mainly eggish direction. That is a branch only of the more rhenish direction of the North Atlantic Ocean. This constellation reminds us of the Sinai, of the Massiv Central, and of other V forms. One of the similar shapes is also India where we suppose a comparable origin. In the Indian Ocean, too, some meridional ridges exist, e. g. the Maledivian and the East Indian Ridge which is very rectilinear and runs 5000 km in nearly exact N—S course. In the P a c i f i c, the 2500 km long Imperator Ridge runs in NNW direction from Kamtschatka southwards. The most spectacular element of this kind at all is the Kermadec — Tonga Ridge between New Sealand and Samoa, with a continuation almost up to Hawaii, an altogether 8000 km long lineament in stressed rhenish direction (Fig. 10).

Within the Pacific Ocean, also other directions are obvious. Hercynian elements can be seen, though very indistinctly. But, in the whole NE part, narrow ridges in ENE direction are very remarkable. We do not know what they definitely mean. The same erzgebirgish course is well marked in the southwestern Indian Ocean, too (Atlantic-Indian Ridge and West Indian Ridge). There is an immense straight line, going from the Easter Islands via Arica — Cape San Roque — Romanche Fracture — Guinea — Gulf of Aden — Kashmir — South of Tarim Basin to the Japonese Sea. STILLE was constructing a "Central Pacific Lineament", also in erzgebirgish direction, from Samoa through the Ocean to the Cocos Ridge where-from we cannot follow it over the Atlantic. But, continuating the direction, we find ENE elements in Europe in the Guadalquivir, in the Cevennes, in SW Germany and in the name-giving Erzgebirge. These are two enormous parallel lines, nearly round the World, in the same direction, I only may mention without any explanation attempt.

We learn that the m e g a l i n e a m e n t s in submeridional direction run not only through oceans. Because they are very old, the d e f o r m a t i o n o f t h e g l o b e seems to be their p r i m a r y m o t i v e. However, their o r i g i n c a n n o t be located in any f l u i d h o r i z o n of the depth, but their rectilinear course over thousands of kilometers speaks for the reaction form of a m a t e r i a l o f h i g h e l a s t i c i t y. As we have to assume liquid movements in the Upper Mantle, the origin of the megalineaments is to l o c a l i z e i n a m o r e i n t e r i o r h o r i z o n.

The ORG is a member of the rhenish striking West European Megalineament, which is a remobilized element of a pre-Paleozoic joint pattern. The whole lineament is not one clear break but split up by numerous subparallel ruptures in

NNW = Egge direction. This may be the reason of the still existent coherence of Europe which could perhaps be divided if a fracture of the same uniformity but of greater length than the Oberrhein Graben would exist.

Zusammenfassung

Es wird versucht, den Mechanismus des ORG nicht ex loco zu lösen. Zu den bekannten Tatsachen der Graben-Stereometrie, der Zerrung, der Süddrift der Westscholle, des Bewegungsalters usw. wird nichts Neues beigetragen. Auf die Rolle der Rheinischen Fuge als präexistierende Scherfläche im varistischen Bewegungsablauf (und auf eine ähnliche Rolle der ebenfalls „uralten" herzynischen Richtung) wird erneut verwiesen, ebenso auf die fundamentale Trennung in der Bruchreaktion des — auf das Rheinische Lineament bezogen — westlichen und östlichen Varistikums. Eine entscheidend wichtige Rolle spielt die „Bruder"-Richtung des NNE-Verlaufs, die Egge-Richtung in NNW. Die Zersplitterung des ORG in der Hessischen Senke wie im Rhodanischen Bereich ist wesentlich auf die Übernahme eines bedeutenden Spannungsausgleichs durch eggisch gerichtete Störungen zurückzuführen. Ein ausgezeichnetes Beispiel für das gleichsinnige Zusammenwirken der beiden meridionalen Richtungen, die sich häufig in V-Form scharen, bietet das französische Central-Massiv. Als nicht regional gebundene Unterschiede beider Richtungen sind stärkere Zerrungsbeträge rheinischer Elemente (Dilatophilie) gegenüber den steileren eggischen Fugen und bedeutend größere Beteiligung von meist hypogenem Vulkanismus (Magmatophilie) auf den Egge-Brüchen festzustellen. Das gilt von den Pyrenäen bis zum mittleren Norwegen.

Das Verhältnis des Bruchmusters zu den orogenen Deformationen ist für das Varistikum eindeutig. Im Alpidischen Orogen sind rheinische und eggische Lineamente im Ostalpin-Paläozoikum ohne Veränderung erhalten bzw. wieder durchgeschlagen, wie ein Vergleich mit der Böhmischen Masse deutlich macht. Sogar herzynische Fugen sind kenntlich. Rheinische Lineamente an der Westalpen/Ostalpen-Grenze, in der Judicarienlinie und im Wiener Becken sind strukturbestimmend. Die rheinische Fuge Wien—Rom dürfte den Deformationsbruch im Apennin beeinflussen, wie sie im Norden Alpen gegen Karpathen absetzt. In Portugal, Spanien, England usw. sind Großbrüche in NNE- und NNW-Richtung deutlich. Das gesamte Rheinische Großlineament in Westeuropa ist jedoch stark zerfasert. Nur eine (unerklärliche) Brücke zwischen etwa Salzburg und Südengland, ca. 300 km breit, ist ziemlich frei von nennenswerten Störungen. Genau in diesem Streifen der tektonischen Ruhe liegt der ORG, der die gesamte Spannung praktisch nur mit den beiden Grabenrand-Verwerfungen löst.

Ähnlichkeiten zwischen dem westeuropäischen Megalineament und dem Riftsystem Ostafrika — Palästina liegen in der Richtung der Einzelstücke (eggisch = Suez, rheinisch = Jordan), im Vulkanismus, in der sinistralen Verschiebung. Mittelafrika zeigt zwischen Tunis und Togo lange rheinische Brüche, welche das Lineament Wien—Rom südwärts fortsetzen. Die afrikanische Ostküste wie der Mittelatlantische Rücken bestehen aus NNW und NNE verlaufenden Elementen. Die

Fortsetzung des Nordatlantik zwischen Nordamerika und Grönland wie zwischen Grönland und Norwegen-Spitzbergen in den Skandik hinein und über den heutigen Nordpol hinaus ist ebenfalls von den beiden meridionalen Richtungen begrenzt. Auch für den jungpaläozoischen Pol verläuft das Megalineament meridional. Die gesamte atlantische Spalte ist somit ein — wahrscheinlich analog zum Rhein-Lineament — uralter Riß in einem gemeinsamen Kontinent, von dem aus dann die Westschollen abdriften. Da es auch im Indik und im Pazifik meridionale Megalineamente gibt, die tausende Kilometer lang sind, muß die Entstehung dieser Bruchlinien auf globale Beanspruchungen zurückgehen. Wohl Kontinentaldrift und Orogenesen, aber nicht das Bruchsystem können mit Strömungen im Oberen Erdmantel erklärt werden. Die Megalineamente stellen vielmehr eine Festkörper-Reaktion dar, die man mit solcher Einheitlichkeit nur in sehr großer Erdtiefe erwarten kann.

Wenn eine Abdrift Westeuropas bisher noch nicht stattgefunden hat, so vielleicht deshalb, weil — im Gegensatz zu dem einzigen Stück ORG — „Unentschlossenheit" zur Ausbildung eines einheitlichen Abrisses besteht. Daß die Kontinentalbrüche hinreichende Tiefe haben, zeigen die hypogenen Magmen auf den eggischen Spalten.

Sorgsame Analysen des zeitlichen Ablaufs im Bewegungsmechanismus der Einzelstücke und damit der Kinematik des gesamten Megalineaments dürften wünschenswerte Erkenntnisse bringen, womit ich neben der Fortsetzung der geophysikalischen Arbeit auch weiterer geologischer Forschung das Wort reden möchte.

References

BECK-MANAGETTA, P.: Geologische Übersichtskarte der Republik Österreich, 1:1 Mill. — Geol. Bundesanst., Wien 1964.
BEDERKE, E.: The Development of European Rifts. — The World Rift System, Report of Symposium, Ottawa 1965.
BROEGGER, W. C. & SCHETELIG, J.: Geologisk Oversiktskart over Kristianiafeltet, 1:250 000. — Norges Geogr. Opmal., Oslo 1923.
BUBNOFF, S. VON: Über die Småländer „Erdnaht". — Geol. Rdsch. 41, 78—90, Stuttgart 1953.
CARLÉ, W.: Bau und Entwicklung der Süddeutschen Großscholle. — Beih. z. Geol. Jb., 16, 1—272, 4 t., Hannover 1955.
Carte géologique détaillée de la France 1:80 000, sheets 145—148, 154—159, 164—168, 173—178, 183—187, 195—198, 206—209, 220—225, 232—237, 247—248. —B.R.G.M., Paris—Orléans.
Carte tectonique internationale de l'Afrique 1:5 Mill. — UNESCO, Paris 1968.
CLOOS, H.: Grundschollen und Erdnähte. — Geol. Rdsch. 35, 133—154, 12 fig., Stuttgart 1948.
DAL PIAZ, G. & ZANETTIN, B. et al.: Carta Geologica d'Italia, 1:100 000, sheets Merano, Bolsano, Riva, Schio etc. — Serv. Geol. d'Italia, Roma 1958—1971.
DIETRICH, G. & ULRICH, J.: Atlas zur Ozeanographie. — Hochschulatlanten, Bibliogr. Inst. Mannheim 1968.
EXNER, CHR.: Geologische Karte der Sonnblickgruppe 1:50 000, m. Erl. — 170 p., 8 t. Geol. Bundesanst., Wien 1964.

FISCHER, A. G., et al.: Geological History of the Western North Pacific. — Science **168**, 1210—1214, N. Y. 1970.
Geolog. Übersichtskarte von Hessen 1:300 000. — Hess. Landesamt f. Bodenforschg., Wiesbaden 1960.
Geolog. Übersichtskarte d. Niederrhein-Westfäl. Karbons 1:100 000. — Geol. Landesamt f. Nordrhein-Westfalen, Krefeld 1971.
Geolog. Übersichtskarte von Nordwestdeutschland 1:300 000. — Amt f. Bodenforschg., Hannover 1951.
Geolog. Übersichtskarte von Südwestdeutschland 1:600 000. — Geol. Landesamt in Baden-Württ., Freiburg 1954.
Geolog. Karte von Thüringen 1:500 000. — Geogr.-Kartogr. Anst. H. Haack, Gotha/Leipzig 1971.
GIRDLER, R. W.: The role of translational and rotational movements . . . The World Rift System, Report of Symposium, Ottawa 1965.
HEEZEN, B. C.: The Rift in the Ocean Floor. — Sci. Amer., New York 1960.
HEEZEN, B. C. & HOLLISTER, C. D.: The Face of the Deep. — Oxford University Press, London 1971.
HILGENBERG, O. C.: Rekonstruktion der Kontinente für Karbon, Perm und Kreide nach palaeomagnetischen Messungen. — N. Jb. Geol. Paläont. Abh. **124**, Stuttgart 1966.
HOBBS, W. H.: Repeating Patterns in the Relief and in the Structure of the Land. — Bull. Geol. Soc. Amer. **22**, 123—176, N. Y. 1911.
ILLIES, H.: Bauplan und Baugeschichte des Oberrheingrabens. — Oberrhein. Geol. Abh. **14**, 1—54, Karlsruhe 1965.
— Kontinentaldrift — mit oder ohne Konvektionsströmungen? — Tectonophysics **2**, 528—557, 10 fig., Amsterdam 1965.
— Development and Tectonic Pattern of the Rhinegraben. — see: ROTHE & SAUER 1967.
— Die großen Gräben. — Geol. Rdsch. **59**, 528—552, Stuttgart 1970.
— Der Oberrheingraben. — Fridericiana, Z. Univ. Karlsruhe, H. **9**, 17—32, Karlsruhe 1971.
KNETSCH, G.: Über ein Strukturexperiment an einer Kugel. — Geol. Rdsch. **54**, 523—548, 13 fig., Stuttgart 1965.
— Über Funktionswechsel des Rheinischen Lineamentes . . . — Z. dt. geol. Ges. **118**, 222—235, 5 fig., Hannover 1969.
KNOTT, S. T. et al.: Red Sea Seismic Reflection Studies. — The World Rift System, Rep. of Sympos., Ottawa 1965.
KODYM, O. et al.: Regional Geology of Czechoslovakia, Atlas; Geolog. map, tecton. map 1:1 Mill., Praha 1968.
KRAUS, E.: Der Abbau der Gebirge. I. Der alpine Bauplan. — Berlin 1936.
LAUGHTON, A. S. & TRAMONTINI, C.: Recent Studies of the Crustal Structure in the Gulf of Aden. — Tectonophysics **8**, 359—376, Amsterdam 1969.
MEYERHOFF, A. A.: Continental Drift, Implications of paleomagnet. Studies. — J. Geol. **78**, Chicago 1970.
MÜLLER, ST. et al.: Crustal Structure beneath the Rhinegraben from Seismic Refraction and Reflection Measurements. — Tectonophysics **8**, 529—542, Amsterdam 1969.
PICARD, L.: Thoughts on the Graben System in the Levant. — The World Rift System, Rep. Sympos., Ottawa 1965.
RASMUSSEN, J. & NOE-NYGAARD, A.: Geology of the Faeroe Islands. — Danm. Geol. Unders. I. **25**, 1—142, 2 t., København 1970.
RICHTER (RICHTER-BERNBURG), G.: Das Rheinische Element im Bilde Westeuropas. — Nachr. Ges. Wiss. Göttingen, Math.-phys. Kl., N. F. **1**, 3, 23—38, 6 fig., Berlin 1934.
— Das Grenzgebiet Alpen—Pyrenäen; Tektonische Einheiten des südostfranzösischen Raumes. — Abh. Ges. Wiss. Göttingen, Math.-phys. Kl., III. F., **19**, Berlin 1939.
RICHTER-BERNBURG, G.: Die Grenze Westalpen—Ostalpen im tektonischen Bilde Europas. — Z. dt. geol. Ges. **102**, 181—187, 6 fig., Hannover 1951.

RICHTER-BERNBURG, G.: Saxonische Tektonik als Indikator erdtiefer Bewegungen. — Geol. Jb. 85, 997—1030, 20 fig., Hannover 1968.
ROTHE, J. P. & SAUER, K.: The Rhinegraben Progress Report 1967. — Abh. geol. Landesamt Baden-Württemberg 6, 1—146, Freiburg 1967.
RUNCORN, S. K.: Paleomagnetic evidence for Continental Drift . . . — Continental Drift. Academic Press, London 1962.
SCHAFFER, F. X.: Geologie der Ostmark. — Wien 1943.
SEDERHOLM, J. J.: Über Bruchspalten und Geomorphologie. — Bul. d. l. Com. geol. de Finlande, 6, Helsingfors 1913.
SONDER, R. A.: Die Lineamenttektonik und ihre Probleme. — Ecl. Geol. Helv. 31, 199—238, 12 fig., Basel 1938.
STILLE, H.: Rheinische Gebirgsbildung im Kristianiagebiet und in Westdeutschland. — Abh. preuß. geol. Landesanst. 95, 110—132, 1 t., Berlin 1925.
— Betrachtungen zum Werden des Europäischen Kontinents. — Z. dt. geol. Ges., 97, 7—29, 5 fig., Berlin 1947.
— Ur- und Neuozeane. — Abh. dt. Akad. Wiss. Berlin, Math.-phys. Kl. Nr. 6, 1—68, 4 fig., 2 t., Berlin 1948.
— Die zirkumpazifische Serotektonik . . . — Abh. dt. Akad. Wiss. Berlin, Kl. III, H. 1, 121—146, 3 fig., Berlin 1960.
Tektonische Karte der Schweiz 1:500 000. — Schweiz. Geol. Komm., Bern 1972.
VENING-MEINEZS, F. A.: The Earth's Crust and Mantle. — Devel. in Solid Earth Geophysics, 1, Elsevier Publ. Comp., Amsterdam 1964.
WEGENER, A.: Die Entstehung der Kontinente und Ozeane. — Die Wissenschaft, 66, Braunschweig 1941.
WYLLIE, P. J.: The Dynamic Earth. Textbook in Geosciences. — New York 1971.

Taphrogenic Lineaments and Plate Tectonics

by

R. B. McConnell

With 3 textfigures

European geologists will recall that the unravelling of Alpine structure in the early part of this century was greatly facilitated by the realization of the post-tectonic epeirogenic uplift of the whole structure, and the segmentation into transverse domes and saddles which caused deep structures to be exhumed by erosion and the higher nappes to be conserved in the synforms of axial pitch. Thus an ideal profile could be constructed from the basal Lepontic nappes to the highest Austro-Alpine sheets. The history of the East African Rift System can also be unravelled owing to an analogous exhumation of deep infracrustal Precambrian structures in Tanzania, Uganda and Malawi which complement the better known sections of Neogene rift volcanics in Kenya and Ethiopia in which the nature of supracrustal graben faulting can be clearly discerned (BAKER, MOHR & WILLIAMS

1972). A preliminary analysis of the evolution of the whole rift system based on a comparative study of the infracrustal and supracrustal structures (McCONNELL 1972) appears to confirm the unexpected conclusion, originally proposed by McCONNELL (1951) and DIXEY (1956), that the original plan of the system was laid out in the early Precambrian.

Intracontinental taphrogenesis is a geological phenomenon and discussion out of context may lead to premature conclusions, it is therefore necessary to give a short summary of the geological setting of the rift system. The African continent between the Sahara Desert and the Cape Ranges in the south consists of a vast Precambrian platform which has not been invaded by the sea since the Silurian. Progress in geological mapping and isotope dating (CAHEN & SNELLING 1966; CLIFFORD 1970) enables the Precambrian history of the continent to be divided into the following major tectonothermal episodes which correspond in age with world-wide orogenic cycles:

1. The formation of early granitic crust prior to c. 3.5 b.y. (3.5×10^9 years before present).
2. Deposition of schist and greenstone belts on a granitic crust, folded and metamorphosed in "anorogenic" (MARTIN 1969) tectonothermal episodes: c. 3.0 b.y.
3. Accumulation of sediments and volcanics followed by a Late Archaean orogeny, the first with linear trends and high-grade metamorphic rocks (granulites, charnockites): c. 2.7—2.3 b.y.
4. Renewed accumulation culminating in a wide-spread Ubendian (Eburnean of West Africa) tectonothermal episode, locally reactivating Late Archaean Africa into mobile belts, domed shields and sunken blocks and basins of early fold belts, as in the type Ubendide Belt, and accentuating the division of Africa into mobile belts, domed shields and sunken blocks, and basins of early Precambrian: c. 2.0 b.y.
5. Formation of intracratonic sedimentary basins, folded and partly metamorphosed — potash metasomatism and granites — in a Kibaran tectonothermal episode: c. 1.0 b.y.
6. Wide-spread molassic and restricted greywacke deposition in tabular cover and turbidite belts followed by the great Pan-African tectonothermal episode at c. 0.7—0.5 b.y. which formed young fold belts (Damarides, Katangides) as well as regenerating earlier orogenies (Mozambique Belt), reviving dislocation zones, and resetting isotope clocks throughout Africa.
7. Intracontinental deposition of Karroo (Carb.-Trias) and Jurassic—Neogene sediments with wide-spread volcanic activity, largely controlled, north of the Zambezi River, by lineaments of the rift system. Rift faulting occurred sporadically, culminating in the Neogene paroxysm north of the Zambezi of which the Gregory and Ethiopian Rift Valleys, with their profuse Neogene volcanism, are outstanding examples.

A glance at the plan of the rift system as shown in Fig. 1 immediately suggests a close relationship with the world-wide Mid-Ocean Rift System exemplified in

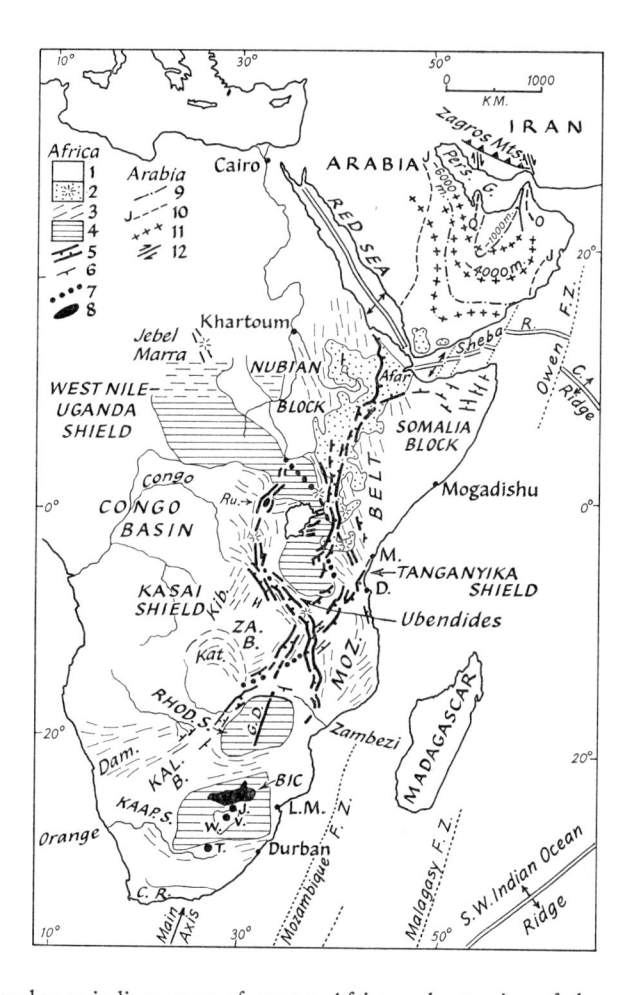

Fig. 1. The taphrogenic lineaments of eastern Africa and extension of the main axis into Arabia. 1, Palaeozoic to Recent; 2, Neogene volcanism; 3, trend lines in Precambrian mobile belts; 4, ancient granitoid shields; 5, Rift System faults; 6, other faults; 7, Precambrian dislocation zones; 8, mafic and ultramafic rocks, 9, top of Mesozoic at sea level and −1000 m; 10, base of Jurassic (ARKELL, 1956); 11, top of Precambrian basement at sea level, −4000 m, and −6000 m; 12, trace of deep structures (FALCON, 1969); Q, Qatar arch; O, Oman arch; C. Ridge, Carlsberg Ridge; Ru, Ruwenzori Mtns.; M., Mombasa; D. Dar es Salaam; Kib., Kibarides; Kat., Katangides; ZA. B., Zambian Block; Rhod. Sh., Rhodesian Shield; Dam., Damarides; Kal. B., Kalahari Basin; G. D., Great Dyke; BIC, Bushveld Igneous Complex; KAAP. S., Kaapvaal Shield; J, Johannesburg hub; V, Vredefort Dome; W, Witwatersrand Basin; T, Trompsburg Igneous Complex; L. M., Lourenço Marques; C. R., Cape Ranges. Structure contours in Arabia after DUBERTRET & ANDRÉ (1969).

the Red Sea and Gulf of Aden branches (HEEZEN 1969; WILSON 1965). This association is undoubtedly justified, and has been supported by geophysical investigations (KHAN & MANSFIELD 1971; GRIFFITHS et al. 1971; GIRDLER et al. 1969; BAKER et al. 1972) which indicate a remarkable resemblance of deep crustal structures to those of mid-ocean ridges and confirms opinions that the Ethiopian rift is underlain by an upwelling of mantle material in the form of penetrative convection (GASS 1970; HARRIS 1970). A great difference, however, from seafloor spreading phenomena is emphasized by KING (1970) and BAKER & WOHLENBERG (1971) who point out that dilation of the Gregory Rift Valley is less by several magnitudes than that of a normal mid-ocean rift, and is comparable with that estimated by ILLIES (1970) for the Upper Rhinegraben, whose crustal structure the Gregory rift also resembles.

A closer look at Fig. 1, however, shows that the great Rift System of Eastern African can be divided roughly into three sectors: (1) the eastern rifts of Kenya and Ethiopia which are mainly cut in Neogene volcanic rocks, and hence represent supracrustal structures; (2) the Western Rift Valley, the eastern rift in Tanzania, and the complicated pattern in Central Africa, all affecting Precambrian basement; and (3) a sector south of the Zambezi River in which the rift directions diverge around the Rhodesian Shield, while the main NNE—SSW axis of the rift system is reflected chiefly by an alignment of Precambrian mafic intrusives (COUSINS 1959) without actual rift faulting. A study of the great amount of information now available for the sectors in which the mechanism of infracrustal deformation can be examined, has led to the conclusion (McCONNELL, in press) that the rift system originated in the Late Archaean cycle (2.7—2.3 b.y.) as a pattern of lineaments, a proto-rift system, associated with deep zones of disturbance in the lithosphere which formed during the breakup of the granitic crust. The complicated pattern would be due to the existence of ancient nuclei already cratonized and not easily penetrated by rifting. Although these taphrogenic lineaments thus formed may coincide locally with orogenies, they represent a fundamentally different type of structure. They tend to be narrow and vertically compressed panels of the crust, affected either by transcurrent or vertical movement, whereas orogenic belts are wider and consist of fold or nappe structures testifying to their different origin. It appears that the tendency in eastern Africa has been for a dextral pattern of transcurrent movement along the lineament (McCONNELL 1972). Any considerable horizontal displacement could have taken place only at a very early stage, and has either been halted by kinking, or become impossible through cratonization, and consequently replaced by predominantly vertical movement. The lineaments are marked by intense crushing and metamorphism, producing such rocks as blastomylonites, phyllonites, flaser gneisses, migmatites and anatectic granites, with locally injected mafic and intermediate igneous rocks. A characteristic of the taphrogenic lineaments in Africa is that they have been rejuvenated at each of the major world-wide cycles of tectonic activity and have hence been designated perennial or persistent lineaments. Where later fold belts have formed, crossing the rifts obliquely, they are seen to be cut by dislocations

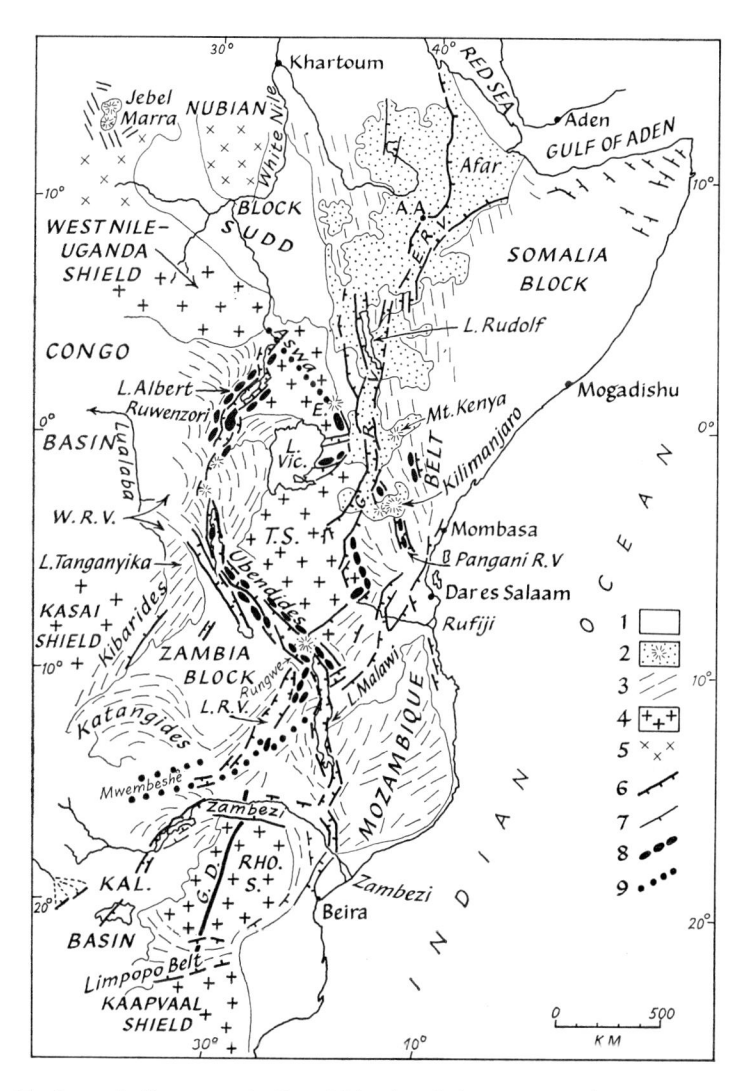

Fig. 2. Taphrogenic lineaments in East Africa in relation to Precambrian cataclastic zones. 1, Palaeozoic to Recent; 2, Neogene volcanism; 3, Precambrian foliation trends; 4, ancient granitoid shields; 5, Precambrian basement, uncorrelated; 6, principal Rift System faults; 7, other faults; 8, Precambrian cataclastic and refoliation-injection zones; 9, Precambrian dislocation zones; AA, Addis Ababa; E. R. V., Ethiopian Rift Valley; G. R. V., Gregory Rift Valley; W. R. V., Western Rift Valley; E, Mt. Elgon; L. Vic, Lake Victoria; T. S., Tanganyika Shield; Rho. S., Rhodesian Shield; G. D., Great Dyke; Kal. Basin, Kalahari Basin; L. R. V., Luangwa Rift Valley.

following the strike of the older taphrogenic lineaments and marked by refoliation, blastomylonites and linear plutonites; as in the Upper Rhinegraben area the Variscan fold belt is cut by Rhenish lineaments (ILLIES 1962).

The presence of migmatites and anatectic granites in the deep lineaments indicates high heat flow, leading to melting and expansion, with consequent reduction in density, isostatic uplift and the arching characteristic of rift valleys (CLOOS 1939). Thus through repeated upwarping and erosion the core of the arch becomes visible at the surface, and the resulting picture is displayed somewhat schematically in Fig. 2, illustrating the evolution of taphrogenic lineaments. The Western Rift Valley is particularly significant as the deepest Late Archaean tectonism is exposed in southwest Tanzania (MCCONNELL 1970) and also in the northern Lake Albert and Edward rifts (CAHEN & LEPERSONNE 1967) which cut across a general latitudinal trend of Precambrian fold belts of Late Archaean, Ubendian and Kibaran ages. The swing of the older strikes into the trend of the Western Rift lineament (LEPERSONNE, in press) is particularly impressive, but is not so pronounced for the younger Kibaran fold belts. This phenomenon is described in more detail elsewhere (MCCONNELL in press) and is taken to confirm the Late Archaean origin of the taphrogenic lineament. The evidence indicates that the lineament may have taken advantage of different pre-existing structure planes of a regmatic pattern (SONDER 1947), but that, once established, it was so modified by a combination of tectonism and heat flow (penetrative convection) that it became conditioned to serve as a preferential channel for all subsequent revivals of activity, culminating in Phanerozoic time in the worldwide rift system of modern global tectonic theory. A similar picture of persistently repeated movement can be deduced from an analysis of the major features of Precambrian structure around the Tanganyika Shield (MCCONNELL 1972).

The nature of the deep mantle disturbance beneath the rift zone of eastern Africa is still problematic, but may be chiefly dependant on a major lineament related to global equilibrium (ILLIES 1972). The main axis (Fig. 1) appears to be directed NNE—SSW, parallel to major fracture zones in the Indian Ocean, and, it is clear that, south of the equator, its reflection in the overlying crust becomes complicated by the existence of early Precambrian nuclei around which it is moulded. The lineament could not crack the pre-existing Tanganyika, Rhodesian, or Kaapvaal Shields, but is manifested in the Great Dyke now dated c. 2.5 b.y., in the Bushveld Igneous Complex (c. 2.0 b.y.), and the subsurface Trompsburg mafic complex (c. 1.3 b.y). To the north of the equator, however, rifting has proceeded more freely and the typical Gregory and Ethiopian Rift Valleys reach the Afar Depression at the junction of the Red Sea and Gulf of Aden which have both been shown to be loci of sea-floor spreading since the Miocene (DRAKE & GIRDLER 1964; LAUGHTON 1966; FALCON et al. 1970). This sector of the rift system has served as a channel for profuse volcanism of per-alkaline, continental type (KING 1970; BAKER et al. 1972), as opposed to the tholeiitic volcanism associated with mid-ocean ridges, and dilation is minimal. MCKENZIE et al. (1970) have indicated that some plate separation in East Africa north of the equator is implied by the slightly divergent spreading directions of the Red Sea and Gulf of Aden, and, although dilation may be very small, this tension vector would account for the exceptional intensity of the rift volcanism. The mantle disturbance appears to continue to the north-northeast beyond the Red Sea-Gulf of Aden junction, causing a downwarp of the Arabian Shield which began in the Jurassic (ARKELL 1956) and continued into the Neogene (DUBERTRET & ANDRÉ 1969), carrying the Precambrian basement to a depth of 6 km south of the Persian Gulf (Fig. 1). The East African rift lineament thus leapfrogs the Afar depression and continues with various structural manifestations (FALCON 1969) up to the frontal thrusts of the Zagros Mountains.

It has long been known that the Red Sea trend is continued in the Horn of Africa (Fig. 1), and NW—SE faults with considerable throws are recorded (Somaliland Oil Exploration Co. Ltd. 1954; MOHR 1963; AZZAROLI & FOIS 1964). This lineament would thus extend beyond Afar. Recent intensive surveying has further shown that the Gulf of Aden trend continues west across Afar as far as the great Ethiopian scarp (BANNERT et al. 1970; GIRDLER 1970; BARBERI et al. 1972). It therefore appears that the Afar Depression is not merely an important node in the world-wide rift system at which the Red Sea and Gulf of Aden spreading ridges join in a single megastructure as shown by TAZIEFF et al. (1972), but lies at the cross-over intersection of three major crustal lineaments. The sea-floor spreading process has utilized the Gulf of Aden ridge only as far west as Afar, the Red Sea lineament only for the length of the present Red Sea basin, and has left the East African lineament relatively untouched. It is postulated therefore that the mid-ocean ridges of global tectonic theory utilize portions only of a great criss-crossing pattern of taphrogenic lineaments already existing in the lithosphere and available for geological investigation on the continents.

The great African taphrogenic lineament is so well developed, and has been the subject of so much investigation that it has been described in some detail. But the situation of the ancestral rifts of the North and South Atlantic Oceans can be compared with that of the nascent Red Sea and Gulf of Aden spreading rifts (MCCONNELL 1969) and is presented schematically in Fig. 3. The early rifts of the three main legs of the Atlantic can be seen to project into West Africa and the Guiana Shield as intracontinental taphrogenic lineaments, and evidence for the Precambrian age of these structures has been summarized from many sources quoted elsewhere by MCCONNELL (1969, in preparation).

We therefore reach the conclusion that taphrogenic lineaments are a fundamental feature of the lithosphere, extending back in time at least as far as 2.7 b.y. At this time in the Late Archaean certain lines of the earth's criss-crossing regmatic pattern were selected for tectonothermal activation which rendered them preferential channels for regeneration during all succeeding orogenic cycles, and, eventually, segments of the lineaments were selected for sea-floor spreading in the great cycle of world-wide global tectonics which began at the opening of the Mesozoic Era and continues to the present day. Although grabens can certainly form in mountain chains they may then be related to shallow fracture zones, whereas the taphrogenic lineaments described in this paper are essentially vertical features, shearing the lithosphere down to the low velocity zone. They are complementary, though of different origin, to the major orogenic fold belts which, in modern theory, are dependent in the first place on horizontal plate motions (DEWEY & BIRD 1970).

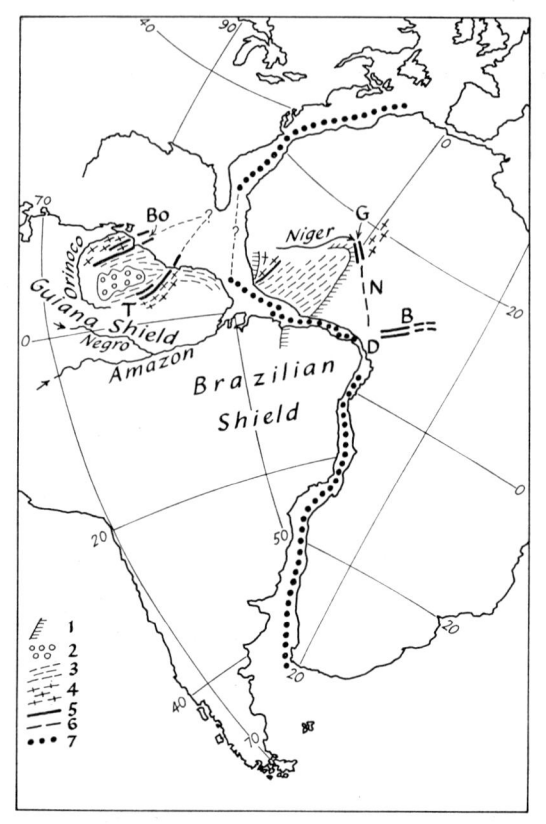

Fig. 3. Taphrogenic lineaments of the Guiana and West African Shields in relation to the ancestral Atlantic rifts. 1, limit of Pan-African reactivation; 2, tabular Proterozoic; 3, Lower Proterozoic fold belts; 4, Archaean fold belts; 5, intracontinental dislocation zones; 6, ditto, presumed; 7, ancestral Atlantic rift; Bo, Bolívar dislocation zone; T, Takutu Rift Valley and dislocation zone; G, Gao Trench; N, middle Niger River valley; Benue Rift Valley and dislocation zone; D, Niger River delta.

References

ARKELL, W. J., 1956: The Jurassic Geology of the World. — 806 p., Oliver & Boyd, London.
AZZAROLI, A. & FOIS, V., 1964: Geological outlines of the northern end of the Horn of Africa. — Proc. 22 Int. Geol. Congr., India, 4, 1197—1211.
BAKER, B. H.; MOHR, P. A. & WILLIAMS, L. A. J., 1972: Geology of the eastern rift system of Africa. — Geol. Soc. Amer. Spec. Pap. 136, 67 pp.

BAKER, B. H. & WOHLENBERG, J., 1971: Structure and evolution of the Kenya Rift Valley. — Nature (Lond.), 229, 538—542.
BANNERT, D.; BRINCKMANN, J.; KÄDING, K. CH.; KNETSCH, G.; KÜRSTEN, M. & MAYRHOFER, H., 1970: Zur Geologie der Danakil-Senke. — Geol. Rdsch., 59, 409—443.
BARBERI, F.; TAZIEFF, H. & VARET, J., 1972: Volcanism in the Afar depression: its tectonic and magmatic significance. — Tectonophysics, 15, 19—29.
CAHEN, L. & LEPERSONNE, J., 1967: The Precambrian of the Congo, Rwanda, and Burundi. — In: K. RANKAMA, Ed., The Precambrian, pp. 143—290, Interscience, New York.
CAHEN, L. & SNELLING, N. J., 1966: The geochronology of equatorial Africa. — North Holland Publishing Co., Amsterdam, 195 pp.
CLIFFORD, T. N., 1970: The structural framework of Africa. — In: T. N. CLIFFORD & I. G. GASS, Ed., African magmatism and tectonics, pp. 1—26, Oliver & Boyd, Edinburgh.
CLOOS, H., 1939: Hebung — Spaltung — Vulkanismus. — Geol. Rdsch., 30, 401—527, 637—640.
COUSINS, C. A., 1959: The structure of the mafic portion of the Bushveld Igneous Complex. — Trans. geol. Soc. S. Afr., 62, 79—201.
DEWEY, J. F. & BIRD, J. M., 1970: Mountain belts and the new global tectonics. — J. geophys. Res., 75, 2625—2647.
DIXEY, F., 1956: The East African Rift System. — Overseas Geology Miner. Resources Bull., Suppl. 1, H. M. Stationery Office, London, 71 pp.
DRAKE, C. L. & GIRDLER, R. W., 1964: A geophysical study of the Red Sea. — Geophys. J. Roy. Astron. Soc., 8, 473—495.
DUBERTRET, L. & ANDRÉ, C., 1969: Péninsule Arabique. — Carte orographie — hydrographie — mers et cartes structurales. — Notes et Mémoires sur le Moyen Orient, X, 285—318. Muséum Nat. d'Hist. Natur., Paris.
FALCON, N. L., 1969: Problems of the relationship between surface structure and deep displacements illustrated by the Zagros Range. — In: KENT, P. E. et al. (eds.), Time and Place in Orogeny, Geol. Soc. Lond., Spec. Publ. 3, 9—22.
FALCON, N. L.; GASS, I. G.; GIRDLER, R. W. & LAUGHTON, A. S. (Eds.), 1970: A discussion on the structure and evolution of the Red Sea and the nature of the Red Sea, Gulf of Aden and Ethiopia Rift junction. — Phil. Trans. Roy. Soc. Lond. Ser. A, 267, 417 pp.
GASS, I. G., 1970: Evolution of the Afro-Arabian dome. — In: T. N. CLIFFORD & I. G. GASS, Eds., African magmatism and tectonics, p. 285—300, Oliver & Boyd, Edinburgh.
GIRDLER, R. W., 1970: An aeromagnetic survey of the junction of the Red Sea, Gulf of Aden and Ethiopian rifts. A preliminary report. — Trans. Roy. Soc. Lond., Ser. A, 267, 359—368.
GIRDLER, R. W.; FAIRHEAD, J. D.; SEARLE, R. C. & SOWERBUTTS, W. T. C., 1969: Evolution of rifting in Africa. — Nature, Lond., 224, 1178—1182.
GRIFFITHS, D. H.; KING, R. F.; KHAN, M. A. & BLUNDELL, D. J., 1971: Seismic measurements in the Gregory rift. — Nature Physical Sci., Lond., 229, 69—71.
HARRIS, P. G., 1970: Convection and magmatism with reference to the African continent. — In: T. N. CLIFFORD & I. G. GASS, Eds., African magmatism and tectonics, p. 419—437, Oliver & Boyd, Edinburgh.
HEEZEN, B. C., 1969: The World Rift System — an introduction to the Symposium. — Tectonophysics 8, 269—289.
ILLIES, J. H., 1962: Oberrheinisches Grundgebirge und Rheingraben. — Geol. Rdsch., 52, 317—332.
— 1970: Graben tectonics as related to crust-mantle interaction. — In: J. H. ILLIES & ST. MUELLER, Eds., Graben Problems, p. 4—27, Schweizerbart'sche Verlagsbuchhandlung, Stuttgart.
— 1972: The Rhine graben rift system — plate tectonics and transform faulting. — Geophys. Surv. 1, 27—60, Dordrecht-Holland.

KHAN, M. A. & MANSFIELD, J., 1971: Gravity measurements in the Gregory rift. — Nature Physical Sci., Lond., **229**, 72—75.

KING, B. C., 1970: Vulcanicity and rift tectonics in Africa. — In: T. N. CLIFFORD & I. G. GASS, Eds., African magmatism and tectonics, p. 263—283, Oliver & Boyd, Edinburgh.

LAUGHTON, A. S., 1966: The Gulf of Aden. — Phil. Trans. Roy. Soc. Lond., Ser. A, **259**, 150—171.

LEPERSONNE, J., in press: Carte géologique du Zaïre. — Musée Roy. de l'Afrique Centrale, Tervuren, Belgium, and Bureau Recherches géol. et minières, Paris.

MARTIN, H., 1969: Problems of age relations and structure in some metamorphic belts of southern Africa. — Geol. Assoc. Canada, Paper **5**, 17—26.

MCCONNELL, R. B., 1951: Rift and shield structure in East Africa. — 18 Int. geol. Congr., **14**, 199—207.

— 1969: Fundamental fault zones in the Guiana and West African Shields in relation to presumed axes of Atlantic spreading. — Geol. Soc. Amer. Bull., **80**, 1775—1782.

— 1970: The evolution of the rift system in eastern Africa in the light of Wegmann's concept of tectonic levels. — In: Graben Problems, p. 285—290, J. H. ILLIES & ST. MUELLER, Eds., Schweizerbart'sche Verlagsbuchhandlung, Stuttgart.

— 1972: Geological development of the Rift System of eastern Africa. — Geol. Soc. Amer. Bull. **83**, 2459—2572.

— (in press): Evolution of taphrogenic lineaments in continental platforms. — Geol. Rdsch., **63**, No. 2.

MCKENZIE, D. P.; DAVIES, D. & MOLNAR, P., 1970: Plate tectonics of the Red Sea and East Africa. — Nature, Lond., **226**, 243—248.

MOHR, P. A., 1963: The geology of Ethiopia. — Univ. Addis Ababa Press, 268 pp. (multigraphed).

Somaliland Oil Exploration Co. Ltd., 1954: A geological reconnaissance of the sedimentary deposits of the Protectorate of British Somaliland. — Crown Agents, Lond., 41 pp.

SONDER, R., 1947: Shar patterns of the earth's crust. Discussion of paper by F. A. VENING MEINESZ. — Trans. Amer. Geophys. Union, **28**, 939—945.

TAZIEFF, H.; VARET, J.; BARBERI, F. & GIGLIA, G., 1972: Tectonic significance of the Afar (or Danakil) depression. — Nature, Lond., **235**, 144—147.

WILSON, J. T., 1965: A new class of faults and their bearing on continental drift. — Nature, Lond., **207**, 343—347.

Martian Tithonius Lacus Canyon and Rhinegraben — a Comparative Morphotectonic Analysis

by

K. Schäfer

With 4 figures in the text

Abstract: A comparative study of morphotectonic features of the Rhinegraben and of the Tithonius Lacus canyon on Mars has been conducted. Similarities as well as coincidences in the physiographies of both systems have been recognized. There is evidence that Mars is at least tectonically and volcanically not a dead planet.

The Rhinegraben has a Martian counterpart, that was the idea of many geoscientists when having been confronted with the "Geotimes" cover photograph of the February 1972 issue. It was a view of Mars, taken on January 12, 1972, by the Mariner 9-spacecraft's wide-angle TV camera from a distance of 1977 kilometers and covering an area of 367 by 480 kilometers (Fig. 1). Now, after the Mars orbiting Mariner 9 has terminated its surface mapping mission by playing back to the Earth several thousands of TV pictures, we know that this photograph turned out to show one of the geologically most interesting landforms on Mars. It is located in Tithonius Lacus, 480 kilometers south of the equator and is part of a gigantic canyon-like system extending east-west across the Aurorae Sinus, Coprates and Tithonius Lacus areas, and covering a distance between the 30° and the 110° meridians of more than the east-west extension of the North American continent.

The goal of this paper is an attempt to outline some significant morphological features of the Martian Tithonius Lacus canyon and to compare those with the physiography of the Rhinegraben area, as it is shown on a relief map (Fig. 3), which has been published many times, e. g. in "Graben Problems" (J. H. ILLIES & ST. MUELLER, eds., 1970).

The most striking morphological feature of both examined areas is indicated by a significant parallelism of the escarpments, which in the two areas are inclined toward the axes of systems of a remarkable bilateral symmetry.

The scarps of the Rhinegraben have been formed along marginal main faults, which enclose a subsided graben wedge. We may infer the same phenomenon from Martian Tithonius Lacus graben since the similarity of that morphological feature is obvious.

Both shoulders of the Rhinegraben have been uplifted considerably with the maximum elevation exactly at the scarps. They dip in opposite directions from the graben axis with 1—4 degrees (ILLIES 1972). Thus, the surface of the shoulders has been predominantly eroded at the scarps, while the remainder of the tilted shoulder plain merely has been incised by erosional processes (Fig. 3).

Erosional degradation of the shoulders and sedimentary accumulation into the subsiding Rhinegraben decreased the tectonical throw of about 4000 m to a difference of altitude of about 1300 m in the southern graben area to some tens of meters in the central part (Fig. 3). The Tithonius Lacus graben, on the other hand, also reveals maximum uplift of the shoulders at the scarps, as it can be seen by the long shadows of the sinking sun (Fig. 1). Sunlight comes from the southwest, so that the southern shoulder is lit and the north dipping northern shoulder is not. From pressure measurements, taken by the ultraviolet spectrometer experiment aboard the Mariner 9-spacecraft which have been converted to relative surface elevation, a gentle 2°-dip of both shoulder plains can be observed. The same pressure measurements revealed a depth of the graben surface relative to the summits of the bordering escarpments of about 6450 m. Hence, the difference of altitude between the summits of the elevated shoulders and the surface of the subsided graben is greater than that of the Rhinegraben. The tectonical downthrow of the Tithonius Lacus graben exceeds that of the Rhinegraben. Also there is no extensive erosion on the shoulders and no remarkable thickness of accumu-

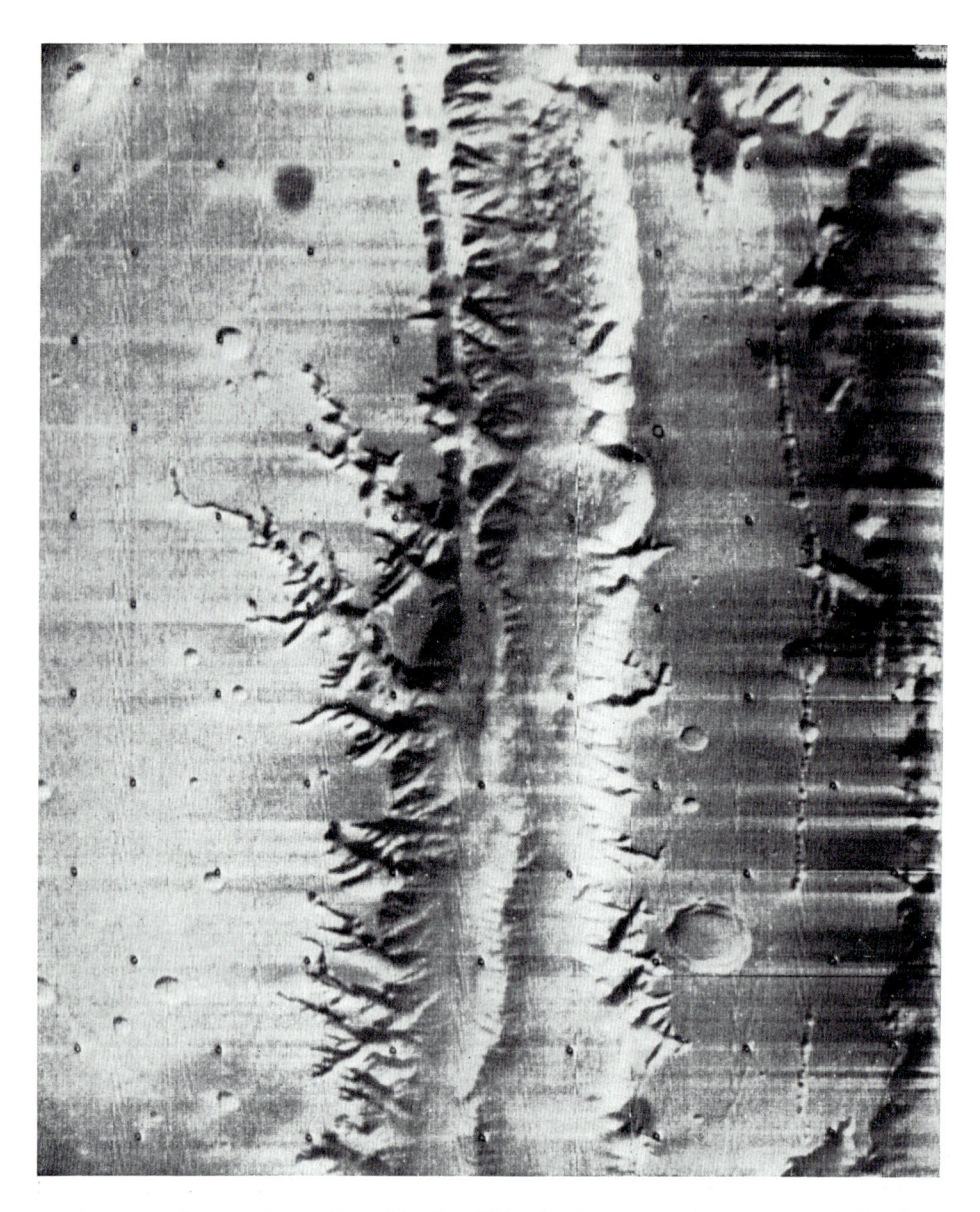

Fig. 1. Mariner 9-picture from Martian Tithonius Lacus covering an area of 367 by 480 kilometers. Graben tectonics is indicated by subsidence of Martian crust along zones of weakness which run strictly parallel. The maximum elevation of the escarpments rises 6450 m above the graben surface. The width of the graben between both marginal master faults is approximately 80 kilometers.

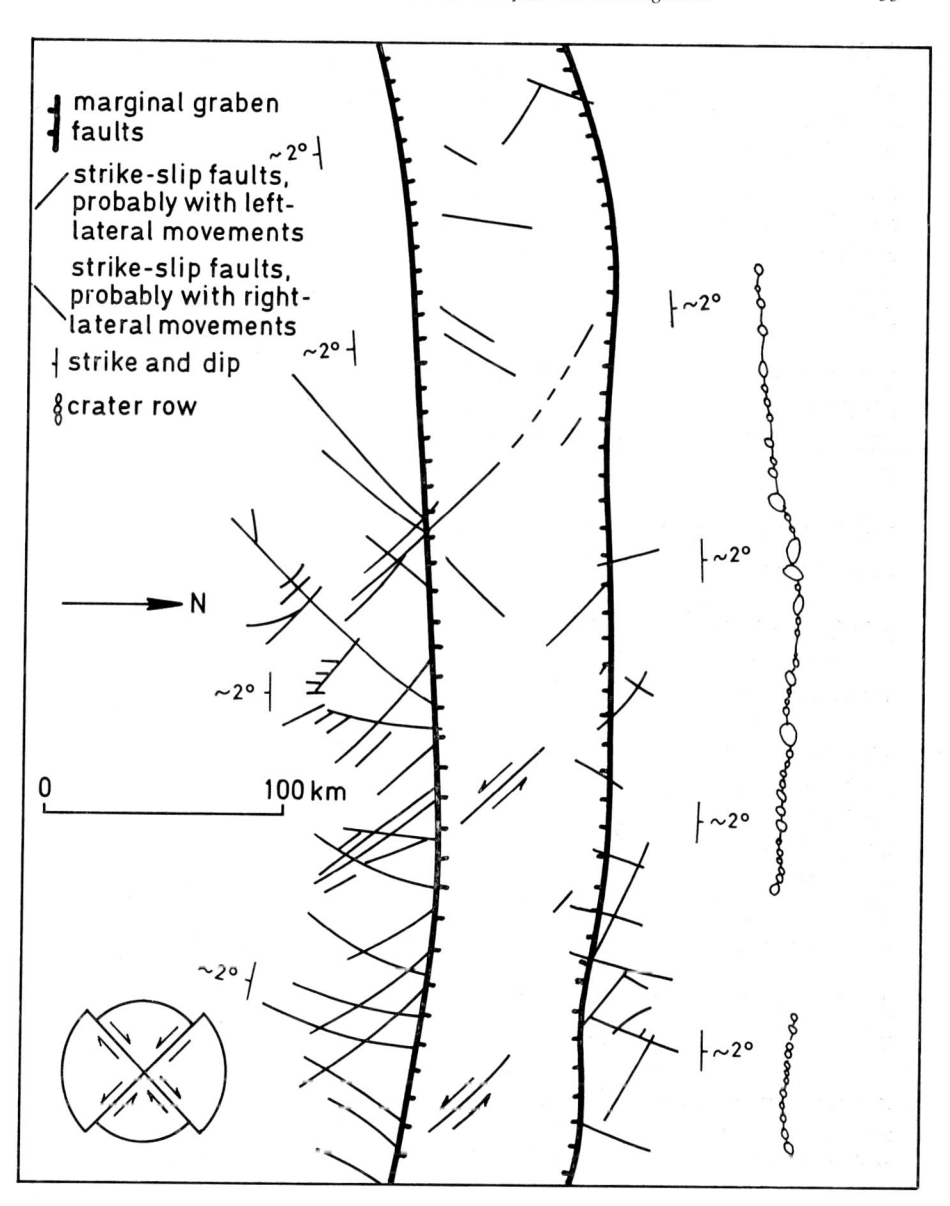

Fig. 2. Branching valleys dissect the escarpments of the Tithonius Lacus graben. The tree-like "stream-system" follows two predominant directions and seems to be entrenched along northeast and northwest striking transcurrent faults.

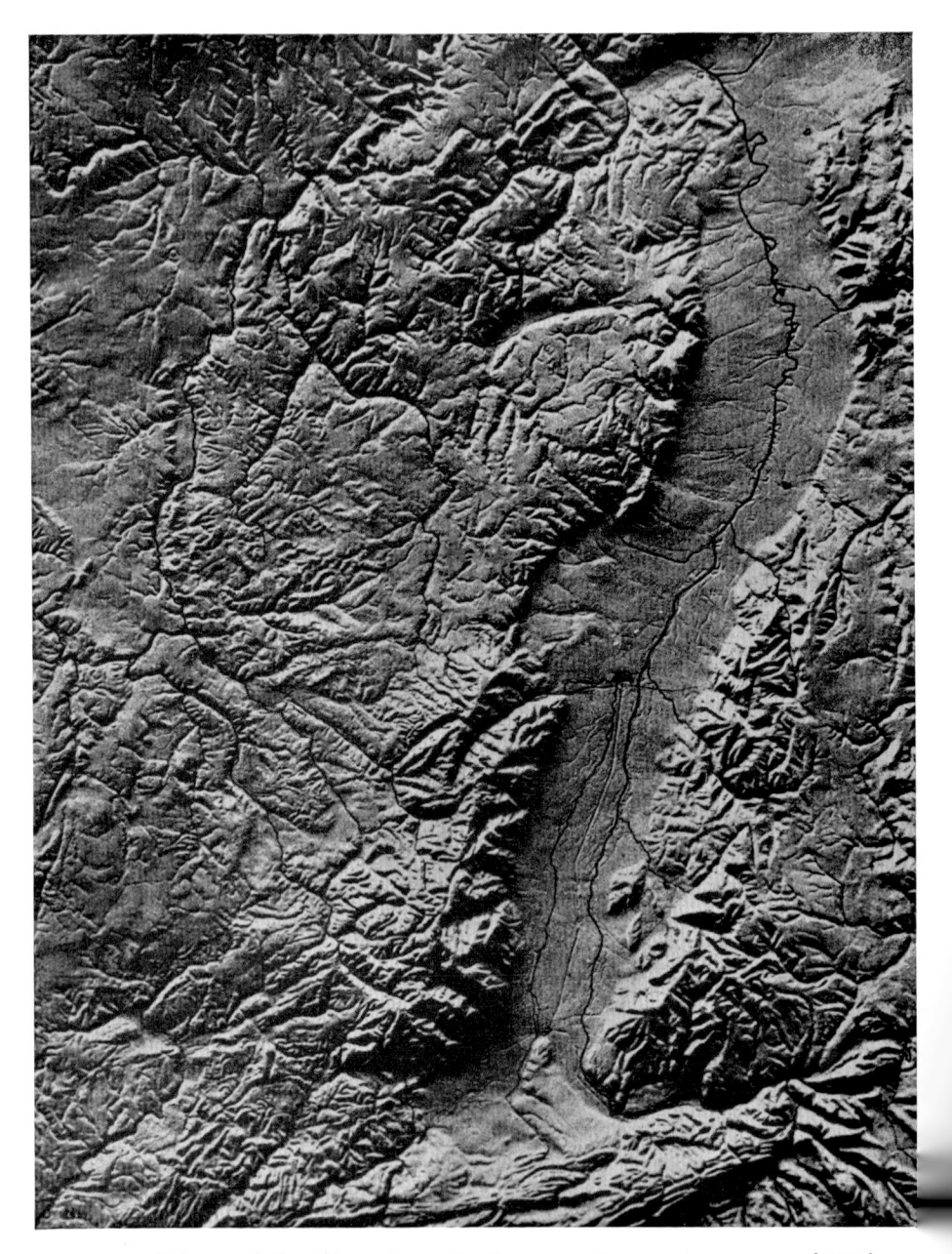

Fig. 3. Relief map of the Rhinegraben and adjacent terrains covering an area of 255 by 325 kilometers. The morphological difference of altitude between the subsided graben and the elevated shoulders attains approximately 1300 m in the southern graben area and decreases to some tens of meters in the central part. The graben width is 36 kilometers.

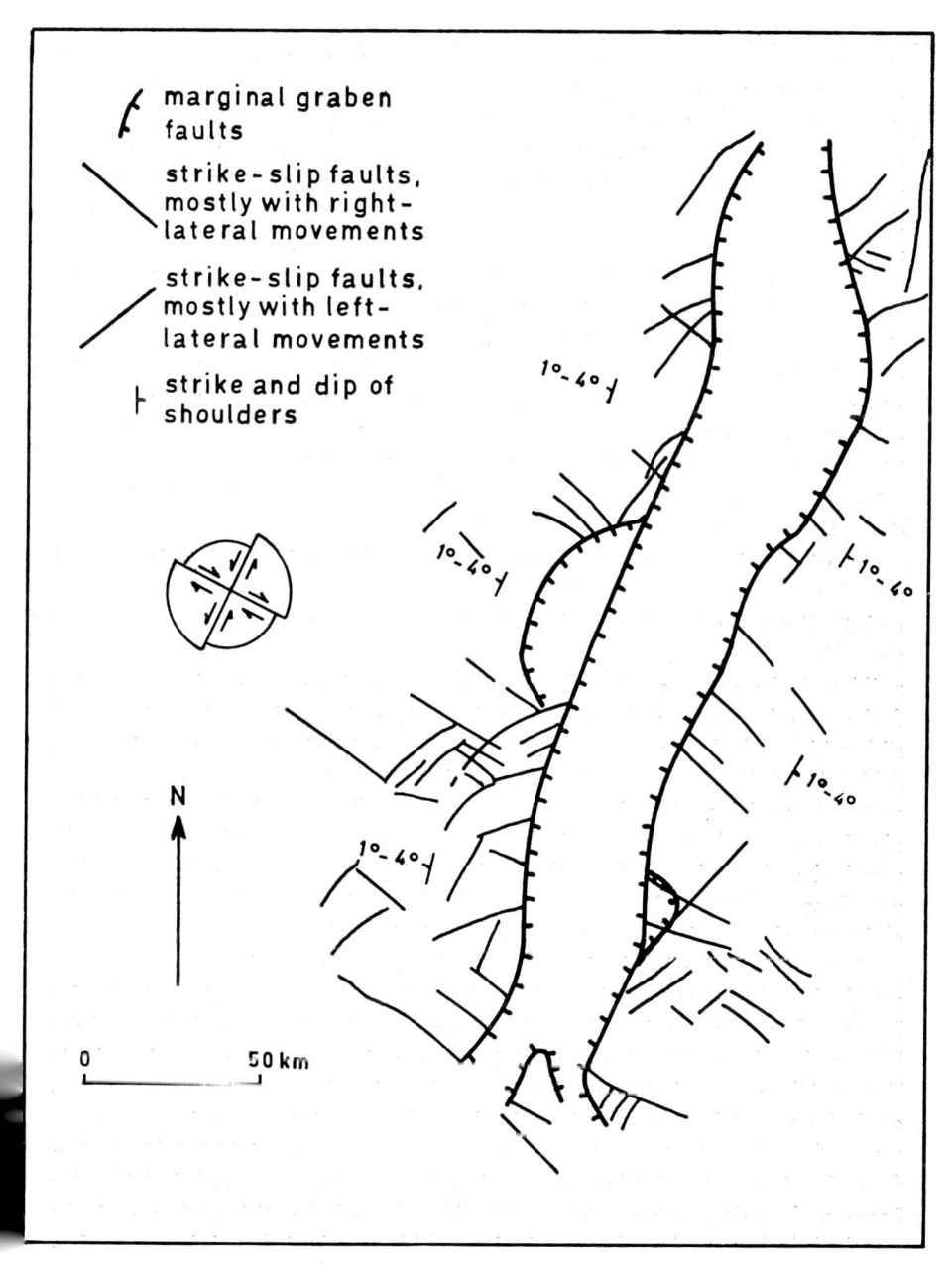

Fig. 4. The Rhinegraben and the Martian Tithonius Lacus graben as well are obliquely traversed by faults which frequently reveal strike-slip movements. Besides the northnortheastern trend of the marginal master faults of the graben, two predominant directions of transcurrent and normal faulting may be distinguished, which are consistent with the present tectonic stress-field.

lated sediments within the graben area. So we are able to identify a mid-graben ridge in the eastern part and numerous tilted blocks in the western part of the graben, both of which are not buried under a sedimentary cover (Fig. 1). Nevertheless, there are or have been erosional forces at work especially along the most elevated part of the southern shoulder and also at the northern shoulder in a less advanced stage. It appears that this type of landform has been developed while there was still water or another fluid matter on Mars. From infrared interferometer spectrometer experiments, taken aboard Mariner 9, only a small amount of water vapor within the Martian atmosphere has been ascertained. But this quantity of water is not sufficient to have formed branching valleys viz. an antecedent river system like the one which is downcutting into the escarpments at both sides of the Rhinegraben (Fig. 3). The valleys of the Tithonius Lacus graben, extending 40—60 kilometers, follow obviously two main directions: one of which strikes north 20° east and the other north 45° west. A subtle examination of the tree-like valley system suggests that activity of Martian winds could have sculptured this unique pattern by eroding predominantly along lines of weakness in the crust, since eolian processes are paramount on the Martian surface.

The Tithonius Lacus graben as well as the Rhinegraben and their adjacent shoulders are traversed by faults which run obliquely to the graben axes (Fig. 2 and 4). From focal mechanism investigations of earthquakes within and around the Rhinegraben area, left-lateral horizontal displacements along the northeast striking faults and dextral strike-slip movements along the northwest running faults may be expected. Transcurrent faults have been observed, which reveal slickensides with shear movements to be consistent with the present tectonic stress-field (Fig. 4) (AHORNER 1968). It seems that the Tithonius Lacus graben fault pattern has originated in the same way. The only difference is that in the Tithonius Lacus area the main tectonic pressure axis runs east-west viz. normal to the marginal main faults of the graben, whereas in Central Europe the stress-field reveals presently a maximum compression in the northwest-southeast direction viz. oblique to the marginal master faults of the Rhinegraben.

As mentioned earlier, both grabens indicate a constant distance between the marginal fault scarps viz. the Rhinegraben of 36 kilometers and the Tithonius Lacus graben of approximately 80 kilometers. The framing normal faults of the Rhinegraben have drifted apart 4,8 kilometers since rifting started in middle Eocene. The master faults reveal a converging dip of about 60°—65° and theoretically would have crossed in a depth of about 31 kilometers, i. e. the crust-mantle boundary in Central Europe, at the moment of initial Rhinegraben formation (ILLIES 1972). If we assume that the crustal spreading of the Tithonius Lacus graben does not exceed considerably that of the Rhinegraben and the marginal faults reveal analogous dipping, which may be inferred from morphological features, a Martian crust-mantle boundary may be expected in about 70 kilometers depth. The morphological characteristics of the Tithonius Lacus area and of other Martian terrains, which can be studied on further Mariner 9-pictures, enable us to suggest that the composition of the Martian crust is sialic rather than oceanic. This is demonstrated by an analogous behaviour of the Martian crust as a consequence of the rifting.

Prominent normal faults along graben margins have not yet been observed bordering the rift valleys of mid-oceanic ridges, though a structural conformity between continental and oceanic grabens has been suggested (van Andel & Bowin 1968).

Analogous to the Rhinegraben we may have an upwarping mantle below the crust in the Tithonius Lacus graben area. This anomalous mantle has been formed after weight reduction of the covering crust by faults and fissure formation. The uppermost mantle has expanded, consequently its volume has increased and its density has decreased. Besides graben tectonics there is other evidence of the occurrence of mobile mantle material underneath the Martian crust. Nix Olympica, a huge central volcano on Mars, is still active, as it has been confirmed by the observation of white clouds over its crater.

There is no doubt that the formation of the crater rows 80 kilometers north of the Tithonius Lacus graben is of volcanic origin and is not caused by impacts (Fig. 2). The fissures, on which the craters are lined up, parallel the graben axis and are supposed to have originated from the same rift process.

From absolute age measurements of lunar rocks, we know that there is an agreement with the ages of the oldest rocks from terrestrial cratons (4,5 eons). If we now follow the assumption that the entire solar system has originated at the same time, we may attain another evidence from Martian graben tectonics. The diameter of Mars is approximately half that of the Earth and twice as much that of the Moon. Its mass is one tenth that of the Earth and eight times that of the Moon. Thus, it seems reasonable, when these three celestial bodies of possibly similar original composition started to cool and solidify at the same time, that for the Earth we may expect a relative thin crust (oceanic and continental), that the Mars' crust is thicker and probably entirely continental, and that the Moon, because of its little mass, became solid in a very early stage, before differentiation of continental material occurred.

References

Ahorner, L., 1970: Seismo-tectonic Relations between the Graben Zones of the Upper and Lower Rhine Valley. — In: Graben Problems, J. H. Illies & St. Mueller, eds., 155—166, (Schweizerbart) Stuttgart.

Illies, J. H., 1972: The Rhine Graben Rift System. — Plate Tectonics And Transform Faulting. — Geophys. Surv., 1, 27—60.

van Andel, T. H. & Bowin, C. O., 1968: Mid-Atlantic Ridge between 22° and 23° North Latitude and the Tectonics of Mid-Ocean Rises. — J. Geophys. Res., 73, 1279—1298.

The Rhinegraben: geologic history and neotectonic activity

Perm, Trias und älterer Jura im Bereich der südlichen Mittelmeer—Mjösen-Zone und des Rheingrabens

von

H. Boigk und H. Schöneich

Mit 6 Abbildungen im Text

The Permian, Triassic and Lower Jurassic within the Range of the Southern „Mittelmeer—Mjösen-Zone" and of the Upper Rhine Graben

Abstract: The concise representation of the paleogeographic conditions from the Permian up to the Dogger inclusively has shown that, at least from the Bunter, the southern part of the „Mittelmeer—Mjösen-Zone" was an area of depression of Rhenish strike. In the Muschelkalk and Keuper it extended northwards as far as into the southern part of the present Rhine Graben. Its central and northern parts, on the other hand, were crossed by elements of Variscan strike. In the „Hessische Senke" they were again replaced by a trough section of a predominantly Rhenish strike. In the Liassic and Dogger the southern part of the zone was subject to a particularly heavy subsidence. At the same time, this part was marked by the miogeosynclinal features of a "marginal Alpine foredeep". With the formation of a minor trough west of the epirogenic main depression, the development of the Bresse Graben was first indicated; its subsidence, and that of the Rhine Graben, took place as late as in the Tertiary.

The zone of Rhenish partial elements is thus the result of processes of subsidence which have repeatedly occurred since the Tertiary. The epirogenic and tectonic movements along the lineament were shifted several times in space and time. But on the whole, they follow a parting plane which intersects various entirely different major units of the older substrata, or is in contact with them, and must therefore have its origin at a greater depth.

Die vorliegende Arbeit knüpft an den Versuch an, die Paläogeographie von Perm, Trias und älterem Jura im Bereich des Rheingrabens zu rekonstruieren (H. Boigk 1967, H. Boigk & H. Schöneich 1970). Sie bezieht in die Untersuchung nunmehr auch die in der südlichen Fortsetzung des Grabens liegenden Gebiete der Mittelmeer—Mjösen-Zone ein und stützt sich dabei auf eine größere Zahl neuerer schweizerischer und französischer Arbeiten. Ihre Ergebnisse wurden aufeinander abgestimmt und durch vergleichende Kompilation (vgl. Schrifttum) zu einer Darstellung der Mächtigkeitsverhältnisse vereinigt. Es wird damit erstmalig der Versuch unternommen, die epirogene Entwicklung des Rheingrabens im Gesamtbild des rheinischen Lineamentes von Südfrankreich bis zur Hessischen Senke wiederzugeben. Die Ausführungen beziehen sich allerdings nur auf größere Zeit-abschnitte, für deren Wiedergabe vielfach gewisse Unsicherheiten in der Abgrenzung und Altersstellung der Sedimente in Kauf genommen werden mußten. Eine Bearbeitung und Berücksichtigung der Mächtigkeitsverhältnisse von Einzelgliedern und Stufen, z. B. des Rhät, des Unteren Muschelkalks oder einzelner Zonen des Lias, wie sie etwa für den Cevennen-Rand, das Morvan-Massiv oder für die östliche Provence vorliegen, hätten zwar ein detaillierteres Bild der vielfältigen Einflüsse auf die Verteilung der Sedimente ergeben, lagen jedoch außerhalb des gesteckten und erreichbaren Zieles. Ebensowenig konnten die durch zahlreiche Spezialuntersuchungen ermittelten Faziesverhältnisse und synsedimentären Erscheinungen Berücksichtigung finden, obwohl sie für das Verständnis der paläogeographisch-

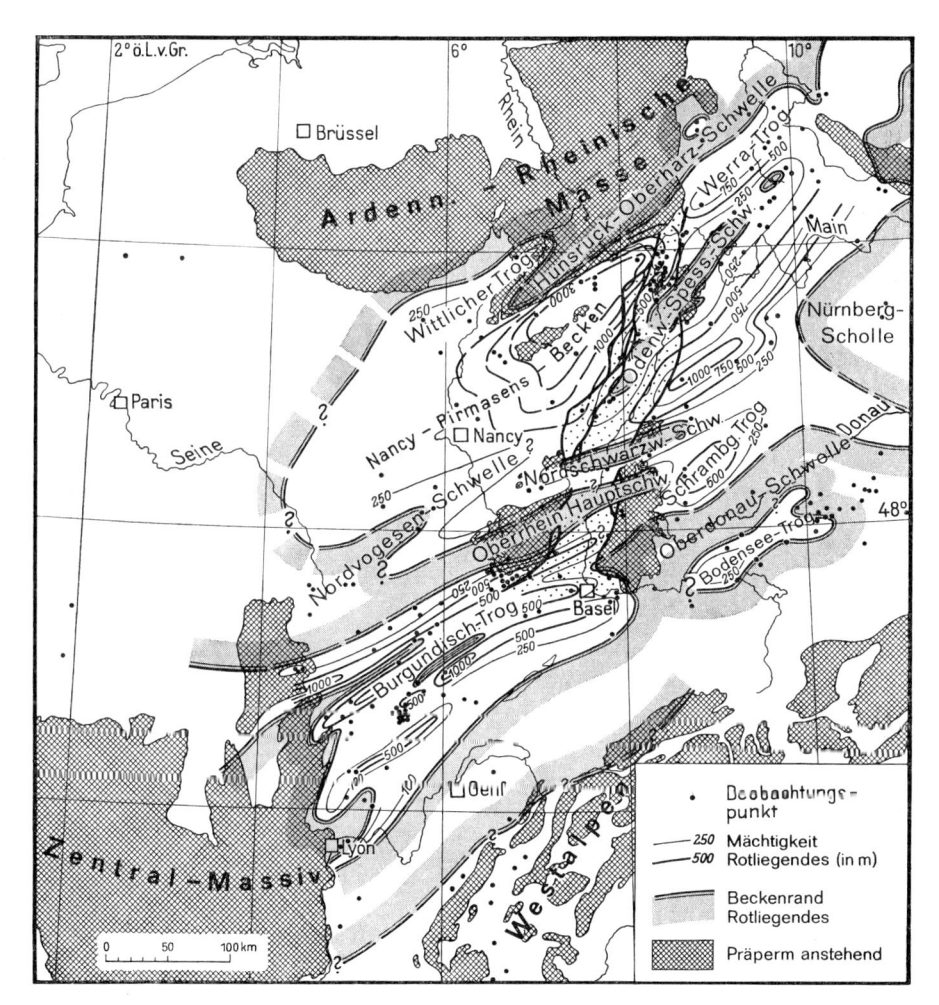

Abb. 1. Paläogeographische Karte des Rotliegenden (u. a. nach H. FALKE, F. NÖRING, H. PORTH, C. SITTLER, N. THEOBALD und E. WINNOCK u. a.).

tektonischen Zusammenhänge von großer Bedeutung gewesen wären. Die Abbildungen geben jedoch, auch wenn sie in Einzelheiten mit Fehlern behaftet sein mögen, ein etwas vereinfachtes großregionales Bild vom epirogenen Werden des rheinischen Lineamentes wieder.

Ob die Mittelmeer—Mjösen-Zone p a l ä o g e o g r a p h i s c h in der variscischen Geosynklinale angedeutet war, ist unbekannt oder hypothetisch. Auch nach der paläozoischen Faltung zeichnen die Becken und Schwellen des R o t l i e g e n d e n (Abb. 1) die posthumen Bewegungen der variscischen Antiklinal- und Synklinalzonen nach. Sie laufen quer über den heutigen Rheingraben hinweg. Auch der Burgundische Trog und der anscheinend isolierte kleinere Bodensee-Trog halten sich an die Streichrichtung des variscischen Gebirges. An beide Senkungszonen grenzt eine mehr oder weniger geschlossene Schwellenregion an, die vom Genfer See bis zum Bodensee zieht. Ablagerungen des Perm (Rotliegenden) sind wieder vom Westrand des Belledonne-Massivs und aus den inneralpinen Senkungsfeldern der Ostschweiz bekannt. Sie nahmen nicht nur die mächtigen Schuttbildungen des Verrucano, sondern auch porphyrische und melaphyrische Ergüsse auf. Wahrscheinlich setzten sich diese Sammeltröge, dem heutigen Streichen der Alpen folgend, in südwestlicher Richtung fort. Demgegenüber bestand zwischen dem südlicheren Teil des Zentralmassivs bzw. der Montagne Noire im Westen und den Massiven von Mercantour und Maures im Osten wieder ein ausgedehnteres Sedimentations-

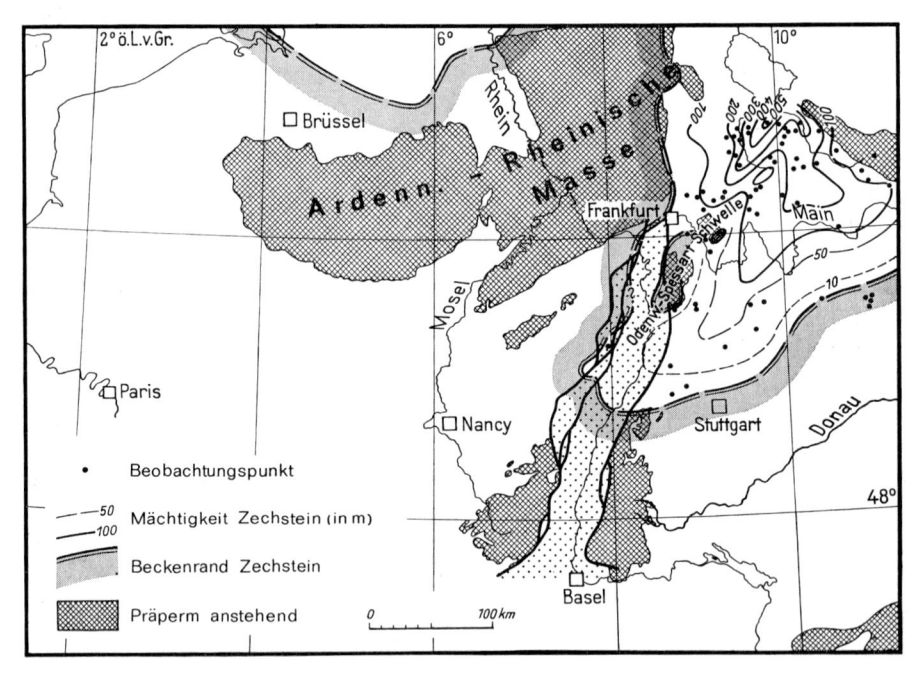

Abb. 2. Paläogeographische Karte des Zechstein (u. a. nach E. BACKHAUS, H. FALKE und F. TRUSHEIM).

gebiet, dessen Umgrenzung und Gliederung jedoch nicht näher festgelegt werden kann. Vielleicht stellt der Sammeltrog klastischer Ablagerungen einen Vorläufer des rheinisch ausgerichteten Rhodanisch-Provençalischen Beckens dar. Auf eine kartenmäßige Darstellung wurde hier wegen kaum zur Verfügung stehender Daten verzichtet.

Das Meer des Zechsteins, das von Nordwestdeutschland über die Hessische Senke bis in den nördlichsten Abschnitt des Rheingrabens vordringt, hinterläßt nur geringmächtige Ablagerungen (Abb. 2). Sie klingen in der südlichen Randfazies an das Rotliegende an und sind deshalb nur schwer von diesem zu trennen. Da zudem nur wenige Beobachtungspunkte bestehen, bleibt die Frage offen, welche epirogenen Elemente die Sedimentmächtigkeiten des Zechsteins im einzelnen bestimmt haben. Aus der Verteilung und Fazies der Ablagerung nördlich und südlich des Mains ist lediglich ersichtlich, daß sich im hessischen Raum und am Niederrhein, also beiderseits der Rheinischen Masse, Meeresbuchten öffnen.

In der Mächtigkeitsentwicklung des Buntsandsteins dokumentieren sich erstmalig beide Richtungen fast gleichberechtigt nebeneinander (Abb. 3). Vindelizisches Land, Morvan—Vogesen-Schwelle, Nancy—Pirmasens-Becken, Kraichgau-Senke und andere kleinere epirogene Elemente spiegeln das Streichen des variscischen Unterbaus wieder. Sie kreuzen z. T., wie mehrfach dargestellt worden ist, den Rheingraben. Auch die gegen Westen vorgreifende Buntsandstein-Bucht, die im N von der Ardennisch-Rheinischen Masse, im S von der Morvan—Vogesen-Schwelle begrenzt wird, ist Ausdruck der posthumen Bewegungstendenzen. Demgegenüber weisen die Hessische Senke, die Trierer und Niederrheinische Bucht wie fast der gesamte Westrand des Großbeckens rheinische Konturen auf. Aber auch am Südende des Rheingrabens ist durch Bohrungen erstmalig eine etwa N—S streichende Senkungszone zwischen Vogesen und Schwarzwald zu belegen. Sie schwenkt jedoch schon wenig südlich von Basel in die präexistente Richtung des burgundischen Rotliegend-Troges ein, um dann erneut wieder mehr den rheinischen Verlauf anzunehmen (vgl. Abb. 1 und Abb. 3). Wahrscheinlich findet diese Senkungszone jenseits einer schmalen Schwelle zwischen Westalpen und Zentralmassiv ihre Fortsetzung im Rhodanisch-Provençalischen Becken, das jedoch nur geringmächtige Ablagerungen germanischer Fazies aufnimmt. Es war im E durch eine in sich gegliederte Schwellenregion begrenzt, in der vor allem die autochthonen und alpinen Kristallin-Massive relative Hochlagen waren. Hier setzt auch der allmähliche Übergang in die alpine Fazies der Trias ein, deren basaler Teil aber noch grobklastisch und kontinental ist.

Im Muschelkalk (Abb. 4) bleibt der Westrand des Beckens seiner Lage nach fast unverändert. Dagegen greift das Meer über den bisherigen Südostrand hinweg und bedeckt Teile des Vindelizischen Festlandes. Zugleich verschiebt sich der Burgundische Trog, der sich mit der allgemeinen Meeresingression sehr stark verbreitert, etwas nach Osten, so daß das Senkungszentrum nunmehr ziemlich genau in der Fortsetzung des heutigen Rheingrabens und unter dem Südostrand des Schweizer Faltenjuras liegt. Der Trog nimmt bis zu 600 m mächtige Ablagerungen germanischer Fazies auf; sie schließen im Mittleren Muschelkalk auch ein Salinar

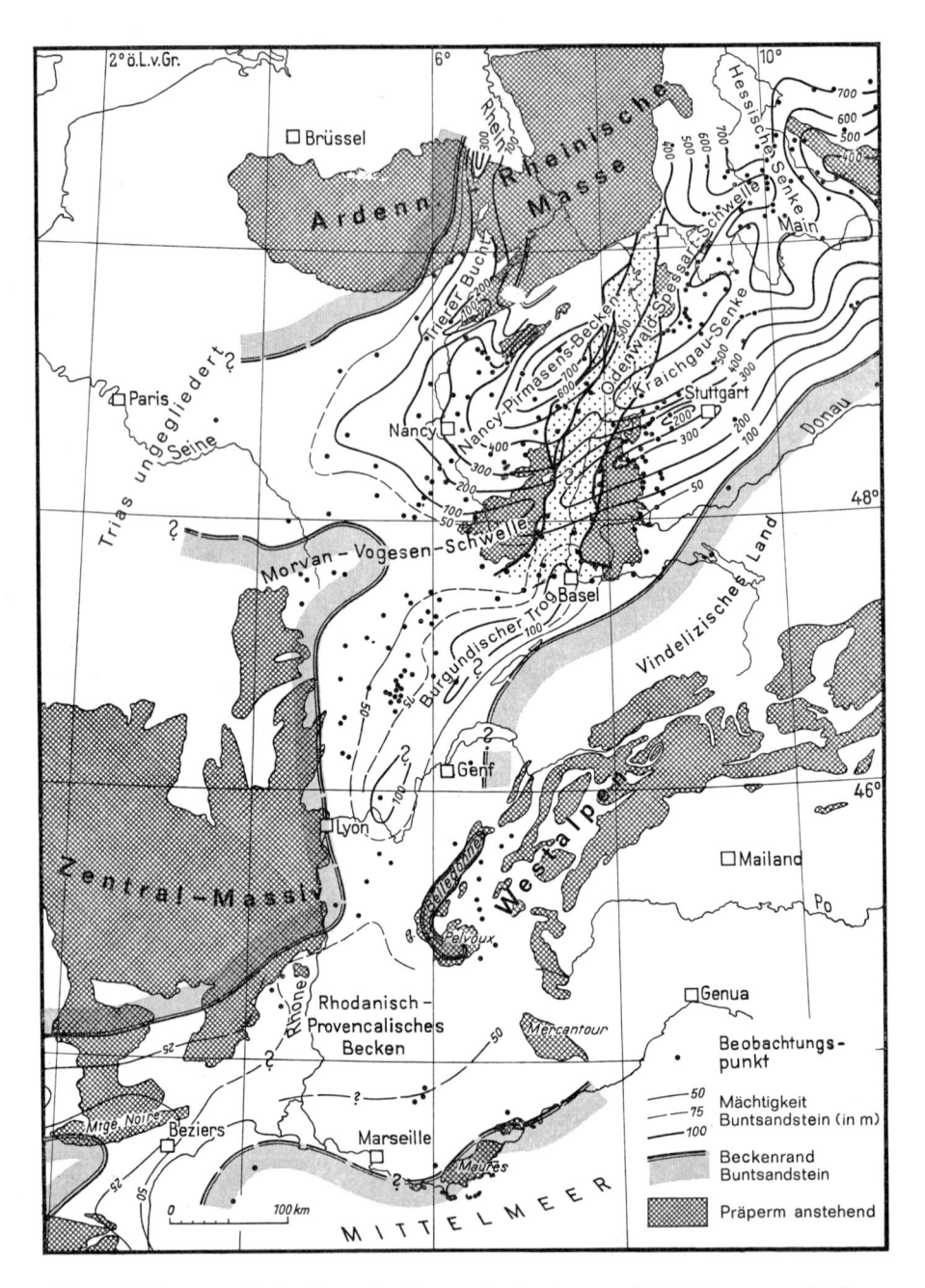

Abb. 3. Paläogeographische Karte des Buntsandsteins (u. a. nach P. F. Burollet, P. Cramer, J. Fourmentraux u. a., E. Müller, J. Ricour, N. Theobald, F. Trusheim, E. Winnock u. a.).

ein. Im Osten reicht der Senkungsraum bis an die subalpinen kristallinen Massive. Da sie sich gegen Südwesten dem Zentralmassiv nähern, bleibt nur eine schmale Verbindung zum Provençalischen Becken offen. Dennoch ähneln die Absätze dieses gegen Süden sich weit öffnenden Troges, von detritischen Randbildungen abgesehen, denen des eigentlichen Germanischen Beckens. Hinweise auf Verbindungen von dort zwischen den voralpinen Massiven (Pelvoux, Mercantour) hindurch und z. T. über sie hinweg zu den miogeosynklinalen Räumen der alpinen Trias ergeben sich besonders für den Ausgang des Zeitabschnittes, in dem die Meerestransgression auch ihre größte Ausdehnung erreicht. In die rheinisch ausgerichtete Senkung wird nunmehr auch der Südteil des Rheingrabens einbezogen. Eine schmale Zone verstärkter Sedimentation trennt die gegen Nordosten endende Morvan—Vogesen-Schwelle von einer schildförmigen Hochlage im Schwarzwald. Dagegen setzen sich in den nördlicheren Bereichen des Rheingrabens die epirogenen Bewegungen vielfach noch auf den eingefahrenen, variscisch ausgerichteten Bahnen fort.

Neben den Undationen in variscischer Richtung, die u. a. im Nancy—Pirmasens-Becken, der Kraichgau-Senke und in der Nordbegrenzung des Vindelizischen Landes zum Ausdruck kommen, lebt im K e u p e r [1] auch das rheinische Element weiter (Abb. 5). Die Subsidenz des Untergrundes ist nicht nur im Südteil des Rheingrabens deutlich erkennbar, sondern besonders groß im Burgundischen Trog. Die Sedimente weisen hier Mächtigkeiten auf, die sonst nur in den halokinetisch besonders mobilen Bereichen Nordwestdeutschlands erreicht und überschritten werden. Sie schließen nach dem Muschelkalk ein weiteres bis zu 700 m mächtiges Salinar ein, das für die Faltung des Schweizer Jura mitverantwortlich ist. Jenseits der Schwellenregion, die das Zentral-Massiv und das externe Kristallin von Belledonne verbindet, öffnet sich auch im Keuper das Rhodanisch-Provençalische Becken. Angaben über die Mächtigkeiten seines Zentrums fehlen. Sie dürften in Anbetracht der starken Beteiligung von Steinsalz an der Bildung von Keuper-Aufbrüchen in der Provence beträchtlich sein. Einflüsse der alpinen Miogeosynklinale lassen auf Verbindungen zwischen den Kristallin-Rücken hindurch zur Tethys schließen. Dagegen ist der Keuper des Maurischen Massivs mit der Germanischen Fazies zu vergleichen und relativ geringmächtig. Den Westrand, den das Zentral-Massiv und die Montagne Noire bilden, zeichnen klastische Saumbildungen aus. Über beide greift das Rhät meist geringfügig hinweg. Demgegenüber scheint sich im Norden die Kraichgau-Senke verbreitet zu haben. Sie öffnet sich in das weite Keuper-Becken Nordwestdeutschlands. Paläogeographische Details sind jedoch am unteren Main nicht rekonstruierbar. Ebenso dehnt sich der Sedimentationsraum zwischen Ardennisch-Rheinischer Masse und Morvan—Vogesen-Schwelle nach Westen zu aus. Der Raum, der sich besonders im Oberen Keuper stark erweitert und das Pariser Becken großräumig umfaßt, unterliegt nunmehr der verstärkten Absenkung und Sedimentation.

---

[1] Im Keuper sind, entgegen der Gepflogenheit französischer Geologen, auch die Lettenkohle und das Rhät einbezogen.

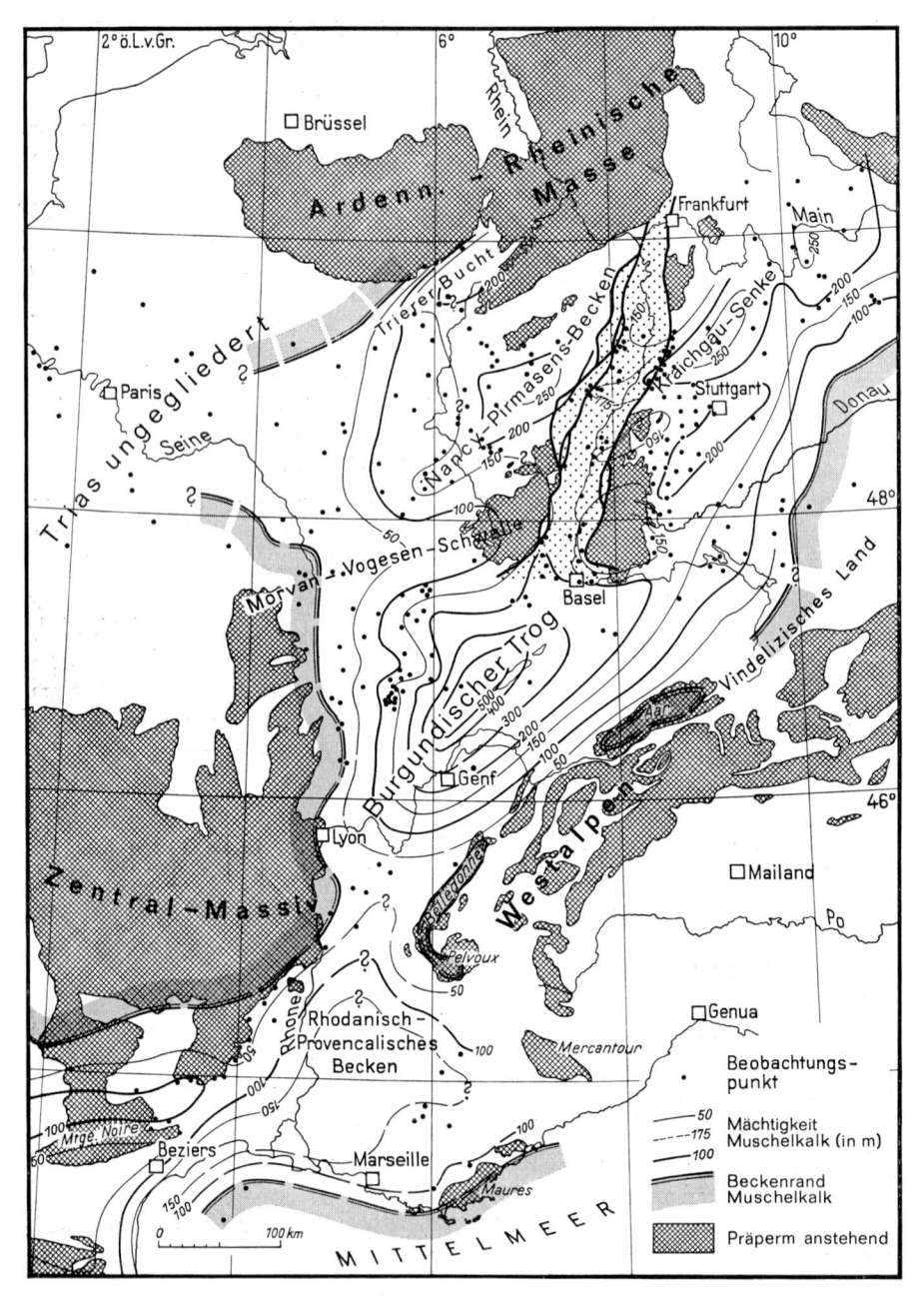

Abb. 4. Paläogeographische Karte des Muschelkalks (u. a. nach U. P. Büchi u. a., U. Emmert, J. Fourmentraux u. a., C. Sittler, E. Winnock u. a.).

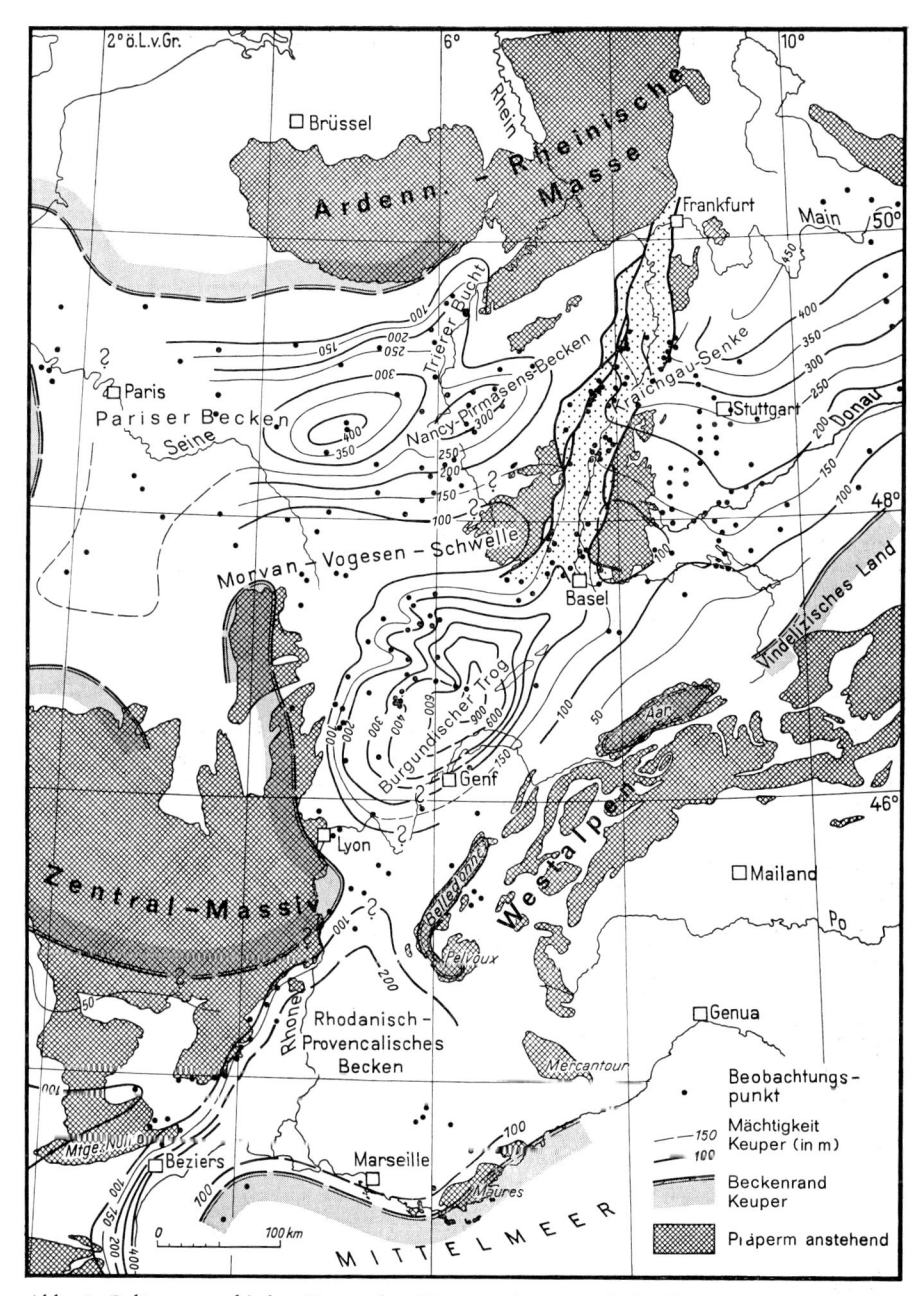

Abb. 5. Paläogeographische Karte des Keupers (u. a. nach U. P. BÜCHI u. a., P. F. BUROLLET, O. F. GEYER & M. P. GWINNER, J. RICOUR, N. THEOBALD, E. WINNOCK u. a.).

Im Lias[2] sind die Mächtigkeiten im Bereich des Rheingrabens ziemlich ausgeglichen (Abb. 6). Sie nehmen generell gegen Südosten in Richtung auf den Rand des Vindelizischen Landes ab. Als variscisch streichendes Element bildet sich nur die Pfalzburg—Kraichgau-Senke ab. Dagegen sinkt südwestlich von Basel die rheinisch streichende Zone des Burgundischen Troges weiter ein. Sie erreicht ihre stärkste Subsidenz im Dauphiné-Becken, das sich gegenüber dem Rhodanisch-Provençalischen Trog der Trias (vgl. Abb. 3—5) beträchtlich verschmälert hat. Die Ablagerungen des nördlichen Teilbeckens sind tonig-mergelig, teilweise auch sandig-kalkig und eisenschüssig wie im lothringischen Raum. Auch der Westrand weist längs des Zentral-Massivs klastische Einflüsse auf. Dagegen breitet sich im zentralen Teil des Dauphiné-Beckens eine bathyale Tiefenfazies aus, deren Mächtigkeit nur annähernd geschätzt werden kann (in Abb. 6 schematisch dargestellt). Sie dürfte etwa 2000 m erreichen. Der vermutlich an Brüchen synsedimentär abgesenkte Raum füllt sich mit marinen, mergelig-kalkigen Sedimenten, die im Hangenden in Tonsteine übergehen. Der Ostrand, der durch die subalpinen Kristallin-Schwellen noch stärker gegliedert ist als zuvor, zeichnet sich durch schmale Durchgänge aus, die zum alpinen Jura-Meer des Briançonnais vermitteln. Im Südosten läuft der Lias am Rand des Maurischen Massivs in der Litoralzone einer heute im Mittelmeer versunkenen epikontinentalen Plattform aus. Bemerkenswert ist schließlich eine schmale, wenig tiefe Senkungszone, die von Lyon ab den Nordteil des Dauphiné-Beckens begleitet. Sie deutet auf die erste Anlage des im Tertiär eingebrochenen Bressegrabens, der gegenüber dem Rheingraben deutlich nach Westen versetzt ist. Nördlich der Morvan-Schwelle senkt sich im Lias auch das Pariser Becken weiter ab. Es weist neben der variscischen Richtung einzelner Teiltröge auch eine rheinische Quergliederung auf. Außerdem öffnet sich die Verbindung nach Nordwesten zu den Britischen Inseln.

Zusammenhängende Untersuchungen über die Paläogeographie des D o g g e r s sind hier nicht mehr durchgeführt worden. Aus der vorliegenden Literatur lassen sich jedoch genügend Anhaltspunkte für eine Rekonstruktion des damaligen Sedimentationsraumes ableiten. Im Bereich des Rheingrabens bestehen Hinweise für das Vorhandensein der Pfalzburg (Zaberner)—Kraichgau-Senke. Sie verläuft über den Rheingraben hinweg. Andererseits ist das rheinische Lineament in synsedimentären Brüchen wirksam. Wahrscheinlich heben sich Schwarzwald- und Vogesen-Schwelle teils submarin, teils subaerisch weiter heraus, während der Rheingraben einsinkt. In seiner südlichen Fortsetzung befinden sich nach einer Darstellung von E. WINNOCK u. a. (1967) nunmehr z w e i ausgeprägte Tröge, die durch ein Schwellengebiet voneinander getretnt sind. Die schmale östliche Senkungszone reicht weit über Genf nach Süden. Der westliche Trog sinkt fast genauso tief wie der östliche ein. Er befindet sich etwa an der Stelle, an der im Tertiär der rheinisch streichende Bressegraben einbricht. Beide Tröge scheinen im Süden zu kommunizieren und in den Dogger des Dauphiné-Beckens überzugehen, dessen Senkungs-

---

[2] Die Darstellung in Abb. 6 gibt die Lias-Mächtigkeiten für das Hettangien bis einschließlich Toarcian wieder. In der französischen Abgrenzung wird auch noch das tiefere Aalenian und das Rhät zum Lias gerechnet (vgl. Fußnote 1).

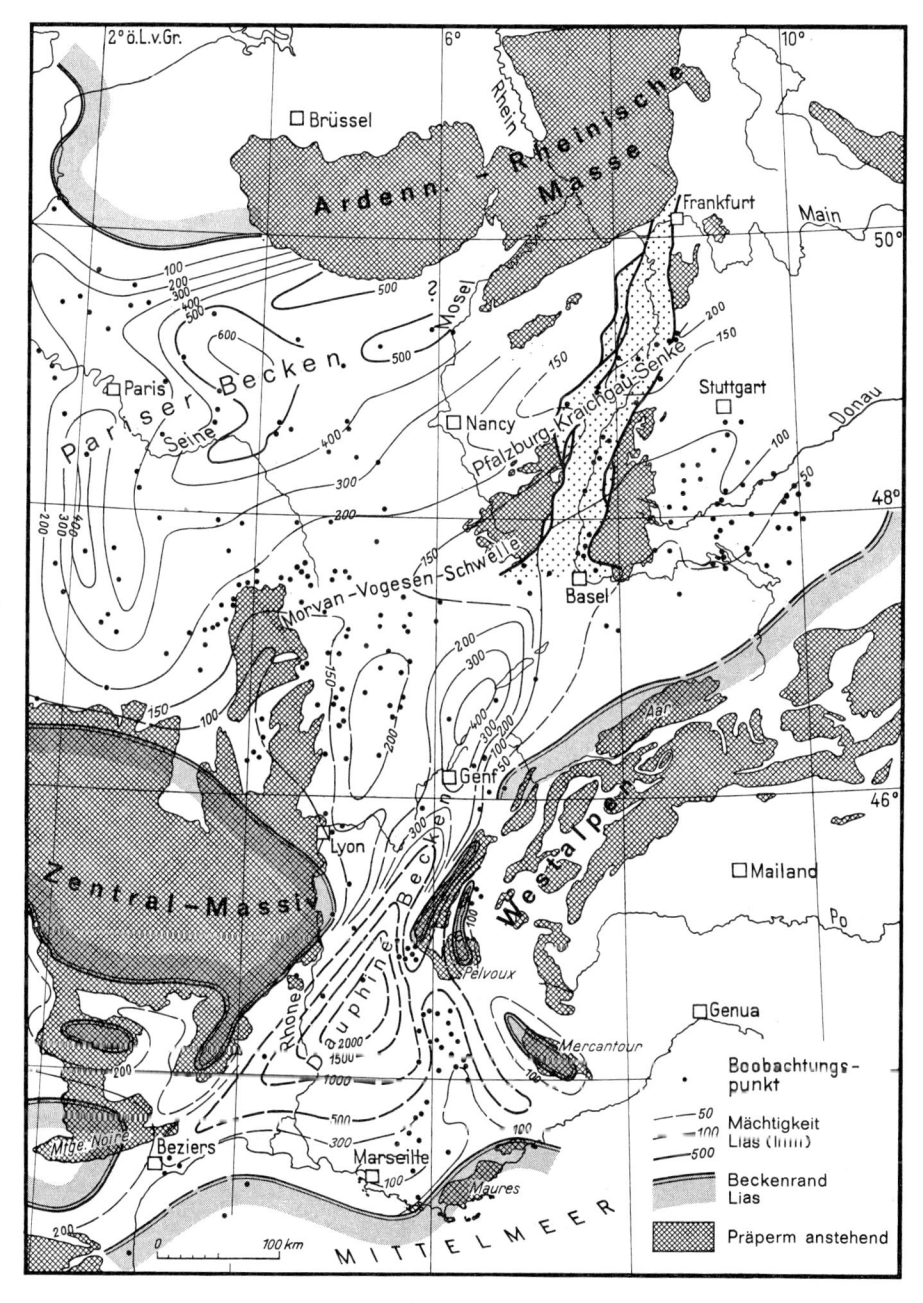

Abb. 6. Paläogeographische Karte des Lias (u. a. nach U. P. BÜCHI u. a., M. GOTTIS, K. HOFFMANN, A. LEFAVRAIS-RAYMOND, E. WINNOCK u. a.).

tendenz auch weiterhin besteht. Sein Westrand weist litorale Fazies auf. Hier bilden sich vor Beginn des Bathonien am Rhône-Rand stellenweise rheinisch streichende Brüche und Flexuren. Auch die Massive von Maures und Esterel gehören mit ihrer litoralen Entwicklung, wie schon im Lias, dem randlichen Epikontinentalbereich an. Demgegenüber nimmt das Zentrum des Dauphiné-Beckens auch im Dogger kontinuierlich sehr mächtige, eintönige Sedimente auf, unter denen Tone vorherrschen.

Literaturverzeichnis

BACKHAUS, E., 1965: Die randliche „Rotliegend"-Fazies und die Paläogeographie des Zechsteins im Bereich des nördlichen Odenwaldes. — Notizbl. hess. L.-A. Bodenforsch., 93, 112—140.

BOIGK, H., 1967: The thickness of the pre-tertiary sediments in the Upper Rhinegraben. — Abh. geol. L.-Amt Bad.-Württ., 6, 39—41.

BOIGK, H. & SCHÖNEICH, H., 1970: Die Tiefenlage der Permbasis im nördlichen Teil des Oberrheingrabens. — In: ILLIES, H. & MÜLLER, ST., Graben Problems, 45—55, Schweizerbart, Stuttgart.

BONTE, A., GOGUEL, J., GREBER, CH., LAFFITTE, P., LIENHARDT, G. & RICOUR, J., 1953: Le bassin houiller de Lons-le-Saunier. — Publ. Bureau Rech. géol. géophys., Nr. 10, 48 S.

BÜCHI, U. P., LEMCKE, K., WIENER, G. & ZIMDARS, J., 1965: Geologische Ergebnisse der Erdölexploration auf das Mesozoikum im Untergrund des schweizerischen Molassebeckens. — Bull. Ver. Schweiz. Petr.-Geol. u. -Ing. 32, Nr. 82, 7—38.

BUROLLET, P., 1963: Présentation de quelques documents d'interprétation du Trias d'Aquitane. — Mém. Bureau Rech. Geol. Min., 15, 309—319.

CALVEZ, Y. LE & LEFAVRAIS-RAYMOND, A., 1961: Lias des sondages de la bordure du Morvan. — Mém. Bureau Rech. Geol. Min., 4, 503—534.

COLLIGNON, M. & DARDENNE, M., 1961: Le Lias dans les sondages de la Société Petrorep. — Mém. Bureau Rech. Geol. Min., 4, 535—542.

CRAMER, P., 1964: Schichtstufenland Nordbayerns, I. Perm, II. Trias. — In: Erl. z. Geol. Karte v. Bayern 1:500 000, 2. Aufl., 55—81.

EINSELE, G. & SCHÖNENBERG, R., 1963: Epirogen-tektonische Grundzüge des süddeutschen Beckens und seiner südöstlichen Randgebiete im Mesozoikum. — Publ. Serv. Geol. Luxembourg, 14, 137—165.

EMMERT, U., 1964: Schichtstufenland Nordbayerns, b. Muschelkalk, c. Keuper. — In: Erl. z. Geol. Karte v. Bayern 1:500 000, 2. Aufl., 81—120.

FALKE, H., 1971: Zur Paläogeographie des kontinentalen Perms in Süddeutschland. — Abh. hess. L.-A. Bodenforsch., 60, 223—234.

FOURMENTRAUX, J., LAVIGNE, J., PONTALIER, Y. & POUJOL, P., 1959: Trias, Jurassique inférieur et moyen de l'est du Bassin de Paris; Présentation des cartes d'isopaques et de lithofacies. — Rev. Inst. Franç. Petrole, 14, Nr. 9, 1063—1090.

GEYER, O. F. & GWINNER, M. P., 1964: Einführung in die Geologie von Baden-Württemberg. — 223 S., Schweizerbart, Stuttgart.

GOTTIS, M., 1961: Le Lias des bordures méridionales du Massif Central, du domaine Pyrénéen et du Bassin Aquitan. — Mém. Bureau Rech. Géol. Min., 4, 59—76.

HABICHT, H., 1966: Die permo-karbonischen Aufschlußbohrungen der Nahe-Senke, des Mainzer Beckens und der Zweibrücker Mulde. — Z. dt. geol. Ges., 115, 631—649.

HÖLDER, H., 1964: Jura. Handbuch der stratigraphischen Geologie, IV. — 603 S., Enke, Stuttgart.

LEFAVRAIS-RAYMOND, A., 1962: Contribution à l'étude géologique de la Bresse d'après les sondages profonds. — Mém. Bureau Rech. Géol. Min., 16, 170 S.

LEFAVRAIS-RAYMOND, A. & HORON, O., 1961: Bassin de Paris (Rapport I). — Mém. Bureau Rech. Géol. Min., **4**, 3—56.

LUCIUS, M., 1953: Quelques aspects de la géologie appliquée dans l'aire de sedimentation luxembourgeoise. — Publ. Serv. Geol. Luxembourg, **9**, 282 S.

NÖRING, F., 1951: Die Fortsetzung der Saar-Senke in Hessen. — Notizbl. hess. L.-A. Bodenforsch., (6) **2**, 22—40.

POUJOL, P., 1961: La série Liassique du Bassin de Paris. Essai de corrélations entre les sondages de la Régie Autonome des Petroles. — Mém. Bureau Rech. Geol. Min., **4**, 577—604.

RICOUR, J., 1962: Contribution à une révision du Trias français. — Mém. pour servir à l'explication de la Carte Geol. détaillée de la France, 471 S., Imprimerie Nationale, Paris.

RICOUR, J., HORON, O. & LIENHARDT, G., 1960: Le Trias du Jura, de la Bresse, de la plaine de la Saône et de la bordure nord du Massif Central. — Compte rendu somm. et Bull. Soc. Geol. France, 7, **2**, 156—167.

SITTLER, C., 1969: Le fossé rhénan en Alsace. Aspect structural et histoire géologique. — Rev. Geogr. Phys. et Geol. Dynam., **11**, 5, 465—493.

THEOBALD, N., 1954: Evolution geologique du Nord-Est de la France, en particulier du Fossé Rhénan, dequis les temps secondaires en relation avec les possibilités pétrolifères. — Ann. sci. Univ. Besançon, 2. Ser. fasc. 1, T. **IX**, 3—92.

TRUSHEIM, F., 1964: Über den Untergrund Frankens, Ergebnisse von Tiefbohrungen in Franken und Nachbargebieten 1953—1960. — Geol. Bavarica, **54**, 92 S.

WINNOCK, E., BARTHE, A. & GOTTIS, C., 1967: Résultats des forages pétroliers français effectués dans la région voisine de la frontière Suisse. — Bull. Ver. Schweiz. Petr.-Geol. u. -Ing., **33**, Nr. 84, 7—22.

An Isobath Map of the Tertiary Base in the Rhinegraben

by

F. Doebl and W. Olbrecht

With 1 figure on a folder

A new isobath map of the Tertiary base of the Rhinegraben is presented. Previous maps considering the same level have been published for the northern part by DOEBL (1967) and for the southern part of the graben by DOEBL (1970). In addition, an isobath map of the pre Tertiary top of the Mainz basin area has been published by SONNE (1970). The presented new map has been compiled by evaluating more than 350 bore-holes which have reached or went close to the base of the Tertiary deposits. About 2000 further bore-holes which penetrated only through parts of the Cenozoic graben succession have been included. Geophysical investigations have not been considered drawing the isobaths. The contour lines referring to the base of the Tertiary succession are indicated in meters below (or above) sea level.

Graben subsidence and graben sedimentary fill started during Middle Eocene times. Tertiary sediments overlie Jurassic beds in the southern part of the graben. Between Landau and Worms Triassic rocks are forming the base whereas north of Worms Permian sediments and volcanics have been found underneath the Cenozoic deposits.

North of Worms the through axis within the Rhinegraben runs close to the western main fault and from there to Heidelberg it shifts to the eastern graben margin. Between Heidelberg and Strasburg the axis of maximum subsidence follows the eastern master fault of the graben. South of Strasburg the western part of the graben again has been affected by preferred subsidence. In the southernmost graben area the through axis splits into two branches, the Dammerkirch graben in the west and the Sierenzgraben in the east, separated by the Muelhausen horst structure, which forms a triangular block. The maximum thickness of the graben fill has been observed northwest of Mannheim. Here the deepest drilling in the Rhinegraben area, the bore-hole Frankenthal 10, has been sunk to a depth of 3335,3 m into Lower Pechelbronn beds of Lower Oligocene age.

To make the map legible, only the marginal master faults and some larger rupture zones of the bordering escarpments have been marked (ANDERLE 1974). The graben itself and its sedimentary fill is traversed by different sets of numerous normal faults, mostly of antithetic character. The scale of the presented map made it impossible to depict these tectonical details.

References

ANDERLE, H. J., 1974: Block tectonic interrelations between northern Upper Rhine Graben and southern Taunus Mountains. — In: Approaches to Taphrogenesis, 243—253.
DOEBL, F., 1967: The Tertiary and Pleistocene sediments of the northern and central part of the Upper Rhinegraben. — Abh. geol. L.-Amt Baden-Württ., 6, 48—54.
— 1970: Die tertiären und quartären Sedimente des südlichen Rheingrabens. — In: ILLIES, H. & MUELLER, ST. (eds.), Graben Problems, Internat. Upper Mantle Proj. Scient. Rep. 27, 56—66.
SONNE, V., 1970: Das nördliche Mainzer Becken im Alttertiär, Betrachtungen zur Paläoorographie, Paläogeographie und Tektonik. — Oberrhein. geol. Abh., 19, 1—28.

Mikrofloristische Untersuchungen zur Altersstellung der jungtertiären Ablagerungen im mittleren und nördlichen Oberrheingraben

von

G. von der Brelie

Mit 1 Abbildung als Beilage und 1 Tabelle im Text

Abstract: Palynological investigations carried out in the northern and central parts of Upper Rhine Graben have proved that the sediments called "Jungtertiär I" respectively "? Obermiozän" are of lower Miocene age (Aquitan to Lower Burdigal). The pollen spectras in these layers contain the same pollen assemblages as the Hydrobia Beds. The lower "Jungtertiär II" and the parts of deeper stratigraphical position in Pliocene belong in the space between uppermost Miocene to Lower Pliocene (Sarmat to Pont). Results of quantitative pollen analysis show that in the "Jungtertiär I" the component of typical Miocene pollen taxa accounts for up to 50% of the total flora. In the "Jungtertiär II", however, these forms amount to 15% or less. The flora of Upper Pliocene (Asti/Piacentin) is quite different and has a floral structure like the typical Pliocene in north western Europe.

Die altersmäßige Einstufung der jungtertiären Schichten des Oberrheingrabens oberhalb der Hydrobien-Schichten wirft wegen des Fehlens echter tierischer Leitfossilien noch immer große Probleme auf. Diese Unsicherheit kommt in den aufgrund petrographischer Kriterien entwickelten lokalstratigraphischen Bezeichnungen für die verschiedenen Schichtglieder (? Obermiozän, Jungtertiär I, Jungtertiär II und Pliozän) deutlich zum Ausdruck (s. u. a. SCHAD 1964) Da die Ablagerungen des jüngeren und jüngsten Tertiärs teilweise terrestrischen bzw. limnisch-fluviatilen Ursprunges sind, bietet sich die Palynologie für den Versuch einer Einstufung der neogenen Ablagerungen in das internationale stratigraphische Gliederungsschema an. Eine Übersicht über das im Rahmen dieser Untersuchungen berücksichtigte Probenmaterial bringt die Tab. 1. In einem Pollendiagramm sind einige der für die verschiedenen lokalstratigraphischen Einheiten typischen Pollenspektren dargestellt. Die Reihenfolge der Sporomorphen richtet sich nach deren stratigraphischer Bedeutung (VON DER BRELIE 1967). Die Probenserien sind angeordnet nach der sich aus der Mikroflorenzusammensetzung ergebenden zeitlichen Abfolge.

1. Die Mikrofloren der jungtertiären lokalstratigraphischen Einheiten im Oberrheingraben

1.1 Die Mikrofloren der Hydrobien-Schichten

In den Pollenspektren aus den Hydrobien-Schichten ist eine deutliche Vorherrschaft von *Engelhardtioipollenites punctatus* und *Engelhardtioidites microcoryphaeus* zu erkennen. Gleichzeitig erreicht, abgesehen von den faziell bedingten Maxima bei *Inaperturopollenites dubius* und den geflügelten Koniferenpollen, *Subtriporopollenites simplex* höhere Werte. Weiterhin von Bedeutung ist das Vorkommen von *Inaperturopollenites emmaensis*, *Cupuliferoipollenites fusus*, *Plicatopollenites plicatus* (eine Formspezies, die im Alttertiär ihre Hauptverbreitung hat), *Quercoidites henrici* und *Rhoipites pseudocingulum*. *Quercoidites microhenrici*, der erst in der oberen Hälfte des Hemmoorium eine größere Verbreitung zeigt, fehlt fast vollständig.

1.2 Die Mikroflora des Jungtertiärs I und des ?Obermiozäns

Die Pollenspektren aus dem Jungtertiär I bzw. ?Obermiozän stimmen untereinander weitgehend überein. Damit ist ihre gleichzeitige Entstehung mikrofloristisch nicht zu bezweifeln. Andererseits zeigt das Diagramm aber auch eine fast vollständige Identität mit den Beispielen aus den Hydrobien-Schichten.

Tab. 1. Das Untersuchungsmaterial.

Lfd. Nr.	Bohrung u. Teufe	Anzahl d. Proben	Labor-Nr. GLA	stratigraph. Einstufung d. Einsenders	Bezeichnung auf dem Diagramm
	Probeneinsender: Gewerkschaft Elwerath				
1	Groß-Gerau 2				GG 2
	454,0—460,0 m	1	50 233	Jungtertiär II	
	485,0—491,0 m	2	50 234 — 235	Jungtertiär I	
2	Wolfskehlen 7				
	465,0—470,0 m	1	50 231	Jungtertiär I	
3	Stockstadt 28				
	503,5—521,0 m	9	50 236 — 244	Jungtertiär II	
4	Stockstadt 102				
	424,5—469,0 m	3	46 091 — 093	Jungtertiär II	
	484,0—557,0 m	4	46 094 — 097	Jungtertiär I	
	599,5—625,0 m	2	46 098 — 099	Hydrobien-Mergel	
5	Stockstadt 103				
	450,0—452,0 m	1	46 100	Jungtertiär II	
	475,0—480,0 m	1	46 101	Jungtertiär I	
6	Stockstadt 116				ST 116
	505,0—597,2 m	4	46 102 — 105	Jungtertiär II	
	597,2—766,3 m	14	46 106 — 119	Jungtertiär I	
7	Stockstadt 128				ST 128
	715,6—827,0 m	8	46 083 — 090	Hydrobien-Schichten	
	Probeneinsender: Deutsche Erdöl AG. (jetzt Deutsche Texaco AG.)				
8	Pfungstadt 6				
	635,6—641,0 m	1	45 463	Obermiozän	
9	Pfungstadt 7 a				PF 7 a
	565,0 m	1	45 464	Pliozän	
	571,6—682,8 m	5	45 465 — 469	Obermiozän	
10	Pfungstadt 9				PF 9
	534,0—546,8 m	5	45 451 — 455	Pliozän*	
	587,5—668,3 m	7	45 456 — 462	Obermiozän	
11	Pfungstadt 13				
	561,0—632,0 m	7	52 567 — 573	Obermiozän	
12	Hähnlein-West 1				HÄ 1
	204,0—209,0 m	1	52 527	„höheres Pliozän"	
	381,8—452,3 m	6	52 528 — 533	„mittleres Pliozän"	
	460,3—463,3 m	2	52 534 — 535	„tieferes Pliozän"	
	735,0 m	1	52 536	„Obermiozän"	
	Probeneinsender: Wintershall AG.				
13	Landau 17				LA 17
	70,0—120,0 m	3	51 987 — 989	Pliozän oder Obermiozän	
14	Landau 91				
	172,5 m	1	55 069	Obere Hydrobienschichten oder jünger	
15	Landau 186				
	125,0—130,0 m	1	58 021	Miozän oder Pliozän	

* Nach dem pollenanalytischen Befund liegt die Grenze zwischen den beiden Schichtgliedern bei 540,8 m Teufe.

Fortsetzung von Tabelle 1

Lfd. Nr.	Bohrung u. Teufe	Anzahl d. Proben	Labor-Nr. GLA	stratigraph. Einstufung d. Einsenders	Bezeichnung auf dem Diagramm
	Probeneinsender: Geologisches Landesamt Rheinland-Pfalz				
16	Mettenheim (Wasserwerk) 242,5—243,5 m	1	60 184		ME
	Probeneinsender: Geologisches Landesamt Baden-Württemberg				
17	Illingen B 5 58,15—59,35 m	1	55 991		IL
18	Steinmauern B 17 67,3— 68,0 m	1	55 969		SM
	Probeneinsender: Dr. Franz Doebl, Landau				
19	Kläranlage Godramstein B I 8,2— 17,0 m	4	63 919 — 922	Globigerinellen-Horizont der Hydrobien-Schichten	GO I

1.3 Die Mikroflora des Jungtertiärs II

Die Schichten aus dem unteren Jungtertiär II unterscheiden sich von den im Vorhergehenden beschriebenen vor allem durch einen erheblichen Rückgang von *Engelhardtioipollenites punctatus* und *Engelhardtioidites microcoryphaeus*. Die miozänen Pollenformen sind noch vollzählig vorhanden, wenn auch jetzt mit geringeren Werten. Größere Anteile, die jedoch vor allem faziell bedingt sind, erreichen *Inaperturopollenites dubius* und *Alnipollenites verus*, wobei beide Formen sich wechselseitig vertreten. Hier zeichnet sich ein Wechsel zwischen trockeneren Erlen-Bruchwäldern *(A. verus)* und feuchteren „Cupressineen"-Beständen *(I. dubius)* ab.

1.4 Die Mikroflora des Pliozäns

Von den mit „Pliozän" bezeichneten Proben enthielten besonders solche aus stratigraphisch tieferen Horizonten Pollenfloren, die in ihrer Zusammensetzung denen aus dem Jungtertiär II vollständig gleichen. Andere pliozäne Proben dagegen lieferten Spektren, in denen die miozänen Leitformen fehlen. Diese Floren werden gekennzeichnet durch höhere Anteile von *Tsuguepollenites igniculus*, *Sequoiapollenites polyformosus*, *Sciadopityspollenites serratus* und *Faguspollenites verus*. Sie gleichen damit vollständig den pliozänen Mikrofloren aus Hessen, der Niederrheinischen Bucht, den Niederlanden und aus Niedersachsen. Diese Florenvergesellschaftung hat im gesamten Oberrheingraben von Frankfurt bis südlich Karlsruhe eine weite Verbreitung. Sie kennzeichnet die jüngsten Sedimente des Tertiärs. Innerhalb der als „Pliozän" bezeichneten Schichtkomplexe treten also zwei stratigraphische Einheiten auf.

Abb. 1. Pollendiagramm der jungtertiären Ablagerungen im mittleren und nördlichen Oberrheingraben. Die genaue Bezeichnung der Bohrungen ist Tabelle 1 zu entnehmen. Die nachstehenden Ergänzungen beziehen sich auf die Eintragungen in den vertikalen Spalten.

Spalte

1	*Plicatopollenites plicatus* (R. POT.) KRU.
2	*Inaperturopollenites emmaensis* MÜRR. & PF.
3	*Engelhardtioipollenites punctatus* (R. POT.) R. POT.
	Engelhardtioidites microcoryphaeus (R. POT.) R. POT., TH. & THIERG.
4	*Quercoidites henrici* (R. POT.) R. POT., TH. & THIERG.
5	*Quercoidites microhenrici* (R. POT.) R. POT., TH. & THIERG.
6	*Tricolporopollenites villensis* (TH. i. R. POT., TH. & THIERG.) TH. & PF.
7	*Rhoipites pseudocingulum* (R. POT.) R. POT.
8	*Cupuliferoipollenites fusus* (R. POT.)
9	*Cyrillaceaepollenites megaexactus* (R. POT.) R. POT.
10	*Tetracolporopollenites* sp. sp.
11	*Sequoiapollenites polyformosus* THIERG.
12	*Sciadopityspollenites serratus* (R. POT. & VEN.) RAATZ
13	*Nyssapollenites* sp.
14	*Periporopollenites stigmosus* (R. POT.) TH. & PF.
15	*Polyatriopollenites stellatus* (R. POT.) PF.
16	*Tsugaepollenites igniculus* (R. POT.) R. POT. & VEN.
	Tsugaepollenites viridifluminipites (WODEHOUSE) R. POT.
17	*Subtriporopollenites simplex* (R. POT.) TH. & PF.
18	*Caprifoliipites microreticulatus* (TH. & PF.) R. POT.
19	*Betulaceoipollenites bituitus* (R. POT.) R. POT.
	Triatriopollenites rurensis TH. & PF.
	Triatriopollenites myricoides (KREMP) PF. i. TH. & PF.
20	*Inaperturopollenites dubius* (R. POT. & VEN.) TH. & PF.
21	*Taxodiaceaepollenites hiatus* (R. POT.) KREMP
22	*Cupuliferoidaepollenites quisqualis* (R. POT.) R. POT.
	Cupuliferoidaepollenites fallax (R. POT.)
23	*Platanoidites gertrudae* (R. POT.) R. POT., TH. & THIERG.
24	*Araliaceoipollenites edmundi* (R. POT.) R. POT.
25	*Cupuliferoipollenites pusillus* (R. POT.) R. POT.
	Cupuliferoipollenites oviformis (R. POT.) R. POT.
26	*Cyrillaceaepollenites exactus* (R. POT.) R. POT.
27	*Faguspollenites verus* RAATZ
28	*Multiporopollenites maculosus* (R. POT.) TH. & PF.
29	*Abietineaepollenites microreticulatus* (R. POT.) R. POT.
	Pinuspollenites labdacus (R. POT.) RAATZ
	Piceaepollenites alatus R. POT.
30	*Fraxinoipollenites confinis* (R. POT.)
31	*Tricolpopollenites asper* TH. & PF.
32	*Ilexpollenites iliacus* (R. POT.) THIERG.
33	*Trivestibulopollenites betuloides* PF. i. TH. & PF.
	Triporopollenites coryloides PF. i. TH. & PF.
34	*Intratriporopollenites instructus* (R. POT.) TH. & PF.
35	*Ulmipollenites undulosus* WOLFF
36	*Alnipollenites verus* (R. POT.) R. POT.
37	*Polyporopollenites carpinoides* PF. i. TH. & PF.
38	*Polyporina multistigmosa* (R. POT.) R. POT.

Stratigraphische Bezeichnung: Plio = Pliozän, OM = Obermiozän, JTII = Jungtertiär II, JTI = Jungtertiär I, Sch = Schichten.

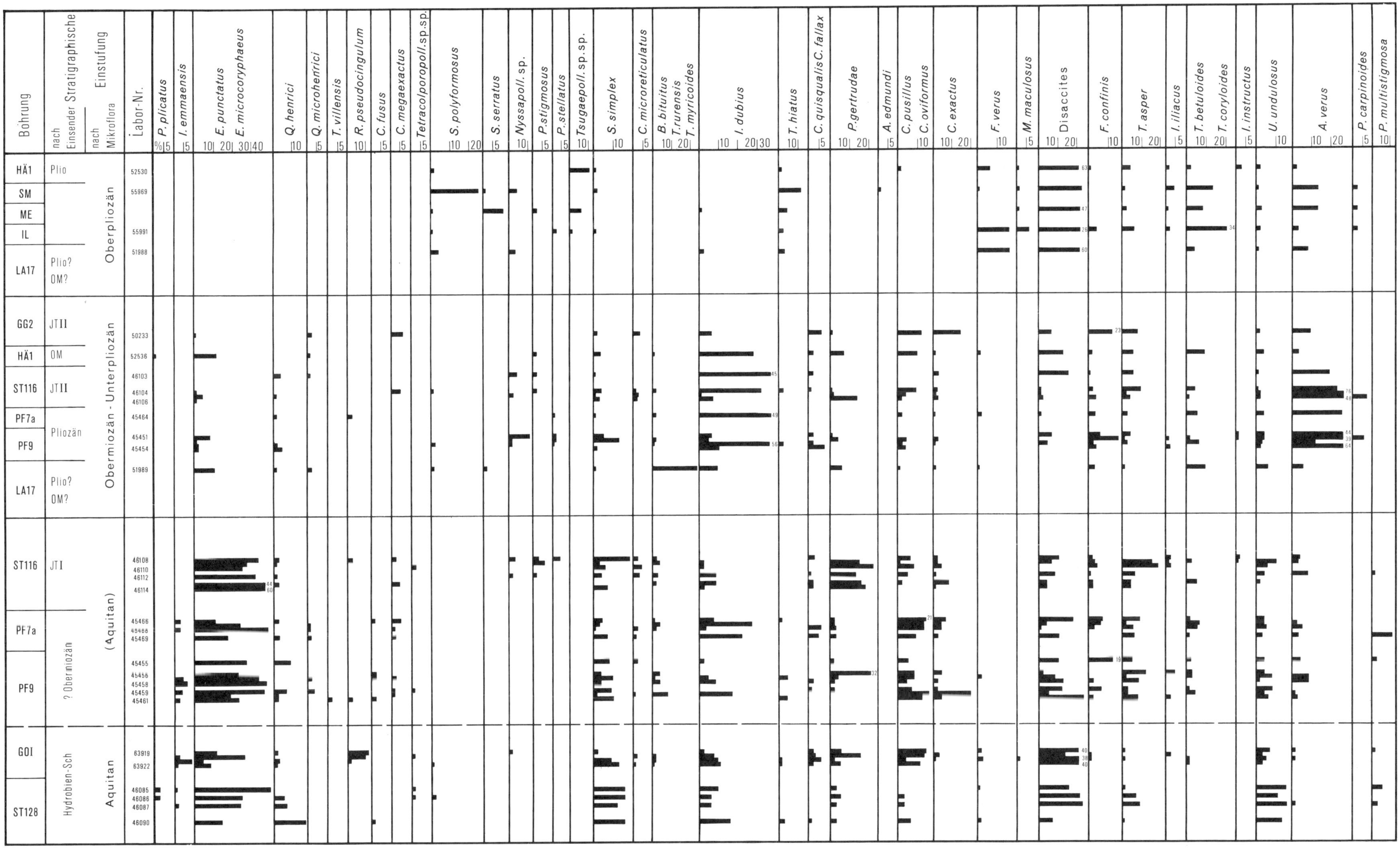

Approaches to Taphrogenesis

Abb. 1

2. Stratigraphische Ergebnisse

Wichtigstes Resultat dieser Untersuchungen ist die Übereinstimmung in der Zusammensetzung der Mikrofloren aus den Hydrobien-Schichten mit denen aus dem Jungtertiär I bzw. ?Obermiozän. Entsprechende durch Maxima von *Engelhardtioipollenites punctatus* und *Engelhardtioidites microcoryphaeus* gekennzeichnete Pollenfloren sind aus den Hydrobien- und den *Corbicula*-Schichten des Mainzer Beckens, aus dem Aquitan der Hanau/Seligenstädter Senke und Oberhessens, der Unterflöz II-Serie der Niederrheinischen Bucht und den Unteren Braunkohlensanden des Vierlandium bekannt. Die dem Jungtertiär I bzw. ?Obermiozän zugeordneten Graben-Sedimente gehören damit zweifelsohne ebenfalls dem Untermiozän an. Dieser Befund deckt sich auch mit der Beobachtung, daß das Jungtertiär I bzw. ?Obermiozän überall konkordant über den Hydrobien-Schichten liegt (SCHAD 1964). Der vielfach geäußerten Vermutung, daß in dem entsprechenden Schichtenkomplex noch jüngere Miozän-Stufen enthalten sein könnten, steht der mikropaläobotanische Befund entgegen. Florengesellschaften mit Maxima von *Quercoidites henrici* und *Rhoipites pseudocingulum*, wie sie in den dem Burdigal zugeordneten Ablagerungen im Raume Frankfurt—Hanau und in der unteren Hälfte der Niederrheinischen Hauptflözgruppe (± tieferes Hemmoorium) angetroffen werden, fehlen im Oberrheingraben. Ebenso wird dort der im oberen Teil der Hauptflözgruppe (oberes Hemmoorium) auftretende Florenwechsel (gekennzeichnet durch eine plötzliche Ausbreitung von *Quercoidites microhenrici, Sciadopityspollenites serratus* und *Sequoiapollenites polyformosus*) nicht angetroffen. Damit scheidet eine Parallelisierung des Jungtertiärs I bzw. ?Obermiozäns schon mit dem Hemmoorium bzw. Burdigal aus.

Die den unteren Abschnitten des Jungtertiärs II bzw. des Pliozäns zugehörigen Ablagerungen enthalten noch bis zu 15 % typisch miozäne Formspezies. Im Jungtertiär I bzw. ?Obermiozän dagegen machen diese Anteile bis zu 50 % aus. Der auffallende Rückgang, selbst wenn sich in ihm gewisse fazielle Verhältnisse widerspiegeln, läßt sich nur durch eine Klimaänderung erklären. Die Pollenfloren aus diesem Bereich zeigen Ähnlichkeit mit den „jüngeren Braunkohlen" im bayerischen Alpenvorland, die in Schichten eingeschaltet sind, die an die Wende Miozän/Pliozän gestellt werden, sowie mit Kohlen des Hausruck, für die teils ein sarmatisches, teils ein pontisches Alter angenommen wird (MEYER 1956). Aber auch in den Indener Schichten der Niederrheinischen Bucht (± Reinbekium) treten vergleichbare Pollenspektren auf. Der heutige Stand unserer Kenntnisse reicht leider noch nicht aus, die Florenzusammensetzung im obersten Miozän und im Unterpliozän genauer zu fassen und damit gegeneinander abzugrenzen. Daher kann für die unteren Abschnitte des „Jungtertiärs II"[1] bzw. des „Pliozäns" im Oberrheingraben gegenwärtig nur eine Einstufung in den Bereich „oberstes Miozän bis unteres Pliozän" vorgenommen werden. Von den typisch oberpliozänen Floren, die jünger als Pont sind, lassen sich diese Schichten gut abtrennen.

---

[1] Unter Jungtertiär II wird der Zeitraum zwischen Jungtertiär I und Quartär verstanden. Es enthält also auch noch höheres Pliozän. Aus diesem lag jedoch kein Material für pollenanalytische Untersuchungen vor.

Im Jungtertiär des Oberrheingrabens sind mikrofloristisch drei Abschnitte zu unterscheiden, denen die lokalstratigraphischen Bezeichnungen wie folgt zuzuordnen sind:

Einstufung nach Mikroflora		lokalstratigraphische Bezeichnungen	
3. Oberpliozän (Asti/Piacentin)		Jungtertiär II bzw. Pliozän	(oberer Komplex)
2. Unterpliozän bis oberstes Miozän	(Pont) (Sarmat)	Jungtertiär II bzw. Pliozän	(unterer Komplex)
1. Untermiozän	(Aquitan bis unteres Burdigal)	Jungtertiär I bzw. ? Obermiozän	

Dank: Der Deutschen Texaco AG., Hamburg, den Gewerkschaften Brigitta und Elwerath Betriebsführungsgesellschaft m.b.H., Hannover, der Wintershall AG., Kassel, sowie den Herren Prof. Dr. J. Bartz (Geologisches Landesamt Baden-Württemberg), Dr. F. Doebl, Landau, und Dr. V. Sonne (Geologisches Landesamt Rheinland-Pfalz) danke ich für die Überlassung der Proben und für die Erlaubnis zur Veröffentlichung der Ergebnisse.

Literaturverzeichnis

Brelie, G. von der, 1967: Quantitative Sporenuntersuchungen zur stratigraphischen Gliederung des Neogens in Mitteleuropa. — Rev. Palaeobotan. Palynol., 2, 147—162.

Doebl, F., 1967: The Tertiary and Pleistocene Sediments of the Northern and Central Part of Upper Rhinegraben. — The Rhinegraben Progress Report 1967, 48—54.

Meyer, B. L., 1956: Mikrofloristische Untersuchungen an jungtertiären Braunkohlen im östlichen Bayern. — Geologica Bavarica, 25, 100—128.

Schad, A., 1964: Feingliederung des Miozäns und die Deutung der nacholigozänen Bewegungen im Mittleren Rheingraben. Eine Auswertung erdölgeologischer Arbeiten. — Abh. geol. L.-Amt Baden-Württ., 5, 1—56.

Die Mächtigkeit des Quartärs im Oberrheingraben

von

J. Bartz

Mit 1 Abbildung als Beilage

An Isopach Map of the Quaternary Deposits of the Rhinegraben

Abstract: As it is shown on the joined map there are two important zones of Quaternary subsidence. These zones of maximum Quaternary thickness are situated in the southern and in the northern graben area, separated by a zone of reduced sedimentation near Rastatt and Karlsruhe. Quaternary strata overlie in the southern part of the graben Oligocene deposits, whereas north of the Kaiserstuhl paleovolcano they follow over sediments of Pliocene age. The Quaternary base is inclined towards the eastern border, so that the trough axis follows just west of the eastern marginal fault zone of the graben. The Quaternary sedimentation was extensively independent of block faulting tectonics, which terminated essentially at the boundary Neogene I/Neogene II.

1. Einleitung

Im Rahmen der Arbeiten für das Upper Mantle Project ist von DOEBL und Mitarbeitern anhand von Erdöl- und Kalibohrungen unter anderem eine Karte der Quartär- und Pliozän-Mächtigkeiten im Oberrheingraben aufgestellt und in den Reports 1967 für das nördliche und 1970 für das südliche Grabengebiet veröffentlicht worden. Die Darstellung umfaßt die Ablagerungen vom Holozän bis ins ältere Pliozän, die einen Zeitraum von mindestens 8 Millionen Jahren einnehmen. Da die pliozänen Ablagerungen im Graben recht ungleich entwickelt sind, z. T. auch fehlen, lassen sich aus den Mächtigkeitsangaben nur generelle Schlüsse auf die jüngsten tektonischen Bewegungen ziehen. Es lag daher der Wunsch nahe, unter Zuhilfenahme weiterer Bohrungen wie Wasser-, Baugrund- oder Kiesbohrungen nur die Mächtigkeit des Quartärs zu erfassen, die einen Zeitraum von etwa 1½ Millionen Jahren einnehmen, und hieraus die jüngsten tektonischen Bewegungen im Oberrheingraben abzuleiten.

Dieses Unterfangen stößt auf nicht beträchtliche Schwierigkeiten. In Gebieten mit geringer Quartär-Mächtigkeit ist die Erfassung und Abgrenzung des Quartärs von pliozänen oder älteren Tertiär-Ablagerungen meist gut möglich. In Gebieten jedoch, in denen das Quartär Mächtigkeiten von 100 und mehr Metern erreicht, liegen — von Ausnahmen abgesehen — nur Erdöl- oder Kalibohrungen vor, deren Spülproben eine Ansprache erschweren und eine sichere Abgrenzung des Quartärs oft unmöglich machen.

Daß ich trotz dieser Schwierigkeiten eine etwa dem heutigen Wissensstande entsprechende Karte der Quartär-Mächtigkeiten im Oberrheingraben fertigstellen konnte, habe ich der freundlichen Unterstützung vieler Kollegen und Freunde zu verdanken. So hat Herr Dr. ANDERLE vom Hessischen Landesamt für Bodenforschung auf meine Anregung hin den nördlichsten Teil des Oberrheingrabens bearbeitet (ANDERLE 1968). Herr Dr. BANGERT vom Geologischen Landesamt Rheinland-Pfalz hat mir eine Manuskriptkarte über die Quartär-Mächtigkeit des Pfälzer Anteils zur Verfügung gestellt. Für das Elsaß habe ich die Karte von SIMLER & THÉOBALD (1970) benutzt, zu der mir Herr SCHWOERER und Mitarbeiter vom Service de la Carte Géologique d'Alsace et de Lorraine, Strasbourg, weitere Hinweise gaben. Den Erdölfirmen Deutsche Texaco AG, Gewerkschaften Brigitta und Elwerath und Wintershall AG danke ich für die gewährte Einsichtnahme in die Profile von Erdölbohrungen, bei deren schwieriger Deutung mich die Herren Dr. DOEBL, Dr. STRAUB und Dr. VEIT in liebenswürdiger Weise unterstützten. Zu besonderem Dank bin ich auch Herrn Dr. VON DER BRELIE vom Geologischen Landesamt Nordrhein-Westfalen verpflichtet, der in den vergangenen Jahren durch die Pollenuntersuchungen von mehreren 100 Torfproben wesentlich zur Gliederung und Abgrenzung des Quartärs beigetragen hat. Schließlich danke ich auch allen Amtskollegen, deren im Archiv des Geologischen Landesamtes Baden-Württemberg niedergelegte Bohrprofile ich auswerten konnte.

2. Die quartären Ablagerungen

Auf der beiliegenden Karte sind als Begrenzung des Oberrheingrabens die Gebirgsrandverwerfungen und soweit faßbar die Vorbergzonen und Randschollen eingetragen worden. Die quartären Ablagerungen sind bis 120 m Mächtigkeit in 20 m, ab 150 m in 25 bzw. 50 m Abständen dargestellt. Erfaßt wurden nur die fluviatilen und limnischen Bildungen. Deckschichten wie Löß, Hangschuttbildungen

u. dgl. wurden i. a. nicht berücksichtigt. Mit Raster wurden Quartär-Mächtigkeiten von 0 bis 60 m, 60 bis 120 m, 120 bis 200 m und über 200 m zusammengefaßt. Gebiete ohne wesentliche quartäre Ablagerungen sind weiß gelassen worden. Zur Mächtigkeit und Ausbildung der quartären Ablagerungen ist folgendes zu bemerken:

a) Basel — Kaiserstuhl

Während im Raum zwischen Basel und Istein die fluviatilen Ablagerungen maximal etwa 30 m, in der Rheinaue unter 20 m mächtig sind, erfolgt nach Norden rasch eine starke Zunahme. So haben nach THÉOBALD (1948) die südwestlich des Kaiserstuhls gelegenen Bohrungen Nambsheim 230 m, Geiswasser 236 m und Heiteren 243 m Quartär-Schotter angetroffen.

Es handelt sich hierbei um grobe, meist sandarme Schotter, bestehend aus Grundgebirgsmaterial aus Schwarzwald und Vogesen, gelegentlich Material aus Mesozoikum und Tertiär der Vorberge, vermischt mit alpinem Material. Randlich und nach der Tiefe nimmt der Grundgebirgsanteil i. a. zu, doch ist alpines Material bis an die Basis der Schotter nachweisbar.

Die oberen 40 bis 100 m bestehen aus frischen Schottern, darunter folgen stark verwitterte und verlehmte Schotter, die z. T. mit frischen Schotterpaketen wechsellagern. Im tieferen Teil der Folge treten häufig Nagelfluh-Bänke auf. An der Basis der Schotter liegt meist eine aus Tonen und Mergeln bestehende Übergangszone, die 10 bis 30 m mächtig ist und aus verwittertem und aufgearbeitetem Tertiär (Oligozän) besteht.

Eine Altersgliederung der mächtigen Schotterablagerung ist nicht möglich, da bislang keinerlei Torfreste aufgefunden wurden. Das obere Schotterpaket mit den frischen Schottern wird der letzten Eiszeit zugerechnet, wofür auch gelegentliche Funde von Molaren von *Mammonteus primigenius* sprechen. Bei den verwitterten Schottern kann es sich z. T. um interglaziale Bildungen handeln. Auch die Nagelfluh-Bänke sind altersmäßig nicht einstufbar, da die Lösung und Ausscheidung des Kalkes weitgehend durch aggressive Grundgebirgswässer und weniger durch klimatische Einflüsse bedingt sind.

Von CORROY & MINOUX (1931) wird von Hüningen ein Molar von *Elephas meridionalis (Archidiscodon meridionalis)* erwähnt, der im Institut von Nancy vorlag, ein weiterer von THÉOBALD (1933) aus dem Industriemuseum von Mülhausen, dessen Fundort nicht bekannt ist. Dies deutet darauf hin, daß wohl im Bereich des Mülhauser Horstes Reste von Ablagerungen vorliegen, die bis in das älteste Pleistozän zurückreichen.

Auf rechtsrheinischem Gebiet greifen die quartären Ablagerungen nördlich Müllheim in größerer Mächtigkeit in den Bereich der Vorbergzone über. Hier wie auch im Bereich der Freiburger Bucht sind die quartären Schotter in über 70 m Mächtigkeit angetroffen worden (SAUER 1967). Es liegen auch hier zuoberst frische Schwarzwaldschotter, die nach der Tiefe zu in stark verwitterte Schotter übergehen. An der Basis folgen Verwitterungsbildungen der liegenden mesozoischen Schichten, die über 10 m mächtig werden können. Es scheinen hier vorwiegend

Rinnenfüllungen vorzuliegen, die auf ein ausgeprägtes Relief vor Ablagerung der Schotter hinweisen. Durch tektonische Bewegungen allein sind diese großen Mächtigkeiten bei den zahlreichen Inseln mesozoischer Gesteine, die aus der Schotterebene auftauchen, jedenfalls nicht zu erklären. Die Höhenlage der Schotter in der Vorbergzone weist auf eine junge Hebung dieses Gebietes hin.

Da nach den vorliegenden Erfahrungen im Raum südlich des Kaiserstuhls keine pliozänen Ablagerungen erhalten sind, habe ich die Quartär-Mächtigkeiten im linksrheinischen Gebiet ohne Änderungen nach der Karte von SIMLER & THÉOBALD (1970) übernommen. Die Schotter erreichen südwestlich und westlich des Kaiserstuhls mit etwa 240 m ihre größte Mächtigkeit. Westlich davon treten die diapiren Strukturen von Mayenheim (100 m-Isopache) und Hettenschlag (20 m-Isopache) als Hochlage deutlich in Erscheinung, während sich auf badischem Gebiet der Weinstetter Diapir durch eine Zunahme der Quartär-Mächtigkeit zu erkennen gibt.

b) Kaiserstuhl — Rastatt

Nördlich des Kaiserstuhls sind die Verhältnisse infolge der großen Mächtigkeit des Quartärs und der geringen Zahl tieferer Bohrungen weniger übersichtlich. Hinzu kommt, daß etwa ab Schlettstadt pliozäne Ablagerungen in größerer Mächtigkeit vorliegen, deren Abgrenzung zum Quartär oft Schwierigkeiten macht.

Auf badischem Gebiet sind die quartären Schotter nördlich des Kaiserstuhls in über 100 m Mächtigkeit nachgewiesen worden. Westlich von Lahr liegt eine vermutlich örtlich begrenzte Hochzone vor. Eine Wasserbohrung bei Nonnenweier hat hier das Quartär nur 47 m mächtig über Pliozän angetroffen. Im Offenburger Raum wurden in zahlreichen Erdölbohrungen (WIRTH 1962) wieder Quartär-Mächtigkeiten von über 100 m festgestellt. Die Schotter liegen auf weißen oberpliozänen Sanden und Tonen, die über 100 m mächtig werden können.

Im Bereich der Vorbergzone liegt wie in der Freiburger Bucht und der südlich anschließenden Vorbergzone wiederum eine überaus starke Eintiefung der aus dem Gebirge austretenden Täler vor. In einer von SCHNARRENBERGER (1933) beschriebenen Bohrung sind im Mündungstrichter der Schutter bei Lahr Kiese, die wohl nur dem Quartär zuzurechnen sind, in einer Mächtigkeit von 101 m über oligozänen Tonmergeln angetroffen worden. Im Mündungstrichter der Kinzig sind sie über 80 m, im Mündungstrichter der Rench über 60 m mächtig erbohrt worden, ohne das Liegende zu erreichen. Eine hier in 40 m Tiefe eingeschaltete Torfschicht hat nach der Pollenuntersuchung von Herrn Dr. VON DER BRELIE (Krefeld) Hinweise auf das letzte Interglazial gegeben. Die Eintiefung und Auffüllung der Täler müssen daher recht jungen Alters sein.

Auf elsässischem Gebiet sind bis in den Schlettstadter Raum die Quartär-Mächtigkeiten nach der Karte von SIMLER & THÉOBALD (1970) dargestellt worden. Nördlich davon habe ich versucht, die Quartär-Mächtigkeiten durch Abzug der auf rechtsrheinischem Gebiet bekannten Pliozän-Mächtigkeiten zu ermitteln, wobei mir Hinweise des Service de la Carte Géologique d'Alsace et de Lorraine, Strasbourg, dienlich waren. Wie auf rechtsrheinischem Gebiet sind auch am Ausgang der Vogesentäler von Breusch, Zorn und Moder nach SIMLER & THÉOBALD tiefere Rinnen nachgewiesen worden, die vornehmlich in Südostrichtung verlaufen.

Nördlich Straßburg geht die Mächtigkeit der quartären Ablagerungen rasch zurück. Im Raum von Hagenau, Sufflenheim und Riedselz liegen die weißen Sande und Tone des Oberpliozäns unter einer geringmächtigen Quartärdecke. Gegen den Rhein hin und östlich des Rheins erreicht das Quartär zwischen Straßburg und Offenburg Werte von über 100 m. Die Trogachse, die bisher auf linksrheinischem Gebiet lag, wechselt im Offenburger Raum auf das rechtsrheinische Gebiet über, wobei gegen Rastatt hin eine laufende Abnahme der Quartär-Mächtigkeit bis auf 60 m festzustellen ist.

Obwohl das Quartär hier stark reduziert ist, scheinen doch sämtliche Quartär-Stufen vorhanden zu sein. Nach Pollenuntersuchungen VON DER BRELIES ließen sich Torfreste von jung/mittelpleistozänem, altpleistozänem und ältestpleistozänem Alter nachweisen. In Übereinstimmung hiermit stehen auch Säugerfunde aus der Kiesgrube bei Wanzenau nördlich Straßburg. Nach WERNERT (1949) wurden im oberen Kies Molaren von *Mammonteus primigenius,* mit der Tieferbaggerung von *Mammonteus trogontherii* und *Palaeoloxodon antiquus* und schließlich von *Archidiscodon meridionalis* angetroffen.

c) Rastatt — Mannheim

Die im nördlichen Elsaß vorliegende geringe Mächtigkeit des Quartärs setzt sich zunächst auch in der Pfalz fort. Am Büchelberg liegt Miozän zutage. Weiter nördlich haben sich die Hardtbäche in oberpliozäne Schichten eingetieft, ihre jungpleistozänen Schotter überlagern diese als dünne Decke. Gegen den Rhein hin sind alt- und ältestpleistozäne Rheinablagerungen erhalten, die als Kiese, Sande und Tone vorliegen. Aus letzteren sind von Jockgrim Säugerreste vom Alter der Mosbacher Fauna bekannt. Solche altpleistozänen Ablagerungen liegen nach Säugerfunden in Kiesgruben auch auf rechtsrheinischem Gebiet in der Niederung in geringer Tiefe vor.

Im Stadtgebiet von Karlsruhe ist das Quartär etwa 55 m mächtig. Nach Norden nimmt es rasch zu und erreicht in der Heidelberger Thermalwasserbohrung mit 382 m seine bisher größte bekannte Mächtigkeit (BARTZ 1951). Zahlreiche Bohrungen im Großraum Karlsruhe lassen eine Gliederung der pleistozänen Schichten zu, die durch viele Pollenuntersuchungen gestützt wird. So läßt sich an einer Bohrung im Hardtwald nördlich Karlsruhe folgendes Profil aufstellen (BARTZ 1959):

0— 36 m obere kiesige Abteilung mit 3 Kiesbänken, die durch Schluff- und Feinsandbänke getrennt werden und dem Jung- und Mittelpleistozän angehören
36— 67 m mittlere sandige Abteilung mit einzelnen Kiesbänken, die dem Altpleistozän zuzurechnen ist
67— 88 m untere sandig-tonige Abteilung, bestehend aus Schluffbänken im Wechsel mit Fein-, Mittel- und Grobsandlagen des ältesten Pleistozäns
88—100 m diese aus weißgrauen Schluffen und Mittel- und Grobsanden bestehende Folge ist nach neueren Pollenuntersuchungen VON DER BRELIES bereits dem Oberpliozän zuzurechnen.

Diese Gliederung in drei Abteilungen läßt sich bis in den Heidelberger Raum verfolgen, sie weist auf eine gleichmäßige Sedimentation in diesem Gebiet hin.

Dem Gebirge ist eine Randscholle vorgelagert, die sich von Ettlingen im Süden über Bruchsal bis gegen Wiesloch verfolgen läßt. Bei Durlach beträgt auf ihr die Mächtigkeit der jung- und mittelpleistozänen Kiese 10 bis 15 m. Sie nimmt bis Bruchsal kontinuierlich auf etwa 60 m zu, weiter nördlich liegt sie unter 10 m.

Im Großraum Mannheim bereitete die Abgrenzung des Quartärs größere Schwierigkeiten. Von THÜRACH (1905) war in einer 175,5 m tiefen Bohrung der Spiegelmanufaktur in Waldhof die Basis des Quartärs bei 146,70 bzw. 168,50 m, auflagernd auf Hydrobien-Schichten, angegeben worden. Von SCHOTTLER (1906), der die Bohrung in den Erläuterungen zu Blatt Viernheim (Käfertal) auch anführt, ist diese Angabe nur indirekt berichtigt worden, indem er schreibt, daß „in keinem der zahlreichen und zum Teil recht tiefen Bohrlöcher bis jetzt das Liegende des Diluviums angetroffen wurde".

Die Auswertung zahlreicher Bohrprofile und die Durchsicht von Bohrproben zweier Tiefbohrungen der BASF in Ludwigshafen/Rhein, die mir liebenswürdigerweise von Herrn Dipl.-Ing. ALTNOEDER zur Verfügung gestellt wurden, ergaben, daß im Niveau der „Hydrobien-Schichten" Schluffe und Mergel vorliegen, die dem älteren Quartär angehören. Nach den regionalen Verhältnissen ist in Mannheim, wo leider keine tieferen Bohrungen vorliegen, die Quartärbasis in über 200 m Tiefe zu erwarten.

d) Mannheim — Frankfurt

Die große Mächtigkeit der quartären Ablagerungen im Mannheim-Heidelberger Raum setzt sich längs dem Odenwaldrande fort. In Erdölaufschlußbohrungen bei Weinheim und Heppenheim sind sie über 250 m mächtig angetroffen worden, von Pliozän mit 500 und mehr m Mächtigkeit unterlagert. Die älteren Quartär-Ablagerungen liegen wie im Raume Karlsruhe—Mannheim als Sande und Tone von grauer, brauner und grüner, meist schmutziger Farbe vor. Die pliozänen Ablagerungen zeigen hingegen nicht mehr die typische weiße Ausbildung, sondern liegen mit ockergelben, grünen und rötlichen Pastellfarben vor. Die Abgrenzung des Quartärs wurde anhand der Spülprobenbeschreibung der zahlreichen dort vorliegenden Erdölbohrungen mit dem Einsetzen von ockergelben Tonen durchgeführt. Torfproben, die liebenswürdigerweise von den Gewerkschaften Brigitta und Elwerath von verschiedenen Bohrungen bei Stockstadt zur Verfügung gestellt und von Herrn Dr. VON DER BRELIE auf Pollen untersucht wurden, stützen diese Grenzziehung.

In der Beschreibung der im Jahre 1939 erfolgten Reichsbohrung 510 b — Nierstein 1 ist von den Bearbeitern HASEMANN und STROBEL die Grenze Quartär/Pliozän bei 325 m gezogen worden. Eine solch große Mächtigkeit läßt sich mit benachbarten neueren Erdölbohrungen nicht in Einklang bringen. Zahlreiche Funde vom Flußpferd, die in der näheren Umgebung der Bohrung bei Hessenaue, Nierstein, Oppenheim, Erfeld und Stockstadt (ADAM 1965) bekannt sind, lassen in geringer Tiefe schon altpleistozäne Ablagerungen erwarten und machen eine wesentlich geringere Quartär-Mächtigkeit wahrscheinlich.

Westlich Darmstadt liegt das Quartär noch in über 100 m Mächtigkeit vor. Nördlich einer Linie Darmstadt—Rüsselsheim nimmt sie rasch ab, ich verweise hier auf die Ausführungen von ANDERLE (1968).

3. Entwicklung des Oberrheingrabens im Quartär und Jungtertiär

Die quartären Ablagerungen im Oberrheingraben zwischen Basel und Frankfurt lassen zwei bedeutende Senkungsgebiete erkennen. Das eine liegt im südlichen Grabengebiet und erstreckt sich südlich vom Kaiserstuhl bis in den Offenburger Raum. Das andere zieht sich südlich von Heidelberg bis in die Höhe von Darmstadt hin und erfaßt wesentliche Teile des nördlichen Grabengebietes. Beide Senkungsgebiete sind durch eine Hochzone im Raum Rastatt—Karlsruhe getrennt.

Im südlichen Senkungsgebiet erreichen die quartären Ablagerungen südwestlich des Kaiserstuhls Mächtigkeiten von über 240 m. Sie bestehen vorwiegend aus groben Kiesen mit geringen Einschaltungen von Feinsandbänken. Die hangenden Schotter werden der Würmeiszeit zugerechnet. Über das Alter der tieferen Schotter sind keine sicheren Aussagen möglich, da Tier- und Pflanzenreste fehlen.

Nördlich des Kaiserstuhls geht die Quartär-Mächtigkeit laufend zurück. Die Trogachse verlagert sich im Offenburger Raum auf das rechtsrheinische Gebiet. Den Kiesen sind in zunehmendem Maße Sande und auch Schluffbänke eingeschaltet. Nach Säugerfunden im Straßburger Raum sind jung-, alt- und ältestpleistozäne Ablagerungen vertreten. Diese Entwicklung setzt sich in der Rastatt-Karlsruher Hochzone fort, wo die Quartär-Mächtigkeit bis auf etwa 55 m zurückgeht. Trotz dieser relativ geringen Mächtigkeit lassen sich auch hier nach Säugerfunden und Pollenuntersuchungen sämtliche Quartär-Stufen nachweisen.

Nördlich Karlsruhe nehmen die Quartär-Ablagerungen wieder rasch an Mächtigkeit zu und erreichen in der Heidelberger Thermalwasserbohrung mit 382 m ihre größte bislang bekannte Mächtigkeit. Mit der Zunahme der Mächtigkeit ist eine wesentliche Erhöhung des Feinkornanteils verknüpft. Das Geröllmaterial der Thermalwasserbohrung stammt ausschließlich vom Neckar. Gegen Mannheim und nach Norden verzahnen sich Rhein- und Neckarablagerungen.

Im Raum Mannheim—Heidelberg nimmt das Senkungsgebiet fast die gesamte Breite des Oberrheingrabens ein, wobei die größten Senkungsbeträge vor dem östlichen Gebirgsrande liegen. In der Höhe von Bensheim ist noch mit Quartär-Mächtigkeiten von 250 m zu rechnen. Nach Westen gehen mit abnehmender Mächtigkeit die Kieseinschaltungen stark zurück. Im Raum Frankental—Worms liegen die quartären Ablagerungen, von geringen gröberen Einschaltungen der Hardtbäche abgesehen, fast nur als Sande und Schluffe vor, deren Abgrenzung zu den die gleiche Ausbildung zeigenden pliozänen Schichten oft recht schwierig ist. Gegen Darmstadt nimmt die Mächtigkeit langsam ab. In der Untermainebene liegen nur noch unbedeutende Mächtigkeiten vor.

Sowohl dem nördlichen wie dem südlichen Senkungsgebiet ist gemeinsam, daß die Senkungsachse nicht in der Grabenmitte verläuft, sondern dem östlichen Gebirgsrande bzw. dessen Vorschollen vorgelagert ist. Im Süden sind die großen Mächtigkeiten nahe dem Kaiserstuhl und vor der bis in den Raum Offenburg ziehenden Hochscholle, die vorwiegend aus mesozoischen Gesteinen aufgebaut ist, entwickelt. Im Norden liegen sie vor dem Odenwaldrande.

Im südlichen Grabengebiet liegen die quartären Ablagerungen auf oligozänen, im Bereich der Vorberge und der Freiburger Bucht auf mesozoischen Schichten auf. Etwa nördlich der Linie Schlettstadt—Lahr schalten sich pliozäne Ablagerungen ein, die im Offenburger Raum über 100 m mächtig werden. Im Bereich der Rastatt—Karlsruher Hochzone sind sie 80 m mächtig nachgewiesen, gegen Heidelberg nehmen sie auf über 640 m zu und liegen auch weiter nördlich in mehreren 100 m Mächtigkeit vor.

Auch südlich der Linie Schlettstadt—Lahr haben einst pliozäne Ablagerungen, wenn auch wohl nur in geringer Mächtigkeit vorgelegen, wie die Funde von *Anancus arvernensis* in bohnerzführenden Spalten des Hauptmuschelkalkes bei Emmendingen, im Hauptoolith bei Herbolzheim und im oligozänen Kalkstein bei Lahr-Dinglingen im Bereich der Vorbergzone beweisen (TOBIEN 1950). Diese Bildungen sind jedoch vor Ablagerung der quartären Schotter abgetragen worden, wobei die Abtragung nicht nur die Randgebiete, sondern auch den Graben selbst betroffen hat.

Nach den weit verbreiteten vorwiegend feinklastischen Bildungen dürfte die Abtragung im jüngeren Pliozän vornehmlich flächenhaft erfolgt sein und nur ein schwaches Relief vorgelegen haben. Im Quartär hingegen weisen die mächtigen Kiesablagerungen auf die Herausbildung eines kräftigen Reliefs hin. Taleintiefungen und -auffüllungen liegen in der Vorbergzone mit über 100 m Mächtigkeit vor. Sie finden sich auch am Ausgang der Vogesentäler. Möglicherweise sind sie den Taleintiefungen im Neckar- und Maingebiet gleichzustellen, die zur Ablagerung der altpleistozänen Sande von Mauer und Randersacker geführt haben.

Die quartären und pliozänen Sedimente zeigen in den Senkungsgebieten relativ gleichmäßige Ausbildung und Lagerungsverhältnisse. Ein Schollenmosaik, wie wir es aus den älteren Tertiär-Schichten kennen, liegt nicht vor, worauf zuletzt ILLIES (1965) hingewiesen hat. Bedeutende Lagerungsstörungen in den älteren Tertiär-Schichten, die durch geophysikalische Untersuchungen und Erdölbohrungen nachgewiesen wurden, geben sich in den pliozänen und quartären Ablagerungen nicht oder nur in ganz unbedeutendem Maße zu erkennen.

Die Entstehung dieser Bruchfelder mit den antithetisch geneigten und gekippten Einzelschollen ist an der Wende Jungtertiär I/Jungtertiär II zum Abschluß gekommen. Hand in Hand damit ging eine wohl pulsierende Heraushebung vor allem des südlichen Grabengebietes, die zu bedeutenden Abtragungen miozäner und älterer Tertiär-Schichten geführt hat. SCHAD (1965) konnte nachweisen, daß allein im Raum von Landau—Bruchsal bis Karlsruhe 400—500 m jungtertiäre Sedimente abgetragen wurden, weiter südlich dürften noch wesentlich höhere Werte erreicht worden sein.

Die ältesten fluviatilen Ablagerungen aus dem südlichen Abtragungsgebiet liegen in den unterpliozänen Dinotherien-Sanden vor, deren Reste im Raum zwischen Worms und Bingen erhalten geblieben sind. Nach deren Ablagerung kam es im jüngeren Pliozän zu den neuen Senkungen, die vor allem das nördliche Grabengebiet in starkem Maße betroffen haben und sich im südlichen Grabengebiet bis etwa Schlettstadt—Lahr nachweisen lassen. Die Hochzone von Rastatt—Karlsruhe macht sich schon hier durch eine geringere Sedimentation bemerkbar.

Bei diesen Senkungen kam es an den Grabenrändern zur Herausbildung eines meist schwachen Reliefs. So liegen vielfach in den randnahen Gebieten an der Basis der pliozänen Ablagerungen Kiesschüttungen vor, im Trogtiefsten sind Grobsande eingeschaltet. Im Norden zeugen die *Arvernensis*-Schotter mit *Anancus arvernensis* von stärkerer fluviatiler Tätigkeit (BARTZ 1950).

Im Gefolge der Faltung des Schweizer Juras werden an der Wende Pliozän/Quartär die Gewässer des Hochrheingebietes, insbesondere der Aare, in den Oberrheingraben umgelenkt. Im Altquartär wird der Südteil des Grabengebietes, der während des gesamten Pliozäns Abtragungsgebiet war, in die Senkung mit einbezogen. Diese erreichte vor allem südwestlich des Kaiserstuhls beträchtliche Ausmaße, die — sofern hier ältestpleistozäne Ablagerungen fehlen — den Senkungsbeträgen im nördlichen Grabengebiet kaum nachstehen.

Die Senkungen im Pliozän und Quartär erfolgten somit weitgehend unabhängig vom älteren Schollenbau. Sie führten zunächst im nördlichen, im Quartär auch im südlichen Senkungsgebiet zu einer Kippung der Auflagerungsfläche der fluviatilen und limnischen Sedimente gegen den östlichen Grabenrand. Während dieser säkularen Bewegungen kam es zu einer relativ gleichmäßigen Ablagerung und zeitlich gleichartigen Ausbildung der Sedimente, die es ermöglichten, die Quartär-Mächtigkeiten in den Senkungsgebieten zu erfassen. Die Tatsache, daß in der Hochzone Rastatt—Karlsruhe sämtliche Pleistozän-Stufen faunistisch und floristisch faßbar sind, berechtigt zu der Hoffnung, auch in den Senkungsgebieten mit den großen Sediment-Mächtigkeiten eine Gliederung des Quartärs durchführen zu können.

Schriftenverzeichnis

ADAM, K. D.: Neue Flußpferdfunde am Oberrhein. — Jh. Geol. L.-Amt Baden-Württ., 7, 621—631, Freiburg i. Br. 1965.

ANDERLE, H. J.: Die Mächtigkeiten der sandig-kiesigen Sedimente des Quartärs im nördlichen Oberrheingraben und der östlichen Untermainebene. — Notizbl. Hess. L.-Amt Bodenforsch., 96, 185—196, Wiesbaden 1968.

BARTZ, J.: Das Jungpliozän im nördlichen Rheinhessen. — Notizbl. Hess. L.-Amt Bodenforsch., VI, 1, 201—243, Wiesbaden 1950.

— Revision des Bohrprofils der Heidelberger Radium-Sol-Therme. — Jber. u. Mitt. Oberrhein. Geol. Ver., NF 33, 1951, 101—125, Freiburg i. Br. 1953.

— Zur Gliederung des Pleistozäns im Oberrheingebiet. — Z. Dt. Geol. Ges., 111, 653—661, Hannover 1959.

BRELIE, G. v. D.: Pollenanalytische Untersuchungen zur Gliederung des Pleistozäns im nördlichen Oberrhein-Graben. — Z. Dt. Geol. Ges. 1963, 115, 902—903, Hannover 1966.

CORROY, G. & MINOUX, G.: Les Mammifères quarternaires de Lorraine. Les Eléphantidés. — Bull. Soc. G. France (5) I, 635—653, Paris 1931.
DOEBL, F.: The Tertiary and Pleistocene sediments of the northern and central part of the Upper Rhinegraben. — Abh. Geol. L.-Amt Baden-Württ., 6, 48—54, Freiburg i. Br. 1967.
— Die tertiären und quartären Sedimente des südlichen Rheingrabens. — In: ILLIES, H. & MUELLER, ST.: Graben Problems, S. 56—66, Stuttgart (Schweizerbart) 1970.
ESSLINGER, G.: Rezente Bodenbewegungen über dem Salinar des südlichen Oberrheintals. — Diss. Berlin 1968, D 83. Offset-Druck Johannes Krause, Freiburg i. Br.
ILLIES, H.: Prinzipien der Entwicklung des Rheingrabens dargestellt am Grabenabschnitt von Karlsruhe. — Mitt. Geol. Staatsinst. Hamburg, 31, 58—122, Hamburg 1962.
— Bauplan und Baugeschichte des Oberrheingrabens. Ein Beitrag zum „Upper Mantle Project". — Oberrhein. Geol. Abh., 14, 1—54, Karlsruhe 1965.
SAUER, K.: Beiträge zur Hydrogeologie der näheren Umgebung von Freiburg i. Br. — Mitt. Bad. Landesver. Naturkunde u. Naturschutz, N. F. 9, 611—637, Freiburg i. Br. 1967.
SCHAD, A.: Abtragungserscheinungen an der Grenze Jungtertiär I/Jungtertiär II im Innern des mittleren Rheintalgrabens. — Senck. leth., 46 a, 363—376, Frankfurt/Main 1965.
SCHMITT, O.: Über pleistozäne Ablagerungen am Rande des Odenwaldes. — Z. Dt. Geol. Ges., 1964, 116, 987—989, Hannover 1966.
SCHNARRENBERGER, C.: Sondages dans les environs de Lahr (Pays de Bade). — C. R. Séanc. du Groupe des géologues petroliers de Strasbourg, 2 ème année No 1 et 2, S. 21—24, Strasbourg 1933—34.
SCHOTTLER, W.: Erläuterungen zur Geologischen Karte des Großherzogtums Hessen, Blatt Viernheim (Käfertal), S. 1—116, Darmstadt (Bergsträßer) 1906).
SIMLER, L. & THÉOBALD, N.: Les alluvions plio-quarternaires du Fossé Rhénan (secteur plaine d'Alsace). — In: ILLIES, H. & MUELLER, ST.: Graben Problems, S. 75—78, Stuttgart (Schweizerbart) 1970.
STRAUB, E. W.: Die Erdöl- und Erdgaslagerstätten in Hessen und Rheinhessen. — Abh. Geol. L.-Amt Baden-Württ., 4, 123—136, Freiburg i. Br. 1962.
THÉOBALD, N.: Carte de la base des formations alluviales dans le sud du Fossé Rhénan. — Mém. Serv. carte géol. Als. Lor. No 9, 1—77, Strasbourg 1948.
— Les restes d'éléphants fossiles conservé au Musée d'histoire naturelle de la Société Industrielle de Mulhouse. — Bull. Soc. Industr. Mulhouse, 5, 324—338, Mulhouse 1933.
THÜRACH, H.: Erläuterungen zur Geologischen Spezialkarte des Großherzogtums Baden, Blatt Mannheim (Nr. 21). — S. 1—24, Heidelberg (Winter) 1905.
TOBIEN, H.: Die Aufzeichnungen H. G. STEHLINS über die pliozänen Säugerreste von Herbolzheim bei Freiburg i. Br. — Mitt.-Blatt Bad. Geol. Landesanst. 1950, 78—84, Freiburg i. Br. 1951.
WERNERT, P.: *Elephas meridionalis Nesti* dans le Bas-Rhin. — Tirage des Cahiers d'Archéologie et d'Histoire d'Alsace, S. 217—222, Strasbourg 1949.
WIRTH, E.: Die geologischen Ergebnisse der Erdölexploration in der Rheinebene zwischen Offenburg und Lahr. — Erdöl u. Kohle, 15, 684—692, Hamburg 1962.

Nivellements und vertikale Krustenbewegungen im Bereich des südlichen Oberrheingrabens

von

H. Mälzer und A. Strobel

Mit 6 Abbildungen im Text

Abstract: In Baden-Württemberg two levellings of high precision were carried out in 1922/39 and 1952/70. The height differences (relative vertical movements) computed by separated adjustments of the two networks are similar to the results of the calculations by MÄLZER (1967, 1972). Small changes in the values (max. 5 mm/30 a) are not significant. To study the vertical movements in active zones of the southern Upper Rhinegraben in 1971/72 the Institut Géographique National (Paris) and the Landesvermessungsamt Baden-Württemberg (Stuttgart) observed two levelling profiles from the Vosges to the Black Forest. The northern profile crosses the Rhine between Strasbourg-Kehl and the southern one near Breisach. In the area Appenweier-Kehl the mean relative vertical movements are —(4 to 7)mm/10 a and between Breisach-Bad Krozingen —(6 to 8)mm/10 a.

Einleitung

Während die Geologie mit ihren Untersuchungsmethoden Aufschluß über die langzeitigen tektonischen Bewegungen im Rheingraben zu geben vermag, ist die Geodäsie dazu berufen, über die gegenwärtig stattfindenden Veränderungen exakte Angaben zu machen. Neben genauesten Messungen ist hierfür vor allem eine sorgfältige Auswahl der Standorte für die Festpunktmarken erforderlich, um lokale Einflüsse auf die Bewegungen der Marken möglichst auszuschalten.

Überprüfung der bisherigen Ergebnisse durch Vergleich der Hauptnivellements

Nachdem die Wiederholungsmessungen im badisch-württembergischen Haupthöhennetz im Jahre 1970 beendet wurden, können die von MÄLZER (1967, 1972) berechneten vertikalen Bodenbewegungen durch einen Vergleich zweier Hauptnivellements überprüft werden. Die beiden Hauptnivellements (1. Nivellement 1922 bis 1939, 2. Nivellement 1952 bis 1970) erfüllen die hierfür notwendigen Genauigkeitsanforderungen. Die mittleren Fehler für ein Doppelnivellement von 1 km Länge betragen

$$m_1 = \pm 0{,}34 \text{ mm} \quad \text{und} \quad m_2 = \pm 0{,}35 \text{ mm},$$

wobei m aus den Differenzen von Hin- und Rückmessung zwischen zwei aufeinanderfolgenden Höhenmarken (durchschnittlicher Abstand 500 m) berechnet wurde.

Dem Vergleich liegen zwangsfreie Ausgleichungen der beiden Nivellements zugrunde. Die Schleifenschlußfehler mußten unter Berücksichtigung der wahren Schwere ermittelt werden, da der Einfluß der Schwereanomalien, besonders im Schwarzwald, nicht vernachlässigt werden durfte. Für die wahren theoretischen

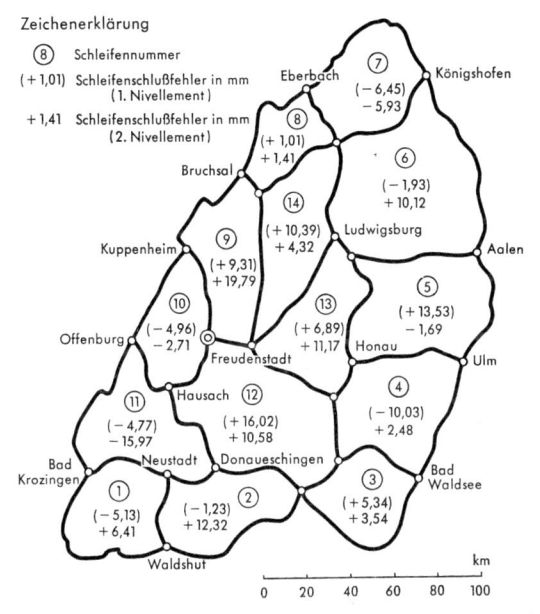

Abb. 1. Schleifenschlußfehler im baden-württembergischen Haupthöhennetz.

Schleifenschlußfehler wurden die von RAMSAYER (1957) berechneten Werte angehalten. Die Ausgleichungskomplexe und der Verlauf der Nivellementlinien sind für beide Nivellements gleich (Abb. 1). Dadurch wird ein Vergleich über viele identische Punkte ermöglicht.

Für ein Doppelnivellement von 1 km Länge ergaben sich aus den Ausgleichungen der Schleifenwidersprüche die mittleren Fehler

$$M_1 = \pm\,0{,}64 \text{ mm} \quad \text{und} \quad M_2 = \pm\,0{,}79 \text{ mm}.$$

Diese mittleren Fehler liegen damit etwa doppelt so hoch wie die unmittelbar aus den Messungen berechneten mittleren Fehler m. Die Unterschiede sind einmal auf systematische Fehleranteile, die sich auf großen Strecken auswirken, und zum anderen auf die lange Beobachtungsdauer für die beiden Nivellements (je 18 Jahre) zurückzuführen; denn es ist innerhalb einer so langen Beobachtungsepoche keine Gewähr gegeben, daß wiederholt verwendete Anschlußpunkte stabil geblieben sind. Um diese Fehlerursachen einzuschränken, sollte man künftig grundlegende Nivellements, die der Untersuchung rezenter Krustenbewegungen dienen, in möglichst kurzer Zeit beobachten.

Die Höhenwertänderungen, die sich aus dem Vergleich der beiden Hauptnivellements ergeben, sind für die Knotenpunkte und die aktivsten Zonen in Abb. 2 dargestellt. Sie können unmittelbar mit den in Abb. 9 von MÄLZER (1967)

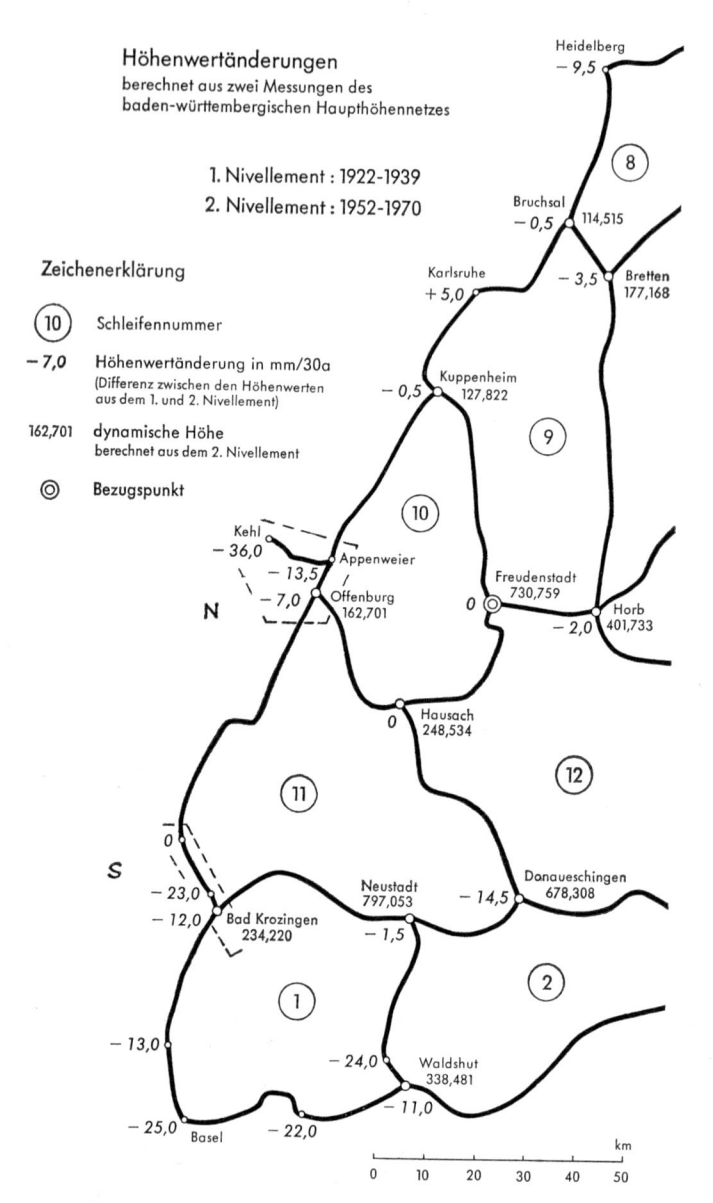

Abb. 2. Höhenwertänderungen im westlichen Teil des baden-württembergischen Haupthöhennetzes. N und S kennzeichnen die Gebiete, in denen 1970 bis 1972 spezielle Präzisionsnivellements zur Untersuchung rezenter vertikaler Bewegungen ausgeführt wurden.

wiedergegebenen Werten verglichen werden. Die Unterschiede zwischen den Höhenwertänderungen betragen im Oberrheingraben bis zu 5 mm/30 a und sind auf die verschiedenen Berechnungsmethoden zurückzuführen. Sie sind gegenüber den Senkungsbeträgen in den markanten Gebieten Appenweier-Kehl und nordwestlich Bad Krozingen nicht signifikant. Insgesamt bringt der Vergleich der beiden zwangsfrei ausgeglichenen Hauptnivellements eine Überprüfung, aber keine Verbesserung des von MÄLZER (1967) angewandten Verfahrens.

Spezielle Nivellements zur Untersuchung rezenter vertikaler Bewegungen

Um gesicherte Grundlagen für künftige Untersuchungen vertikaler Bewegungen zu schaffen, kamen das Institut Géographique National, Paris, und das Landesvermessungsamt Baden-Württemberg im Jahre 1971 überein, mit zwei Nivellementprofilen den Oberrheingraben in etwa west-östlicher Richtung von den Vogesen bis zum Schwarzwald zu durchqueren. Die Profile verlaufen durch Zonen erhöhter Bewegungsintensität: das Nordprofil führt über Straßburg—Kehl—Appenweier—Offenburg und das Südprofil über Colmar—Breisach—Bad Krozingen. Bei der Erkundung und Festlegung der Profillinien und Höhenpunkte auf deutscher Seite hat Prof. Dr. ILLIES durch geologische Beratung mitgewirkt.

Nachdem bereits 1970 im Nordprofil die rechtsrheinischen Linien Offenburg—Appenweier, Appenweier—Kehl und Offenburg—Kehl vom Landesvermessungsamt Baden-Württemberg neu beobachtet worden waren, wurden im Herbst 1971 diese durch Verbindungslinien hoher Genauigkeit zu einem Netz flächenhaft ergänzt (Abb. 3). Das Netz wurde an unterirdisch im Grundgebirge eingebrachte Höhenbolzen bei Nußbach und Ohlsbach (UB) sowie an eine Punktgruppe bei Durbach unterhalb des Mooskopfes angeschlossen. Zeitlich wurden die Messungen innerhalb eines Jahres ausgeführt. Insgesamt hat das Netz auf deutscher Seite eine Linienlänge von 87 km; der Punktabstand beträgt im Mittel 0,3 bis 0,4 km und ist an geologisch interessanten Stellen verdichtet. Der aus den Schleifenwidersprüchen errechnete mittlere Fehler beträgt $\pm$ 0,4 mm/$\sqrt{\text{km}}$; er entspricht somit einer feldmäßigen Präzisionsmessung. Die Maßstabsprüfung der verwendeten Nivellierlatten erfolgte unabhängig durch das Landesvermessungsamt und das Geodätische Institut der Universität Karlsruhe und ergab übereinstimmende Ergebnisse.

Bislang können in dem neu angelegten Untersuchungsnetz nur längs der bereits bestehenden Linien 1. Ordnung relative Höhenwertänderungen für identische Punkte abgeleitet werden. Auf dem Linienabschnitt Appenweier—Offenburg zeichnen sich nur im lokalen Bereich von Offenburg für die Zeitintervalle 1929/60 und 1960/70 gleichgerichtete Bewegungstendenzen von 3 bis 4 mm/10 a ab. Dagegen deuten im Regionalfeld zwischen Appenweier und Kehl sowohl die Beobachtungen für den Zeitraum 1929/64 als auch für 1964/70 auf analoge vertikale Geschwindigkeiten hin (Abb. 4). Nach Westen zu, wo auch größere pleistozäne Mächtigkeiten (bis zu 120 m) auftreten, erreichen die relativen vertikalen Ab-

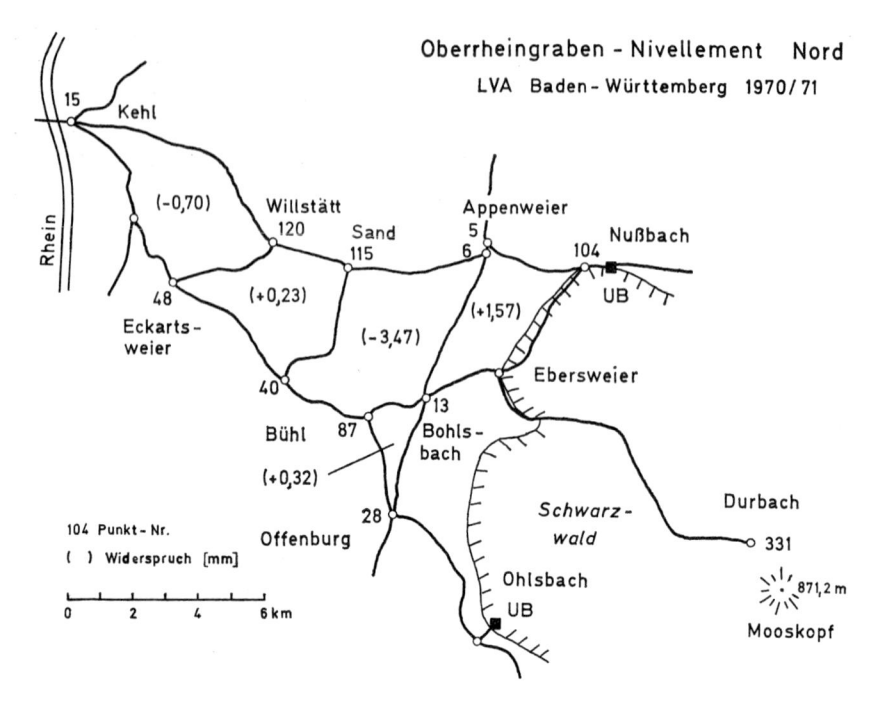

Abb. 3. Verlauf der Nivellementlinien zur Beobachtung rezenter vertikaler Bewegungen im Gebiet Kehl—Appenweier—Offenburg (Nordprofil).

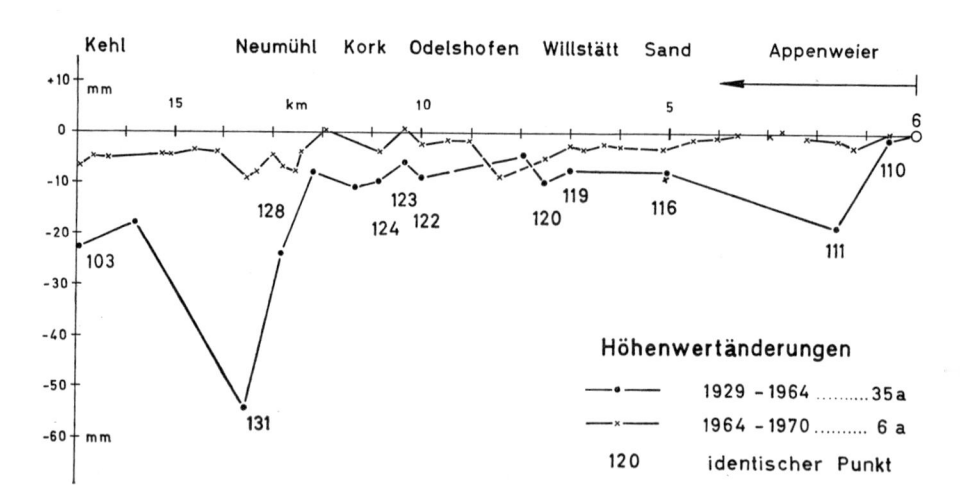

Abb. 4. Relative Höhenwertänderungen zwischen Appenweier und Kehl, abgeleitet aus den drei Beobachtungsepochen 1929, 1964 und 1970.

senkungen 15 mm/10 a. Einem Hinweis von ORTLAM (1970) zufolge überquert die Nivellementlinie bei dem Höhenpunkt 131 eine Verwerfung, die in westlicher Krümmung von der Kinzig bis zur Murg im Oberrheingraben verläuft. Im Mittel liegen die relativen Geschwindigkeiten auf dem rund 17 km langen Profil bei —(4 bis 7) mm/10 a. Die neu 1970/71 beobachteten Linien im Netz Kehl—Appenweier—Offenburg fallen mit Linien niederer Ordnung zusammen, so daß hier Höhenwertänderungen bislang nicht gesichert abgeleitet werden können.

Das S ü d p r o f i l erstreckt sich rechtsrheinisch von Breisach in südöstlicher Richtung entlang einer schon bestehenden Linie 1. Ordnung bis Bad Krozingen und endet bei Untermünstertal am Fuße des Belchenmassivs (Abb. 5). Im südlichen Bereich des Tuniberges ist das Profil durch zwei Schleifen erweitert, um die zu erwartenden Bewegungsvorgänge über die hier vorhandenen Bruchstufen zu erfassen. Auf deutscher Seite ist dieses Nivellementprofil bei einem mittleren Punktabstand von 0,4 km insgesamt etwa 40 km lang. Die Präzisionsmessungen wurden im Herbst 1972 vom Landesvermessungsamt Baden-Württemberg ausgeführt. Be-

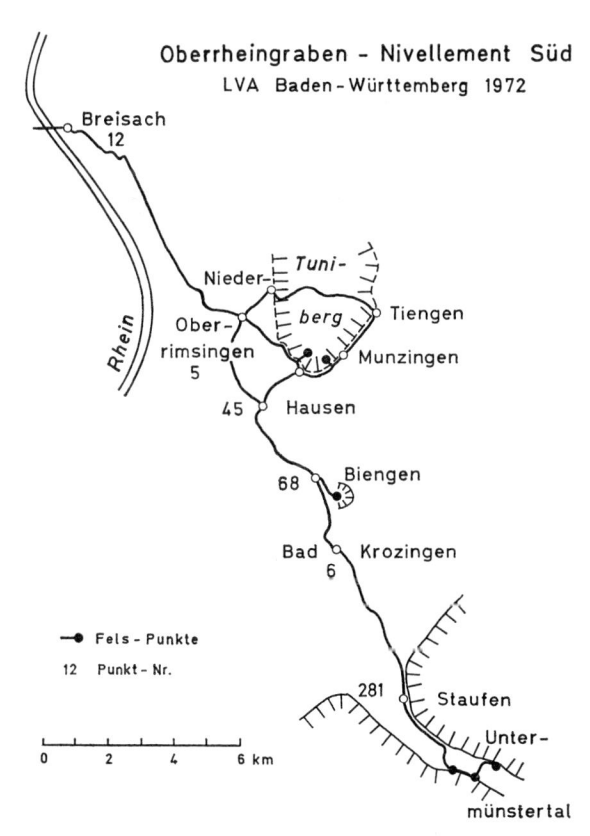

Abb. 5. Verlauf der Nivellementlinien zur Beobachtung rezenter vertikaler Bewegungen zwischen Breisach und Staufen (Südprofil).

zogen auf Breisach betragen im Profilabschnitt Oberrimsingen—Bad Krozingen die vertikalen Geschwindigkeiten im Mittel —(6 bis 8)mm/10 a. Nach den bisherigen Beobachtungen wird das Geschwindigkeitsmaximum von —13 mm/10 a im Raume Biengen angetroffen (Abb. 6).

Unter der Voraussetzung der bisher erreichten Genauigkeiten und bei gleichbleibenden relativen vertikalen Geschwindigkeiten erscheinen in frühestens 10 bis 12 Jahren Wiederholungsmessungen sinnvoll, um zu statistisch gesicherten Aussagen zu gelangen.

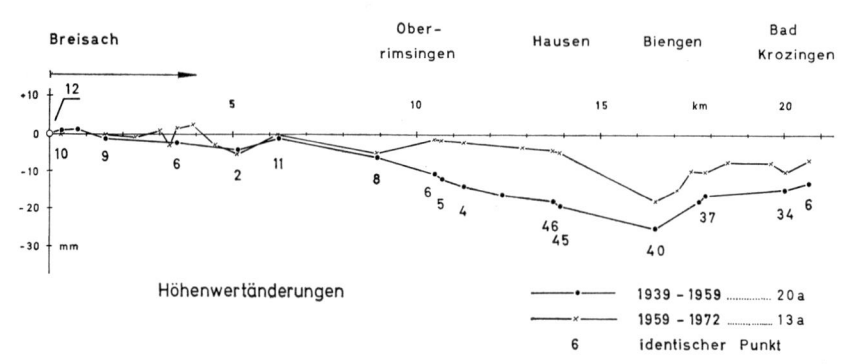

Abb. 6. Relative Höhenwertänderungen zwischen Breisach und Bad Krozingen, abgeleitet aus den drei Beobachtungsepochen 1939, 1959 und 1972.

Literatur

MÄLZER, H., 1967: Untersuchungen von Präzisionsnivellements im Oberrheingraben von Rastatt bis Basel im Hinblick auf relative Erdkrustenbewegungen. — 23 S., Dt. Geodät. Komm., B 138, München.
— 1972: Die Höhenwertänderungen im Oberrheingraben. — In: KERTZ, W. et al. (eds.), Unternehmen Erdmantel, 220—223, Steiner, Wiesbaden.
ORTLAM, D., 1970: Interferenzerscheinungen rheinischer variszischer Strukturelemente im Bereich des Oberrheingrabens. — In: Graben Problems, 91—97, Schweizerbart, Stuttgart.
RAMSAYER, K., 1957: Schwerereduktionen des Badisch-Württembergischen Haupthöhennetzes. — 33 S., Dt. Geodät. Komm., A 22, München.

Geodätische Messungen im Rheingraben

von

Gh. Bozorgzadeh und E. Kuntz

Mit 3 Abbildungen im Text

Geodetic Measurements in the Rhinegraben

Abstract: In order to investigate modern electronic distance measuring techniques and to study the influences of topography and atmosphere on their results the Geodetic Institute of the University of Karlsruhe has established a test-net based on three points on the western and four points on the eastern border of the Rhinegraben. Since 1967 all distances have been measured several times with microwave instruments and also with a laser equipment. With the present best obtainable precision of 1.5 cm for 40 km, the horizontal movements of the borders estimated at 0.5 mm p. a. can only be proved in some decades. Furthermore, the deflections of the vertical along an east-west profile across the graben had been determined by astrogeodetical procedures with an accuracy of 0.2"—0.4". Their rapid changes are correlated with the strong variations of the Bouguer-anomalies from maximum values at the borders to a minimum near Landau.

1. Elektronische Entfernungsmessungen

Das Testnetz Karlsruhe (Abb. 1) wurde ursprünglich zur Untersuchung elektronischer Entfernungsmeßgeräte und äußerer Einflüsse auf die elektronische Entfernungsmessung angelegt. Es überspannt die Rheinebene nordwestlich von Karlsruhe und damit den auf dem ersten Oberrheingraben-Kolloquium Wiesloch 1966 als geologisch besonders interessant herausgestellten Profilstreifen I (DFG-Protokoll 1966). Auf Veranlassung der Geologen wird seit Jahren versucht, die Genauigkeit des Netzes so zu steigern, daß es Grundlage für spätere Untersuchungen über horizontale Krustenbewegungen in diesem Raum werden kann.

Das Netz bestand anfangs aus den sechs Punkten Turmberg, Michaelsberg, Letzenberg auf dem badischen und Kalmit, Madenburg, Stäffelsberg auf dem pfälzischen Grabenrand. Später wurde der Punkt Herxheim in der Grabensohle hinzugenommen, um vor allem den Einsatz elektrooptischer Entfernungsmeßgeräte effektiver gestalten zu können. Schließlich haben wir in den Jahren 1970/71 das Netz auf der badischen Seite im Süden um den Punkt Mahlberg erweitert, der im Gegensatz zu den nördlichen Punkten Michaelsberg und Letzenberg auf geologisch gesichertem Untergrund liegt.

Das Netz wurde wiederholt mit verschiedenen Mikrowellengeräten und neuerdings auch elektrooptisch mit einem Laser-Geodimeter durchgemessen (KUNTZ 1971). Die mittleren Punktunsicherheiten der freien Netzausgleichung der Mikrowellenmessungen liegen im Durchschnitt bei etwa 4 cm, wie die Fehlerellipsen in Abb. 1 zeigen. Hierbei handelt es sich um die innere Genauigkeit der Messungen. In Wirklichkeit sind die Punktunsicherheiten größer, weil mit nicht erfaßten systematischen meteorologischen Effekten gerechnet werden muß. Bei der Mikro-

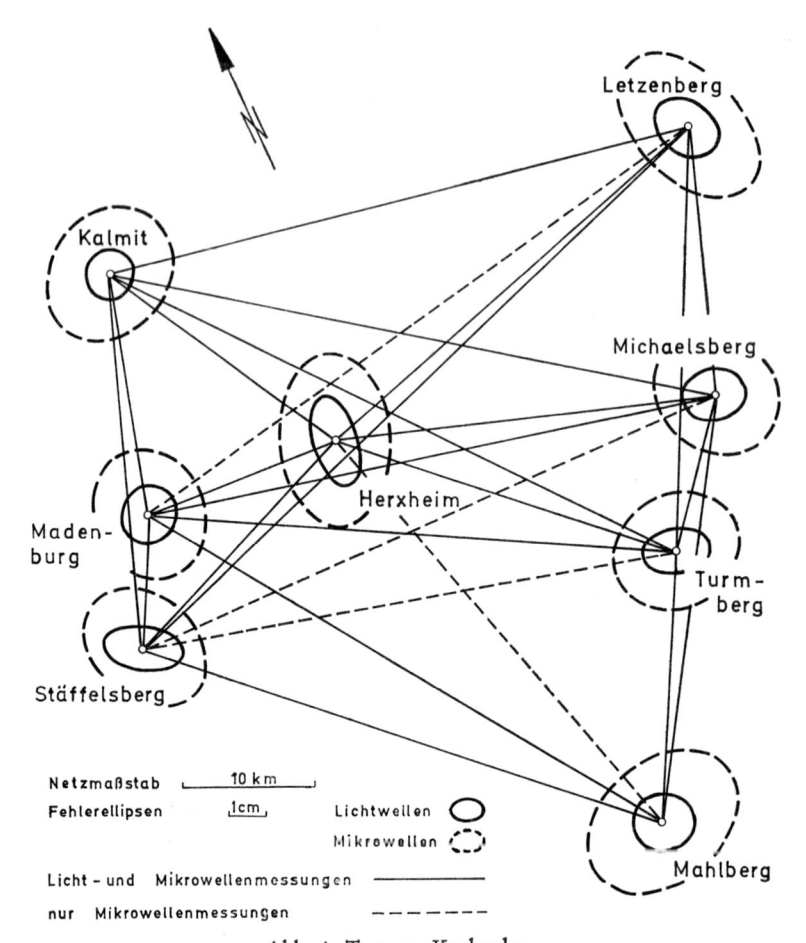

Abb. 1. Testnetz Karlsruhe.

wellenmessung ist es besonders die Luftfeuchte, die einen sehr starken Einfluß auf die Geschwindigkeit der Wellen ausübt.

Im Gegensatz hierzu ist die Geschwindigkeit der Lichtwellen von der Luftfeuchte nahezu unabhängig. Das ist auch der Grund dafür, daß mit Lichtwellen höhere innere Genauigkeiten erreicht werden können. Die Fehlerellipsen in Abb. 1 der freien Ausgleichung der Laser-Geodimetermessungen zeigen im Durchschnitt nur noch mittlere Punktunsicherheiten von 1 bis 2 cm. Für den mittleren Streckenfehler m ergab sich nach der Ausgleichung die Beziehung

$$m = \pm (5 \text{ mm} + 2,6 \cdot 10^{-7} \cdot S). \quad (S = \text{Streckenlänge}).$$

Aber auch hier ist die äußere Genauigkeit geringer. Auch elektrooptische Messungen können durch falsche Erfassung der für den Meßstrahl repräsentativen Lufttemperatur systematisch verfälscht sein. In KUNTZ & MÖLLER (1971) wird diese Fehlerquelle näher untersucht und zu ihrer Erfassung ein Verfahren der gleichzeitigen Messung mit Licht- und Mikrowellen vorgeschlagen.

Für das Studium horizontaler Krustenbewegungen sind solche systematischen Verfälschungen nur dann von untergeordneter Bedeutung, wenn auch die späteren Wiederholungsmessungen unter gleichartigen meteorologischen Bedingungen durchgeführt werden und Fehler gleicher Art erwartet werden können. Die Differenzen sind dann weitgehend frei von systematischen Fehlereinflüssen. Trotzdem bleibt ein Unsicherheitsfaktor, der zu Fehlschlüssen Anlaß geben könnte. Um ihn abschätzen zu können, wird gegenwärtig der südliche und wichtigste Teil des Testnetzes nach dem erwähnten Verfahren der gleichzeitigen Messung mit Licht- und Mikrowellen neu vermessen.

Nehmen wir sehr optimistisch als derzeit maximal erreichbare Streckengenauigkeit über den ca. 40 km breiten Graben einen Wert von 1,5 cm an, so könnten horizontale Bewegungen der Grabenränder mit einiger Sicherheit erst dann nachgewiesen werden, wenn sie die Größenordnung von ca. 5 cm erreichen. Nach Ansicht der Geologen (ILLIES 1972) liegt die zu erwartende Bewegung unter 0,5 mm/Jahr. Das bedeutet, daß nach dem heutigen Stand der geodätischen Meßtechnik nahezu ein Jahrhundert vergehen muß, bevor im Rheingraben Bewegungen dieser Größe festgestellt werden können.

2. Astronomische Bestimmung von Lotabweichungen

Das vorliegende Lotabweichungsprofil (Abb. 2), welches nach Absprache mit Geowissenschaftlern der Universität Karlsruhe und mit Rücksicht auf vorhandene Bougueranomalien angelegt wurde, verbindet die Dreieckspunkte I. Ordnung Karlsruhe—Durlach, Turmberg im Osten und Maikammer, Kalmit im Westen des Rheingrabens. Das Profil führt durch Gebiete positiver Bougueranomalien an den Grabenrändern und negativer Anomalien im Grabenbereich selbst, wobei minimale Werte im Nordosten von Landau auftreten. Der durchschnittliche Punktabstand von etwa 3 km entspricht dem Vorschlag von WOLF (1967).

Von den gebräuchlichen Verfahren der genauen astronomischen Ortsbestimmung wurde die Methode der gemeinsamen Bestimmung von Länge und Breite nach dem Höhenstandlinienverfahren mit Prismen-Astrolabium benutzt. Die Beobachtungen wurden in den Jahren 1969—1970 auf 19 trigonometrischen Punkten durchgeführt. Um einen Einblick in die innere und äußere Genauigkeit des Meßverfahrens zu gewinnen, wurden für jeden Punkt Messungen in zwei Nächten in verschiedenen Jahreszeiten von zwei Beobachtern mit zwei Astrolabien gleichzeitig und parallel ausgeführt. Im Verlauf der gesamten Meßreihe wurde dabei das Instrumentarium zyklisch zwischen den Beobachtern vertauscht. Der erreichte mittlere Fehler für die Breite und für die auf die Universität Karlsruhe bezogene Länge bewegt sich im Rahmen der bei Simultanbestimmung zu erwartenden Größenordnung von 0.2"—0.4" (BOZORGZADEH & SCHNÄDELBACH 1971).

Die auf das Besselellipsoid bezogenen vorläufigen Lotabweichungen sind in Abb. 3 dargestellt. Topographische Reduktionen sind noch nicht angebracht.

7 Approaches to Taphrogenesis

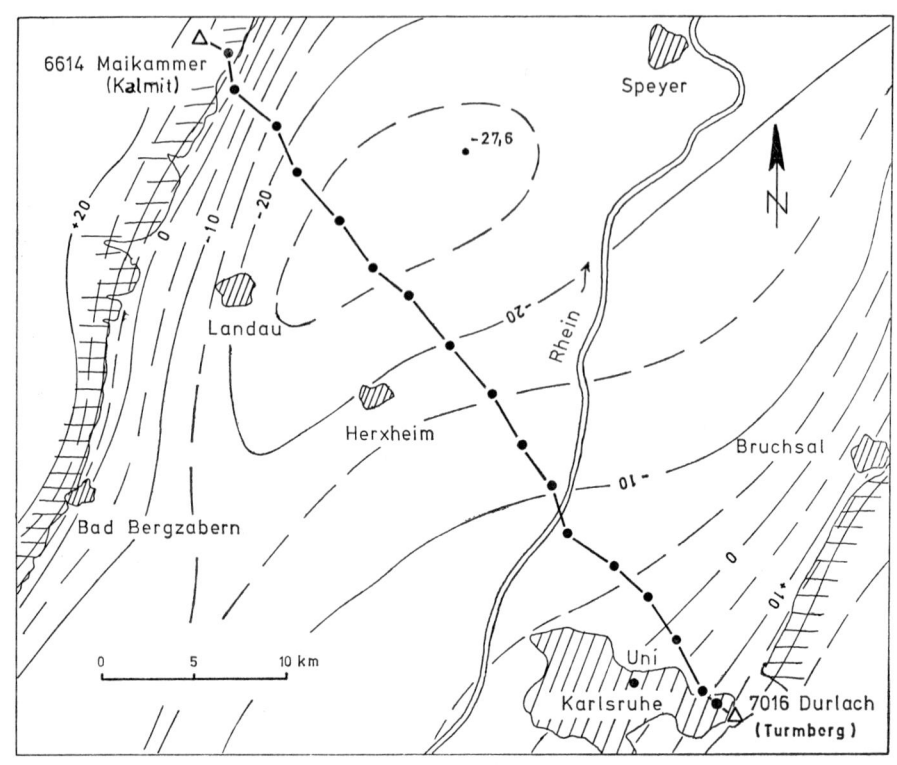

Abb. 2. Übersichtsskizze des Rheingrabens im Bereich nördlich von Karlsruhe, Bougueranomalien und Lotabweichungsprofil Turmberg-Kalmit.

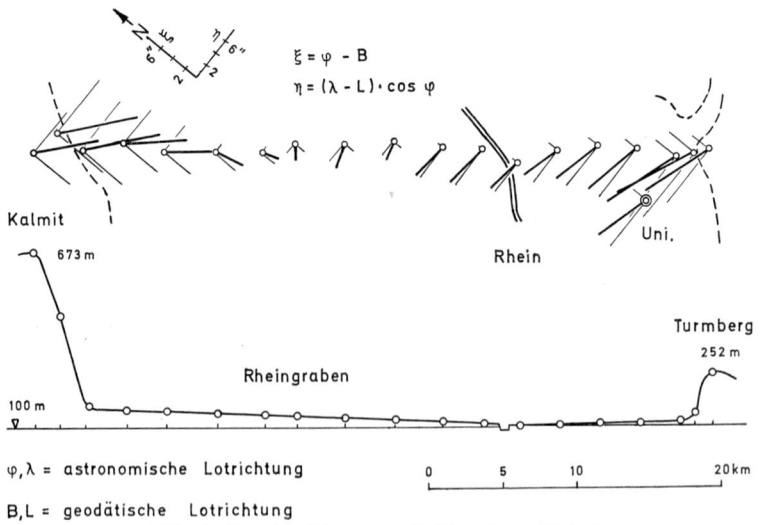

$\xi = \varphi - B$
$\eta = (\lambda - L) \cdot \cos \varphi$

φ, λ = astronomische Lotrichtung
B, L = geodätische Lotrichtung

Abb. 3. Lotabweichungsprofil Turmberg-Kalmit.

Neue astronomische Beobachtungen sind im Jahre 1971 in diesem Gebiet durchgeführt worden, eine weitere Verdichtung ist beabsichtigt, so daß eingehende Interpretationen und Geoidstudien möglich werden.

Literaturverzeichnis

BOZORGZADEH, GH. & SCHNÄDELBACH, K., 1971: Determination of Deflection of the Vertical by Simultaneous Observations with two Ni-2 Astrolabes. — Vorgelegt auf der IUGG Generalversammlung 1971, Moskau.
ILLIES, H., 1972: Der Oberrheingraben. — Unternehmen Erdmantel, 35—56, Steiner, Wiesbaden.
KUNTZ, E.: 1971: Elektronische Entfernungsmessungen auf Teststrecken und im Testnetz Karlsruhe. — 55 S., Dt. Geodät. Komm., Reihe B, Heft Nr. 182, München.
KUNTZ, E. & MÖLLER, D., 1971: Gleichzeitige elektronische Entfernungsmessungen mit Licht- und Mikrowellen. — AVN, 7, 254—266.
WOLF, H., 1967: Die Bedeutung der Lotabweichungen für das Upper-Mantle-Project. — Dt. Geodät. Komm., Reihe B, Heft Nr. 153, 109—112, München.

Microearthquake-Activity observed by a Seismic Network in the Rhinegraben Region[*]

by

K.-P. Bonjer and K. Fuchs

With 5 figures in the text

From 1966 to 1970, a homogeneous short-period and highly sensitive telemetric network has been installed by the Geophysical Institute of Karlsruhe University, including the following seismic stations: Buehlerhoehe (BUH), Feldberg (FEL), Kalmit (KLT) and a receiving unit for the French station Welschbruch (WLS). This network has been supplemented in 1972 by a short-period station in the Mine-Observatory Schiltach and the new telemetric station Tromm in the Odenwald.

The increase in station sensitivity and density has lowered the detection level within the Rhinegraben area especially in the southern part to magnitudes of about one. This resulted in a strong increase of the number of earthquakes detected per year by at least a factor of ten. The seismicity map for 1971/72 is given in Fig. 1. It does not include events from the Swabian Jura. Within these two years more than 160 events have been located. This figure becomes comparable to the number of earthquakes with magnitudes ≥ 3.0 included in the seis-

---

[*] Contribution No. 92, Geophysical Institute, University Karlsruhe.

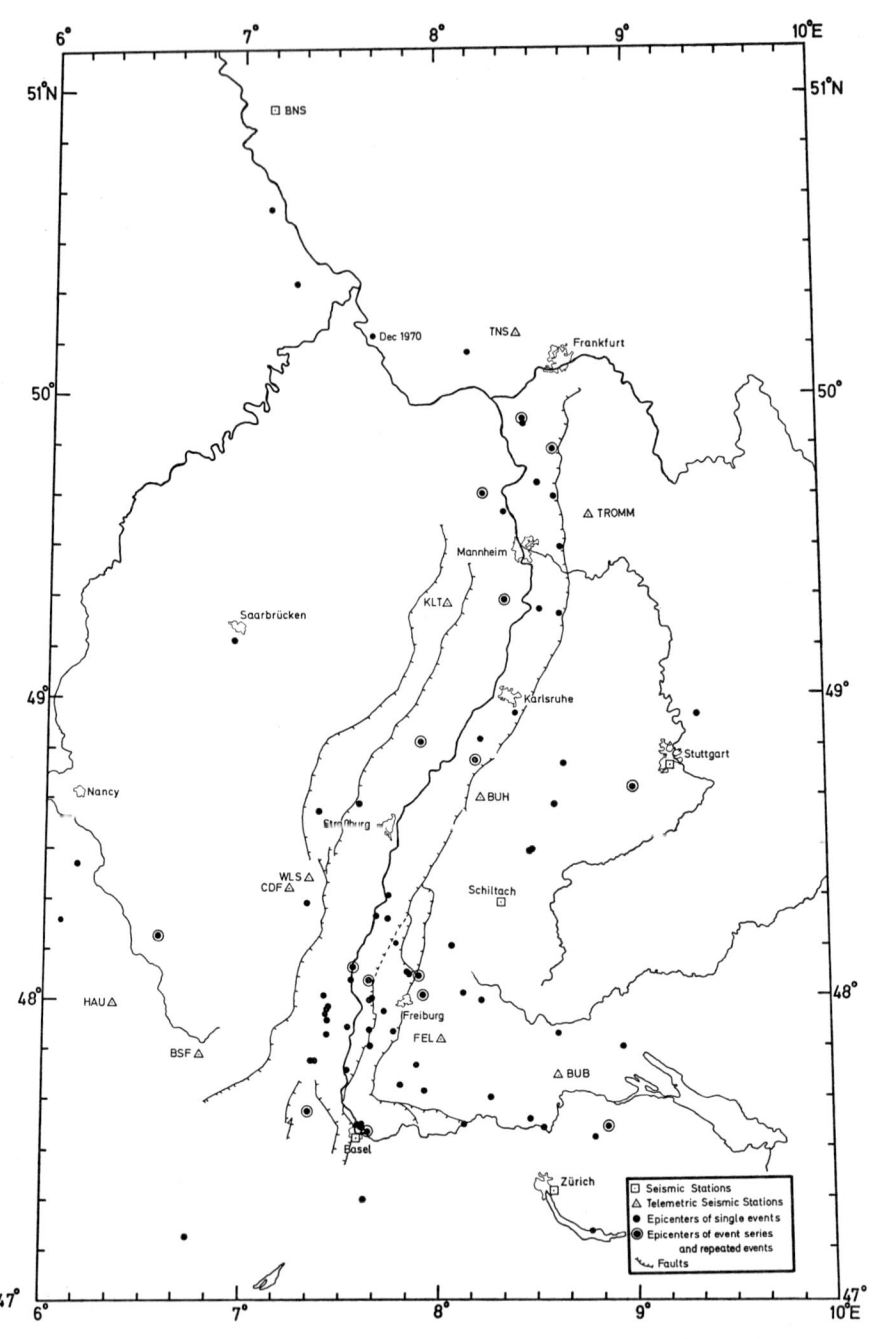

Fig. 1. Seismicity map for the years 1971 and 1972 for the Rhinegraben region. Events from the Swabian Jura are not included.

micity maps of several other authors (e. g.: HILLER et al. 1967; AHORNER 1970) for the last decades. The main trends of seismic activity in Fig. 1 correlate closely with these former maps: the strong activity in the southern part of the graben, its predominance on the eastern as compared to the western shoulder and a seismic gap in the Kraichgau and Pfaelzer Wald region north of Karlsruhe.

It is not to be expected that the observations of only two years show already all the details obtained in a period of decades. However, some details should be noted, especially the absence of activity in the regions of Strasbourg/Kehl and

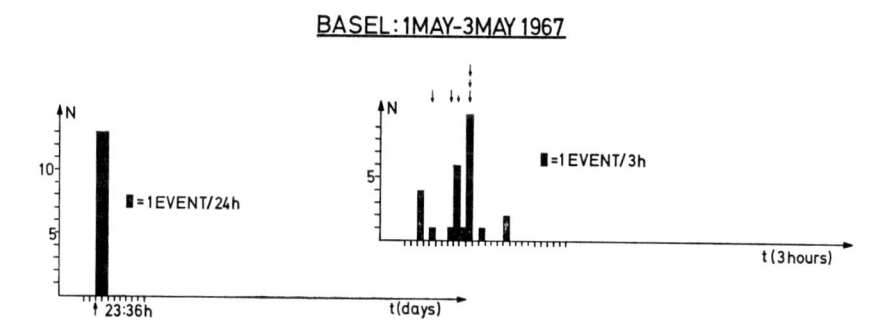

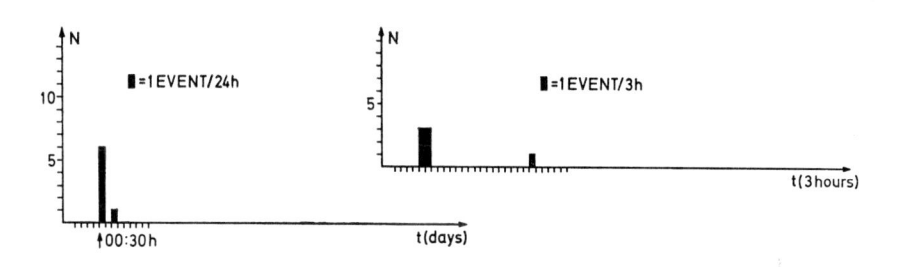

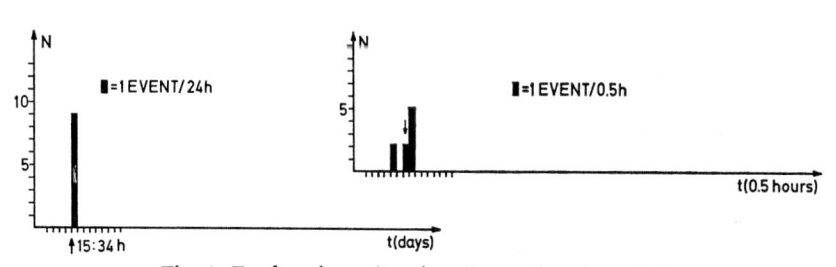

Fig. 2. Earthquake series plotted as a function of time.

Thann. Compared with the older maps there seems to be a lack of activity within the Kaiserstuhl, but the region around and its extension to the south appears to be very active.

Observations with the new station Tromm will allow us to decide whether the observed seismic activity of the northern part of the graben is really smaller or is an artifact imposed by large station separations.

An important finding is the frequent occurrence of earthquake series of different types possibly related to the tectonics of the graben. Some of these earthquake series are presented in Fig. 2 and 3 as a function of time. We can distinguish three types. The events of Basel (1967), not shown in Fig. 1, and Freiburg (1972) form a sequence within a short time interval without any defined main shock. Following MOGI (1967) and several other authors, the occurrence of these earthquake swarms can be related to an extremely heterogeneous crustal structure and a very concentrated distribution of external stress. The events of Denzlingen (1972) show two foreshocks preceding a remarkable main shock one order of magnitude larger which is followed by six aftershocks. This phenomenon can be associated with the crustal structure which is partly heterogeneous and where the external stress is not uniform. The events of Baden-Oos (1971) and Speyer (1972), Fig. 3, start with the main shock which is followed by a sequence of aftershocks. This can be explained by a relatively homogeneous structure and uniform external stress. The frequent occurrence of such series may prove to be a valuable tool in delineating seismo-tectonic relations. If a rapid detection and locating method allows the use of mobile stations, an improvement of hypocenter location and focal mechanism is to be expected.

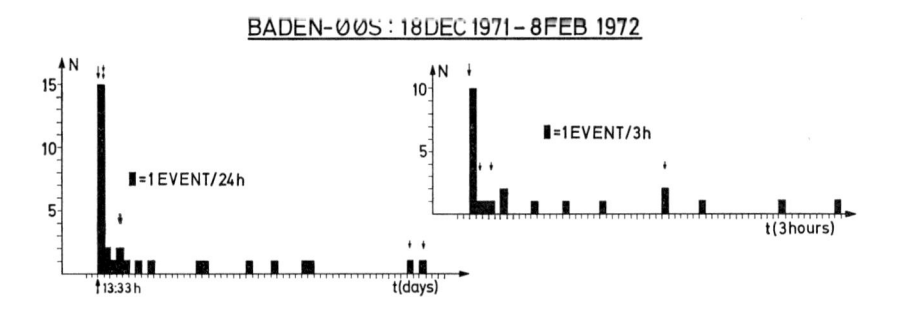

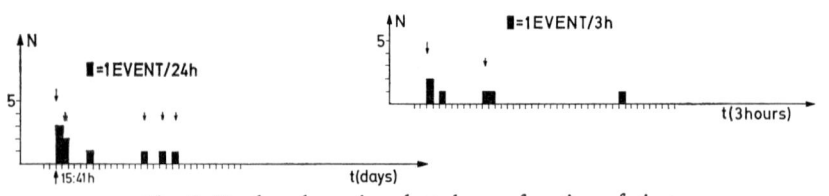

Fig. 3. Earthquake series plotted as a function of time.

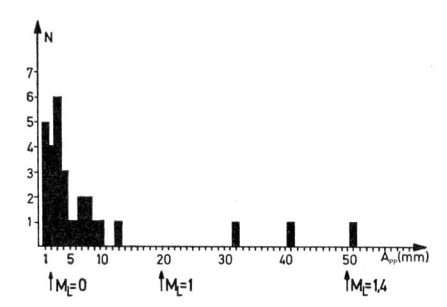

Fig. 4. Number of aftershocks of the Baden-Oos event versus maximum amplitude (peak/peak) and magnitude. Mainshock magnitude amounts to $M_L = 2.4$.

New conclusions may also be drawn regarding the magnitude-frequency relationship in the Rhinegraben area. SCHNEIDER (1971) deduced a maximum in this relationship at a magnitude of about 3. The strongly increased number of detected events requires a further increase of frequency with decreasing magnitude. The numerous events in the aftershock series of Baden-Oos permit an estimate of the magnitude frequency relationship (Fig. 4) which clearly shows an increase down to the detection threshold.

Teleseismic events recorded in the new network show that the crustal structure strongly influences the amplitudes and frequency content of seismic signals. One

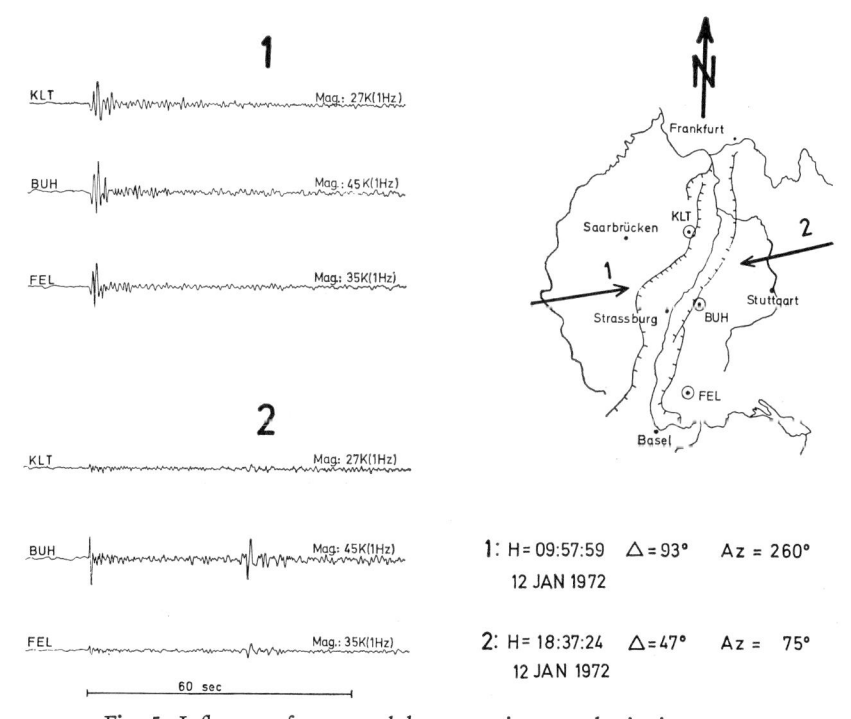

Fig. 5. Influence of structural heterogeneity on teleseismic events.

of numerous examples is given in Fig. 5. It shows the vertical components of two events recorded the same day from a different azimuth and different epicentral distance. While the amplitudes of the records of the three stations for event No. 1 are roughly proportional to the magnification, the record of BUH shows unusual amplification for event No. 2, which must be attributed to local structural heterogeneity. Therefore, station corrections have to be made calculating body wave magnitudes. Probably, this must also be carried out for local shock magnitudes.

This study has been sponsored by the Deutsche Forschungsgemeinschaft (German Research Association).

Literature

AHORNER, L., 1970: Seismo-tectonic relations between the graben zones of the Upper and the Lower Rhine Valley. — In: ILLIES, H. & MUELLER, ST.: Graben Problems. Schweizerbart, Stuttgart.

HILLER, W., ROTHÉ, J.-P. & SCHNEIDER, G., 1967: La Séismicité du Fossé Rhénan. — Abh. geol. L.-Amt Baden-Württ., 6, 98—100.

MOGI, K., 1967: Earthquakes and Fractures. — Tectonophysics, 5, 35—55.

SCHNEIDER, G., 1971: Seismizität und Seismotektonik der Schwäbischen Alb. — Ferdinand Enke, Stuttgart.

Herdmechanismen von Erdbeben im Oberrhein-Graben und in seinen Randgebirgen

von

L. Ahorner und G. Schneider

Mit 5 Abbildungen und 1 Tabelle im Text

Focal Mechanisms of Earthquakes in the Upper Rhine Graben and its Bordering Regions

Abstract: Fault-plane solutions of 14 earthquakes (time interval 1933—1972, magnitude range $2.8 < M_L < 5.4$) which had their epicenters in the Upper Rhine Graben and its bordering regions were analysed and the seismological results compared with the tectonic situations near the epicenters. Seismotectonic strike-slip mechanisms are dominating in and around the Upper Rhine Graben. The relative motion direction is left-lateral along the graben axis and right-lateral perpendicular to it. Dip-slip mechanisms are mostly of tensional type with fault-planes parallel to the graben axis. The principal horizontal stress field within the earth's crust of Southwest Germany as deduced from seismotectonic strike-slip mechanisms is directed NW—SE, i. e. diagonally to the general trend of the Upper Rhine Graben. The same regional stress field has been found within larger parts of the Alps and the Lower Rhine area.

I. Einleitung

Herdmechanische Untersuchungen sind für Erdbeben, die ihren Herd im Oberrhein-Graben und in seinen Randgebirgen hatten, bisher nur in wenigen Fällen ausführbar gewesen. Einmal war die Zahl der stärkeren Beben seit 1900 gering. Zum anderen bildete die bis vor wenigen Jahren sehr ungünstige Verteilung der seismologischen Stationen ein weiteres Hindernis, um zuverlässige Angaben über den Herdvorgang gewinnen zu können. Erst die Einrichtung spezieller Beobachtungsnetze im Bereich des Oberrhein-Grabens durch das Geophysikalische Institut der Universität Karlsruhe, das Institut de Physique du Globe der Universität Strasbourg und der Ecole Normale Superieur in Paris ermöglichte es, daß neuerdings auch für verhältnismäßig schwache Beben der Magnitude $M_L < 4$ herdmechanische Analysen durchgeführt werden können. Für die Zukunft eröffnen sich dadurch neue Wege zur Erforschung des seismotektonischen Geschehens im Bereich der größten und tektonisch am besten bekannten Grabenstruktur Europas.

Die erste Untersuchung über den Herdvorgang eines Grabenbebens wurde von HILLER (1934) für das Beben vom 3. Februar 1933 bei Rastatt veröffentlicht. Neuere herdmechanische Daten über Erdbeben des Oberrhein-Grabens und seiner Randgebirge haben SCHNEIDER, SCHICK & BERCKHEMER (1966), AHORNER (1967, 1972) u. a. publiziert. RANALLI & SCHEIDEGGER (1967), SCHNEIDER (1968, 1971) und AHORNER (1970) haben die seismotektonische Bedeutung der Ergebnisse diskutiert.

In der vorliegenden Arbeit wird der Versuch unternommen, das bis heute verfügbare herdmechanische Beobachtungsmaterial nach modernen Gesichtspunkten neu auszuwerten und einheitlich darzustellen. Während bei den früheren Untersuchungen zumeist nur zweidimensionale Ergebnisse bezüglich der Geometrie des Herdvorganges erhalten wurden, hat unsere Arbeit zum Ziel, ein vollräumliches Bild von der Orientierung der Herdflächen und des Spannungsellipsoides im jeweiligen Herdvolumen zu gewinnen.

Ausgangspunkt unserer Untersuchungen ist die initiale Bewegungsrichtung der P-Wellen. Für die Darstellung der Herdflächen-Lösungen (engl. fault-plane solutions) wird das von RITSEMA (1955) u. a. vorgeschlagene Verfahren benutzt: Man denkt sich um den Herd eine Kugelfläche gelegt und betrachtet die untere Hälfte dieser Herdkugel in stereographischer Projektion im Wulff'schen Netz. Herdmechanische Knotenebenen (Verschiebungsfläche, Hilfsfläche) treten in dieser Art der Projektion als Großkreise in Erscheinung, während die vom Herd abgehenden Wellenstrahlen entsprechend ihrem Azimut- und Abtauchwinkel als Durchstichpunkte auf der Kugeloberfläche aufgetragen werden. Der Abtauchwinkel eines Wellenstrahles läßt sich bei bekannter Herdtiefe, Herdentfernung und Krustenstruktur im allgemeinen mit ausreichender Genauigkeit (auf ± 10°) bestimmen.

Die gewonnenen Herdflächen-Lösungen sind grundsätzlich zweideutig. Man kann zunächst nicht unterscheiden, welche Herdfläche der seismischen Verschiebungsfläche und welche der hierzu senkrecht liegenden Hilfsfläche entspricht. Erst die Berücksichtigung zusätzlicher Kriterien, so der spektralen Amplitudenverteilung der seismischen Wellen, der Streckung makroseismischer Isoseisten oder der geologischen Lagerungsverhältnisse im Herdgebiet, ermöglicht eine Flächenauswahl (vgl. z. B. SCHICK 1970, 1972, AHORNER, MURAWSKI & SCHNEIDER 1972). Ohne zusätzliche Entscheidungshilfen sind aus den Herdflächen-Lösungen dagegen die Hauptspannungsachsen P, B und T (größte, mittlere und kleinste Hauptspannung im Herdvolumen zum Zeitpunkt der Spannungsentlastung) bestimmbar.

II. Beschreibung der Herdflächen-Lösungen

Im Rahmen der vorliegenden Arbeit werden 14 Erdbeben der Jahre 1933—1972 betrachtet. Die allgemeinen Daten dieser Beben sind in der Tab. 1 zusammengestellt. Die Lage der Herde geht aus der Übersichtskarte Abb. 1 hervor. In den Abb. 2, 3 und 5 sind die Herdflächen-Lösungen abgebildet. Es handelt sich etwa zur Hälfte um ältere Beben, die schon früher herdmechanisch untersucht worden sind und für die jetzt eine Neubearbeitung durchgeführt wurde. Die übrigen, zumeist jüngeren Beben wurden zum ersten Mal herdmechanisch untersucht.

Die Herdflächen-Lösungen werden im folgenden einzeln besprochen. Die Reihenfolge geht von Norden nach Süden, wobei die gleiche Zahlenbezeichnung angewandt wird wie in der Tab. 1 und den Abb. 1—5.

Tab. 1. Verzeichnis der untersuchten Erdbeben.

Nr.	Gebiet	Datum	Herdzeit GMT	Epizentrum Lat.	Long.	Magnitude M_L
1	Boppard	1. 12. 1970	$10^h49^m08^s$	50°13′N	7°42′E	3,8
2	Birkenfeld	17. 8. 1960	$15^h28^m06^s$	49°42′N	7°15′E	4,1
3	Worms	24. 2. 1952	$21^h25^m34^s$	49°37′N	8°22′E	4,9
4	Heidelberg	12. 7. 1971	$07^h10^m52^s$	49°31′N	8°41′E	3,7
5	Speyer	25. 2. 1972	$15^h41^m11^s$	49°19′N	8°26′E	3,6
6	Karlsruhe	7. 6. 1948	$07^h15^m18^s$	48°58′N	8°20′E	4,6
7	Weißenburg	8. 10. 1952	$05^h17^m15^s$	48°53′N	8°01′E	4,8
8 a	Rastatt	8. 2. 1933	$07^h07^m16^s$	48°51′N	8°12′E	5,4
8 b	Rastatt	18. 12. 1971	$13^h33^m11^s$	48°48′N	8°12′E	2,8
9	Hornisgrinde	30. 12. 1935	$03^h36^m14^s$	48°37′N	8°13′E	5,0
10	Boofzheim	4. 9. 1959	$08^h56^m54^s$	48°23′N	7°43′E	4,1
11	Südschwarzwald	19. 9. 1965	$08^h10^m44^s$	47°57′N	8°16′E	4,1
12	Südschwarzwald	28. 4. 1961	$20^h48^m49^s$	47°43′N	7°53′E	4,9
13	Schwäb. Alb	22. 1. 1970	$15^h25^m17^s$	48°18′N	9°02′E	5,3

(1) Erdbeben bei Boppard 1970

Die Herdflächen-Lösung für dieses im nördlichen Vorfeld des Oberrhein-Grabens in 10 km Tiefe ausgelöste Erdbeben ist bei AHORNER (1972) abgebildet. Der Herdvorgang entspricht einer seismischen Schrägaufschiebung mit vorherrschender horizontaler Bewegungskomponente. Die Verschiebungsfläche streicht entweder NW—SE und fällt mit 55° nach SW ein (linkshändige Bewegung, Verhältnis der horizontalen zur vertikalen Bewegungskomponente H/V = 1,6) oder sie streicht SW—NE und fällt mit 60° nach SE ein (rechtshändige Bewegung, Verhältnis H/V = 1,4). Die erstgenannte Fläche folgt der allgemeinen Richtung der seismisch aktiven Mittelrhein-Zone und der zahlreichen Querverwerfungen des Rheinischen Schiefergebirges, so daß die Deutung als Verschiebungsfläche naheliegt. Eine Isoseistenstreckung ist in dieser Richtung allerdings nicht festzustellen. Vielmehr ergibt sich nach der makroseismischen Karte von AHORNER (1972) eine deutliche Vorzugsrichtung in SSW—NNE, was aber mit keiner der beiden Herdflächen übereinstimmt.

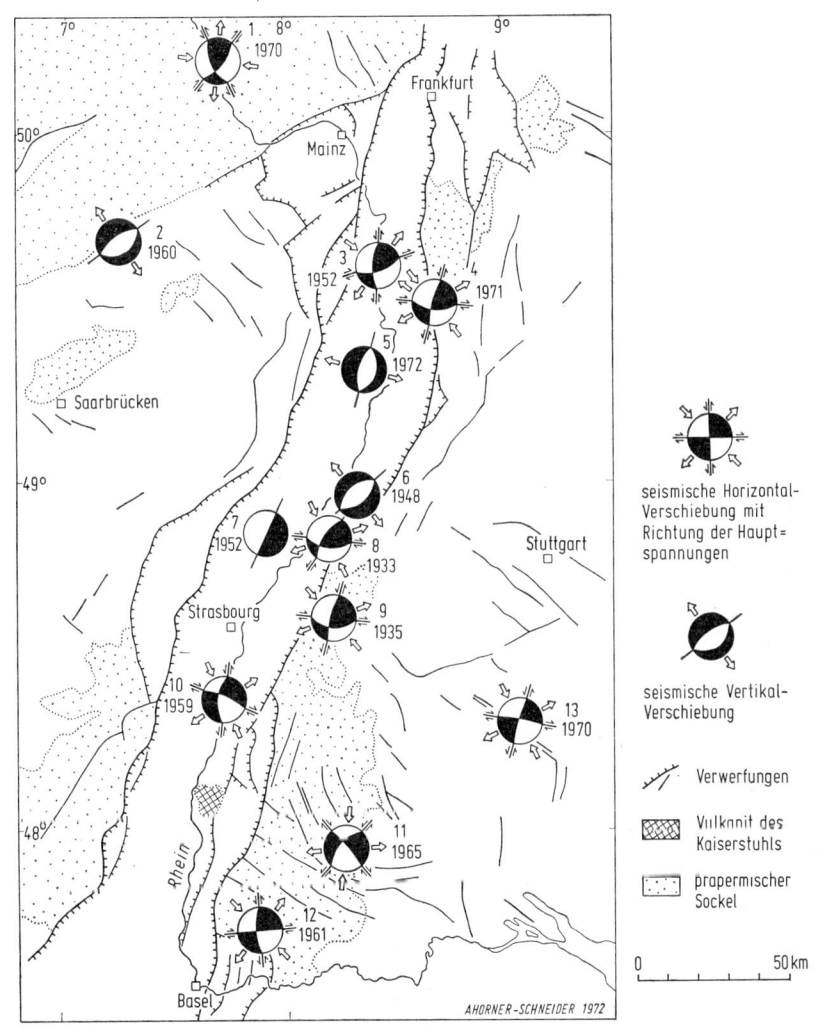

Abb. 1. Übersicht der bisher untersuchten Erdbeben-Herdmechanismen im Bereich des Oberrhein-Grabens und seiner näheren Umgebung. Bezüglich der Herdkugeldarstellung vgl. Abb. 2.

108 L. Ahorner und G. Schneider

Birkenfeld 17.Aug.1960 M=4,1 h~10 km

2

P N 145°E 75°SE B N 55°E 0°
T N 325°E 15°NW ● Kompr. ○ Dilat.

Worms 24.Febr.1952 M=4,9 h=15 km

3

P N 127°E 0° B N 217°E 44°SW
T N 37°E 46°NE ● Kompr. ○ Dilat.

Heidelberg 12.Juli 1971 M=3,7 h~5 km

4

P N 321°E 12°NW B N 215°E 51°SW
T N 61°E 46°NE ● Kompr. ○ Dilat.

Speyer 25.Febr.1972 M=3,6 h~5 km

5

P N 105°E 80°SE B N 15°E 0°
T N 285°E 10°NW ● Kompr. ○ Dilat.

Karlsruhe 7.Juni 1948 M=4,6 h=5 km

6

P N 0°E 90° B N 50°E 0°
T N 140°E 0° ● Kompr. ○ Dilat.

Weißenburg 8.Okt.1952 M=4,8 h=6 km

7

(P N 115°E 45°SE) (B N 25°E 0°)
(T N 295°E 45°NW) ● Kompr. ○ Dilat.

Hornisgrinde 30.Dez.1935 M=5,0 h=22 km

9

P N 318°E 20°NW B N 208°E 40°SW
T N 70°E 41°NE ● Kompr. ○ Dilat.

Boofzheim 4.Sept.1959 M=4,1 h=1-2 km

10

(P N 156°E 42°SE) (B N 316°E 47°NW)
(T N 57°E 10°NE) ● Kompr. ○ Dilat.

Südschwarzwald 19.Sept.1965 M=4,1 h=18 km

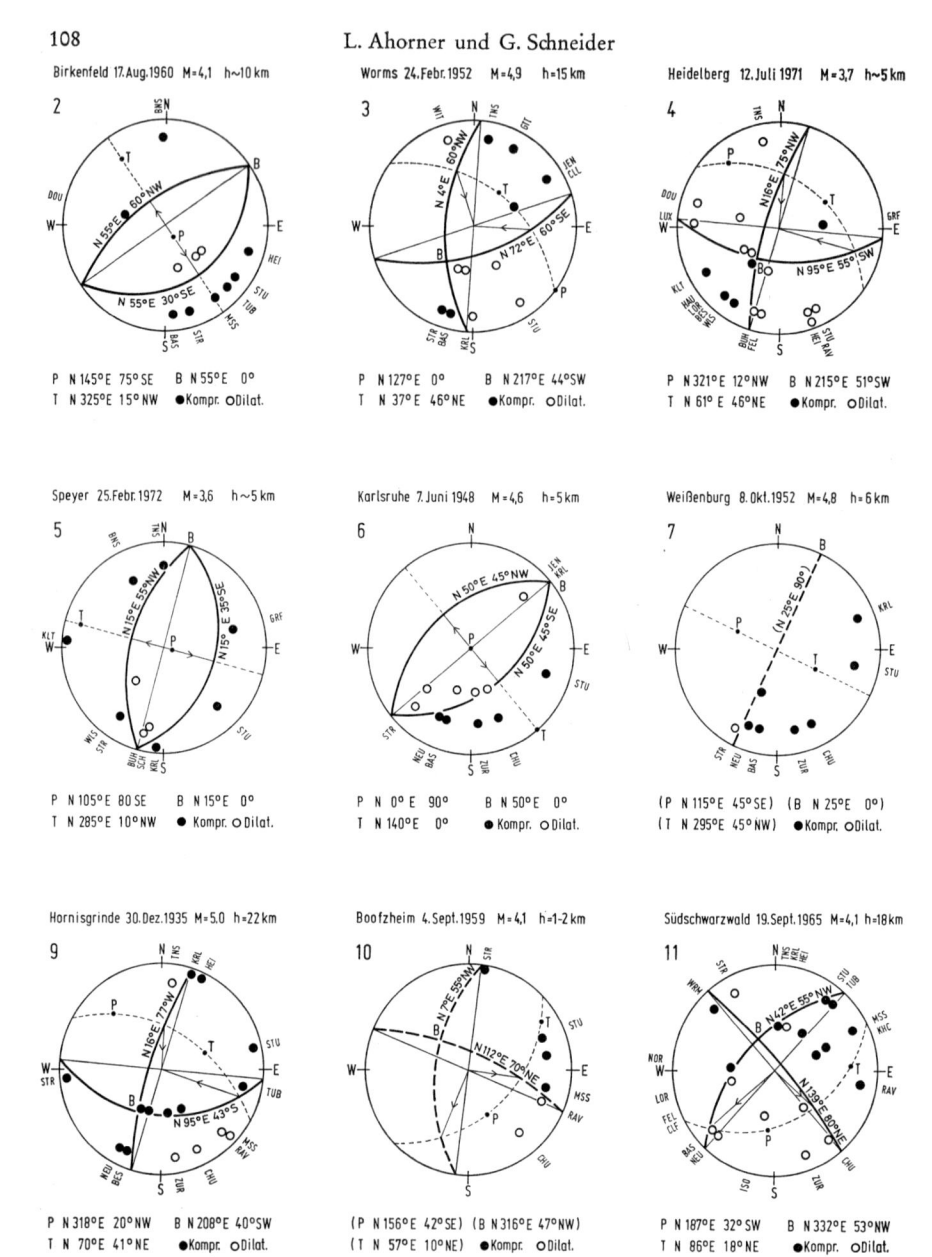

11

P N 187°E 32°SW B N 332°E 53°NW
T N 86°E 18°NE ● Kompr. ○ Dilat.

Abb. 2. Herdflächen-Lösungen von Erdbeben der Jahre 1933—1972. Projektion der unteren Hälfte der Herdkugel im Wulff'schen Netz. Das Streichen und Fallen der Herdflächen (Verschiebungsfläche und Hilfsfläche) und die Richtungen der Spannungsachsen P, B und T (größte, mittlere und kleinste Hauptspannung im Herdvolumen) sind angegeben. Die mit Pfeilen versehenen Linien stellen die Bewegungsspur der Hangendscholle auf der jeweiligen Herdfläche dar (Slip-Winkel). An den Durchstichpunkten der Wellenstrahlen ist die initiale Bewegungsrichtung der P-Wellen vermerkt (Kompression bedeutet Bewegung vom Herd weg, Dilatation Bewegung auf den Herd zu). Stationsabkürzungen nach der internationalen Norm.

(2) Erdbeben bei Birkenfeld 1960 (Abb. 2)

Obgleich die Herdflächen-Lösung nicht sehr gut durch Beobachtungen belegt ist, läßt sich sagen, daß es sich um eine seismische Abschiebung entlang einer SW—NE streichenden Verschiebungsfläche handelt, das heißt um eine vorwiegend vertikal gerichtete Bewegung. Ein nennenswerter horizontaler Bewegungsanteil ist nicht festzustellen. Das Streichen der Herdverwerfung folgt dem Südrand des Rheinischen Schiefergebirges gegen die Saar-Nahe-Senke. Hiermit paßt gut zusammen, daß das Epizentrum in der unmittelbaren Nähe dieser altangelegten tektonischen Grenzzone liegt (vgl. AHORNER 1962).

(3) Erdbeben bei Worms-Ludwigshafen 1952 (Abb. 2)

Dieses Beben wurde schon früher herdmechanisch bearbeitet und als seismische Horizontalverschiebung gedeutet (AHORNER 1967). Die neue Herdflächen-Lösung bestätigt die frühere Deutung. Wir haben es mit einer Schrägaufschiebung mit vorherrschender horizontaler Bewegungskomponente zu tun. Das ist bemerkenswert, weil der Herd innerhalb des Oberrhein-Grabens nahe an dessen westlichen Rand liegt, wo man derartige Bewegungsformen an sich nicht erwartet. Die seismische Verschiebungsfläche streicht entweder SSW—NNE und fällt mit 60° nach W ein (linkshändige Bewegung, Verhältnis H/V = 1,7) oder sie streicht WSW—ENE und fällt mit 60° nach SE ein (rechtshändige Bewegung, Verhältnis H/V = 1,7). Welche der beiden Möglichkeiten zutrifft, ist schwer zu entscheiden. Die starke Streckung der inneren Isoseisten in Grabenlängsrichtung ist wohl hauptsächlich ein Ausbreitungseffekt und erlaubt keine herdmechanischen Schlüsse (vgl. WECHSLER 1955). Die geologischen Verhältnisse legen allerdings eine Deutung der SSW—NNE, also grabenparallel verlaufenden Herdfläche als Verschiebungsfläche nahe.

(4) Erdbeben bei Heidelberg 1971 (Abb. 2)

Die durch Beobachtungen gut belegte Herdflächen-Lösung stimmt weitgehend mit der eben besprochenen überein, obgleich der Herd diesmal am Ostrand des Grabens liegt. Wiederum ergibt sich als Bewegungstyp eine seismische Schrägaufschiebung mit vorherrschender horizontaler Bewegungskomponente. Die eine Herdfläche streicht grabenparallel in SSW—NNE-Richtung und fällt mit 75° nach NW ein (linkshändige Bewegung, Verhältnis H/V = 1,5), die andere streicht quer dazu in WNW—ESE-Richtung und fällt mit 60° nach SE ein (rechtshändige Bewegung, Verhältnis H/V = 3,7). Wegen ihrer richtungsmäßigen Übereinstimmung mit den tektonischen Strukturen im Epizentralgebiet ist wohl die erstgenannte Fläche als Verschiebungsfläche zu deuten. Die seismologischen Daten allein erlauben keine Flächenauswahl.

(5) Erdbeben bei Speyer 1972 (Abb. 2)

Mit diesem Beben begegnet uns im Grabenbereich erstmals eine seismische Abschiebung mit vorwiegend vertikal gerichteter Schollenbewegung. Die Verschiebungsfläche streicht grabenparallel in SSW—NNE-Richtung und fällt entweder 55° nach NW oder 35° nach SE ein. Bewegungsformen dieser Art sind in einer auf Krustenausweitung zurückgehenden Grabenstruktur wie dem Oberrhein-Graben zu erwarten und bedeuten keine Überraschung.

(6) Erdbeben bei Karlsruhe-Forchheim 1948 (Abb. 2)

Auch hier könnte es sich um eine seismische Abschiebung entlang einer grabenparallel streichenden Bewegungsfläche handeln. Das Verteilungsbild von Kompression und Dilatation ist ganz ähnlich wie beim Birkenfelder Beben 1960 (2). Die Herdflächen-Lösung ist allerdings unsicher, weil aus dem Gebiet westlich und nördlich vom Herd keine seismometrischen Daten vorliegen. Die makroseismische Karte von WECHSLER (1955) zeigt eine deutliche Streckung der Isoseisten in Grabenlängsrichtung, was aber ein Ausbreitungseffekt sein kann.

(7) Erdbeben bei Weißenburg-Seltz 1952 (Abb. 2)

Wegen der einseitigen Stationsgruppierung ist diese Herdflächen-Lösung noch unsicherer als die vorangehende. Möglicherweise liegt eine seismische Vertikalverschiebung entlang einer grabenparallel verlaufenden Fläche vor. Die Karte von ROTHÉ & DECHEVOY (1967), die den Verlauf wichtiger Verwerfungen und die Ausbildung der Isoseisten im Epizentralgebiet zeigt, widerspricht dieser Deutung nicht.

(8a, 8b) Erdbeben bei Rastatt 1933 und 1971 (Abb. 3)

Das Rastatter Erdbeben des Jahres 1933, eines der stärksten Beben im Oberrhein-Graben in historischer Zeit, wurde bereits von HILLER (1934) und später von SCHNEIDER, SCHICK & BERCKHEMER (1966) und AHORNER (1967) als horizontaler Scherungsbruch gedeutet. Ein schwaches Erdbeben, das 1971 in fast genau dem gleichen Gebiet stattfand, hat die frühere Interpretation bestätigt. Die kombinierte Herdflächen-Lösung der beiden fast 4 Jahrzehnte auseinander liegenden Ereignisse läßt keine Abweichung in der Verteilung von Kompression und Dilatation erkennen. Auch in anderen Herdgebieten, so auf der Schwäbischen Alb, in Oberschwaben, im Wallis und in Nordtirol, konnte eine über Jahrzehnte anhaltende Konstanz der Bewegungsform nachgewiesen werden (SCHNEIDER 1968).

Der Rastatter Herd entspricht dem Bewegungstyp nach einer seismischen Schrägaufschiebung mit vorherrschender horizontaler Bewegungskomponente. Die eine Herdfläche streicht SW—NE und fällt mit 60° nach NW ein (linkshändige Bewegung, Verhältnis H/V = 1,2), die andere streicht WNW—ESE und fällt mit 50° nach S ein (rechtshändige Bewegung, Verhältnis H/V = 1,7). Eine Flächenauswahl ist nicht sicher zu treffen. HILLER (1934) hat aus den Laufzeitbeobachtungen auf eine Bruchfortpflanzung von SW nach NE entlang einer 10—15 km langen „Herdlinie" geschlossen. Dies spricht für die Deutung der grabenparallelen Herdfläche als Verschiebungsfläche. Auch die geologisch-tektonischen Verhältnisse im Epizentralgebiet legen diese Deutung nahe. Infolge der geringen Herdtiefe von h = 4 km ist gut zu erkennen, daß der Herd in unmittelbarer Nähe einer durch die Erdölexploration bekannt gewordenen Vorstörung des östlichen Grabenrandbruches liegt (Abb. 3 links; BREYER & DOHR 1967). Diese Vorstörung be-

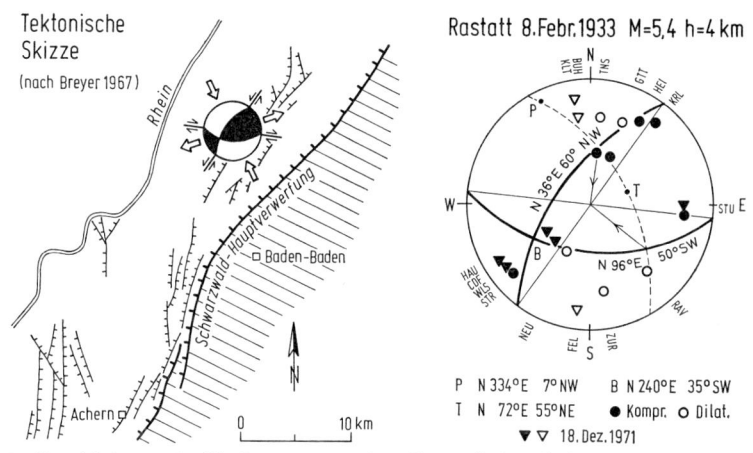

Abb. 3. Kombinierte Herdflächen-Lösung für die Erdbeben bei Rastatt 1933 und 1971 (8 a und 8 b) und geologisch-tektonischer Bau des Epizentralgebietes. Geologie vereinfacht nach BREYER (in BREYER & DOHR 1967).

grenzt die „Rastatter Mulde" gegen eine nach NE anschließende Zwischenscholle, auf der die Tertiärbasis rund 2000 m höher liegt. Auch im Quartär haben hier noch erhebliche Schollenverstellungen stattgefunden (BARTZ 1967). Das Streichen und Einfallen der Vorstörung entspricht genau der SW—NE streichenden Herdfläche des Rastatter Herdes. Auf der anderen Seite zeigen die von LACOSTE (1934) und SCHMIDT-ZITTEL (in HILLER 1934) entworfenen Isoseistenkarten eine auffällige Streckung des Gebietes stärkster Erschütterungen nicht nur in Grabenlängsrichtung, sondern auch quer dazu in Richtung der WNW—ESE verlaufenden Herdfläche (Abb. 4). Man kann daher nicht völlig ausschließen, daß auch in dieser Richtung Bruchvorgänge wirksam waren. Durch eine Bruchfortpflanzung in Richtung nach W ließen sich die starken Erschütterungswirkungen auf elsässischer Seite zwanglos erklären. Beiden Möglichkeiten gerecht würde die Annahme eines multilateralen Bruchvorganges, welcher sich auf beiden Herdflächen abspielte.

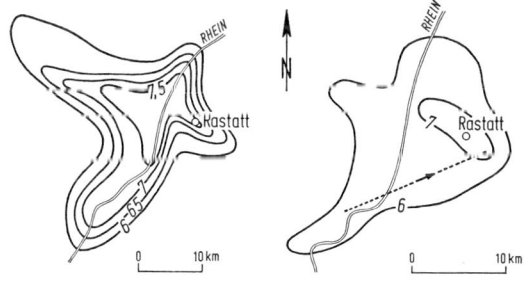

Abb. 4. Makroseismische Karten für das Erdbeben bei Rastatt 1933 (8 a). Dargestellt sind nur die inneren Isoseisten. Links nach LACOSTE (1934), rechts nach SCHMIDT-ZITTEL (in HILLER 1934). Das mikroseismische Epizentrum liegt bei Rastatt. Die von HILLER vermutete Herdlinie ist punktiert eingetragen.

Solche Bruchvorgänge sind theoretisch zu erwarten. In der Praxis hat man sie bisher aber nur selten angetroffen.

(9) Erdbeben im Gebiet der Hornisgrinde 1935 (Abb. 2)

Vom Hornisgrinde-Gebiet im nördlichen Schwarzwald ging am 30. Dezember 1935 im Abstand von etwa einer halben Stunde ein Doppelbeben aus. Die Herdflächen-Lösung bezieht sich auf den zweiten, etwas stärkeren Stoß. HILLER (1936) hat bereits herdmechanische Überlegungen zu diesem Ereignis angestellt und auf eine Aufschiebung entlang einer SW—NE streichenden Fläche geschlossen. Unsere Auswertung ergab als Bewegungstyp eine Schrägaufschiebung mit vorherrschender horizontaler Bewegungskomponente. Die eine Herdfläche streicht SSW—NNE und fällt mit 77° nach W ein (linkshändige Bewegung, Verhältnis H/V = 1,1), die andere streicht WNW—ESE und fällt mit 43° nach S ein (rechtshändige Bewegung, Verhältnis H/V = 4,3). Welche der beiden Flächen der Verschiebungsfläche entspricht, ist nicht eindeutig zu entscheiden. Aufgrund der relativen Amplituden der S-Wellen an den Stationen MSS und NEU (Amplitudenverhältnis 10:1 nach Entfernungsreduktion) müßte die Wahl zwischen den beiden möglichen Bewegungsrichtungen zugunsten der transversal zum Oberrhein-Graben verlaufenden Herdfläche getroffen werden. Der Isoseistenverlauf läßt nach der Karte von HILLER (1936) keine deutliche Streckung in einer der beiden Richtungen erkennen. Nach der geologischen Übersichtskarte des Nordschwarzwaldes von METZ (in GEYER & GWINNER 1968) sind jedoch im Bereich der Hornisgrinde vorwiegend SSW—NNE streichende Brüche vorhanden, was mehr auf eine grabenparallel angeordnete Verschiebungsfläche hindeutet.

(10) Erdbeben bei Boofzheim 1959 (Abb. 2)

Von diesem inmitten des Oberrhein-Grabens etwa 20 km südlich von Strasbourg ausgelösten Erdbeben liegen nur wenige zur Auswertung geeignete Registrierungen vor, so daß die Herdflächen-Lösung sehr unsicher ist. Es könnte sich um eine seismische Schrägabschiebung entlang einer SSW—NNE streichenden und mit 55° nach NW einfallenden Herdfläche (linkshändige Bewegung, Verhältnis H/V = 2,7) oder einer WNW—ESE streichenden und mit 70° nach NE einfallenden Herdfläche handeln (rechtshändige Bewegung, Verhältnis H/V = 2,7). Aber auch andere Mechanismen sind mit den Beobachtungen verträglich. Die makroseismische Karte von ROTHÉ & DECHEVOY (1967) zeigt eine undeutliche Streckung der inneren Isoseisten in SSW—NNE-Richtung. Dies stimmt mit der bevorzugten Streichrichtung bekannter geologischer Verwerfungen im Epizentralgebiet überein (SIMLER & MASCLAUX 1967). Vielleicht hat sich die seismische Verschiebung in der gleichen Richtung abgespielt. Bemerkenswert ist die geringe Herdtiefe des Bebens (h = 1—2 km).

(11) Erdbeben im Südschwarzwald bei Neustadt 1965 (Abb. 2)

Während SCHNEIDER, SCHICK & BERCKHEMER (1966) bei diesem Beben eine seismische Aufschiebung entlang einer WNW—ESE streichenden Fläche annahmen, zeigt die im Rahmen der Neubearbeitung vorgenommene Darstellung der Richtungsdaten im Wulff'schen Netz, daß sich die Beobachtungen auch widerspruchslos durch eine Schrägabschiebung mit vorherrschender horizontaler Bewegungskomponente erklären lassen. Die eine Herdfläche streicht dann SW—NE und fällt mit 55° nach NW ein (linkshändige Bewegung, Verhältnis H/V = 5,9), die andere streicht NW—SE und fällt mit 80° nach NE ein (rechtshändige Bewegung, Verhältnis H/V = 1,4). Die neue Deutung hat den Vorteil, daß die Orientierung der Hauptspannungen im Herdvolumen nun besser in das großräumige Spannungsfeld Südwestdeutschlands paßt als vorher. Für die Flächenauswahl ist von Bedeutung, daß die geologischen Bruchstrukturen im Epizentralgebiet vorwiegend NW—SE streichen. Möglicherweise folgt die seismische Verschiebungsfläche der gleichen Richtung. Großtektonisch befindet sich der Herd im nördlichen Randbereich der Freiburg—Bonndorf-Bruchzone (CARLÉ 1955).

(12) **Erdbeben im Südschwarzwald bei Schopfheim-Zell 1961** (Abb. 5)

Das Beben des Jahres 1961 im mittleren Wiesental zeigt in seiner Herdflächen-Lösung, die sich nur wenig von der schon früher von SCHNEIDER, SCHICK & BERCKHEMER (1966) gegebenen unterscheidet, den gleichen Charakter wie die Beben auf der westlichen Schwäbischen Alb (vgl. das folgende Beben und SCHNEIDER 1971). Die seismische Dislokation besteht vorwiegend in horizontalen Bewegungen auf nahezu senkrecht zur Erdoberfläche stehenden Herdflächen. Die eine Herdfläche streicht N—S und fällt mit 65° nach W ein (linkshändige Bewegung, ± horizontal), die andere streicht W—E und steht saiger (rechtshändige Bewegung, Verhältnis H/V = 2,1). Nach der bei METZ & REIN (1958) veröffentlichten Karte der geologisch-tektonischen Baueinheiten im Südschwarzwald herrschen innerhalb der Scholle des Dinkelberges und der Weitenauer Vorberge die

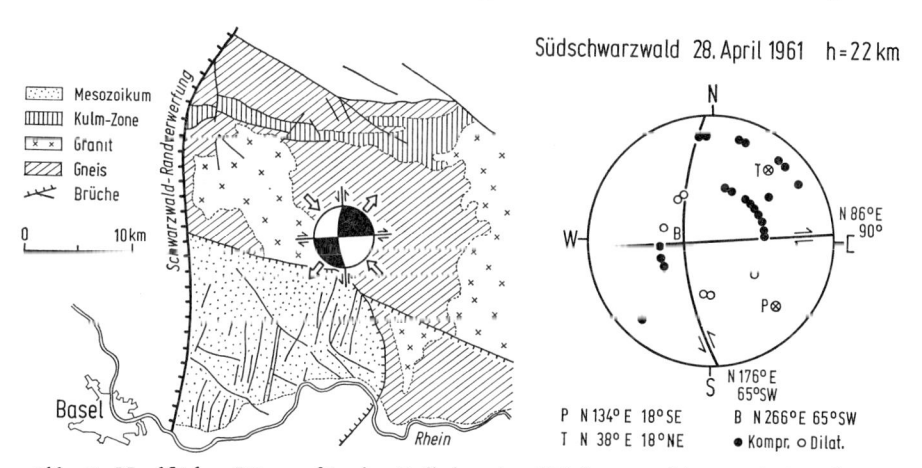

Abb. 5. Herdflächen-Lösung für das Erdbeben im Südschwarzwald 1961 (12) und geologisch-tektonischer Bau des Epizentralgebietes. Geologie vereinfacht nach METZ & REIN (1958).

rheinisch streichenden Brüche gegenüber allen anderen tektonischen Bruchrichtungen stark vor. Es besteht die Möglichkeit, daß sich diese Brüche auch innerhalb der oberen Erdkruste des Südschwarzwälder Granitgebietes fortsetzen und die Bewegungsflächen für dieses Erdbeben gebildet haben. Andererseits wird die Grenze des Südschwarzwälder Granitgebietes gegen die Weitenauer Vorberge durch eine große W—E verlaufende Störung geprägt. Man kann daher auch eine W—E-Orientierung der seismischen Verschiebungsfläche nicht ausschließen.

(13) Erdbeben auf der Schwäbischen Alb 1970

Dieses Beben, dessen Herdflächen-Lösung bei AHORNER, MURAWSKI & SCHNEIDER (1972) abgebildet ist, wird stellvertretend für eine größere Zahl von Erdbeben der westlichen Schwäbischen Alb angeführt, die alle den gleichen Herdvorgang erkennen lassen, nämlich eine linkshändige Horizontalverschiebung entlang einer SSW—NNE streichenden und nahezu saiger stehenden Herdfläche (vgl. auch SCHNEIDER, SCHICK & BERCKHEMER 1966, SCHNEIDER 1971). Die klare Entscheidung über die Richtung der Verschiebungsfläche ist im Herdgebiet der Schwäbischen Alb aufgrund von Spektralvergleichen zwischen Raumwellenimpulsen sowie der Epizentrenaufreihung und sehr auffälligen Isoseistenstreckung in SSW—NNE-Richtung möglich (SCHICK 1970, SCHNEIDER 1971, AHORNER, MURAWSKI & SCHNEIDER 1972). Dabei zeigt sich, daß zwischen den seismotektonischen Bewegungsvorgängen in der Tiefe und den an der Erdoberfläche sichtbaren Bruchstrukturen, z. B. dem Hohenzollern-Graben, keine einfachen Beziehungen bestehen.

III. Diskussion der Ergebnisse

Für das gesamte Untersuchungsgebiet ergibt sich aus den bisher vorliegenden Herdflächen-Lösungen in bezug auf die seismotektonische Formung ein deutliches Vorherrschen der horizontalen Bewegungskomponente. Von den betrachteten 14 Erdbeben gehen 10, also mehr als zwei Drittel, auf seismische Schrägaufschiebungen bzw. Schrägabschiebungen mit vorherrschender horizontaler Bewegungskomponente oder auf reine Horizontalverschiebungen zurück.

Seismisch wirksame horizontale Schollenverschiebungen kommen sowohl im Oberrhein-Graben als auch an seinen Rändern vor. Sie weisen in der Grabenlängsrichtung eine linkshändige, quer zum Graben eine rechtshändige Bewegungstendenz auf. Die im geologischen Lagerungsbild dominierenden Abschiebungen mit vorwiegend vertikaler Bewegungsrichtung treten in der Seismotektonik weniger in Erscheinung. Offenbar neigt die auf Krustendehnung zurückzuführende Abschiebungstektonik mehr zu aseismischen Kriechbewegungen. Hierfür liefert der Vergleich der Mohr'schen Spannungsdiagramme für die Verwerfungstypen Abschiebung, Aufschiebung und Horizontalverschiebung einen Hinweis: Die absolute Größe der Hauptspannungen, die sich bis zum Erreichen der Grenzkurve ansammeln kann, ist bei der Abschiebung als Ausdruck der Dehnung des Materials wesentlich geringer als bei den anderen Verwerfungstypen (vgl. JAEGER & COOK 1969).

Besonders hervorzuheben ist, daß drei der im Grabenbereich ausgelösten Erdbeben (die Beben 3, 4 und 8) auf eine seismische Schrägaufschiebung zurückgehen, also auf eine zumindest örtliche Krusteneinengung. Teilschollen des Grabens werden offenbar gegenüber ihrer Umgebung herausgehoben. Dies scheinen auch geodätische Wiederholungsnivellements anzudeuten, welche im Raum Kandel—Karlsruhe—Bruchsal einen Aufstieg der Grabenregion gegenüber dem Nordschwarzwald erkennen lassen (KUNTZ, MÄLZER & SCHICK 1970). MÜLLER (1970) hat hierfür die Erklärung gegeben, daß bei einer Horizontalverschiebung längs der Grabenachse an den Knickpunkten des Grabens lokale Pressungserscheinungen auftreten können. Versuche von MURAWSKI (1969) zur Grabentektonik zeigen ebenfalls, daß Differentialbewegungen in Form von Hebungen innerhalb der Grabenscholle kinematisch möglich sind.

Die aus den seismischen Horizontalverschiebungen im Bereich des Oberrhein-Grabens und seiner Umgebung abzuleitenden Spannungsrichtungen ergeben ein sehr einheitliches Bild. Die größte Hauptspannung (Druckspannung) ist vorwiegend NW—SE orientiert, die kleinste Hauptspannung (Zugspannung) vorwiegend SW—NE. Diese Richtungsangaben gelten streng genommen nur für die Spannungsentlastung im jeweiligen Herdvolumen. Die Richtungsbeständigkeit bei den über das ganze Untersuchungsgebiet verteilten Beben ist aber kaum anders zu erklären als durch ein großräumiges, mehr oder weniger einheitlich ausgerichtetes Spannungsfeld innerhalb der Erdkruste von Südwestdeutschland, wie es schon früher von RANALLI & SCHEIDEGGER (1967), AHORNER (1967) und SCHNEIDER (1968) vermutet wurde. Auch im größten Teil der Alpen sowie im Mittel- und Niederrhein-Gebiet läßt sich aufgrund der seismischen Horizontalverschiebungen ein Spannungsfeld mit einer NW—SE orientierten horizontalen Hauptspannung nachweisen (AHORNER, MURAWSKI & SCHNEIDER 1972). Für den Oberrhein-Graben ist von Bedeutung, daß die größte horizontale Hauptspannung diagonal zur Grabenrichtung verläuft. Hieraus resultiert eine linkshändige Scherbeanspruchung, welche durch die Erdbeben-Herdmechanismen sowie durch geologische Beobachtungen an den Grabenschultern (MURAWSKI 1960, ILLIES 1965) nachzuweisen ist.

Bei den seismischen Vertikalverschiebungen, die im Grabenbereich in geringer Zahl festgestellt wurden, scheint die kleinste Hauptspannung im Herdvolumen NW—SE orientiert zu sein. Dies weist auf eine Dehnungsbeanspruchung quer zum Graben hin, was mit den geologischen Ansichten über die Grabenentstehung im Einklang steht. Gegenüber dem großräumigen Spannungsfeld, wie man es aufgrund der seismischen Horizontalverschiebungen ableitet, ergibt sich allerdings ein Widerspruch. Da die mitgeteilten Herdflächen-Lösungen für die Vertikalverschiebungen zumeist sehr unsicher sind, wird man genauer auswertbare künftige Ereignisse abwarten müssen, bevor man diesem Richtungsunterschied in der Spannungsentlastung, der vielleicht auf einer unterschiedlichen Stockwerksreaktion beruht, genauer nachgeht.

Allgemein läßt sich zu den Herdflächen-Lösungen sagen, daß der Zusammenhang zwischen den an der Erdoberfläche erkennbaren geologisch-tektonischen

Strukturen und den Bewegungsvorgängen im Erdbebenherd selten ganz einfach ist. Häufig entspricht die aus den Herdflächen-Lösungen resultierende Dislokationsform in ihrem Charakter nicht dem geologischen Bild. Hierfür können folgende Gründe maßgeblich sein:

a) Die geologischen Lagerungsformen sind zumeist das Resultat eines langdauernden Bewegungsprozesses, der sich über viele Jahrmillionen erstreckt. Demgegenüber umfassen die Beobachtungen über die seismotektonischen Bewegungsvorgänge insgesamt nur den Zeitraum von einigen Jahrzehnten. Das zur Verfügung stehende Zeitfenster ist also so klein, daß eine eindeutige Korrelation zwischen den geologischen Lagerungsformen und den seismotektonischen Bewegungsvorgängen kaum zu erwarten ist.

b) Die Dislokationstypen Abschiebung, Aufschiebung und Horizontalverschiebung zeigen eine unterschiedliche seismische Effektivität. Der Anteil der aseismischen Kriechbewegungen ist bei den auf Krustendehnung zurückgehenden Abschiebungen offensichtlich größer als bei den anderen Dislokationstypen. Dies führt dazu, daß die Abschiebungen im geologischen Lagerungsbild zwar eine große Rolle spielen, in der Seismotektonik aber nur wenig in Erscheinung treten, während es bei den Aufschiebungen und Horizontalverschiebungen umgekehrt ist.

c) Untersuchungen an Gesteinen unter hohen Drucken und Temperaturen haben ergeben, daß die Reaktion eines der tektonischen Belastung ausgesetzten Materials stark vom Umgebungsdruck und der Temperatur abhängt (z. B. GRIGGS & HANDIN 1960). Selbst wenn man voraussetzt, daß sich das tektonische Spannungsfeld innerhalb der Erdkruste regional wie zeitlich sehr konservativ verhält, muß der Verlauf der physikalischen Zustandsgrößen und der Materialeigenschaften mit der Tiefe einen wesentlichen Einfluß auf die Reaktion des Materials ausüben und eine Stockwerkstektonik bewirken, welche den Vergleich der seismischen Dislokationen in der Tiefe mit geologischen Bewegungsabläufen an oder in der Nähe der Erdoberfläche erschwert.

Dank: Die Deutsche Forschungsgemeinschaft unterstützte das Forschungsvorhaben durch eine finanzielle Beihilfe.

Literatur

AHORNER, L., 1962: Das Erdbeben im Saar-Nahe-Becken vom 17. August 1960. — Sonderveröff. Geol. Inst. Köln, 7, 24 S.
— 1967: Herdmechanismen rheinischer Erdbeben und der seismotektonische Beanspruchungsplan im nordwestlichen Mittel-Europa. — Sonderveröff. Geol. Inst. Köln, 13, 109—130.
— 1970: Seismo-tectonic Relations between the Graben Zones of the Upper and Lower Rhine Valley. — In: ILLIES, H. & MÜLLER, ST. (Hrsgb.): Graben Problems, 155—166, Schweizerbart, Stuttgart.
— 1972: Erdbebenchronik für die Rheinlande 1964—70. — Decheniana, 125, 259—283.
AHORNER, L., MURAWSKI, H. & SCHNEIDER, G., 1972: Seismotektonische Traverse von der Nordsee bis zum Apennin. — Geol. Rdsch., 61, 915—942.

BARTZ, J., 1967: Recent movements in the Upper Rhinegraben between Rastatt and Mannheim. — In: ROTHÉ, J. P. & SAUER, K. (Hrsgb.): The Rhinegraben progress report 1967. Abh. geol. L.-Amt Baden-Württ., 6, 1—2.
BREYER, F. & DOHR, G., 1967: Bemerkungen zur Stratigraphie und Tektonik des Rheintalgrabens zwischen Karlsruhe und Offenburg. — In: ROTHÉ, J. P. & SAUER, K. (Hrsgb.): The Rhinegraben progress report 1967. Abh. geol. L.-Amt Baden-Württ., 6, 42—43.
CARLÉ, W., 1955: Bau und Entwicklung der Südwestdeutschen Großscholle. — Beih. Geol. Jb., 16, 272 S., Hannover.
GEYER, O. F. & GWINNER, M. P., 1968: Einführung in die Geologie von Baden-Württemberg. 2. Aufl. — 228 S., Schweizerbart, Stuttgart.
GRIGGS, D. & HANDIN, J., 1960: Rock deformation. — 382 S., Geol. Soc. Amer., Memoir, 79, Baltimore.
HILLER, W., 1934: Der Herd des Rastatter Bebens am 8. Februar 1933. — Gerl. Beitr. z. Geophys., 41, 170—180.
— 1936: Das Hornisgrinde-Beben. — Seism. Ber. württembg. Erdbebend. Jg. 1935, Stuttgart 1936, Anhang, 10—23.
ILLIES, H. 1965: Bauplan und Baugeschichte des Oberrheingrabens. — Oberrh. geol. Abh., 14, 1—54.
JAEGER, J. C. & COOK, N. G. W., 1969: Fundamentals of rock mechanics. — 513 S., Methuan, London.
KUNTZ, E., MÄLZER, H. & SCHICK, R., 1970: Relative Krustenbewegungen im Bereich des Oberrheingrabens. — In: ILLIES, H. & MÜLLER, ST. (Hrsgb.): Graben Problems, 170—177, Schweizerbart, Stuttgart.
LACOSTE, J., 1934: Les tremblements de terre en France 1933. — Ann. Inst. Phys. du Globe, Univ. de Strasbourg, 2ième partie, Séismologie, 8—10.
METZ, R. & REIN, G., 1958: Erläuterungen zur Geologisch-petrographischen Übersichtskarte des Südschwarzwaldes. — 134 S., Schauenburg, Lahr.
MÜLLER, ST., 1970: Geophysical Aspects of Graben Formation in Continental Rift Systems. — In: ILLIES, H. & MÜLLER, ST. (Hrsgb.): Graben Problems, 27—37, Schweizerbart, Stuttgart.
MURAWSKI, H., 1960: Das Zeitproblem bei der Tektogenese eines Großgrabensystems. — Notizbl. Hess. Landesamt f. Bodenforschg., 88, 294—342.
— 1969: Bruchtektonik mit modifizierter Bruchbildung. — Geol. Rdsch., 59, 193—212.
RANALLI, G. & SCHEIDEGGER, A. E., 1967: Tectonic stress field in Central Europe. — Z. Geophys., 20, 193—201.
RITSEMA, A. R., 1955: The fault-plane technique and the mechanism in the focus of the Hindu Kush earthquakes. — Indian J. Met. and Geophys., 6, 1—10.
ROTHÉ, J. P. & DECHEVOY, N., 1967: La séismicité de la France de 1951 à 1960. — Ann. Inst. Phys. du Globe, Univ. de Strasbourg, Nouv. Série, 3ième partie, Geophysique, 8, 19—95.
SCHICK, R., 1970: A method for determining source parameters of small magnitude earthquakes. Z. Geophys., 36, 205—224.
— 1972: Erdbeben als Ausdruck spontaner Tektonik. — Geol. Rdsch., 61, 896—914.
SIMLER, L. & MASCLAUX, R., 1967: Cartes des profondeurs de deux horizons sedimentaires dans la plaine d'Alsace. — In: ROTHÉ, J. P. & SAUER, K. (Hrsgb.): The Rhinegraben progress report 1967. Abh. geol. L.-Amt Baden-Württ., 6, S. 68.
SCHNEIDER, G., 1968: Erdbeben und Tektonik in Südwestdeutschland. — Tectonophysics, 5, 459—511.
— 1971: Seismizität und Seismotektonik der Schwäbischen Alb. — 79 S., Enke, Stuttgart.
SCHNEIDER, G., SCHICK, R. & BERCKHEMER, H., 1966: Fault-plane solutions of earthquakes in Baden-Württemberg. — Z. Geophys., 32, 383—393.
WECHSLER, H., 1955: Die Erdbebentätigkeit in Südwestdeutschland in den Jahren 1938—1954. — Geol. Diplomarbeit, TH Stuttgart, 110 S.

In situ Stress Measurements in Southwest Germany — First Results

by

Gerhard Greiner

With 1 textfigure

In summer 1972 some experimental work dealing with in situ stress determination in rock had been done in the area of the "Schwäbische Alb"-mountains. These measurements were carried out for research in rock mechanics. Some results are interesting enough to be presented within this book.

Position of measurements

The coordinates of the points of measurements are 9° 01' E / 48° 17' N and 9° 05' E / 48° 10' 27" N located in the region near the town of Ebingen, about 65 km SSW of Stuttgart. This area is known to have the highest rate of seismicity in SW-Germany (SCHNEIDER 1971), a factor that encouraged experiments in stress determination there.

The points of measurements were carefully examined for not having some topographical stress concentration effects. Hence the investigations were carried out in a region where there was a lack of important relief energy.

The experiments occurred in two quarries of Upper Jurassic limestones. This type of rock has been found to have the necessary elastic properties when it was tested in situ and in laboratory.

Method used

At the test sites the "doorstopper" borehole strain gauge device, developed in South Africa by LEEMAN (1969), was used to measure absolute stresses in the interior of rock masses.

The principle of the method is to drill a borehole, smooth the bottom with diamond bits until it is quite flat, glue a resistance strain gauge rosette orientated at the center of the bottom and take initial readings. During overcoring a relief in stress at the location where the strain gauge is glued occurs. This de-stressing of the rock indicates a specific strain within the rock core which is determined by the resistance strain gauge after having recovered it from the borehole. The stresses present on the end of the borehole before overcoring may be calculated from the strain gauge readings together with the already determined modulus of elasticity and Poisson's ratio. Very important in this context is the consideration of stress concentration factors that compare the regional stresses to those acting at the end of the borehole. VAN HEERDEN (1969) furnished some good solutions of the latter problem.

The state of stress at a point is defined by nine components of stress which constitute a second order tensor, known as the "stress tensor". It has been shown that the shear stresses are not all independent. Thus, the state of stress at a point is completely specified by six independent components (JAEGER & COOK 1969). In cases where there are no informations about the configurations of the stress field, only measurements in three boreholes pointing in different known directions can give the necessary six independent equations for determining the six unknown values.

Results (Fig. 1)

The measurements have been carried out in six boreholes, three at each point of the experiments. One of these points was situated in the "Hohenzollerngraben" which is a graben structure that runs in a ESE to WNW direction a distance of about 30 km. The other measurements have been taken at a place about 10 km from this tectonic structure (see Fig. 1).

The values of the rock's modulus of elasticity measured at these locations range between $2{,}56 \cdot 10^6$ and $5{,}50 \cdot 10^6$ psi as determined in the borehole by the NX borehole plate bearing device of GOODMAN et al. (1969). Poisson's ratio of the rock was shown to be near 0.20, calculated from static compressional tests.

Within the series of investigations it was possible to get 60 simultaneous independent values of strain.

The stresses calculated of these values show some characteristic features:

None of the absolute values of the three principal stresses are equal. This creates an anisotropy of lateral stresses acting in this region. The most interesting result is that the axis of maximum compression points in a direction of nearly 50° NW within the graben and 28° NW out of the graben.

This is in good agreement with the picture of the seismo-tectonic stress field of this area, induced by the Alpine orogenic belt as first presented by RITSEMA (1970).

The author's results also agree with those of SCHNEIDER (1971) who used fault-plane solutions of shallow earthquakes to determine the direction of maximum compressive stress.

Recently ILLIES (1972) discussed the interaction between the alpine orogenic belt and the Rhine Graben Rift System, revealing a stress pattern similar to the one observed by the author.

The maximum principal stresses are directed with a slight inclination to the surface. Their absolute values (234 psi and 271 psi) are, naturally, not as high as those found by other authors from basement-measurements in Scandinavia (HAST 1969), Canada (EISBACHER & BIELENSTEIN 1971), the USA (NORMAN 1970) or South Africa (CAHNBLEY 1970). However, it must be considered that our measurements took place in soft and sometimes fractured sedimentary rocks near the surface. For these circumstances the results are satisfactory (see RALEIGH et al. 1972).

Nevertheless, the absolute values of lateral stresses are more than three times higher than was expected, considering the weight of the overburden. The large anisotropy of lateral stresses creates high shear stresses in the earth's crust which

120 Gerhard Greiner

are important if stresses in greater depths are extrapolated from near-surface measurements.

The deductions dealing with the actual distribution of stresses in correspondence to the tectonic pattern of this region will be discussed later.

This paper was prepared in the "Sonderforschungsbereich 77 (Felsmechanik)" at the University of Karlsruhe. The measurements in the field have been done in co-operation with Interfels GmbH, Salzburg—Bentheim.

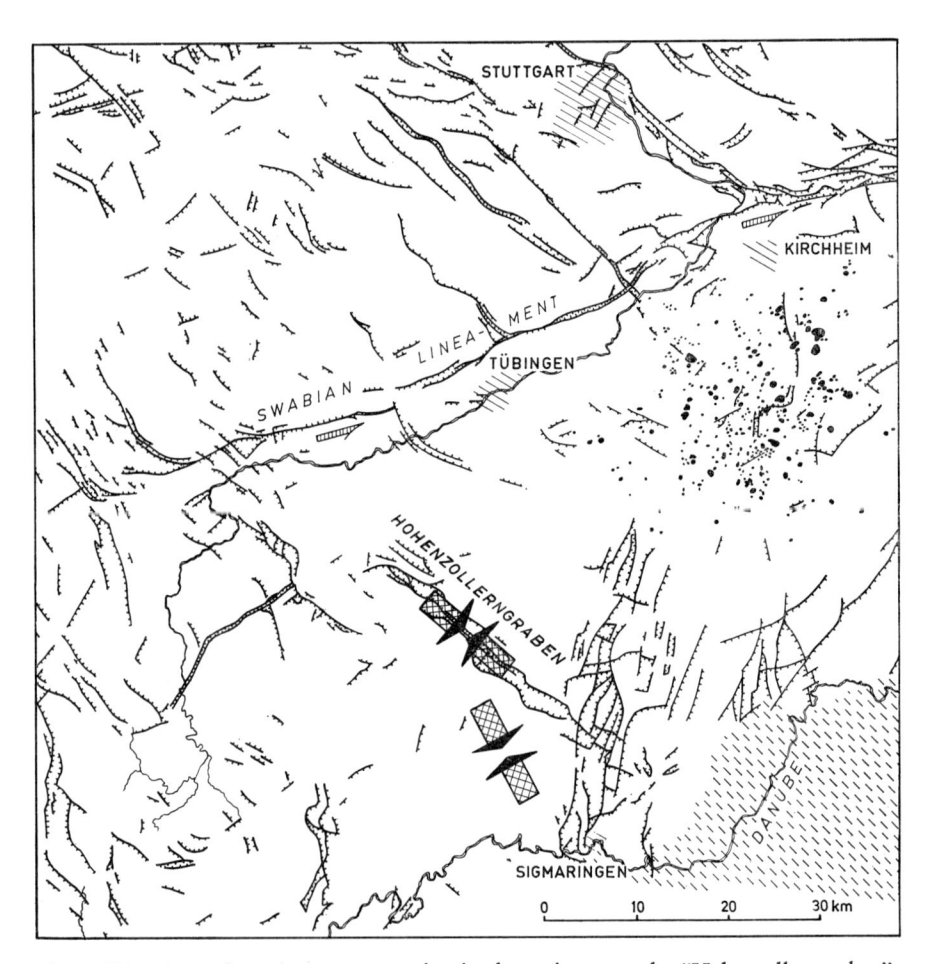

Fig. 1. Directions of maximum compression in the region near the "Hohenzollerngraben". Compare the agreement of the directions of principal compressive stresses to those estimated by other authors (see the text). Furthermore it is important to recognize that there is no great difference in stress directions within the tectonic structure and those measured outside of it. (The results are inserted in a tectonic map, kindly furnished by ILLIES, see p. 445 of this book).

References

CAHNBLEY, H., 1970: Grundlagenuntersuchungen über das Entspannungsbohrverfahren während des praktischen Einsatzes in großer Teufe. — Diss. TU Clausthal, 166 pp.

EISBACHER, G. H. & BIELENSTEIN, H. U., 1971: Elastic strain recovery in Proterozoic rocks near Elliot Lake, Ontario. — J. Geophys. Res., 76, 2012—2021.

GOODMAN, R. E. et al., 1968: The measurement of rock deformability in bore holes. — Paper presented at the 10th Symposium on Rock Mechanics, 1968, Univ. of Texas.

HAST, N., 1969: The state of stress in the upper part of the earth's crust. — Tectonophysics, 8, 169—211.

HEERDEN, W. L. VAN, 1969: Stress concentration factors for the flat borehole end for use in rock stress measurements. — Eng. Geol., 3, 307—323.

ILLIES, J. H., 1972: The Rhine Graben Rift System — plate tectonics and transform faulting. — Geophys. Surv., 1, 27—60.

JAEGER, J. C. & COOK, N. G. W., 1969: Fundamentals of rock mechanics. — 513 pp., Methuen & Co. Ltd., London.

LEEMAN, E. R., 1969: The "doorstopper" and triaxial rock stress measuring instruments developed by the C.S.I.R. — J. S. Afr. Inst. Min. Metall., 69, 305—339.

NORMAN, C. E., 1970: Geometric relationship between geologic structure and ground stresses near Atlanta, Ga. — US Bur. Mines Rept. of Invest., 7365, 23 pp.

RALEIGH, C. B. et al., 1972: Faulting and crustal stress in Rangely, Colorado. — In: HEARD, H. C. et al. (eds.), Flow and fracture of rocks, 275—284, geophys. monograph 16, Amer. Geophys. Union, Washington.

RITSEMA, A. R., 1970: Seismo-tectonic implications of a review of European Earthquake Mechanisms. — Geol. Rdsch., 59, 36—56.

SCHNEIDER, G., 1971: Seismizität und Seismotektonik der Schwäbischen Alb. — 79 pp., Enke, Stuttgart.

The Rhinegraben rift system: deep structure and tectonic features

The 1972 Seismic Refraction Experiment in the Rhinegraben — First Results [1,2,3]

by

Rhinegraben Research Group for Explosion Seismology [4]

With 10 figures and 2 tables in the text and on 1 folder

Abstract: Continuing a detailed study of the structure of the crust and uppermost mantle of the Rhinegraben, a specially designed seismic-refraction investigation was carried out in the southern part of the Rhinegraben area during September 1972. The main reversed profile is located in the Rhinegraben proper. This paper describes the interpretation of the three main phases: P_g — refracted arrivals from the Variscan basement; P_n — refracted arrivals from the Mohorovičić discontinuity; and P_MP — reflected phases from the crust-mantle boundary. Sedimentary corrections were applied. Contour maps showing the depth of the crust-mantle boundary are presented; these are based on the interpretation of both the P_MP and the P_n data. In the southern part of the Rhinegraben the crust-mantle boundary uplifts to a depth of 24—25 km. The interface dips rather sharply southwards to the south of Mulhouse, and northwestwards to the northwest of the Vosges mountains. The dip is less towards the east. From the reversed P_n data a true velocity close to 8.1 km/s is derived for the uppermost mantle. The apparent velocities of the P_n traveltime curves on other profiles can be explained in terms of the topography of the Mohorovičić discontinuity.

---

[1] Contribution I.P.G.-N.S. No. 78 of the Institut de Physique du Globe, Université de Paris VI.
[2] Contribution No. 163 within a joint research program of the Geophysical Institutes in Germany sponsored by the Deutsche Forschungsgemeinschaft (German Research Association). — Contribution No. 98 Geophysical Institute, University Karlsruhe.
[3] Contribution No. 81, Institute of Geophysics, Swiss Federal Institute of Technology, Zürich.
[4] J. Ansorge[5], B. Edel[6,7], D. Emter[8], K. Fuchs[6], C. Gelbke[6], A. Hirn[9], St. Mueller[5], E. Peterschmitt[7], C. Prodehl[6], L. Steinmetz[9], P. Streicher[7].
[5] Institut für Geophysik, Eidgenössische Technische Hochschule, Postfach 266, CH-8049 Zürich, Switzerland.
[6] Geophysikalisches Institut, Universität, Hertzstr. 16, D-75 Karlsruhe 21, Germany.
[7] Institut de Physique du Globe, Université, 5 Rue René Descartes, F-67 Strasbourg, France.
[8] Institut für Geophysik, Universität, Richard-Wagner-Str. 44, D-7 Stuttgart-O, Germany.
[9] Institut de Physique du Globe, Université-VI, 11 quai St.-Bernard, F-75 Paris-V, France.

Zusammenfassung: In Fortsetzung der intensiven Erforschung von Erdkruste und oberem Mantel unter dem Rheingraben wurde im September 1972 im südlichen Teil des Rheingrabens eine refraktionsseismische Spezialuntersuchung ausgeführt. Das gegengeschossene Hauptprofil liegt im Gebiet des eigentlichen Grabens. Die gegenwärtige Untersuchung beschreibt die Auswertung der drei Hauptphasen: P_g — die an der Oberfläche des variskischen Kristallins geführte Welle; P_n — die an der MOHOROVIČIĆ-Diskontinuität geführte Welle; und P_MP — die von der Krusten-Mantel-Grenze reflektierte Welle. Zu Beginn der Auswertung wurden Sedimentkorrekturen angebracht. Es werden Tiefenlinienpläne von der Krusten-Mantel-Grenze vorgestellt, die auf der Interpretation sowohl der P_MP- als auch der P_n-Einsätze beruhen. Im südlichen Teil des Oberrheingrabens wölbt sich die Krusten-Mantel-Grenze bis zu Tiefen von 24—25 km auf. Die Grenzfläche fällt im Bereich südlich von Mülhausen nach Süden und nordwestlich der Vogesen nach Nordwesten stark ein, während das Abtauchen nach Osten nicht so kräftig ausgeprägt ist. Aus den gegengeschossenen P_n-Daten auf dem Hauptprofil folgt für den obersten Mantel eine wahre Geschwindigkeit von etwa 8,1 km/s. Die Scheingeschwindigkeiten der P_n-Laufzeitkurven auf anderen Profilen lassen sich durch die Topographie der MOHOROVIČIĆ-Diskontinuität erklären.

Résumé: Faisant suite a l'étude intensive de la croûte terrestre et du manteau supérieur sous le Fossé Rhénan, une campagne de sismique réfraction a été réalisée en Septembre 1972 dans la partie Sud du Fossé. Le profil principal, inversé, suit l'axe du Fossé. La présente étude décrit l'interprétation des trois phases principales: P_g — onde propagée dans la partie superficielle du socle cristallin; P_n — qui suit la discontinuité de MOHOROVIČIĆ; P_MP — onde réflechie par la limite croûte-manteau. L'influence des sédiments sur les temps de propagation a été corrigée. Des cartes d'égales profondeurs de la discontinuité croûte-manteau sont présentées. Elles reposent à la fois sur l'interprétation des arrivées des ondes P_n et des ondes P_MP. Dans la partie Sud de la zone étudiée, la profondeur est minimale entre 24—25 km. La discontinuité plonge rapidement vers le Sud, au Sud de Mulhouse, et vers le Nord-Ouest, au Nord des Vosges. Vers l'Est, la pente est plus faible. L'étude des ondes P_n sur le profil principal indique une vitesse vraie d'environ 8,1 km/s dans le manteau supérieur. Les vitesses apparentes des ondes P_n sur les autres profils peuvent s'expliquer par la topographie de la discontinuité de MOHOROVIČIĆ.

Introduction

Since 1966, the structure of the crust and uppermost mantle in the area of the Rhinegraben has been investigated in detail by explosion seismology (see, e. g., ANSORGE et al. 1970; MEISSNER et al. 1970). The results obtained prior to April 1972 for the central and southern Rhinegraben are summarized by ANSORGE et al. (1974). During the period of 19th—28th September 1972 a specially designed seismic-refraction experiment was carried out in the southern part of the Rhinegraben. This investigation was a joint experiment of French, German, and Swiss Geophysical institutes.

The experiment had four different aims:

1) Investigation of the crustal and upper-mantle structure in the area of the Rhinegraben,
2) calibration of earthquake stations in the area of the Rhinegraben, and to record,
3) a long-range profile in the direction of Böhmischbruck in the Bohemian massif, and
4) two profiles in Switzerland.

This paper presents the data relevant to the first aim, i. e., the investigation of crustal and upper-mantle structure in the area of the Rhinegraben. The main phases P_g, P_MP, and P_n have been used in the inversion process; more detailed results will be presented in a future publication.

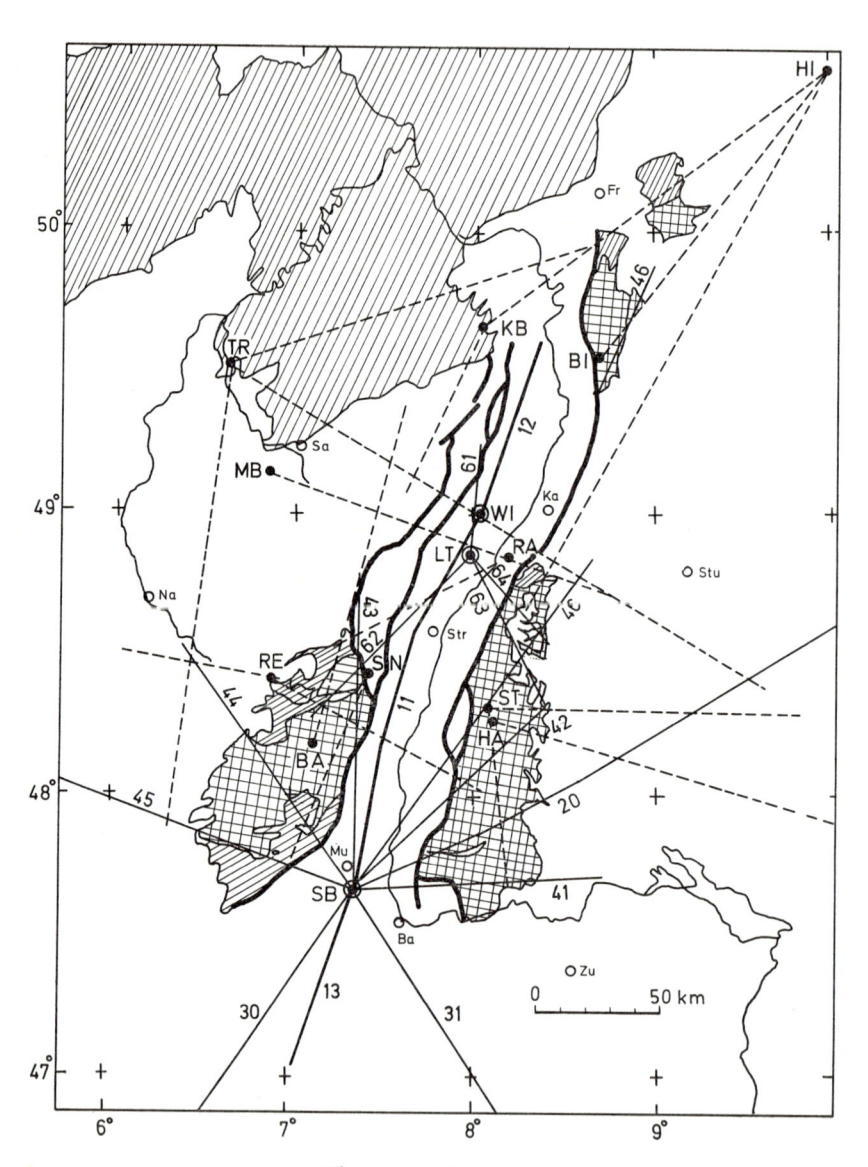

Fig. 1. Location map.

Description of the experiment

From borehole shots, ranging in size from 400 kg to 2800 kg TNT, at shotpoints near Steinbrunn (south of Mulhouse), Wissembourg, and Leutenheim (south of Wissembourg), 15 profiles with maximum distances ranging from 70 to 400 km were recorded (Fig. 1). The main profile in the Rhinegraben proper was reversed. Each shot was recorded by 80 similar portable seismic-refraction stations which recorded on magnetic tape the frequency modulated and multiplexed output (BERCKHEMER 1970) from 2 Hz three-component geophones. The time signal was transmitted by station HBG (75 kHz) near Geneva, Switzerland. 30 stations of the "Institut national d'Astronomie et de Géophysique" were operated by the "Instituts de Physique du Globe" of the Universities of Paris and Strasbourg, 40 stations by the German Geophysical Institutes of the Universities of Clausthal, Frankfurt, Göttingen, Hamburg, Karlsruhe, Kiel, München, and Stuttgart, and by the Bundesanstalt für Bodenforschung, Hannover, and 10 stations were operated by the Institute of Geophysics of the Swiss Federal Institute of Technology, Zürich. Headquarters at Strasbourg, Schiltach (Black Forest), and Zürich provided the necessary communication with shotpoints and stations, and checked continuously the quality of the recordings and instruments by daily data collection and playback throughout the whole experiment.

All profiles recorded in September 1972 are shown by solid lines in Fig. 1 (see also Tab. 1).

The main profile (11, 12, 13) is located in the Rhinegraben proper, parallel to the graben axis, and was recorded as a reversed profile (line 11 only). At each end the profile was extended beyond the reversing shotpoint, up to a maximum distance of 190 km towards the north (11 and 12), and up to 215 km towards the south (11 and 13). On profile 12 the shots from Wissembourg were also recorded. No recordings were made on profile 13 during the shots at Steinbrunn. The station

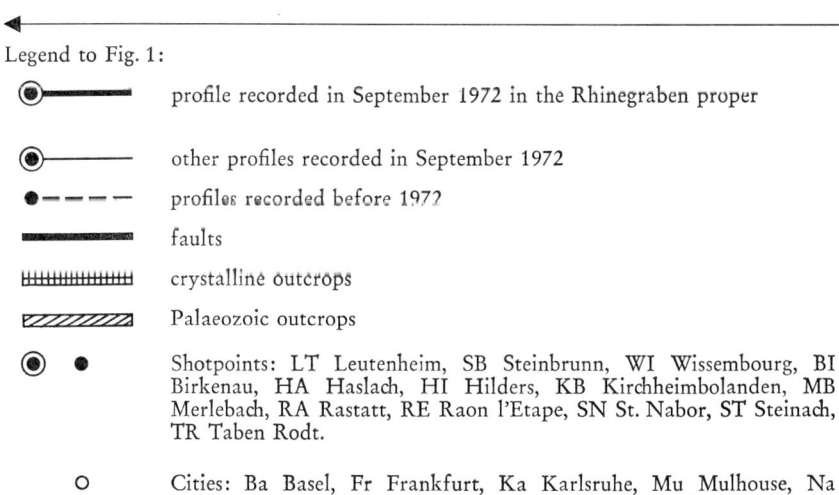

◀

Legend to Fig. 1:

⦿━━━━━ profile recorded in September 1972 in the Rhinegraben proper

⦿━━━━   other profiles recorded in September 1972

●– – – – profiles recorded before 1972

▬▬▬▬▬   faults

⊞⊞⊞⊞⊞⊞ crystalline outcrops

▨▨▨▨▨   Palaeozoic outcrops

⦿ ●   Shotpoints: LT Leutenheim, SB Steinbrunn, WI Wissembourg, BI Birkenau, HA Haslach, HI Hilders, KB Kirchheimbolanden, MB Merlebach, RA Rastatt, RE Raon l'Etape, SN St. Nabor, ST Steinach, TR Taben Rodt.

○   Cities: Ba Basel, Fr Frankfurt, Ka Karlsruhe, Mu Mulhouse, Na Nancy, Sa Saarbrücken, Str Strasbourg, Stu Stuttgart, Zu Zürich.

Tab. 1. List of shotpoints in September 1972.

N°	Date	Shot time h mn s	Shotpoint	φ ° N	λ ° E	Charge T
01	19	18 01 01,630	Steinbrunn	47,67321	7,33171	0,4
02	20	19 01 01,459	Steinbrunn	47,67351	7,33171	0,5
04	21	19 01 01,407	Steinbrunn	47,67263	7,33171	1,0
05	23	18 00 59,422	Wissembourg	49,01115	8,03778	0,5
06	24	09 01 01,183	Steinbrunn	47,67382	7,33218	2,8
10	25	18 01 01,365	Wissembourg	49,01070	8,03731	2,0
07	26	19 01 01,338	Leutenheim	48,86316	7,98266	0,35
11	27	19 01 01,471	Steinbrunn	47,67224	7,33140	0,25
12	28	19 01 01,100	Steinbrunn	47,67423	7,33134	0,9

interval varies; on the profile from Steinbrunn it averages 2 km and on the profile from Wissembourg it is 3 km up to 140 km distance, then 6 km to the end of the profile.

The shotpoint at Steinbrunn is located within Tertiary limestones on the horst of Mulhouse. The drill holes in which the shots were fired, did not reach the water table. On the other hand, the shotpoint at Wissembourg was located within Quaternary sands beneath the water table, and the efficiency of these shots was much higher than that of the shots at Steinbrunn. It should be pointed out that the profile between Wissembourg and Steinbrunn is not only the first reversed profile in the area, but also that it follows the strike of the Rhinegraben and reaches within the graben a recording length (200 km) not attained on previous profiles.

The shotpoints Steinbrunn and Leutenheim are both located in areas of local earthquake epicentres (see, e. g., AHORNER et al. 1972; BONJER & FUCHS 1974). The profiles 41, 42, 43, 44, 45, and 46 from Steinbrunn, and the profiles 61, 62, 63, 64, 65 from Leutenheim, head towards existing or planned earthquake stations in the area of the Rhinegraben. On most profiles recordings were obtained up to sufficient distances to give information on both crustal and upper mantle structure.

Profile 20 crosses the Black Forest and the Suebian and Franconian Jura and heads towards Böhmischbruck in the Bohemian massif at a distance of 420 km, thus reversing a long-range profile from Böhmischbruck towards Basel (EMTER 1971). Adequate energy was recorded out to a distance of 350 km.

The profiles 30 and 31 reverse existing profiles in Switzerland: Profile 30 was recorded parallel to the axis of the Swiss Jura mountains towards Lyon and reverses profiles towards Basel from shotpoints in the Rhônegraben. Profile 31 crosses the Swiss Molasse basin and heads towards Lago Bianco (Gotthard massif) within the western Alps and reverses a profile from Lago Bianco towards the Vosges mountains (see, e. g., BEAUFILS 1967).

The earlier profiles, shown as dashed lines in Fig. 1, are presented in detail by ANSORGE et al. (1970, 1974), MEISSNER et al. (1970). The corresponding shotpoint data are published in part by STEIN & SCHRÖDER (1974). Profiles from the shotpoints shown in Tab. 2 are included in the present interpretation.

Tab. 2. List of shotpoints from which profiles were recorded before 1972

Shotpoint	longitude	latitude
Raon l'Etape	6°.8712	48°.4058
Col des Bagenelles	7°.1105	48°.1976
St. Nabor	7°.4123	48°.4438
Steinach	8°.0606	48°.2921
Haslach	8°.1178	48°.2662
Merlebach	6°.7988	49°.1627
Taben Rodt	6°.6115	49°.5493
Kirchheimbolanden	7°.9598	49°.6751
Birkenau	8°.717	49°.550
Büdingen	9°.183	50°.300
Hilders	10°.033	50°.533

Correlation and identification of phases

All seismograms obtained during the 1972 experiment were digitized at a rate of 200 samples/second at the Institut de Physique du Globe of the University of Paris. Except for some seismograms showing noise with frequencies below 20 Hz, the digital data were then filtered with a 3—20 Hz bandpass filter. Finally record sections were plotted for the vertical components, using different reduction velocities: 6 km/s (see, e. g., Figs. 2, 3, 5, and 6) and 8 km/s (see, e. g., Fig. 4).

Figs. 2 and 3 show the record sections of the reversed profile between Wissembourg and Steinbrunn. A correlation of the most conspicuous phases is shown. In the first arrivals, the phases P_g and P_n are correlated; in later arrivals, the main phase P_MP can be defined on almost all profiles. Apart from these three main phases, of which the inversion into depth-velocity structure will be discussed in this paper, the correlation of some additional clear phases is also shown in Figs. 2 and 3.

P_g can be traced on both profiles to a distance of 60—70 km. This wave is interpreted as a refracted wave from the Variscan basement. The arrivals cannot be combined by a straight traveltime curve; rather their deviations from a straight line indicate that the Variscan basement is not a plane surface and that the different sedimentary layers may vary in thickness and composition. For example, on the profile from Steinbrunn (Fig. 3) the P_g arrivals are relatively early at a distance of about 40 km which is located within the Potash basin of the Upper Alsace and can be correlated with the position of a salt dome. The relatively large reduced traveltimes of 0.9—1.5 sec on the profile from Steinbrunn and 1.0—1.6 sec on the profile from Wissembourg reflect the great thickness of the sediments within the graben area proper and clearly indicate that the thickness varies between the southern and the central part of the Rhinegraben.

Due to sufficiently large recording distances, P_n arrivals can be well correlated on both profiles. In order to facilitate the correlation, record sections with a reduction velocity of 8 km/s were also plotted. An example is shown in Fig. 4. The P_n phases are interpreted as a refracted wave travelling within the uppermost mantle

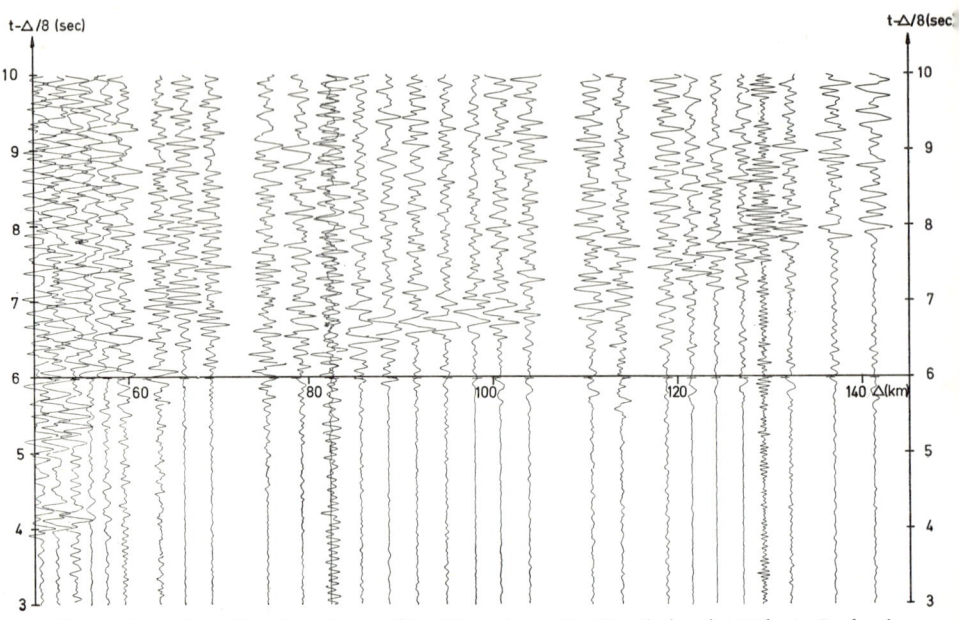

Fig. 4. Record section for the profile Wissembourg-S (50 < Δ < 145 km). Reduction velocity 8 km/s.

beneath the MOHOROVIČIĆ-discontinuity. On the profile from Wissembourg (Figs. 2 and 4) these arrivals are visible from a distance of about 80 km onwards and have good energy out to a distance of 140 km. At greater distances, the energy becomes weaker, but the arrivals can still be correlated. Although on the reversed profile from Steinbrunn (Fig. 3) the signal-to-noise ratio is poor, the P_n wave can clearly be recognized between 105 and 120 km distance and can be followed out to the end of the profile. On the profile from Wissembourg, at some locations within the distance ranges over which the P_n arrivals can be clearly defined, the scatter of the arrivals is quite considerable and may be due to the presence of salt domes or some other unknown cause. On the profile from Steinbrunn no scatter of the arrivals is observed and they can be well correlated with a straight traveltime curve.

For the P_n arrivals the two profiles reverse each other (between Sélestat and Strasbourg) over a certain segment of the crust-mantle boundary within which the corresponding phases are observed in both directions. Within this area apparent velocities are measured as follows: 8.12 km/s from Wissembourg and 8.06 km/s from Steinbrunn. This indicates that a true velocity close to 8.1 km/s satisfies the P_n traveltimes. On other profiles (20, 42, 45, 46, see also Fig. 6) the correlation of the P_n arrivals gives values which can be interpreted as apparent velocities corresponding to a true velocity of 8.1 km/s (see also Figs. 8 and 9).

On almost all profiles, in later arrivals a clear wave group can be correlated which is termed $P_M P$ and which is interpreted as a reflected wave from the crust-mantle boundary (MOHOROVIČIĆ-discontinuity). On the profile from Wissembourg (Fig. 2), in contrast to most other profiles, this phase $P_M P$ is well pronounced along only a short distance range between 90 and 110 km. On the reversed profile from Steinbrunn (Fig. 3) it can be clearly distinguished from 60 to 120 km distance. Whilst the $P_M P$ phase is relatively weakly developed within the graben proper, it is very well pronounced on the profiles which cross the flanks of the Rhinegraben. Figs. 5 and 6 present two examples showing parts of the record sections of the profiles 45 (distance range 85—140 km) and 46 (distance range 70—125 km).

On the profiles in the area of the Rhinegraben the critical distance is about 55—60 km. In general the $P_M P$ traveltime curve is tangential to the prolongation of the P_n traveltime curve towards smaller distances. The profile from Wissembourg to the south is an exception; here, in the distance range of 60—80 km, higher apparent velocities must be assumed locally for the P_n phase in order for it to be tangential to the observed $P_M P$ phase. This indicates that the structure of the crust changes considerably within this area.

Apart from the phases P_g, P_n, and $P_M P$, some other phases are correlated in the record sections shown in Fig. 2 and 3; these will not be discussed in detail. On the profile from Wissembourg (profile 11, Fig. 2) several phases appear as later arrivals within the distance range from 0 to 60 km and are most likely to be interpreted as multiples of the P_g phase. This explanation is supported by record sections showing seismograms subjected to a digital polarization filter with the aid of the horizontal components.

In addition to the $P_M P$ phase, a second phase, here termed $P_b P$, can be recognized as a later arrival beyond a distance of 60 km and this is interpreted as a reflection from a boundary within the earth's crust. This interface lies at a depth of approximately 19 km, and the mean velocity between this boundary and the MOHOROVIČIĆ-discontinuity is about 6.6—7.0 km/s.

Results of depth calculations

In this section the depth of the crust-mantle boundary is derived using all the $P_M P$ and P_n data available on profiles within the area of investigation shown in Fig. 1. A detailed interpretation, and depth calculation, using the $P_M P$ data is complicated by the fact that some recording sites are located on the sediments in the graben proper, the thickness of which may reach up to 4 km, and some are located on the flanks of the graben on basement rocks or on a relatively thin sedimentary cover. Difficulties also arise for profiles which originate north of the Vosges mountains and the Black Forest and cross other areas with sedimentary cover, e. g., the profiles from Taben Rodt (Fig. 1).

An initial, and approximate, picture of variations in the crustal structure can be obtained by contouring the reduced traveltime of the $P_M P$ wave at a fixed distance of 60 km which is close to the critical distance in the area of investigation.

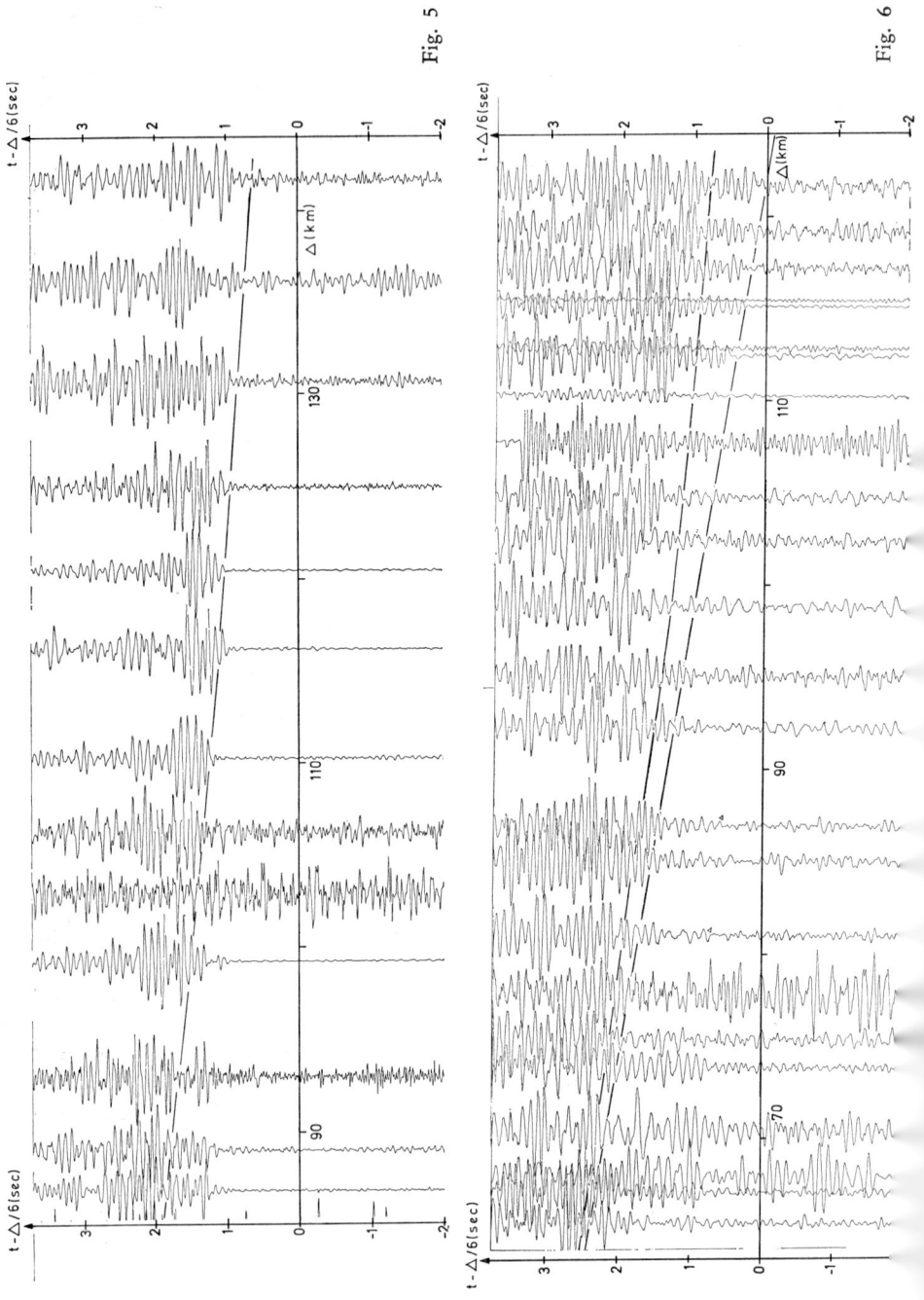

Fig. 5. Record section for the profile Steinbrunn-W (profile 45, $85 < \Delta < 145$ km) with correlation of P_MP. Reduction velocity 6 km/s.

Similar contour maps were presented for other areas by CHOUDHURY et al. (1971), GIESE & STEIN (1971), and PRODEHL (1970). The resulting variations are due either to mean velocity variations within the crust or to changes in crustal thickness. To eliminate traveltime delays caused by varying thicknesses of sediments and other near-surface influences, the P_MP traveltimes were corrected by subtracting the P_g traveltimes observed at the same distance. The resulting corrected traveltimes $t_{P_MP} - t_{P_g}$ at 60 km are contoured in Fig. 7. They correspond to the structure of the crust below the sedimentary layers which cover the Variscan basement in large parts of the area of investigation.

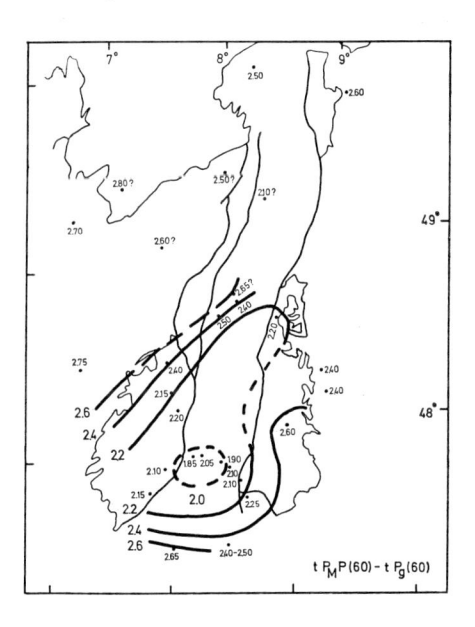

Fig. 7. Contour map of the traveltime difference $t_{P_MP} - t_{P_g}$ at $\Delta = 60$ km.

The contour lines are shown only for the southern part of the Rhinegraben, where a sufficient amount of data is available. The low values in the area of the graben between Strasbourg and Basel, indicate an apparent uplift of the crust-mantle boundary. Crustal thickness increases to the south, to the northwest, and to the east.

For a more detailed depth determination, the P_MP and the P_n data were used independently in the following manner.

In order to be able to interpret all P_MP data homogeneously all arrival times were reduced to a reference depth of 3 km below the earth's surface in the whole graben. The depth to the "base of the Tertiary" within the graben is reported in some detail by DOEBL & OLBRECHT (1974). The thicknesses and velocities of the sediments between the base of the Tertiary and the basement can only roughly be

Fig. 6. Record section for the profile Steinbrunn-NNE (profile 46, $65 < \Delta < 125$ km) with correlation of P_MP and P_n. Reduction velocity 6 km/s.

estimated (BOIGK & SCHÖNEICH 1974); in particular, the thickness of the Carboniferous sediments is unknown. Similar problems arise for the areas north of the Vosges mountains and the Black Forest; here knowledge of sedimentary thickness is even poorer.

Every station and every shotpoint was projected to the reference level along the ray path of the P_MP reflection, taking into account individual mean velocities for the Quaternary, Tertiary, Mesozoic, and Palaeozoic sediments and for the basement above the reference depth of 3 km. The ray geometry in the overburden was derived assuming that below the reference depth lies an upper crust of 16 km thickness with a velocity of 6.1 km/s and a lower crust of 6 km thickness with a velocity of 6.8 km/s. Thus for every shotpoint-station pair the new P_MP traveltime and distance at the reference level was derived.

Using these corrected values of distance and traveltime, the mean velocity and the average crustal thickness were determined for each profile by the T^2-X^2-method. The resulting mean velocities for the crust below the reference depth of 3 km vary between 6.1 and 6.4 km/s; on average a mean crustal velocity of 6.25—6.30 km/s results. Using 6.25 km/s and 6.30 km/s as mean crustal velocities, the corresponding depths to the crust-mantle boundary were calculated from the P_MP traveltimes. The resulting depth values, with the reference depth added, are shown in the contour maps in Fig. 8 (V = 6.25 km/s) and Fig. 9 (V = 6.30 km/s). All depths in these maps are measured from the earth's surface.

The P_n data provide additional information on depth, and local variations in depth, of the crust-mantle boundary. Figs. 8 and 9 show solid lines for those parts of the profiles along which P_n phases were recorded.

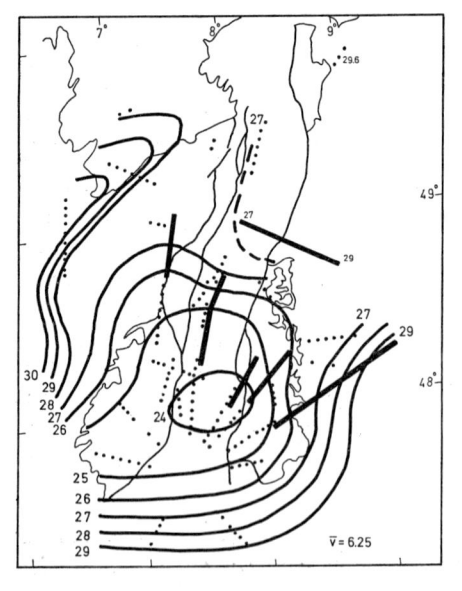

Fig. 8. Contour map of the depth to the crust-mantle boundary for the area of the southern Rhinegraben, using a mean crustal velocity of 6.25 km/s. Points show locations for which depths were calculated from the P_MP data. Solid lines show the distance range in which P_n phases were recorded.

As shown above, from the reversed profile Wissembourg—Steinbrunn a true velocity of 8.1 km/s is obtained for the uppermost mantle. Based on this value, the depth and inclination of the MOHOROVIČIĆ-discontinuity is estimated from the P_n data for each profile, using either 6.25 or 6.30 km/s as a mean crustal velocity below the reference depth of 3 km and estimating a mean velocity for the uppermost 3 km from the available data on the sediments. The resulting depths are included in Fig. 8 and 9.

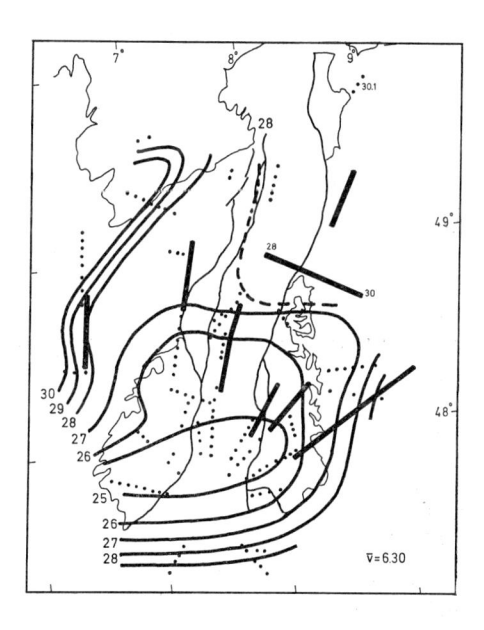

Fig. 9. Contour map of the depth to the crust mantle boundary for the area of the southern Rhinegraben, using a mean crustal velocity of 6.30 km/s. Further explanations see Fig. 8.

The differences in depth between the two resulting contour maps are in general not greater than 1 km. As could be suggested from the contour map $tp_MP - tp_g$ in Fig. 7, a rise of the crust-mantle boundary to a minimum depth of 24 km is found in the area of the Vosges mountains — southern Rhinegraben — Black Forest. The maximum elevation of this discontinuity is reached near Mulhouse.

The boundary dips relatively abruptly towards the south and reaches a depth of 28 km only 40 km south of Mulhouse. Northwest of the Vosges mountains the crust-mantle boundary also dips steeply towards the northwest. The largest gradient is located approximately parallel to the line from Nancy to Saarbrücken; here the boundary dips from about 28 to 31 km depth within a distance of 15 km. To the east the gradient is smaller; along the line NE from Zürich to Stuttgart an approximate depth of 28 km is reached.

The presently available data did not allow the contour lines to be continued into the Rhinegraben, and its surroundings, north of Karlsruhe.

Discussion

As shown in the previous sections, more extensive and better explosion seismic data are now available in the southern part of the Rhinegraben. Consequently, the evaluation of this much more reliable data leads to a reinterpretation of the previous observations upon which the former models of the crust and upper mantle in this area had been based. Which aspects of the structure under consideration could be confirmed and which had to be modified? To facilitate the answer to this question we have, for comparison, included in this paper the cross section of the crustal structure (Fig. 10) in the southern Rhinegraben which was previously presented in Ansorge et al. (1974) and Mueller et al. (1973). The eastward extension of this section is based primarily on results published by Emter (1971). The uplift of the crust-mantle boundary or upper side of the "rift cushion", fairly well established previously, remains essentially unchanged as can be seen by comparison of Figs. 8 and 9 with Fig. 10. Also the general topography of this interface with its asymmetric steep slope to the west and northwest and its smaller dip to the east was confirmed. The strike of the contour lines of the crust-mantle boundary in Figs. 8 and 9 seems to have a Variscan component. In addition the uplift of this interface around the shotpoint Steinbrunn continues towards the west-southwest which coincides with the direction of tectonic features mentioned by Illies (1973). The magnetic anomalies in this area show similar features (Edel & Lauer 1974).

Although the fine structure of the upper crust is not discussed in this paper some details of the new observations correlate well with the earlier results. For example, the phase P_bP mentioned in a previous section is related most probably to the layer above the crust-mantle boundary with a velocity between 6.7 and 6.8 km/s as shown in Fig. 10.

In summary, the uplift of the crust-mantle boundary in the area of the Rhinegraben has been confirmed in general and defined in detail by the new seismic observations. The mean crustal structure above the elevated upper mantle remains the same as in previous models, although a detailed analysis of the earlier phases may lead to some modifications. However, the rift cushion hypothesis itself and, consequently, ideas regarding the function of this intermediate layer had to be changed considerably. The low velocity of 7.6—7.7 km/s in the "rift cushion" presented earlier was based mainly on the unreversed observations of apparent velocities along the profiles Taben Rodt-SE and Bagenelles-N. Since the reversed line Steinbrunn—Wissembourg (profile 11) gave now a true velocity of 8.1 km/s a reinterpretation of the earlier data was necessary. In the new interpretation of the earlier observations, together with those of 1972, both the low and typical values of P_n-velocities can be explained in terms of an 8.1 km/s refractor with a varying topography as shown in Figs. 8 and 9. The lower boundary of the "rift cushion" in previous models for the area south of Karlsruhe (Ansorge et al. 1970) was based on the observations of 8.1 km/s for the P_n-phase on the profile Taben Rodt-SE beyond a distance of 210 km. Since this phase is interpreted now also as originating from the upper interface of the "cushion" the existence of a lower

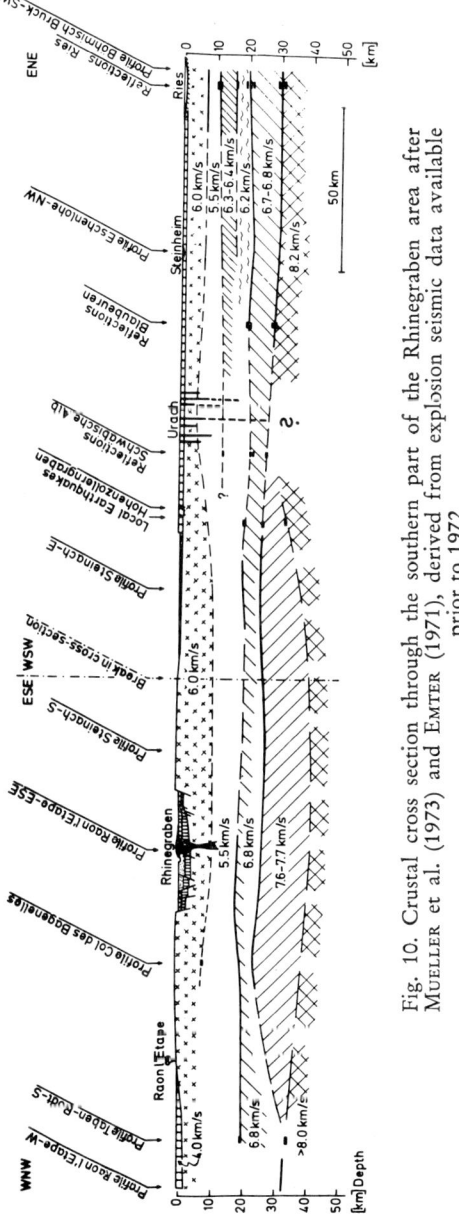

Fig. 10. Crustal cross section through the southern part of the Rhinegraben area after MUELLER et al. (1973) and EMTER (1971), derived from explosion seismic data available prior to 1972.

boundary is not confirmed. In the earlier cross section of the southern Rhinegraben, the presence of this interface was questionable as shown by a dashed line at a depth of about 40 km in Fig. 10. There are some independent indications for further changes below the uplifted crust-mantle boundary. DOHR (1970), during a special reflection survey in the central part of the Rhinegraben, observed echoes between 7.0 and 9.5 seconds which could be associated with the transition structure above the "cushion" and the upper boundary of it. Later reflections at 13.5 seconds and around 17.5 seconds may indicate more or less abrupt velocity changes below the main interface. EMTER (1971) has introduced an additional interface between 50 and 60 km depth about 80 to 100 km east of the graben axis. But since at present we do not have enough long range data in the graben area itself we can make a clear statement neither about the deeper structure of the graben system nor about whether or not there is a lower boundary of the "cushion". Consequently, this leaves the "rift cushion" hypothesis itself unsupported by the present data.

An uplifted structure with a minimum depth of 24 km, a half width of about 150 km and a velocity of 8.1 km/s should also be visible in a pronounced gravity high which is not observed. So far, all gravity investigations have dealt only with the narrow gravity minimum which is caused by the sedimentary filling of the graben itself and perhaps the structure of the low velocity zone in the upper crust, as shown by MUELLER & RYBACH (1974). The missing gravity anomaly leaves us with another unsolved problem which may also be related to the deeper structure in that area.

Acknowledgements

The experiment was a joint project of French, German, and Swiss geophysical institutes. The project was enabled by the financial support of the Institut National d'Astronomie et Géophysique of France, the German Research Association, and the Fond National Suisse de la Recherche Scientifique.

The authors are indebted to all participants who have contributed to the success of the experiment by their personal engagement. Dr. BAMFORD kindly read the manuscript.

Literature

AHORNER, L., MURAWSKI, H. & SCHNEIDER, G., 1972: Seismotektonische Traverse von der Nordsee bis zum Apennin. — Geol. Rdsch. 61, 915—942.
ANSORGE, J.; EMTER, D.; FUCHS, K.; LAUER, J. P.; MUELLER, ST. & PETERSCHMITT, E., 1970: Structure of the Crust and Upper Mantle in the Rift System around the Rhinegraben. — In: ILLIES, H. & MUELLER, ST.: Graben Problems, 190—197, Schweizerbart, Stuttgart.
ANSORGE, J.; EMTER, D.; FUCHS, K.; LAUER, J. P.; MUELLER, ST.; PETERSCHMITT, E. & PRODEHL, C., 1974: Seismic-refraction investigations in the central and southern part of the Rhinegraben. — In: GIESE, P. & STEIN, A.: Results of deep seismic sounding in the Federal Republic of Germany and some other areas (in press).
BEAUFILS, Y., 1967: Expérience du Lac Blanc. — Proc. 9th Gen. Ass. Europ. Seismol. Comm. (Copenhagen 1966), 257—263, Akademisk Forlag, København.

BERCKHEMER, H., 1970: MARS 66 — a magnetic tape recording equipment for deep seismic sounding. — Z. Geophys. **36**, 501—518.
BOIGK, H. & SCHÖNEICH, H., 1974: Perm, Trias und älterer Jura im Bereich der südlichen Mittelmeer-Mjösen-Zone und des Rheingrabens. — In: ILLIES, H. & FUCHS, K.: Approaches to Taphrogenesis, 60—71, Schweizerbart, Stuttgart.
BONJER, K. & FUCHS, K., 1974: Microearthquake-Activity Observed by a Seismic Network in the Rhinegraben Region. — In: ILLIES, H. & FUCHS, K.: Approaches to Taphrogenesis, 99—104, Schweizerbart, Stuttgart.
CHOUDHURY, M.; GIESE, P. & DE VISINTINI, G., 1971: Crustal structure of the Alps: Some general features from explosion seismology. — Boll. Geofis. Teor. ed Appl. **13**, 211—240.
DOEBL, F. & OLBRECHT, W., 1974: An Isobath Map of the Tertiary Base in the Rhinegraben. — In: ILLIES, H. & FUCHS, K.: Approaches to Taphrogenesis, 71—72, Schweizerbart, Stuttgart.
DOHR, G., 1970: Reflexionsseismische Messungen im Oberrheingraben mit digitaler Aufzeichnungstechnik und Bearbeitung. — In: ILLIES, H. & MUELLER, ST.: Graben Problems, 207—218, Schweizerbart, Stuttgart.
EDEL, J. B. & LAUER, J. P., 1974: Magnetic Anomalies, Thermal Metamorphism and Tectonics in the Rhinegraben Area. — In: ILLIES, H. & FUCHS, K.: Approaches to Taphrogenesis, 155—160, Schweizerbart, Stuttgart.
EMTER, D., 1971: Ergebnisse seismischer Untersuchungen der Erdkruste und des obersten Erdmantels in Südwestdeutschland. — Dissertation, Univ. Stuttgart, 108 p.
GIESE, P. & STEIN, A., 1971: Versuch einer einheitlichen Auswertung tiefenseismischer Messungen aus dem Bereich zwischen der Nordsee und den Alpen. — Z. Geophys. **37**, 237—272.
ILLIES, H., 1973: The Rhinegraben rift system — plate tectonics and transform faulting. — Geophys. Surv. **1**, 27—60.
MEISSNER, R., BERCKHEMER, H., WILDE, R. & POURSADEG, M., 1970: Interpretation of seismic refraction measurements in the northern part of the Rhinegraben. — In: ILLIES, H. & MUELLER, ST.: Graben Problems, 184—190, Schweizerbart, Stuttgart.
MUELLER, ST.; PETERSCHMITT, E.; FUCHS, K.; EMTER, D. & ANSORGE, J., 1973: Crustal structure of the Rhinegraben area. — In: MUELLER, ST.: Crustal structure based on seismic data. Tectonophysics (in press).
MUELLER, ST. & RYBACH, L., 1974: Crustal Dynamics in the Central Part of the Rhinegraben. — In: ILLIES, H. & FUCHS, K.: Approaches to Taphrogenesis, 379—388, Schweizerbart, Stuttgart.
PRODEHL, C., 1970: Seismic refraction study of crustal structure in the western United States. — Geol. Soc. Amer. Bull. **81**, 2629—2646.
STEIN, A. & SCHROEDER, H., 1974: Shotpoint data of quarry blasts 1958—1971. — In: GIESE, P. & STEIN, A.: Results of deep seismic sounding in the Federal Republic of Germany and some other areas (in press).

Beiträge der Reflexionsseismik zur Frage der Schwarzwaldrandstörung

von

L. Erlinghagen und G. Dohr

Mit 3 Abbildungen im Text

Abstract: Reflection seismic measurements during oil exploration in the Upper Rhine Graben after the second world war have given a better insight into the nature of the fault zone at the edge of the graben. Modern seismic work for deep seismic sounding sponsored by the Deutsche Forschungsgemeinschaft 1968 and 1971 as well as vibroseis measurements of Prakla-Seismos near the city of Baden-Baden show a rather clear main fault zone the inclination of which is indicated to be approximately 50—60°.

1. Einführung

Die Randzonen des Oberrheingrabens waren in den Jahren nach dem 2. Weltkrieg bevorzugtes Ziel der seismischen Untersuchungen der Erdölindustrie. Umfangreiche reflexionsseismische Messungen dienten nicht nur der Erkundung der generellen Lagerungsverhältnisse der tertiären und z. T. auch der mesozoischen Schichten, sondern in vielen Fällen speziell der Auffindung bohrwürdiger Strukturen in den Randzonen des Grabens. Durch diese Aufgabenstellung ergab sich schon frühzeitig die Notwendigkeit einer sehr detaillierten und sorgfältigen Bearbeitung der Störzonen.

2. Frühere Messungen

2.1. Alte Messungen der Erdölindustrie

Aus den genannten Arbeiten resultieren zahlreiche Kartierungen der erdölgeologisch wichtigen Horizonte, die — durch die Ergebnisse von Tiefbohrungen gestützt — ein recht gutes Bild der tektonischen Verhältnisse in den Randzonen des Oberrheingrabens lieferten.

In der Abb. 1 ist als Beispiel für die Ergebnisse dieser älteren Messungen die Auswertung einer Profillinie im mittleren Teil des Grabens südlich von Rastatt dargestellt. Das Profil quert die Randstaffeln des Grabens. Bereits die damalige Interpretation der Meßergebnisse ließ einige recht wichtige Aussagen über den Charakter der Störungen in diesem Bereich, ihre Einfallsrichtungen sowie die Auf-

Abb. 1. W—E-Profil im mittleren Teil des Oberrheingrabens. Diese Teufendarstellung basiert auf der Auswertung eines älteren, mit konventioneller Einfachüberdeckung, geschossenen Profils (1960).

Fig. 1. Result of reflection seismic measurements in the middle part of the Upper Rhine Graben. The single cover profile crosses the forest border of the graben. This interpretation — dating from 1960 — assumed that the main faults had an inclination at about 60 degrees.

Beiträge der Reflexionsseismik 139

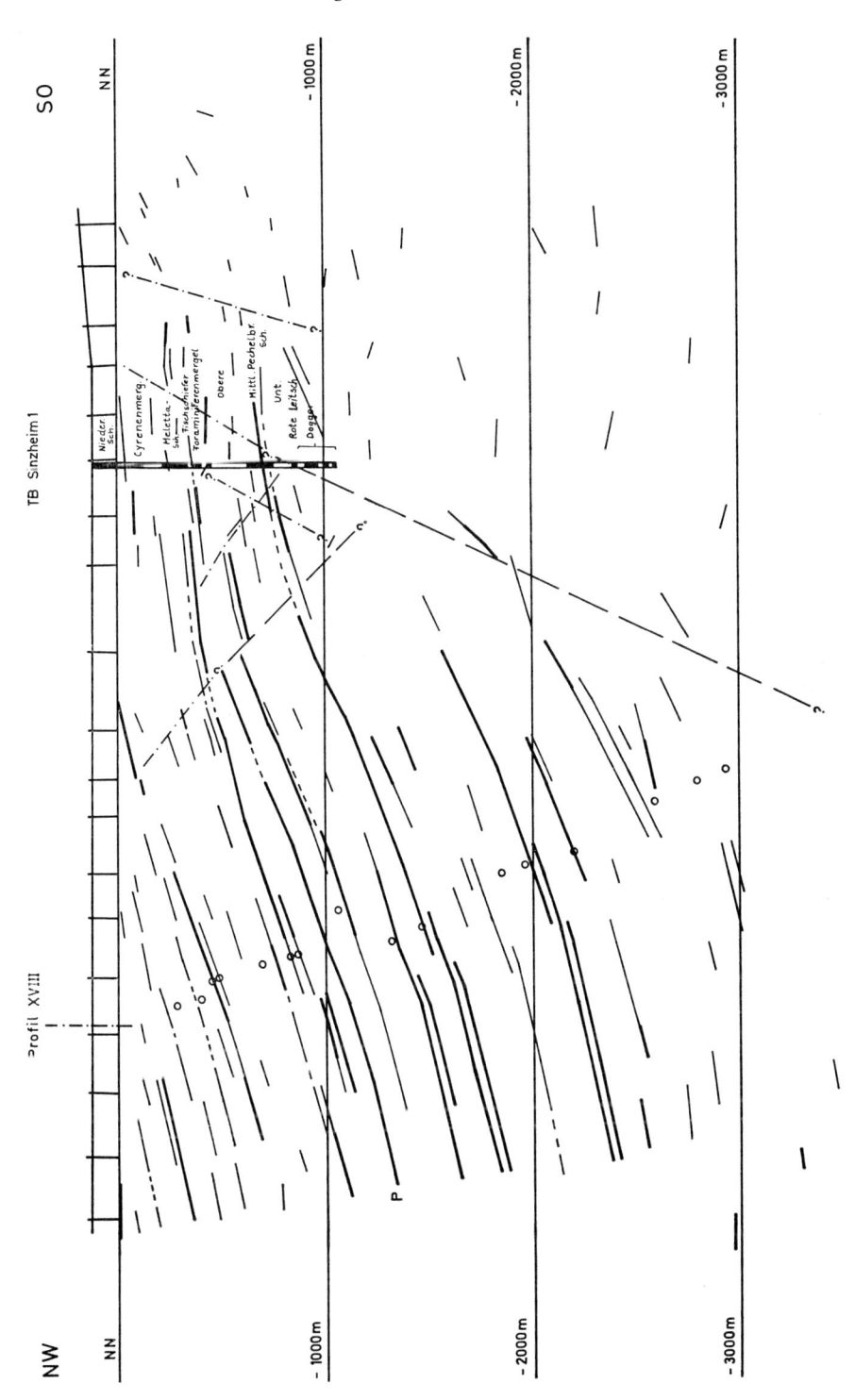

gliederung der randlichen Schollen zu. Man erkennt in der Abb. 1 eine „Hauptstörzone", die sich mit einiger Sicherheit noch bis in den Teufenbereich von 2000 bis 3000 m erkennen läßt. Ihr Einfallen ist mit ca. 60°—70° gedeutet. Bemerkenswert erscheinen ferner Y-förmige Störungssysteme, d. h. der Hauptbruch als Syntheter wird von kleineren antithetischen Brüchen begleitet.

Eine ähnliche Erscheinung deutet sich übrigens bei den Störungssystemen der ungefalteten Molasse Oberbayerns an.

2.2. Arbeiten im Auftrage der Deutschen Forschungsgemeinschaft

Die seismischen Arbeiten der Erdölindustrie sind leider Mitte der sechziger Jahre zum Erliegen gekommen. Erst im Jahre 1968 und 1971 sind erstmalig zwei speziell für die Aufgaben der Tiefensondierung angelegte reflexionsseismische Profile im Raum Rastatt im Auftrage der Deutschen Forschungsgemeinschaft geschossen worden.

Die Abb. 2 zeigt das 1971 gemessene W—E verlaufende Profil. Es schneidet im östlichen Teil die Randstörungen des Grabens südlich von Rastatt. Damit gewinnt es in seiner modernen Form der Flächenschriftdarstellung zugleich Bedeutung für das Problem der Schwarzwaldrandstörungen.

Bei diesem Profil fallen relativ kräftige und zeitlich annähernd sölig liegende Reflexe bei 2,2—2,8 sec Reflexionszeit im östlichen Teil des Profils auf. Eine geologische Deutung dieser Erscheinung kann hier nicht diskutiert werden. Sofern es sich nicht um Reflexe aus seitlichen Ebenen handelt, käme solchen Beobachtungen im Hinblick auf die tektonischen Verhältnisse am Grabenrand große Bedeutung zu.

Nach früheren Messungen (1968) wurde das „Basement" bei einer Reflexionszeit von ca. 2,6 sec vermutet.

3. Neue Vibroseis-Messungen der Prakla-Seismos 1971/72

3.1. Allgemeines zum Vibroseis-Verfahren

Zur Jahreswende 1971/72 wurde im Oberrheingraben erstmals das Vibroseis-Verfahren angewendet, das mit der Schallerregung an der Erdoberfläche arbeitet. Hierbei treten als Schallquelle an die Stelle der Sprengungen aus mehr oder weniger tiefen Bohrlöchern Vibrationen von Vibratoren, die auf Lkw montiert sind. In der Regel arbeiten drei Vibratoren synchron miteinader, die über Funk mit bestimmten Signalen von Meßwagen aus gesteuert werden. Die Vibratoren setzen die

---

Abb. 2. W—E-Profil südlich von Rastatt. Diese Sektion zeigt das Ergebnis der 6-fach Stapelung eines im Auftrage der DFG geschossenen Profils. Bei der Diskussion der Lage der Haupt-andstörung ist die Verschwenkung bei teufenrichtiger Darstellung zu berücksichtigen!

Fig. 2. W—E reflection profile in the middle part of the Upper Rhine Graben near Rastatt. The eastern part crosses the fault zone at the eastern border of the graben and indicates a main fault zone, the position of which corresponds to that in Fig. 1 (measurements sponsored by the Deutsche Forschungsgemeinschaft).

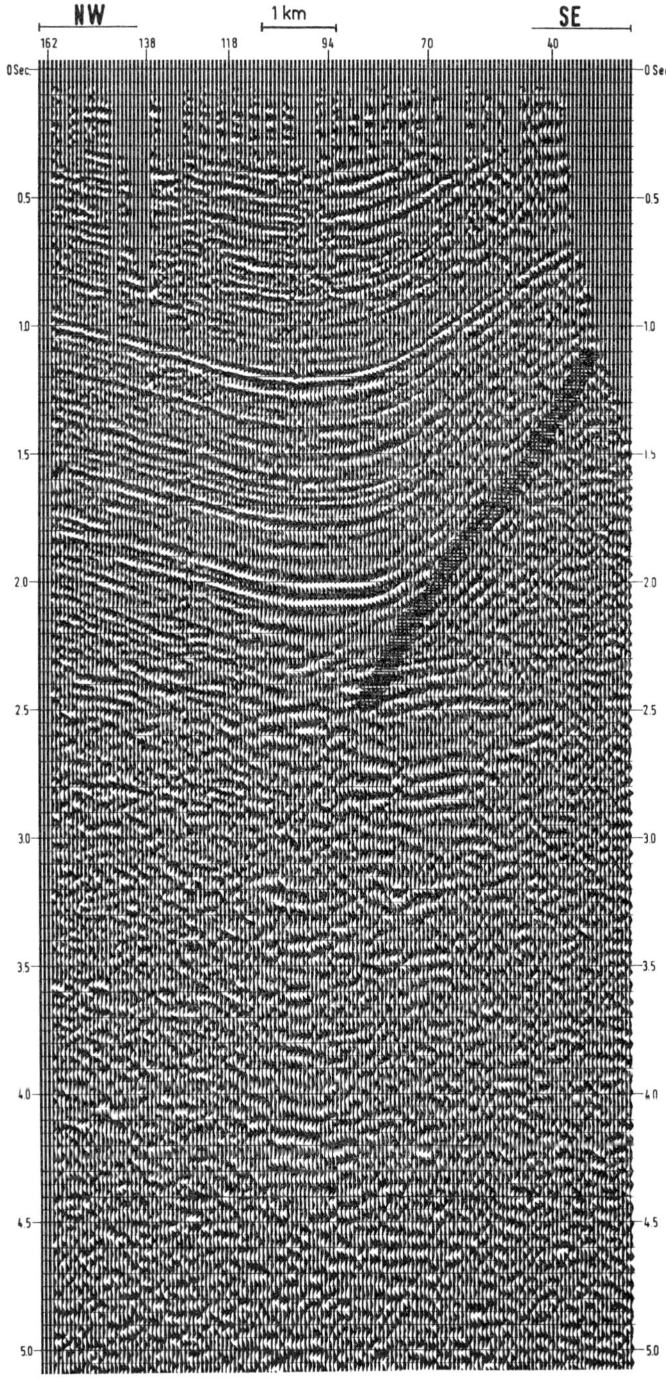
NW
1 km
SE
162
138
118
94
70
40
0 Sec.
0.5
1.0
1.5
2.0
2.5
3.0
3.5
4.0
4.5
5.0

Erdoberfläche in die Schwingungen, die ihnen durch das übermittelte sinusartige Signal vorgegeben werden. Diese Signale — Steuersignale oder Sweeps genannt — haben Frequenzbereiche zwischen 6 und 100 Hz. Dabei wird eine bestimmte Bandbreite, z. B. 15 bis 50 Hz, in einer bestimmten Zeitdauer, z. B. 7 Sekunden, in den Frequenzen kontinuierlich durchfahren, und zwar entweder mit steigender oder fallender Frequenzfolge. Dieses an der Erdoberfläche ausgestrahlte lange Signal verbreitet sich in den Untergrund, wird — wie der Meßimpuls der Sprengseismik — an Schichtgrenzen refraktiert bzw. reflektiert und beim Durchgang durch die geologische Schichtenfolge entsprechend den physikalischen Eigenschaften der Gesteine gefiltert. Die von den verschiedenen Horizonten reflektierten langen Wellenzüge treffen entsprechend ihren verschieden langen Laufwegen bzw. Laufzeiten an den Geophonen zu verschiedenen Zeiten nacheinander ein und überlagern sich gegenseitig. Durch Wiederholung von Einzelmessungen und später durch elektronische Summierungsverfahren in der Meßapparatur wird die statistische Unruhe, hervorgerufen durch Straßenverkehr, Windunruhe usw. weitgehend geschwächt.

Gemessen wurde im Oberrheingraben mit einer 12fachen Überdeckung des Untergrundes. Bei dieser Meßanordnung wird ein Punkt im Untergrund 12mal angemessen, und zwar mit verschiedenen Entfernungen zwischen Schallquelle und Schallempfänger an der Erdoberfläche. Dadurch wird die Energie pro gemessenen Untergrundpunkt gesteigert, und es werden multiple Einsätze, hervorgerufen durch wiederholtes Reflektieren von Schallwellen zwischen den Schichten, wirksam unterdrückt.

Die von den Geophonen kommenden Daten werden mit Hilfe digitaler Apparaturen auf Magnetbänder übertragen, und diese Daten werden im Datenzentrum in Hannover weiter digital verarbeitet.

Zunächst werden die langen Signale durch einen sogenannten Korrelationsprozeß in kurze Wellenzüge transformiert, die dem seismischen Impuls, erzeugt durch Sprengungen, ähnlich sind. In diesem Prozeß wird durch Produktsummenbildung zwischen dem ausgesendeten Ursprungssignal und der am Geophon empfangenen Ergebnisspur der Korrelations- bzw. Kohärengrad zwischen diesen beiden Funktionen gemessen. Das Ergebnis ist das Korrelationssignal bzw. das Korrelogramm als Ergebnis einer Messung an einem Meßpunkt. Die nachfolgenden digitalen Prozesse werden genauso wie bei den durch Sprengseismik erhaltenen Daten angebracht, wie z. B. die Anbringung statischer und dynamischer Korrekturen, der Stapelprozeß zur Herstellung einer seismischen Sektion, Filterprozesse u. a.

Die Vorteile des Verfahrens liegen in folgenden Punkten begründet:
1. es entfallen Bohrungen und Sprengungen und damit entsprechende Kosten. Insbesondere in Gebieten mit schwierigen Bohrbedingungen — wie in Schottern und harten Kalken — bringt das Vibroseis-Verfahren kostenmäßig erhebliche Vorteile;
2. die Leistung ist in der Regel höher und zum Teil sogar erheblich höher als bei der Sprengseismik;
3. es können Messungen auch in bewohnten Gebieten und Schutzgebieten durchgeführt werden, die für die Sprengseismik praktisch ausfallen.

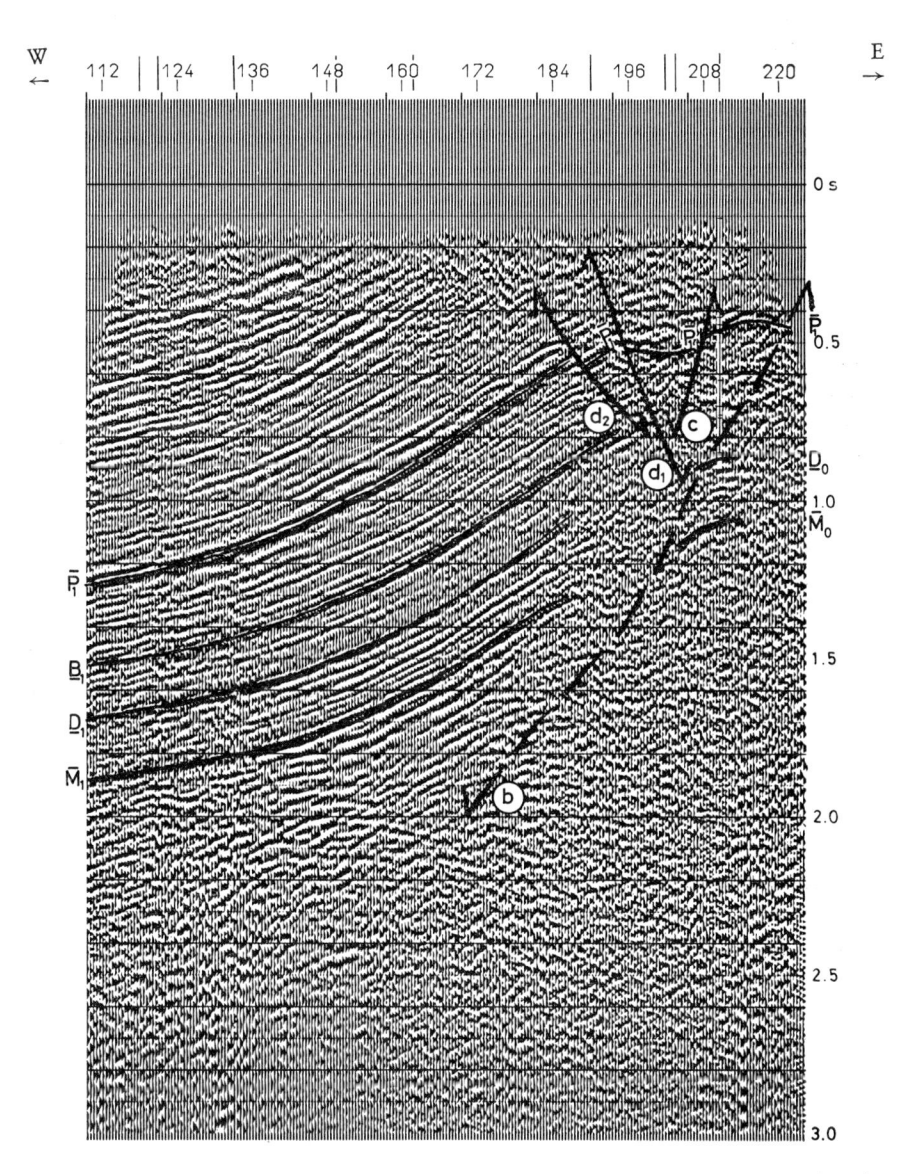

Abb. 3. W—E-Profil im mittleren Teil des Oberrheingrabens, gemessen mit dem Vibroseis-Verfahren (Aufnahme: Prakla-Seismos). Bemerkenswert sind die Aufgliederungen der Störzone und die antithetischen Vorstörungen im oberen Teil.

Fig. 3. Vibroseis profile near the city of Baden-Baden, measured by Prakla-Seismos in 1972. In this section, too, a major fault zone can be seen in the rigth part of the profile. This profile and its faultzone are similar to that in Fig. 2. (This section has been given with the kind permission of the Vorstand der Bäder- und Kurverwaltung Baden-Baden.)

Der Kilometerpreis liegt in der Regel um etwa 20—30% unter den Kosten sprengseismischer Arbeiten. Natürlich läßt sich dieses Verfahren nicht einsetzen in Moorgebieten und bei Geländeneigungen, die ein Befahren der Meßstrecken mit den Vibratoren nicht mehr zulassen.

3.2. Profile über die Randverwerfungen des Oberrheingrabens

Im Rahmen seismischer Messungen im Oberrheingraben wurden mit dem Vibroseis-Verfahren auch Profile vom Graben aus über die Randstörungen hinaus nach Osten gemessen. Die Abb. 3 zeigt ein seismisches Profil, das im Bereich der Randstörungen über eine Tiefbohrung verläuft und somit eine Interpretation der seismischen Reflexionshorizonte erlaubt.

Auch hier erkennt man die mit b bezeichnete Hauptrandstörung, deren Einfallswinkel sich durch den Abriß der tieferen Reflexionsgruppen zu etwa 60° abschätzen läßt. Ebenfalls lassen sich im oberen Teil antithetische Vorstörungen erkennen. Die Reflexionszeit von etwa 2,0 sec, bis zu welcher die Hauptstörung erkennbar ist, dürfte einer Teufe von ca. 3500 m entsprechen.

4. Zusammenfassung, Ergebnisse

Faßt man die Ergebnisse der hier gezeigten und diskutierten Profile — die zugleich für die Gesamtheit des vorliegenden Meßmaterials stehen — zusammen, so resultieren folgende Feststellungen:

— Alle Profile zeigen eine Hauptstörung oder Hauptstörzone. Ihr Einfallswinkel läßt sich zu ca. 60—70° abschätzen.
— Diese Hauptstörzone läßt sich über längere Strecken verfolgen und kartieren. Es handelt sich jedoch offenbar nicht um eine durchlaufende Störung. Vielmehr scheint der Hauptstörzug mehrfach abzusetzen, der Sprung wird vielfach von einer ± parallel versetzten Störung übernommen.
— Die Hauptstörzone ist von annähernd parallel verlaufenden Vorstörungen begleitet, die geringeren Versatz, aber gleiche Verwurfsrichtung zeigen.
— Vielfach treten Y-förmig angelegte, antithetische Vorstörungen auf.

Structure and Development of the Southern Part of the Rhine Graben According to Geological and Geophysical Observations

by

F. Breyer

With 3 figures in the text

1. Delimitation of the area

The area dealt with in this paper extends from the region south of Karlsruhe as far as to Kandern north of Basel. The framework was automatically widened when this part of the Graben on the right hand of the River Rhine was fitted in the structural pattern of the entire Graben. Obliged to be brief, I had to renounce quoting the authors whose papers and maps have been used or considered. This will be done in a more detailed report to be published in "Geologisches Jahrbuch", Reihe E, Geophysik.

2. Observational fundamentals

The material used has been taken from three different groups of sources. The first group, which is the most extensive one as regards the newly explored areas, is formed by the seismic-reflection measurements. The results were made available by the oil companies working in this region. The second group is formed by oil wells, some potash drillings and thermal drillings the major part of which has not been published yet. The third group consists of the detailed geological maps the one part of which covers important areas of the Schwarzwald and was published some time ago. The other part was provided by the geological institutes of the universities of Freiburg i. Br. and Heidelberg.

To a minor extent gravity measurements were also taken into consideration. But as the network of measuring points was hardly dense enough, they played a minor rôle.

To reduce all data obtained to a common denominator, a contour map was first compiled of the Bunter base, and then a similar one of the Tertiary base.

The subsurface contours of the Bunter were determined according to the geological maps on which the Bunter and/or younger beds are shown to be exposed. This was in particular the case in the northern Schwarzwald, i. e. in the farther surroundings of the Hornisgrinde. The area of Baden-Baden was rather neglected, but ever more emphasis was laid on the regions adjoining the Hornisgrinde in the south. In this region, i. e. between the Rench and Kinzig rivers, a number of structures became obvious which have hitherto been unknown and which are of importance for a better information about the Graben flanks. To obtain even more details of the Triassic base and to fill existing gaps, the subsurface contours of the Rotliegend were investigated and, finally, the orographic altitudes of the basement were considered.

In the foothills of the Middle Schwarzwald, between the Kinzig and Elz rivers, the most recent mapping was evaluated. Unlike the above-mentioned northern region the boundaries between the Bunter stages and the Muschelkalk were additionally determined and, after compiling schematic sections, referred to the Triassic base.

The reports on the seismic-reflection measurements do not give the subsurface contours of the Triassic and/or Tertiary base. In case a connection with deep drillings was possible, both horizons were plotted in the depth profiles. Neither of the horizons is continuously represented by a characteristic reflection. This phenomenon is mainly due to the similar facies of the Lower Bunter and the Upper Rotliegend. More important, however, is the fact that from oil-geological aspects there was no reason to penetrate into such a depth with corresponding charges. The base of the Tertiary is partly hidden by physically equal properties of the rocks, partly it lies at shallow depths unfavourable for the recording of reflections.

In some regions the contours were only obtained by converting the plotted depths of the reflections after it had become obvious that the travel-time curve had to be changed. This was the case when it was not known in advance which geological sequence had to be expected and which travel-time curve was the suitable one.

The thickness of the Triassic stages was obtained from some drillings, determined from the geological maps and or taken from published material. The reflexion measurements mostly were of an excellent quality, and there was thus no difficulty in tracing a number of boundaries over large distances.

When determining the Tertiary base other problems arose. It is not in all places that there is a transgression of the Lymnaea marls or of the thin, even older Eocene beds. In some places a transgression of the Pechelbronn beds occurs, in others the Pliocene directly overlies the Mesozoic, and this not only near the Graben margin but also on certain structures near the River Rhine. In addition to the observations made so far, the Mesozoic beds were found to reach into the Dogger or Malm to a varying extent. As a result of these observations, a geological map of the pre-Tertiary transgression plain was drafted.

3. Tectonic elements of the Graben and its flanks

To describe the structural forms in detail, a division into three parts will be made which is more or less evident from the structural conditions. The old division into Rheinebene (Rhine plain), Vorbergzone (foothills), and Hochschwarzwald (Schwarzwald mountains) has no longer been satisfactory. The Quaternary, often together with the Pliocene, covers formerly elevated blocks comparable to submerged foothills with differences in depth of a few 100 m. Moreover, it had earlier been found out that there were also faulted block systems behind the foothills.

A tripartition in N—S direction is obtained with a northern part extending as far as to the transverse zone of Variscan strike, which starts in the Alsace near Barr, runs via Erstein and Appenweier and can be followed up to the Hornisgrinde. The middle part extends from the transverse zone to the tectonically less distinct line which begins with the southern slope of the Tuniberg, running into the Elz valley. The southernmost third part, finally, extends from this line to the Swiss Jura. — To obtain a further, more convenient division I have specially marked those faults from among the lot of known ones which have got a vertical displacement of 500 m at the least. These major faults separate mainly two kinds of blocks from one another, for one the elongated, partly very deep troughs, for the other the horst-like regions fractured by numerous minor faults.

Proceeding from the north, there is at first the extremely deep Rastatter Mulde (Rastatt Trough) in which the Triassic base sinks down to 4.700 m below m. s. l., whereas the Tertiary base lies at a depth of 3,700 m. On its eastern margin the trough is accompanied by the Kuppenheim block. The separating fault has got a vertical displacement of 1,000 m. Some minor ones follow in the direction of the trough. The block itself is sinking down to the North. For this reason the vertical displacement of the fault which follows in eastern direction and which leads to the Schwarzwald amounts to different heights, on the River Murg to about 2,000 m.

Farther southwards, the Rastatt Trough approaches the Schwarzwald. The equivalents of the Kuppenheim block are not even half as wide as this one. Its tectonic height changes in southern direction in a positive sense. In the foothill zone of Baden-Baden the Pechelbronn beds, the Meletta beds, and Cyrena marls are exposed at the surface; in the foothills south of Bühl Triassic and Liassic beds are outcropping, in the Achern marginal zone of the Schwarzwald Variscan crystalline rocks are exposed. The axis of the Rastatt Trough rises to the same degree. The trough ends within the region of Achern.

In the east, the Kuppenheim block and the foothills of Baden-Baden are adjoined by the well explored region of Baden-Baden; in this region which is of a highly complex structure, the Bunter occurs in isolated areas, and the Rotliegend base could hardly be determined. Therefore the areal investigation only starts at the Hornisgrinde (see below).

Between the Rastatt Trough and the Kehl Trough following towards SW a shallower block is intercalated which is cut by numerous faults striking N—S. Many of these faults are of a marked antithetic character. All partial blocks rise in southern direction, leading to the uplift region of Appenweier. In northern direction this fault system extends beyond the River Rhine into the Lower Alsace where the antithetic faults occur as far as to the eastern margin of the Zabern fault zone. — In the Kehl Trough the Triassic base reaches down to 3,200 m.

In the uplift region of Appenweier the Paleogene sequence ever more decreases in thickness towards SE. It is already on its western margin that the

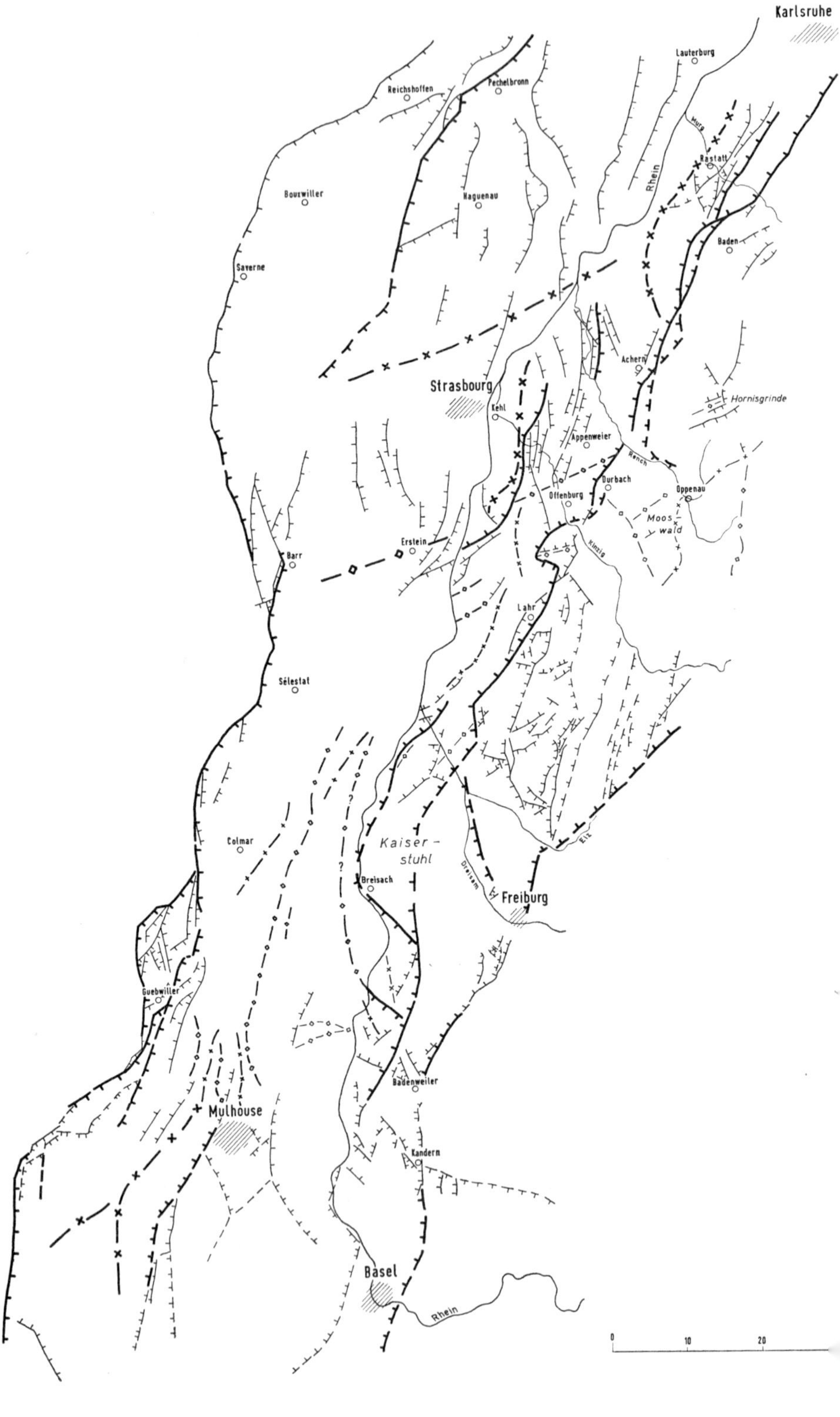
Karlsruhe
Lauterburg
Reichshoffen
Pechelbronn
Murg
Rastatt
Bouxwiller
Haguenau
Rhein
Baden
Saverne
Achern
Strasbourg
Kehl
Hornisgrinde
Appenweier
Rench
Durbach
Oppenau
Offenburg
Moos wald
Erstein
Barr
Kinzig
Lahr
Sélestat
Colmar
Kaiser-stuhl
Breisach
Dreisam
Elz
Freiburg
Guebwiller
Badenweiler
Mulhouse
Kandern
Basel
Rhein
0 10 20

Lymnaea marls are no longer observed. The various beds of the Upper and Middle Oligocene gradually disappear under the Pliocene transgression; finally the Tertiary as a whole is no longer to be found. In a region near the mountains striking about N—S — the town of Offenburg is situated in its centre — Liassic and Triassic rocks presumably exist under the Quaternary. On the eastern margin, along the Schwarzwald, the Bunter appears in some places.

The regions of Appenweier and Offenburg are already located on the transverse swell striking SW—NE. This swell is continued beyond the River Rhine with the Erstein swell. In northeastern direction the Hornisgrinde dome is encountered.

At a distance of 10 km there is another series of mostly anticline- or horst-like structures, which are, however, not so distinctly connected with one another. They are, for one, a group of anticlinal axes, which quite probably continue beyond the River Rhine; for the other, the Triassic horst of Niederschopfheim is concerned which is still quite a problem from tectonic aspects; and finally there is a group of structures within the region between Rench and Kinzig the forms of which have been discovered by the combined study of the Triassic base, the Rotliegend base and the surface of the crystalline basement. This group is formed by the Durbach Trough and the Mooswald and Oppenau Troughs together with the intercalated swells. Mention be made in this context of the Wälderhof swell.

These three troughs form a deep-lying area which reaches far into the Schwarzwald. In this place, at a distance of 12 km from the Schwarzwald margin, the Triassic base lies 500 m above m. s. l. at the most, whereas the dome of the Hornisgrinde which is adjacent in the north reaches an altitude of almost 1,000 m.

In the middle part, which starts south of the transverse swell, there is, in contrast to the region north of the swell, no indication of a particularly deep graben or a deep trough on the right hand of the River Rhine up to the Tuniberg south of the Kaiserstuhl. The deepest troughs of the Rhine Graben here lie left of the River Rhine.

Faulted foothills and marginal blocks may be traced up to the Elz valley fault. The following division is possible: contiguous to the River Rhine and separated from the Alsace by a continuous chain of ridges or anticlines with Dogger covered by Pliocene there is an elongated, shallow trough in which, in the deepest places, the Tertiary base reaches down to 400 m below m. s. l. at the most. Adjacent in the east there is the Lahr—Emmendingen foothill zone which is subdivided into the Tertiary foothills, Dogger to Muschelkalk occurrences, and a

◄

Fig. 1. The fracture system and the structural lines of the southern Rhine Graben. On the right side of the river Rhine all essential fractures are given. Fractures with displacements of 500 meters or more have been drawn with thick lines, all others with thin lines. The teeth show to the sunken block. — The structural lines have been drawn in various thicknesses according to their importance. Anticlines are shown by small squares, synclines by small crosses. — On the left side of the river mostly the structural elements of importance are given only, e. g. in the region of fractures of Saverne.

wide and elongated Bunter region. Finally, a third zone must be mentioned which is formed by hardly subsided blocks partly characterized by caps of the Rotliegend formation. This block system extends up to the Elz valley fault behind which the Schwarzwald mountains rise with a maximum difference in altitude of 600 m, decreasing to 300 m and less.

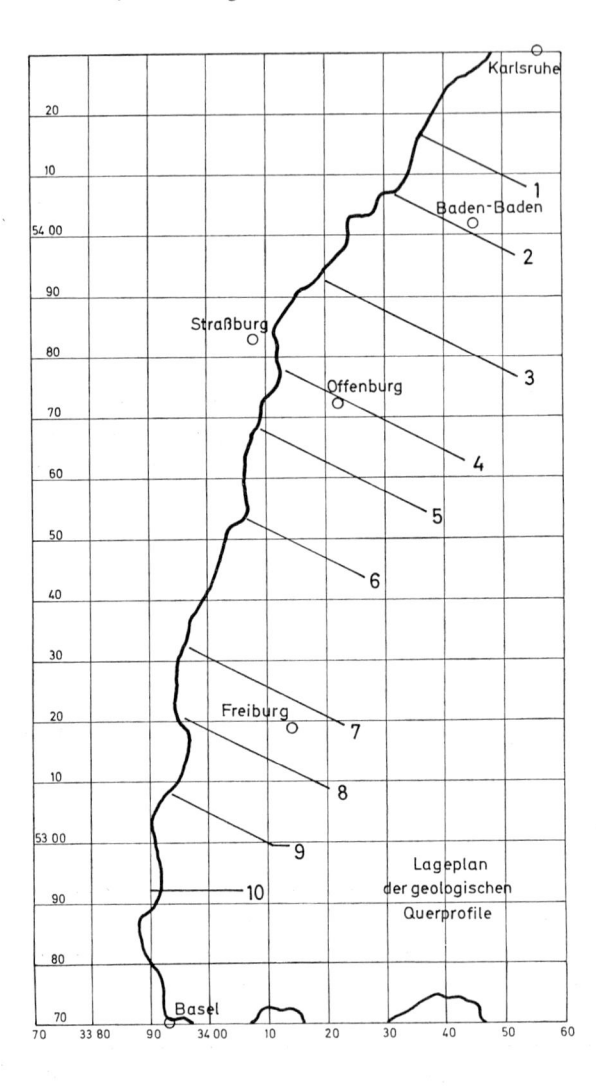

Fig. 2. Map showing the situation of the geological cross sections of Fig. 3.

Between the flat trough just mentioned and the Lahr—Emmendingen foothill zone there is a fault with a considerable vertical displacement; it is the straight-lined continuation of the fault system separating the Northern Schwarzwald from the Rastatt Trough. Where the Kinzig leaves the mountains the exact course has not been proved yet. The main marginal fault which has hitherto been thought to lie between the Bunter hills and the crystalline basement with its occasional Rotliegend caps and actually is formed by a number of minor faults now markedly looses in importance if compared with the first one. It runs in southern direction to the Kaiserstuhl. There it is quite obviously identical with the fault the existence of which has been postulated earlier in this region. On the western margin of the Tuniberg it appears again.

The "Freiburger Bucht" is in its central part cut by a major fault from N to S, which runs across the Nimberg. It branches off the aforesaid north of Riegel and, south of Freiburg, joins the eastern marginal fault of the "Freiburger Bucht", which appears as the southern continuation of the Elz valley fault. There does not seem to be a fault of Tertiary or younger age which crosses the Freiburger Bucht in the direction of the Kaiserstuhl.

So the third, southern part of the Rhine Graben section under discussion has been reached.

Under the western half of the Kaiserstuhl there is quite probably the continuation of the flat Eocene trough which extends from Offenburg between the foothills and the sub-Pliocene Dogger anticlines aligned along the River Rhine. For lack of adequate measurements its western margin cannot be given. But it is quite probable that immediately west of the River Rhine there is a trough with thick Paleogene deposits. Its existence is indicated by the nortern continuation of the anticlinal structure of Weinstetten.

SW of the Kaiserstuhl a fault with a great vertical displacement runs from Breisach towards the southern part of the Tuniberg, joining here the fault which comes from the Kaiserstuhl in the north.

In this place a southeastern spur of the thick Alsacian Tertiary basin changes over to the right-hand side of the River Rhine, forming the well-known potash basin of Buggingen. This basin is cut into two parts of almost equal size by the Weinstetten structure. Here we are in quite a well-known region.

Yet, as regards the southern half of the "Freiburger Bucht" and its continuation as far as to Heitersheim roughly, much progress has not been made so far in the analysis. The only exception are the Mesozoic blocks which occur on the margin of the Schwarzwald, starting with the Schönberg and running to Badenweiler and farther southwards almost without interruptions. Here seismic measurements have not been made yet. In the places where they have been started again a number of structural elements have been observed which gradually rise in southern direction and are separated from one another by faults striking NW—SE. So the existence of the traditional Klemmbach fault has in fact become dubious.

South of Badenweiler we reach the final southern part of the area in discussion, a region which has become well-known by means of geological maps, many prospect drillings, seismic-reflection measurements, and relevant publications.

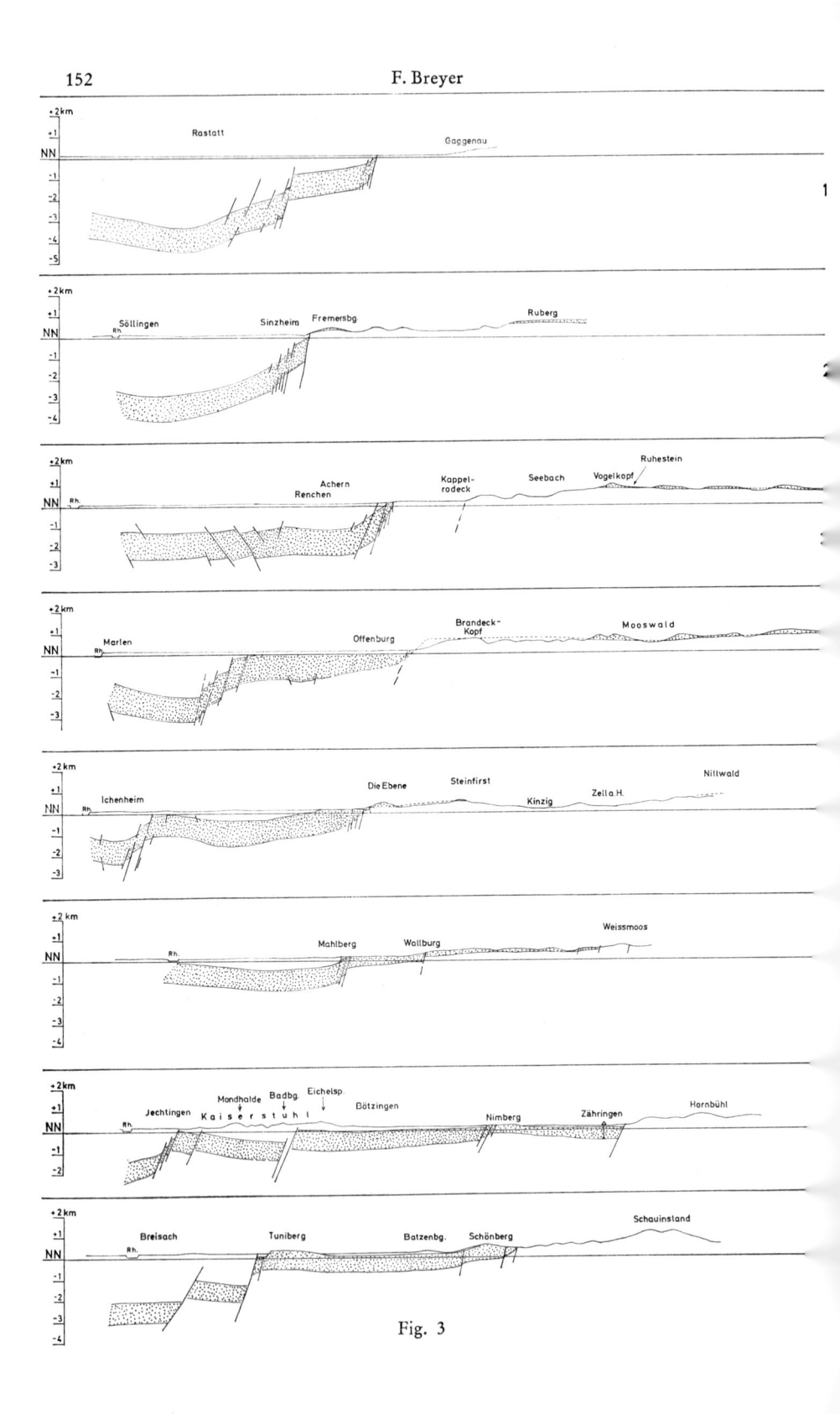

Fig. 3

4. The development of the Graben and its framing mountain ranges

The following discussion of the development of the Graben is not meant to be complete. With the most recent observations it is merely tried to illustrate those aspects which have become more evident now.

The Graben has its origin in the formation of loosely connected depressions in the Upper Eocene. It is rather uncertain whether a depression indicated far in the south during the Bunter is part of the Rhine Graben which developed at a much later time. — The surface covered by the Tertiary transgression does not show any traces of a previous upwarping which might have led to the formation of a "Scheitelgraben". Otherwise older beds must be encountered along the Graben centre, younger ones on either side. Instead a swell region is observed along the Erstein—Appenweier line described above, which still follows the old Variscan main strike. Moreover, there are a great number of Oxfordian and Rauracian occurrences along the eastern margin, which partly exist in the form of pebbles and may only have developed farther eastwards. But any correspondingly young beds are not to be found in the west.

The depressions with thick Upper Eocene and Lower Oligocene deposits do not cover the whole width of the present Graben. The arrangement of the depressions seems to be caused by a local sagging of the underground of a varying degree. By and large, this sagging is of the well-known Rhenish trend; but it is modified by the distribution of more or less mobile "crustal" parts in the underground which have their origin in the Variscan era and were still evident in the distribution of the sedimentary basins in both the Rotliegend and the Triassic. The zone of "weakness" of Rhenish trend is only gradually getting prevalent. The Appenweier uplift region is inundated as late as in the Lower Oligocene, but it is never subject to such rapid sagging movements as the Kehl and Rastatt

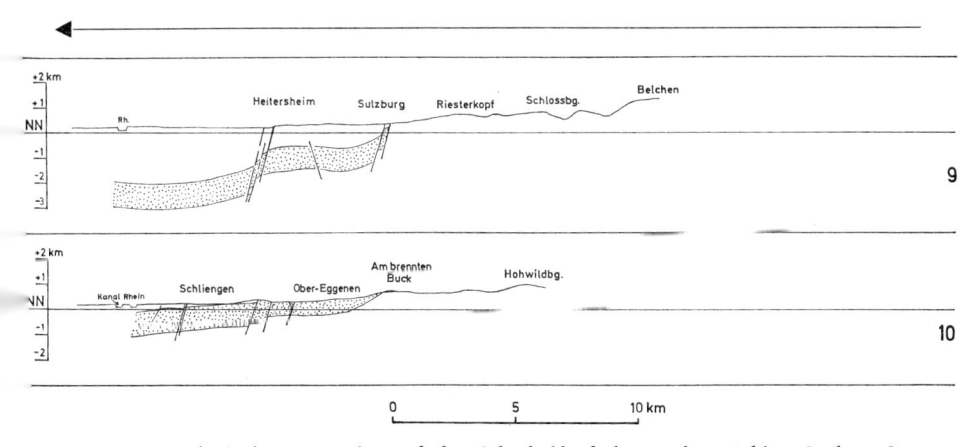

Fig. 3. 10 geological cross sections of the right half of the southern Rhine Graben. Surface and base of the Mesozoic are shown only. Tertiary and Quaternary and Permian and Variscian basement are not separated. Where reflexion seismic was available, i. e. in the plains, surface and base are given, otherwise the base only.

Troughs. — Between this uplift region and the Kehl Trough one of the faults described above is running; it appears as a synsedimentary faulting. Likewise the major fault encountered in the underground of the Weinstetten anticline has turned out to be synsedimentary. SW of the fault the Lymnaea marls and Pechelbronn beds are of a markedly greater thickness than in the NE.

Thus it has moreover become obvious that the primary depressions were already bordered by major faults when the flanks had not been risen yet. South of the transverse swell, i. e. from Appenweier on, the transgression of the Pechelbronn beds over the equally arranged Dogger stages takes place in a similar way as in the primary depressions. The homogeneity of the present Upper Rhine Graben is merely simulated by Pliocene and Pleistocene sediments covering the previously existing variety.

If the thickness of beds in the northern Graben and the southern Graben is compared not only on the basis of drill logs but also on the basis of seismic-reflection results, there is, as regards the Rastatt Trough, a 3:1 ratio between north and south for the Upper Eocene and Oligocene, whereas the ratio is vice versa, i. e. 1:3, for all younger beds. The transition is to be found within the region between Karlsruhe and Speyer.

After the Lower Miocene there is a folding in the south which is stressed by fractures. This phenomenon is clearly to be seen in the seismic profiles, but it also becomes evident from the drillings, especially in the Alsace, and in the distribution of the beds underlying the transgressing Pliocene and/or Pleistocene beds.

The fact that a folding is concerned has in particular become obvious from the drilling Weinstetten 1. Based on this drilling the problem of the salt domes in the Alsace may be re-considered. These salt anticlines have for long been looked upon as actual domes caused by the salt formations. Obviously rather broad saddles with gently dipping flanks and steep anticlines of the Weinstetten type are alternating. It is supposed that there is a connection between these saddles and older faults, as is the case in Weinstetten. But apart from Weinstetten this has nowhere been proved up to now.

5. Relations to the Alps

There is a remarkable anti-symmetry with respect to the structural events in the Alpine region adjacent in the south. In the southern Rhine valley thick sediments develop in the Upper Eocene and Lower Oligocene, whereas in the deposition area of the Molasse and the Helvetian basin there is at the same time either no sedimentation at all or such a minor one that merely a few remains have been left. Thick sediments of an Alpine geosyncline have not become known of this time.

In the Middle and Upper Oligocene up to the Aquitanian a sedimentation of a minor degree occurs in the Rhine Graben, while a geosyncline of several kilometers' depth is being formed on the northern margin of the Alps.

In the Lower Miocene the sedimentation in the Molasse region is quite considerable, whereas in the southern part of the Rhine Graben the rate of sedimentation is low, in the northern part, however, high.

In the Upper Miocene, the folding of the Molasse begins; parallel to it there is no sedimentation at all in the southern part of the Rhine, but merely a folding, uplifting and erosion of the deposits. In the northern part, the sagging movements and sedimentation continue.

In the Pliocene and Pleistocene the interrelations are growing more complex and rather vague owing to the subsidence in the northern Graben, which takes place in the immediate neighbourhood, and the uplifting of the flanks in the north and south, and owing to the lack of adequate geological time marks.

Although such reservations must be made, the question arises whether this contradictory behaviour was balanced at greater depths, e. g. by means of mass displacements. It would be quite imaginable that thus the results of the seismic-reflection measurements and other geophysical methods could be brought into line with one another.

Magnetic Anomalies, Thermal Metamorphism and Tectonics in the Rhinegraben Area

by

J. B. Edel and J. P. Lauer

With 3 figures in the text

Abstract: Two belts of magnetic anomalies which can be followed across the Rhinegraben are related, within the crystalline basement of the flanks, to a thermal metamorphism of Hercynian age and generally not to any basic volcanism. At least in the graben, they seem to correspond to higher positions of the basement. Used as markers, they can give information on eventual post-hercynian horizontal movements.

Aeromagnetic maps for a flight altitude of 3,000 meters show, West of the Rhine and peculiarly above the crystalline Vosges mountains, two belts of positive anomalies which traverse the massif in an East—West direction; where the crystalline basement outcrops, they correspond:

— in the North ($+ 130\ \gamma$), to the Champ-du-Feu — Hohwald and Etival granodiorites
— in the South ($+60\ \gamma$), to the southern border of the granitic Ballons-massif.

A systematic study of magnetic susceptibilities of rock samples collected all over the crystalline Vosges on a 1 km-grid (ROCHE 1967) reveals that high susceptibilities (300.10^{-6} emuGGS/cm^3) are precisely encountered within the basement of the two former belts.

Magnetic measurements at ground level

In these two regions we have undertaken a ground level study which includes tight measurements of the field along NS profiles and of numerous susceptibilities. Microscopic observations of thin rock sections and thermomagnetic analyses in a strong field supplement our work. In both belts we have met, on the ground, important magnetic field anomalies which correlate especially well with "granitic" massives and their contact zones.

Remanent magnetizations are negligible. The variation of susceptibilities along profiles is the same as for anomalies of the magnetic field: thus the magnetized structures are outcropping.

Thermomagnetic analysis shows that, in these sedimentary or volcano-sedimentary rocks near to the metamorphic contact, magnetite is created (T_c about 580° C) out of non-magnetic iron oxides and hydroxides. The growth of susceptibility is paralleled by a growth of grain size of ferromagnetic crystals. In the granite itself, endomorphism produces a perpheric more basic and magnetic crown.

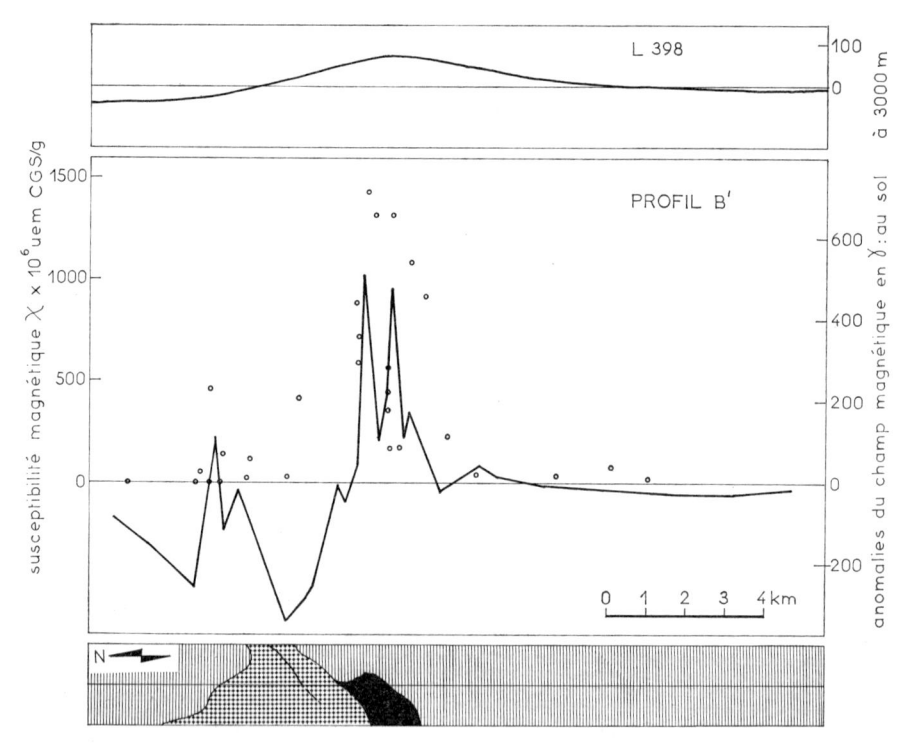

Fig. 1. Southern belt. Anomalies of the total magnetic field at 3,000 m compared with anomalies on the ground and magnetic susceptibilities (○) measured along this profile; the latter are maximum above borders of the granitic body.

The phenomenon responsible for the creation of high susceptibilities seems to be thermal metamorphism (= endomorphism + exomorphism) which accompanies the settlement of these massives: susceptibility is maximum close to the contact and then decreases both towards the interior and the exterior, within sedimentary or volcano-sedimentary rocks.

For the southern belt, for instance, Fig. 1 shows that the most magnetized zones are the border facies (monzonitic and syenodioritic) of the Ballons-granite (GUERIN 1967) Devoniandinautian volcan-sedimentary terranes (Culm) close to the contact. The aeromagnetic anomalies show the same general pattern as the anomalies on the ground and the variation of susceptibilities (EDEL 1972).

In the northern belt we obtain similar curves. There, anomalies of the field are bound to the Champ-du-Feu—Hohwald granodiorites and both of the bordering diorites of Neuntelstein (VON ELLER 1970) and the well-known schists of Steige transformed into hornfelses (ROSENBUSCH 1877; LAUER & TAKTAK 1971). That correlation can be followed over a distance of at least 50 km. Magnetic structures which cause the anomalies in a height of 1,200 m are outcropping (BRGM 1971).

Extension into the Schwarzwald

On the eastern extension of our two belts within the basement of the Schwarzwald, there are hornfelses in the North near Baden-Baden and, in the South, within the piched Culm near Badenweiler.

Furthermore, magnetic mapping o n t h e g r o u n d of the Rhinegraben SAIAGH 1966; RAFAMANTANANTSOA 1969; SPREUX 1971; REICH 1941) shows that two belts of anomalies join these two regions to the corresponding ones in the Vosges (Fig. 2 and 3). The subsidence of the graben being of about 2 to 3 km as an average, the amplitudes of anomalies on the ground in the graben and at a height of 3,000 m above the Vosges are comparable.

Finally, in the two concerned regions of the Schwarzwald, the existence of magnetic anomalies has been shown at 1,200 m (BOSUM & HAHN 1970) and on the ground (GREINER & SITTIG 1971) near Baden-Baden, and by us in the South (+ 40 γ).

So there appear two markers going through the Rhinegraben which are bound to buried ridges (Erstein and Mulhouse) where sinking of the graben seems to have been of a lower extend. Their extension into both of the Vosges and the Schwarzwald massives are associated with a hercynian thermal metamorphism.

Such markers may guide further research on possible horizontal displacements, for instance: along the border master faults.

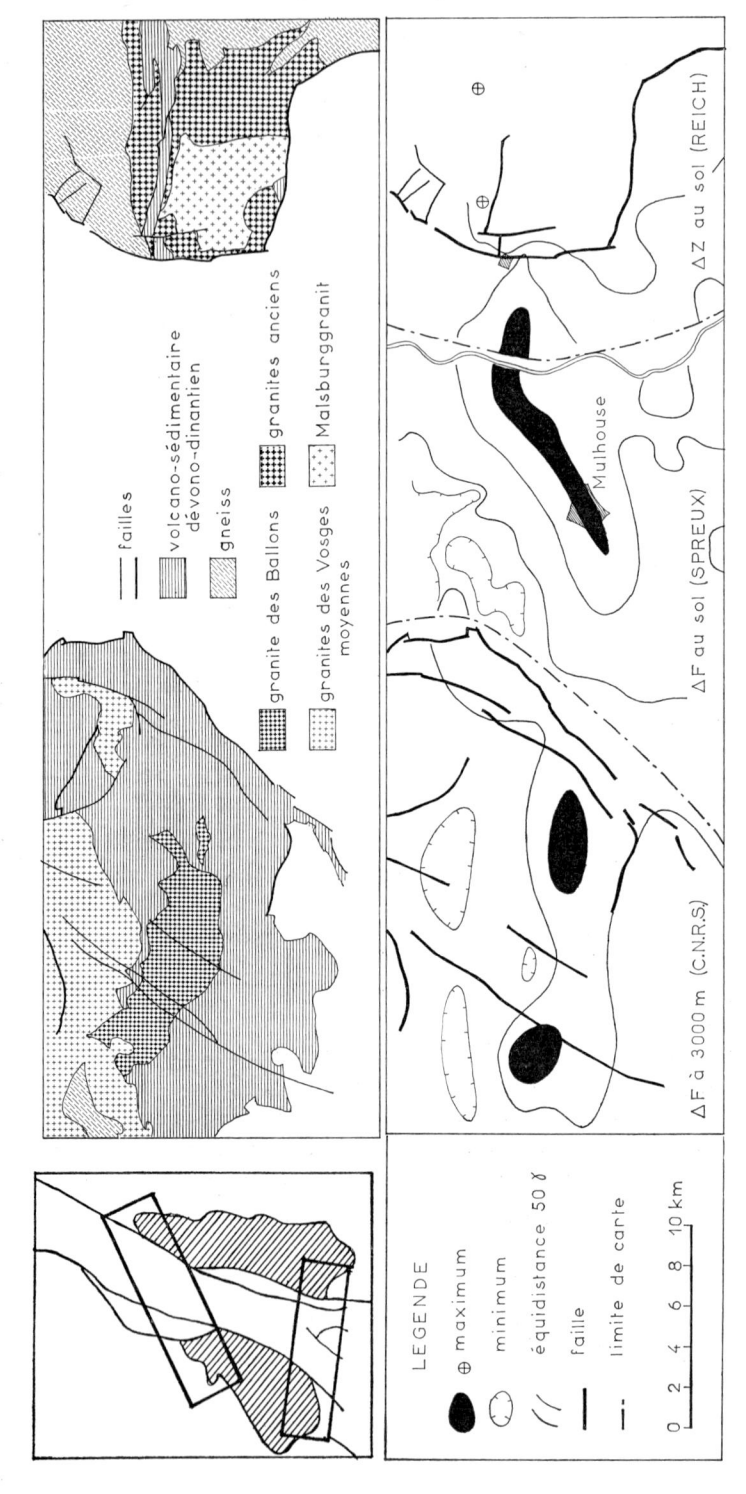

Fig. 2. Southern belt. Comparison between geology and magnetic anomalies at 3,000 m (on the left, above the Vosges massif and on the ground in the Rhinegraben).

Magnetic Anomalies, Thermal Metamorphism and Tectonics

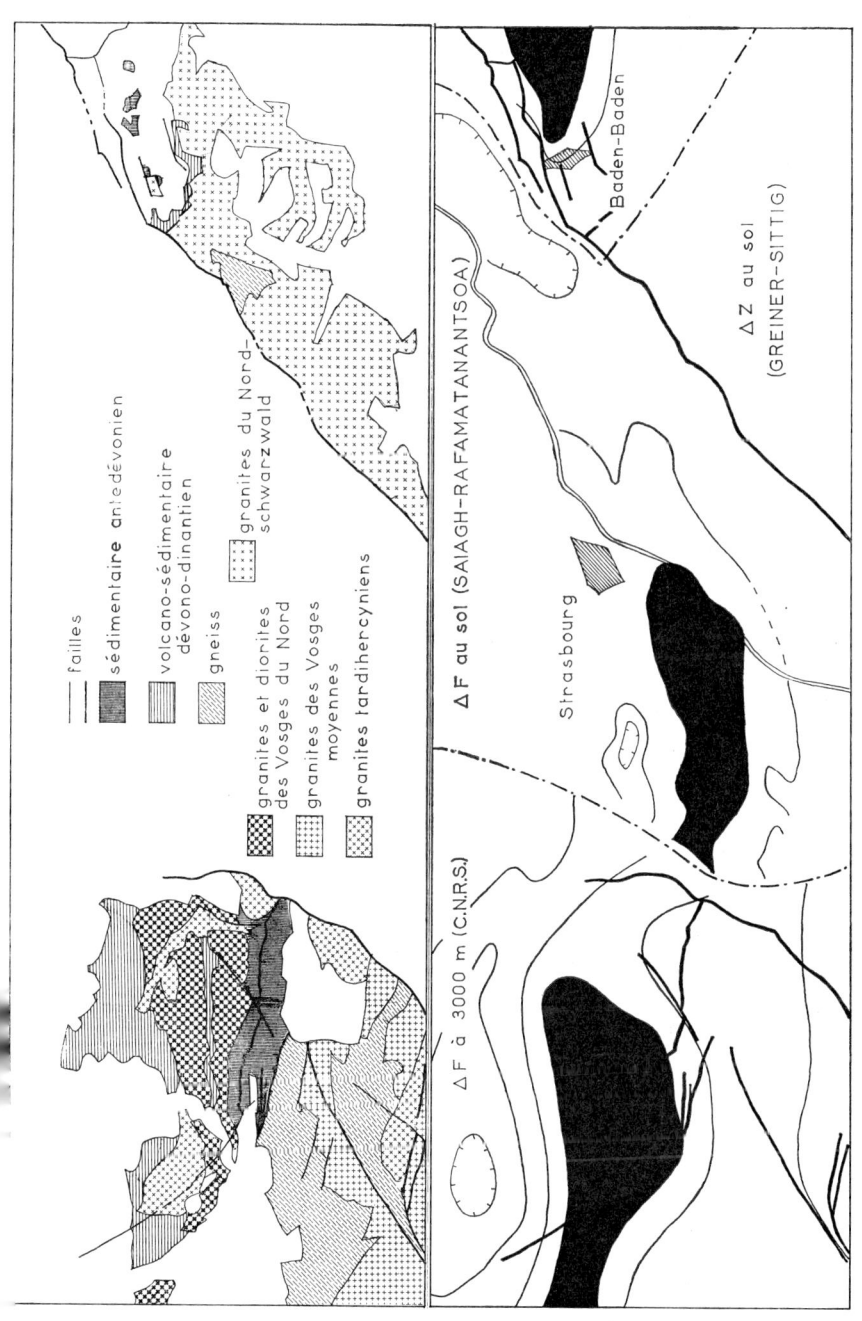

Fig. 3. Northern belt. Comparison between geology and magnetic anomalies at 3,000 m and on the ground (scale and contour intervall are the same as in Fig. 2).

Bibliography

BOSUM, W. & HAHN, A., 1970: Interpretation der Flugmagnetometervermessung des Oberrheingrabens. — In: ILLIES, J. H. & ST. MUELLER (eds.), Graben Problems, 219—223, Schweizerbart, Stuttgart.

BRGM, 1971: Carte magnétique de la France au 80 000e, feuille de Strasbourg. — Notice explicative, 71, 18 p.

EDEL, J. B., 1972: Diplome Ing. Géophys., 92 p., Inst. Phys. Globe, Strasbourg.

ELLER, J. P. VON, et al., 1970: Carte géologique du socle vosgien, partie septentrionale. — Bull. Serv. Carte géol. Als. Lorr., 23, 5—28.

GREINER, G. & SITTIG, E., 1971: Personal communication.

GUERIN, H., 1967: Faciès de bordure du granitic des Ballons etc. . . . — Bull.Serv. Carte géol. Als. Lorr., 20, 37—57.

LAUER, J. P. & TAKTAK, G., 1971: Propriétés magnétiques des roches au voisinage du contact métamorphique etc. . . . — Comptes Rendus Ac. Sci., 272, 924—927.

RAFAMATANANTSOA, M., 1969: Diplome Ing. Géophys., 70 p., Inst. Phys. Globe, Strasbourg.

REICH, H., 1941: Maps for Z. Unpublished.

ROCHE, A., 1967: Magnetic researches on the alsatian part of the Rhinegraben. — In: ROTHE, J. P. & SAUER, K. (eds.), The Rhinegraben Progress Report 1967. — 1967. — Mém. Serv. Carte géol. Als. Lorr., 26, 116—120.

ROSENBUSCH, H., 1877: Die Steiger Schiefer und ihre Kontaktzone etc. . . . — Abh. geol. Spezialkarte Els. Lothr., 1, 79—274.

SAIAGH, H., 1966: Diplome Ing. Géophys., 95 p., Inst. Phys. Globe, Strasbourg.

SPREUX, A., 1971: Diplome Ing. Géophys., 94 p. Inst. Phys. Globe, Strasbourg.

Die Tektonik des nördlichen Schwarzwaldes und ihre Beziehung zum Oberrheingraben

von

D. Ortlam

Mit 4 Abbildungen im Text

Abstract: On the basis of a contour map (texture) related to the boundary Middle/Upper Bunter the new tectonic conditions and the age of the tectonic pattern are represented in the region of the uplift of the Northern Black Forest. A large transversal fault (with a transversal amount of 7 km) striking in SE—NW direction was proved for the first time. In the central part of the Upper Rhine Graben the studies enabled a division of the earth's crust in three tectonic levels.

Im Anschluß an die Arbeiten zur Fertigung einer Streichkurvenkarte des Kraichgaues (D. ORTLAM 1970 a) wurde zum gleichen Zweck für das Gebiet des Nordschwarzwald-Gewölbes (Ettlingen—Pforzheim—Freudenstadt) eine zweite Streichkurvenkarte (Abb. 1) angelegt. Sie ist einheitlich auf die Grenze Mittlerer/ Oberer Buntsandstein — nämlich der Oberfläche des Violetten (Karneol-) Horizontes 2 (VH 2, D. ORTLAM 1967), einem fossilen Bodenkomplex — bezogen worden, nachdem der VH 2 oder andere in bestimmtem Abstand zu ihm befindliche Horizonte auf insgesamt 19 Meßtischblättern kartiert wurden[1]. Der Genauigkeitsgrad der Streichkurvenkarte dürfte im Schnitt ± 5 m betragen, so daß Verwerfungen ab 10 m Sprunghöhe erkannt wurden. Um die Übersichtlichkeit nicht zu beeinträchtigen, wurden jedoch die 10 m-Linien und viele Verwerfungen unbedeutender Sprunghöhe und Länge aus der Karte wieder entfernt.

Der Verlauf der Streichkurvenlinien zeigt als dominierendes Element die variszisch (SW—NE) streichende Nordschwarzwälder Schwelle (B. v. FREYBERG 1935, D. ORTLAM 1966, 1967), deren Achse von SW nach NE eintaucht. Der VH 2 erreicht im Bereich der Hornisgrinde Höhen um 1200 m und fällt nach NE auf Werte um 100 m ab. Die Nordschwarzwälder Schwelle wird im NW vom Pforzheimer Becken (D. ORTLAM 1968) und im SE vom Offenburger Becken (B. v. FREYBERG 1935, D. ORTLAM 1967) begrenzt. Während das Pforzheimer Becken eine Undationstiefe zwischen 400 m und 500 m aufweist (= Becken I. Ordnung), erreicht das Offenburger Becken nur Werte zwischen 200 m und 300 m (= Becken II. Ordnung).

Insgesamt lassen sich 5 verschiedene Störungssysteme beobachten, wovon 4 bereits bei den Untersuchungen im Kraichgau (D. ORTLAM 1970 a) beschrieben wurden:

1. Das hercynisch-krummschalige System
2. Das rheinisch-krummschalige System
3. Das streichende System
4. Das radiale System
5. Das variszisch-krummschalige System

Das bedeutendste Störungssystem ist das hercynisch-krummschalige, dessen Verwerfungen den Bereich der gesamten Streichkurvenkarte durchziehen. Seine wichtigsten Verwerfungskomplexe sind die nordwestlichen Ausläufer des Freudenstädter Grabens (Abb. 1, Nr. 1), die Nagolder Verwerfungszone (Nr. 2), die Wildberg—Herrenalber Verwerfungszone (Nr. 3), die Verwerfungszonen von Weil der Stadt (Nr. 4 und 5) und von Enzklösterle (Nr. 6).

Das rheinisch-krummschalige System ist hauptsächlich im westlichen Teil des Untersuchungsgebietes anzutreffen: die Ostverwerfung des Oberrheingrabens (Nr. 7), die Bühlertäler (Nr. 8) und Gaggenauer Verwerfung (Nr. 9) sowie die Südausläufer des Pfinztalgrabens (Nr. 10).

Das streichende System verläuft mehr oder weniger parallel zu den Streichkurvenlinien. Störungen dieses Typs sind die Pforzheimer (Nr. 11), die Neuenbürger (Nr. 12) und die Gernsbacher Verwerfungszonen (Nr. 20).

Das radiale System steht im allgemeinen senkrecht auf den Streichkurvenlinien, wobei direkte Übergänge zwischen dem radialen und dem streichenden System zu beobachten sind (Nr. 12). Beispiele für Störungen des radialen Systems

---

[1] Für finanzielle Unterstützung danke ich der Deutschen Forschungsgemeinschaft, Bad Godesberg.

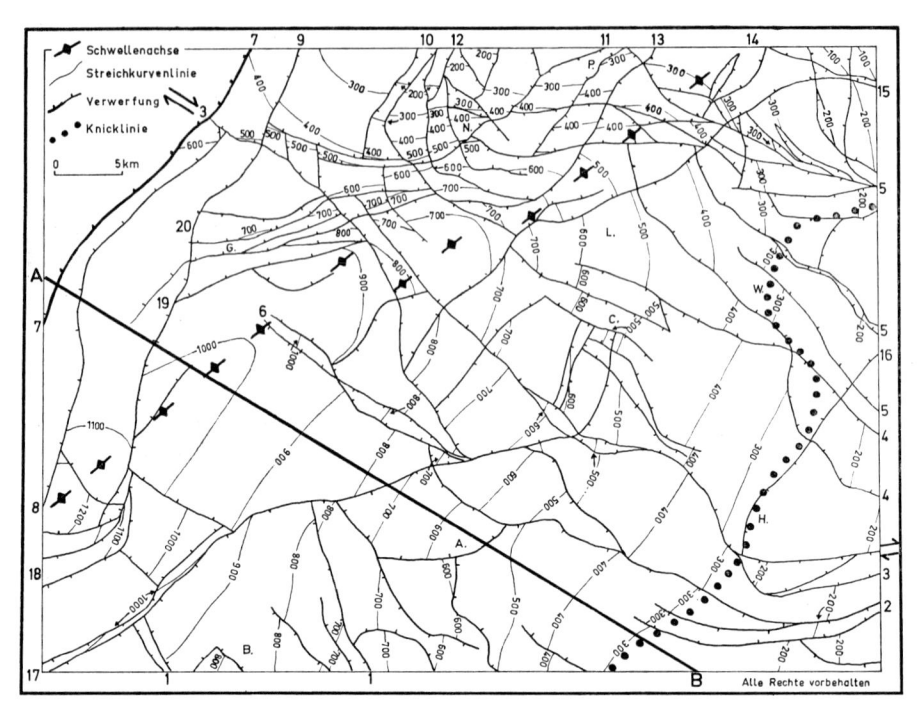

Abb. 1. Streichkurvenkarte des nördlichen Schwarzwaldes, bezogen auf die Oberfläche des Violetten (Karneol-)Horizontes 2 (VH 2) = Grenze Mittlerer/Oberer Buntsandstein in m zu NN (A. = Altensteig, B. = Baiersbronn, C. = Calw, G. = Gernsbach, H. = Herrenberg, L. = Bad Liebenzell, N. = Neuenbürg, P. = Pforzheim, W. = Weil der Stadt).

sind die Verwerfungszonen von Würm (Nr. 13), Wurmberg (Nr. 14) und Weissach (Nr. 15).

Die Verwerfungen des variszisch-krummschaligen Systems konnten im Nordschwarzwald zum ersten Mal in bedeutender Zahl und Länge beobachtet werden. Beispiele dieses Typs sind die Verwerfungszonen von Simmersfeld (Nr. 17), der Hornisgrinde (Nr. 18) und von Lichtental (Nr. 19). Auch beim variszisch-krummschalig streichenden System zeigen sich Übergänge und Beziehungen sowohl zum rheinisch-krummschaligen (Nr. 18 und 9) als auch zum hercynisch-krummschaligen System (Nr. 2 und 3).

Betrachtet man den Achsenverlauf der Nordschwarzwälder Schwelle (Abb. 1), so kann man in zwei Bereichen des mittleren Abschnittes deutliche Horizontalverschiebungen der Achse im Verlauf der hercynisch-krummschalig streichenden Wildberg—Herrenalber Verwerfungszone feststellen. Die horizontalen Versetzungsbeträge belaufen sich auf 5 km an der westlichen Horizontalverschiebung und 2 km an der östlichen Horizontalverschiebung, so daß daraus ein Gesamtversetzungsbetrag von 7 km resultiert. Dies ist **die erste große Blattverschiebung des Schwarzwaldes**. Sie ist rechtsdrehend und stimmt damit in ihrem Bewegungssinn mit dem Kräftediagramm rezenter Erdbeben wie z. B. jenem von Rastatt 1933 (W. HILLER 1934) überein.

Zum Alter der verschiedenen Störungssysteme können folgende Aussagen gemacht werden:

1. Die erste vertikale Bewegungsphase des hercynisch-krummschaligen Systems läßt sich im Gebiet Herrenalb auf das Perm festlegen, da die Schichtmächtigkeiten des Rotliegenden im Bereich von Verwerfungen sprunghaft zunehmen. Die zweite vertikale Bewegungsphase findet im Tertiär im Zusammenhang mit dem Einbruch des Oberrheingrabens statt. Schließlich wird eine jüngste horizontale Bewegungsphase eingeleitet, die die Ostverwerfung des Oberrheingrabens rechtsdrehend verformt (Abb. 1) und rezent durch Erdbeben nachgewiesen werden kann.

2. Die erste vertikale Bewegungsphase des variszisch-krummschaligen Systems muß an die Wende Karbon/Perm gestellt werden, da die Simmersfelder Verwerfung (Nr. 17) im Untergrund Granit gegen Gneis verwirft (Abb. 2), wobei die Gneisdecke auf der NW-Scholle nach dem Verwerfungsakt aber vor der Ablagerung des Oberrotliegenden bereits abgetragen wurde. Die zweite vertikale Bewegungsphase erfolgte im Tertiär und läuft mit dem Einbruchsvorgang des Oberrheingrabens konform.

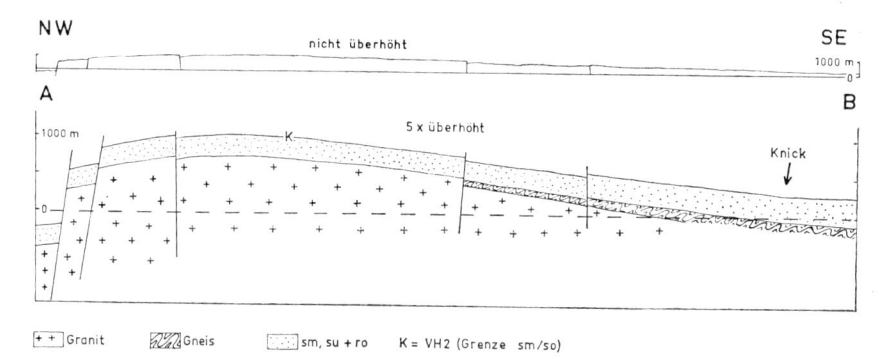

Abb. 2. Profilschnitt A—B (Abb. 1) durch das Nordschwarzwald-Gewölbe.

3. Die Vertikalbewegungen des rheinisch-krummschaligen, des streichenden und des radialen Systems fanden zur Zeit der Entstehung des Oberrheingrabens und der tertiären Heraushebung der Nordschwarzwälder Schwelle statt. Die Vertikalbewegungen sind bis in jüngste Zeiträume zu beobachten, da an einigen Verwerfungen wie z. B. der Pforzheimer Verwerfung (Nr. 11, F. RÖHRER 1919) und des Pfinztalgrabens (Nr. 10, D. ORTLAM 1970 a, Taf. 58) Würm-Lösse gegen Oberen Buntsandstein verworfen werden. Außerdem belegen die Erdbeben längs der Ostverwerfung des Oberrheingrabens und deren Begleitstörungen die rezenten Vertikal-Bewegungen, die durch geodätische Feinnivellements nachgewiesen sind (E. KUNTZ, H. MÄLZER & R. SCHICK 1970 und H. PRINZ & E. SCHWARZ 1970).

Auf der Ostabdachung des Nordschwarzwald-Gewölbes fallen die Schichten mit 1—2° nach SE hin ein. In einem Abstand von 45—55 km von der Ostrandverwerfung des Oberrheingrabens ist eine Knicklinie zu beobachten (Abb. 1). Östlich von ihr verflacht sich der Einfallswert der Schichten von 1—2° (westlich der Knicklinie) auf die Hälfte mit 0,5—1°. Diese Knicklinie, die ebenfalls auf dem Profilschnitt A—B (Abb. 2) festzustellen ist, wird durch den tertiären bis rezenten Aufstieg variszischer Strukturen (z. B. im Bereich der Nordschwarzwälder Schwelle) deutlich nach NE ausgelenkt. Die Anlage der Knicklinie hängt vermutlich direkt mit dem Einbruch des Oberrheingrabens zusammen und wird durch jüngere Bewegungen überprägt.

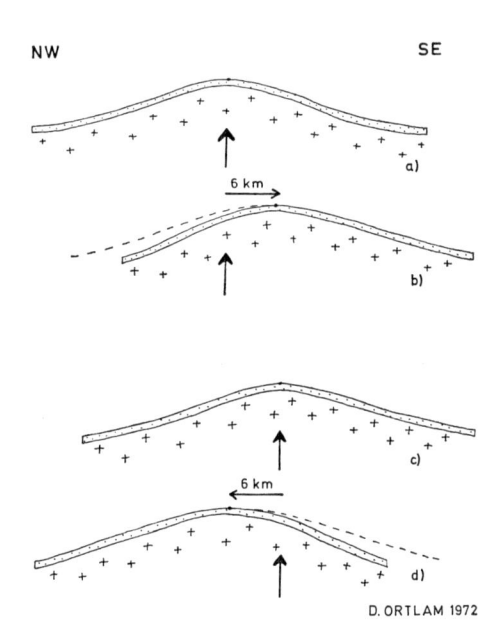

Abb. 3 a—d. Schematischer geologischer Schnitt zur Entwicklung der Nordschwarzwälder Schwelle im Bereich der Wildberg—Herrenalber Blattverschiebungszone. Erläuterung im Text.

Bei der Betrachtung der Dichte der Streichkurvenlinien auf beiden Seiten der Wildberg—Herrenalber Blattverschiebungszone stellt man fest, daß die Leehanglagen in bezug auf die Zerscherungsrichtung besonders steil ausgebildet sind, wie z. B. im Gebiet von Altensteig mit steiler Ausbildung im Gegensatz zum Gebiet von Calw mit flacherem Anstieg, oder das Gebiet von Neuenbürg mit steilem Anstieg und das Gebiet von Gernsbach mit flacherem Einfallen nach NW. Die Ursache der Versteilung der Leehanglagen ist schematisch in Abb. 3 dargestellt. Abb. 3 a und c zeigen schematisch die Nordschwarzwälder Schwelle beiderseits der Blattverschiebungszone im ursprünglichen Zustand mit dem zentral darunterliegenden epirogenen Aufstiegsvektor (Pfeil). Wird nun das obere Stockwerk durch die Blattverschiebung um einen bestimmten Betrag nach SE versetzt, verändert der in einem tieferen Stockwerk liegende Aufstiegsvektor seinen Ort nicht und versteilt dadurch den Leehang der Schwelle in zunehmendem Maße (Abb. 3 b). Das gleiche Bild ergibt sich in umgekehrter Richtung auf der anderen (südwestlichen) Seite der Blattverschiebungszone im Bereich der Nordschwarzwälder Schwelle (Abb. 3 d).

Aus den o. g. Fakten ergibt sich eine Gliederung der Erdkruste im zentralen Teil des Oberrheingrabens in drei tektonische Stockwerke (Abb. 4):

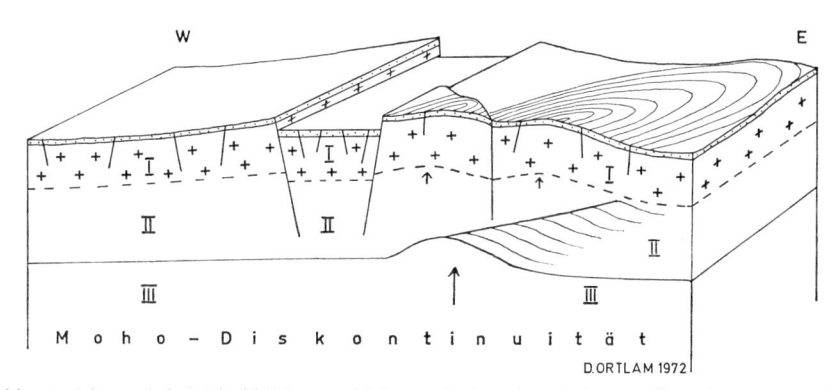

Abb. 4. Schematisches Blockbild zur Ableitung dreier tektonischer Stockwerke im zentralen Teil des Oberrheingrabens.

1. Das erste Stockwerk umfaßt ungefähr 80 % der vorhandenen Verwerfungen, die als Ausgleich der inneren Spannungen im obersten Teil der Erdkruste entstanden sind. Die Verwerfungen weisen vermutlich Tiefgänge von 2—4 km auf.
2. Das zweite Stockwerk wird lediglich von den großen Störungssystemen wie den rheinischen und variszischen Bruchlinien sowie den hercynischen Blattverschiebungen erreicht. Es ist bemerkenswert, daß hauptsächlich auf diesen drei Störungssystemen Erzlösungen im Laufe der Erdgeschichte aufgestiegen sind. Zur Festlegung der Untergrenze dieses Stockwerkes dient folgende Überlegung: Die großangelegte hercynische Blattverschiebung von Wildberg—Herrenalb verlangt im Untergrund eine mehr oder weniger horizontale Abscherungsfläche. Im sedimentären Oberbau (Buntsandstein und Rotliegendes) läßt sich diese Gleitfläche nicht beobachten. Da die Granite und Gneise im erkalteten Zustand keine Gleitflächen auszubilden vermögen, muß man die Abscherungsfläche zwangsläufig in tieferen Erdkrustenteilen suchen. Die Abscherungsfläche und damit die Untergrenze des zweiten Stockwerkes ist wahrscheinlich dort anzunehmen, wo die feste Erdkruste in zähplastische Teile der unteren Erdkruste übergeht, d. h. in Tiefen zwischen 15 und 20 km.
3. Im dritten Stockwerk befinden sich schließlich in zähplastischem und flüssigem Material die epirogenen Motoren der variszisch ausgerichteten Strukturen, deren Undationen seit dem Jungpaläozoikum bis heute nachgewiesen werden können (D. ORTLAM 1970 a, b). Weder der Einbruch des Oberrheingrabens noch die großangelegte Wildberg—Herrenalber Blattverschiebung konnten diesen wahrscheinlich im Bereich der Moho-Diskontinuität liegenden „Motor"

beeinflussen. Vielmehr beeinflußte dieser säkular wirkende „Motor" die sukzessive Entstehung und Auslenkung des Para-Oberrheingrabens aus dem Ortho-Oberrheingraben als Interferenzerscheinung zwischen variszischer und rheinischer Richtung (D. ORTLAM 1970 b).

Schrifttum

FREYBERG, B. v., 1935: Zur Paläogeographie des Jungpaläozoikums in Deutschland. — Z. dt. geol. Ges. **87**, 193—209.

HILLER, W., 1934: Der Herd des Rastatter Bebens am 8. Februar 1933. — Gerland **41**, 170—180.

KUNTZ, E.; MÄLZER, H. & SCHICK, R., 1970: Relative Krustenbewegungen im Bereich des Oberrheingrabens. — In: ILLIES, H. & MÜLLER, ST.: Graben Problems, 170—177.

ORTLAM, D., 1966: Fossile Böden und ihre Verwendung zur Gliederung des höheren Buntsandsteins im nördlichen Schwarzwald und südlichen Odenwald. — Jber. u. Mitt. oberrhein. geol. Ver., N. F. **48**, 68—78.

— 1967: Fossile Böden als Leithorizonte für die Gliederung des höheren Buntsandsteins im nördlichen Schwarzwald und südlichen Odenwald. — Geol. Jb. **84**, 485—590.

— 1968: Neue Ergebnisse aus dem höheren Buntsandstein des nördlichen Schwarzwaldes und des Kraichgaues. — Geol. Jb. **86**, 693—750.

— 1970 a: Eine Strukturkarte des südlichen Kraichgaues. — Geol. Jb. **88**, 553—566.

— 1970 b: Interferenzerscheinungen rheinischer und variszischer Strukturelemente im Bereich des Oberrheingrabens. — In: ILLIES, H. & MUELLER, ST.: Graben Problems, 91—97.

PRINZ, H. & SCHWARZ, E., 1970: Nivellement und rezente tektonische Bewegungen im nördlichen Oberrheingraben. — In: ILLIES, H. & MUELLER, ST.: Graben Problems, 177—183.

RÖHRER, F., 1919: Eine Verwerfung diluvialen Alters im Untergrund von Pforzheim. — Jber. u. Mitt. oberrh. geol. Ver. N. F. **8**, 58—61.

Le rôle des décrochements dans le socle vosgien et en bordure du Fossé rhénan

par

Michel Ruhland

Avec 4 figures dans le texte

The Role of Strike-Slip Faults in the Vosgian Basement and along the Border of the Rhinegraben.

Abstract: The Hercynian basement of the southern Vosges is traversed by some N 40° E striking sinistral strike-slip faults. The displacement along the Thur fault is of about 8 to 9 km causing a dextral rotation of about 30° of the Markstein block (Fig. 1).

The Tertiary tectonics produced fracture zones at the western border of the Rhinegraben which incorporate additional strike-slip movements. The horizontal displacement along the Tertiary fault zone of Steinbach is of about 900 m. Thus, the total sinistral displacement along the Rhinegraben axis reveals 15 to 18 km between the Vosges and the Black Forest (Fig. 3).

Introduction

Le socle vosgien est sillonné, surtout dans sa partie méridionale, par de grands accidents, qui ont fonctionné à la fin des temps hercyniens en tant que décrochements. Leur rôle au Tertiaire est plus délicat à mettre en évidence. Ils ont pourtant participé, lors de l'effondrement du Fossé, à la déformation de la surface post-hercynienne des Vosges.

Les grands décrochements du socle vosgien

Les grandes zones de décrochements échelonnées selon une maille de 5 à 10 km sont essentiellement de type sénestre (Fig. 1). Deux accidents, la faille de Ste-Marie-aux-Mines—Retournemer—Ognon et la faille de Boenlesgrab—Lautenbach, orientées N 30—40, sont parallèles entre eux. Greffée sur l'accident majeur de Ste-Marie-aux-Mines, la faille de la Thur déplace de 8 à 9 km la «ligne des klippes» Markstein—Sauwas, dont on retrouve la trace rive droite au Thalhorn et au Drumont. De plus la forme incurvée du décrochement a entraîné une rotation dextre d'environ 30° de l'ensemble de la série du Markstein. En effet lorsque l'on réajuste la ligne des klippes sur l'affleurement du Thalhorn on retrouve pour la série du Markstein une orientation axiale N 100—110 conforme à l'ensemble du Culm des Vosges méridionales (série d'Oderen—Bussang—Malvaux) et semblable à l'orientation structurale des terrains paléozoïques de la zone de Badenweiler—Lenzkirch du Sud de la Forêt-Noire. L'extension régionale de ce caractère fondamental permet de l'utiliser comme repère, dans l'analyse de relation Vosges—Schwarzwald.

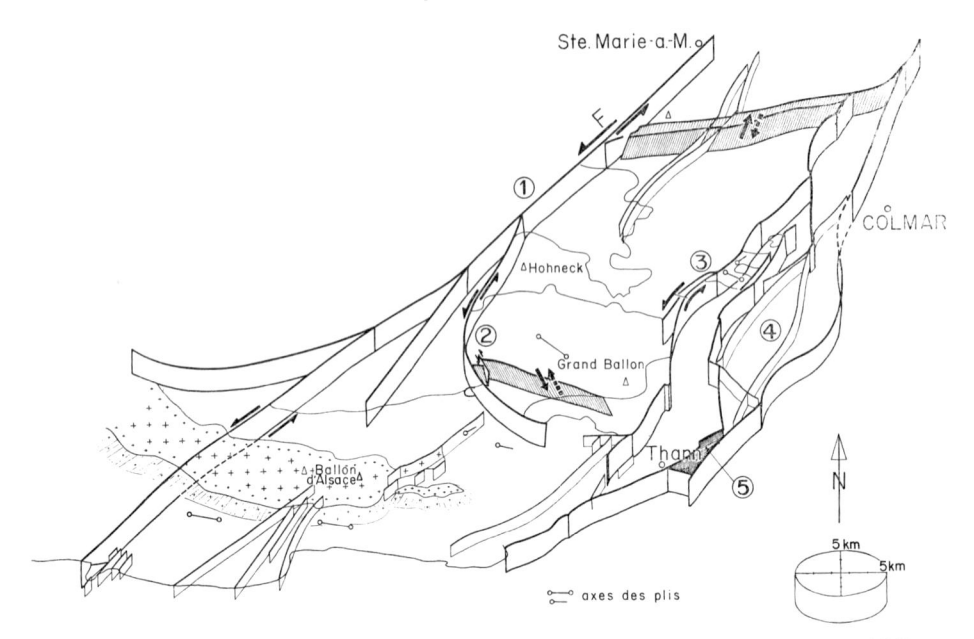

Fig. 1. Représentation schématique des grands accidents des Vosges moyennes et méridionales. (1) faille de Ste-Marie-aux-Mines—Retournemer—Ognon; (2) faille de la Thur; (3) faille Lautenbach—Boenlesgrab—Hunsrück; (4) champ de fractures de Guebwiller—Rouffach avec le fossé de Wintzfelden; (5) brèche tectonique tertiaire de Steinbach.

Les jeux de coulissage ont persisté ultérieurement car les décrochements affectent également la couverture triasique qui subsiste, dans la région des Trois-Epis, comme buttes témoins sur le socle. Dans ces cas les grès ont un même comportement physique—miroirs de failles incurvés, débit amygdalaire, paraclases à stries horizontales en systèmes associés — que le matériel granitique du socle.

Tectonique des champs de fractures

Le fossé de Wintzfelden

La particularité du champ de fractures de Guebwiller—Rouffach est de comprendre un fossé interne, appelé Fossé de Wintzfelden. Long de 10 km et large de 1 à 2 km il est constitué de bandes de terrain d'orientation subméridienne; la partie centrale est affaissée de 600 m par rapport aux extrémités effilées. Les terrains les plus récents qui y affleurent sont le Keuper et les premiers niveaux du Lias. La nature de ces terrains ne permet aucune observation de tectonique fine. En revanche celles-ci sont parfaitement réalisables dans les roches qui bordent le fossé, granite du socle à l'Ouest, Buntsandstein de la zone de horst à l'Est. Les miroirs de faille, couverts de striations horizontales, y sont très abondants; leur sens des déplacements ainsi que la distribution des diaclases dans les grès permettent de conclure à un mouvement relatif sénestre.

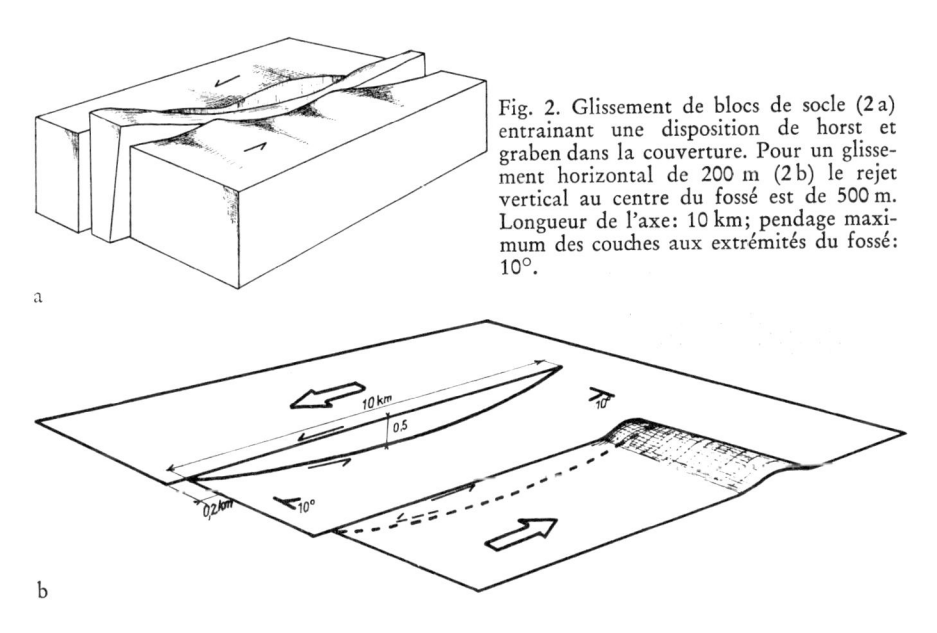

Fig. 2. Glissement de blocs de socle (2 a) entrainant une disposition de horst et graben dans la couverture. Pour un glissement horizontal de 200 m (2 b) le rejet vertical au centre du fossé est de 500 m. Longueur de l'axe: 10 km; pendage maximum des couches aux extrémités du fossé: 10°.

Faute de points de repère la valeur du déplacement horizontal ne peut être apprécié. Celui-ci tout en restant faible provoque, par le jeu de décrochement des lanières et selon leur rayon de courbure (Fig. 2), un important effondrement de la partie centrale, 3 à 5 fois supérieur au rejet horizontal.

Le champ de fractures de Vieux-Thann

L'étroit champ de fracture de Vieux-Thann est bordé le long de la faille vosgienne, orientée N 40, par une très importante brèche de faille minéralisée, d'âge tertiaire. Un lever cartographique de détail, effectué récemment par G. HIRLEMANN a montré que la brèche de Steinbach se trouve disloquée et décrochée en plus d'une dizaine de tronçons principalement décalés par un mouvement relatif sénestre. Le décalage transversal subit est de 900 m pour une longueur de 5 300 m, soit plus de 15%. Si l'on détermine le déplacement relatif selon la direction rhénane celui-ci est de l'ordre de 40% par rapport à la dimension de la zone considérée.

Relations Vosges — Schwarzwald

Les niveaux repères qui peuvent être pris en considération pour analyser les relations entre Vosges et Schwarzwald sont à rechercher dans les niveaux plissés du Culm. A l'étroite zone de Badenweiler—Lenzkirch, d'orientation axiale N 100—110, semble correspondre les séries de Malvaux—Giromagny, abritées et préservées des complications tectoniques ultérieures par le môle du granite du Ballon d'Alsace (Fig. 3). En outre ces terrains sont le siège d'une importante anomalie magnétique que EDEL & LAUER ont suivi à travers le Fossé rhénan. Tous

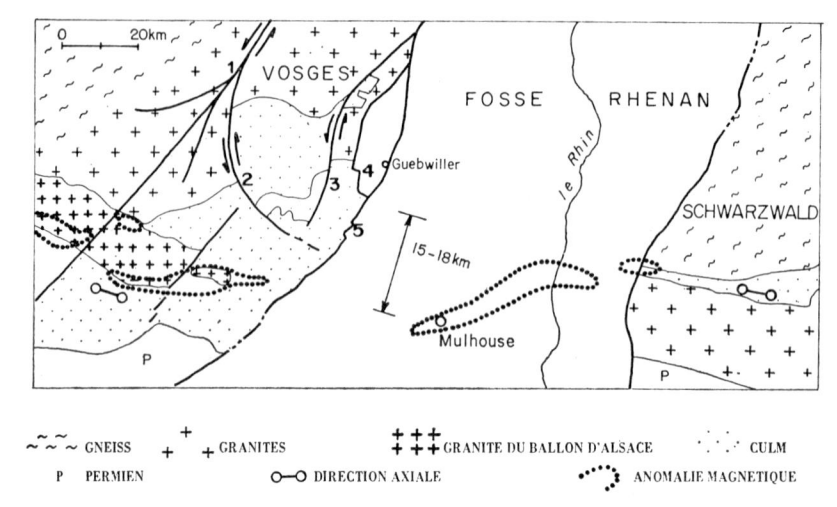

Fig. 3. Relations Vosges — Schwarzwald. Niveaux repères: Culm des Vosges méridionales et zone Badenweiler — Lenzkirch du Sud Schwarzwald; anomalies magnétiques.

ces critères conduisent à la conclusion suivante: les déplacements horizontaux sénestres post-hercyniens entre Vosges et Schwarzwald sont de l'ordre de 15 à 18 km. Selon la largeur du Fossé rhénan considérée le décalage relatif est de 40%, c'est-à-dire analogue à celui déterminé dans les brèches tertiaires de Steinbach. Il reste cependant à préciser l'emplacement des décrochements majeurs sous le Fossé rhénan.

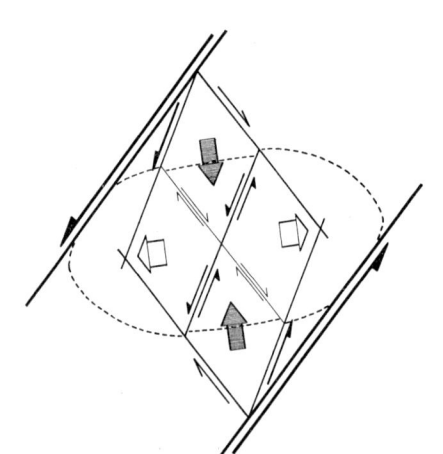

Fig. 4. Modèle de déformation: Décrochement sénestre majeur et cisaillements secondaires et zone de tension subméridienne.

D'après l'ensemble des faits observés, le mécanisme de déformation résulte de décrochements majeurs N 40, de type sénestre (Fig. 4), générateurs de cisaillements annexes N 25 et 145 et de zones de distensions subméridiennes.

Bibliographie

AHORNER, L., 1970: Seismo-tectonic Relations between the Graben Zones of the Upper and Lower Rhine Valley. — Graben Problems, 155—166, E. Schweizerbart'sche Verlagsbuchhandl., Stuttgart.

EDEL, J. B. & LAUER, J. P., 1974: Magnetic Anomalies, Thermal Metamorphism and Tectonics in the Rhine-graben Area. — Approaches to Taphrogenesis, 155—160, E. Schweizerbart'sche Verlagsbuchhandl., Stuttgart.

ELLER, J. P. VON, FLUCK, P., HAMEURT, J. & RUHLAND, M., 1972: Présentation d'une carte structurale du socle vosgien. — Sci. Géol. Bull., 25, 3—19, Strasbourg.

ILLIES, H., 1965: Bauplan und Baugeschichte des Oberrheingrabens. — Oberrh. Geol. Abh., 14, 1—54.

RUHLAND, M. & HIRLEMANN, G., 1973: Tectonique du socle vosgien et de la bordure du Fossé Rhénan au Tertiaire. — Réunion Sc. de la Terre, p. 373, Paris.

THEOBALD, N., 1952: Structure du champ de fractures de Guebwiller. — Bull. Carte Géol. France, 50, 237, 1—53.

Block-faulting at a Rhinegraben scarp and a comparable recent landslide structure at the Korinthian graben

Le dispositif structural du champ de fractures
de Ribeauvillé (Haut-Rhin, France)
Un exemple de fragmentation tectonique (block-faulting)
en bordure occidentale du Fossé rhénan

par

Georges Hirlemann

Avec 2 figures dans le texte

The Fracture Field of Ribeauvillé (Haut-Rhin, France). An Example of Block-Faulting at the Western Border of the Rhinegraben

Abstract: The fracture field of Ribeauvillé is considered to be a promontory area between the Vosges scarp and the Rhinegraben trough. A block-diagram of this area is presented. The base of the Buntsandstein (Lower Triassic) has been used as a reference horizon. The tectonic pattern reveals a set of parallel elongated blocks which are separated additionally by transverse fractures. The total amount of throw between the top of the granitic basement on the elevated shoulder and the base of the Buntsandstein underneath the nearby graben fill is about 2900 meters.

Two pairs of related fault systems can be distinguished: one system is composed of the N 10 — N 30 and N 90 — N 110 directions, and another one reveals a N 40 — N 70 and a N 120 — N 150 strike. The first fault system prevails in the southern part of the described area, and the latter predominates in the northern region where the Bilstein granite outcrops. The tectonic framework in this area seems to be related to the N 80° structure of the Hercynian granite.

Pre-existent fracture zones within the Hercynian basement have considerably influenced the Tertiary taphrogenesis of the Rhinegraben.

Introduction

Le champ de fractures de Ribeauvillé, situé en Alsace moyenne à quelques kilomètres au NW de Colmar, fait face aux formations volcaniques du Kaiserstuhl en pays de Bade (Fig. 1). Il se développe au pied des Vosges moyennes constituées par des terrains cristallophylliens antévarisques (gneiss perlés de Ribeauvillé, migmatites de Kaysersberg et des Trois-Epis) et des granites varisques (granites de Thannenkirch, du Brézouard et du Bilstein). A la hauteur de Ribeauvillé le Fossé rhénan est comblé par 800 à 900 m de Tertiaire et 100 à 120 m de Quaternaire.

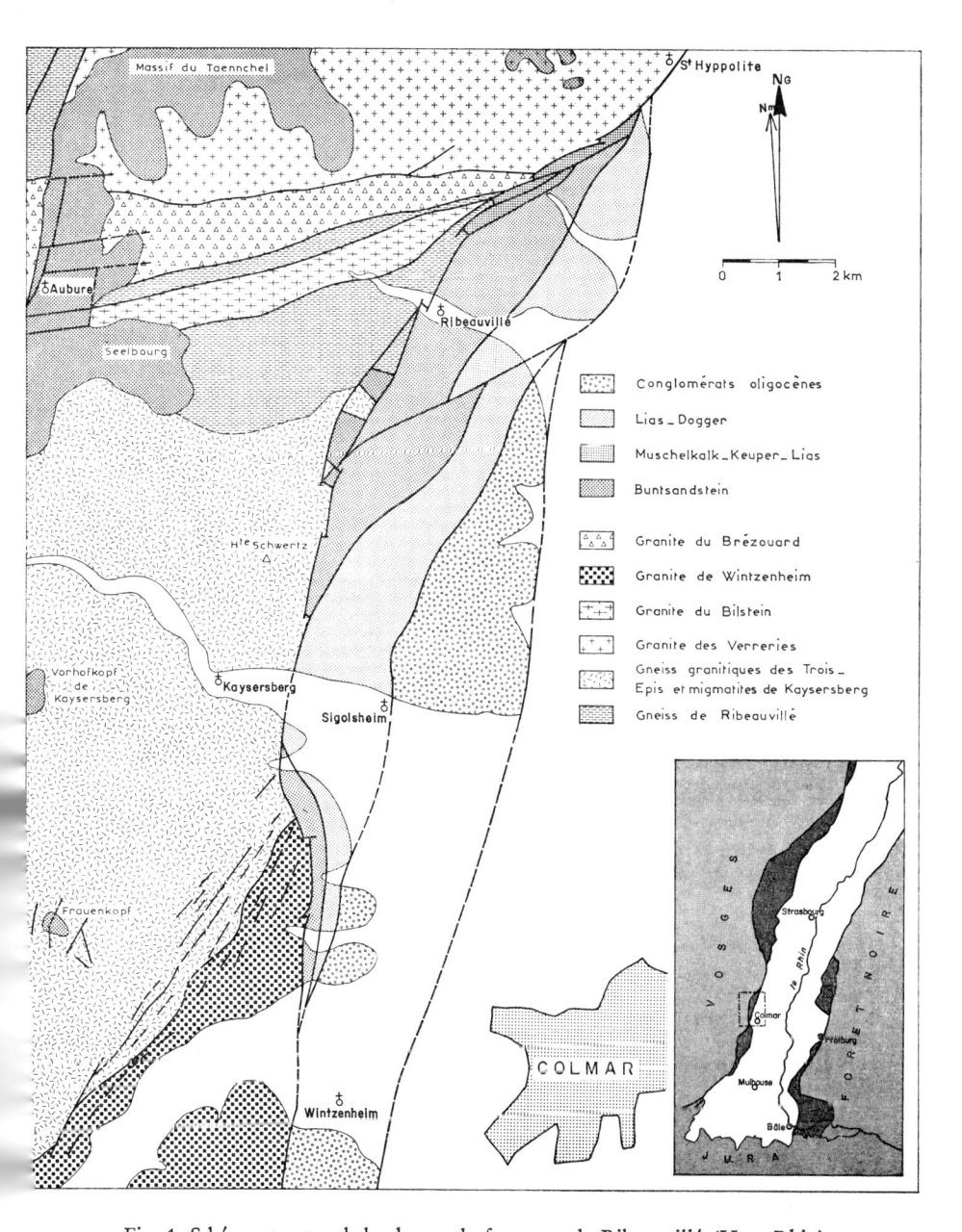

Fig. 1. Schéma structural du champ de fractures de Ribeauvillé (Haut-Rhin).

Dans le champ de fractures, la structure en mosaïque et l'effondrement différentiel des compartiments permettent à tous les niveaux stratigraphiques (du Buntsandstein au Dogger) d'affleurer.

Le bloc-diagramme

Le dispositif structural du champ de fractures de Ribeauvillé et de ses abords immédiats est illustré par le bloc-diagramme de la Fig. 2.

Il a été construit à l'aide du pantographe à parallélogramme PERSPEKTOMAT P 40. L'angle de projection choisi est de 35° 15′. La surface de référence correspond au mur du grès vosgien (Buntsandstein moyen); les rejets sont calculés à partir du massif triasique du Taennchel (+ 700 m d'altitude) et par extrapolation verticale de la série triasico-jurassique dans le sondage d'Ostheim (—1600 m d'altitude). L'image ainsi obtenue est en quelque sorte l'écorché du socle débarrassé de sa couverture sédimentaire.

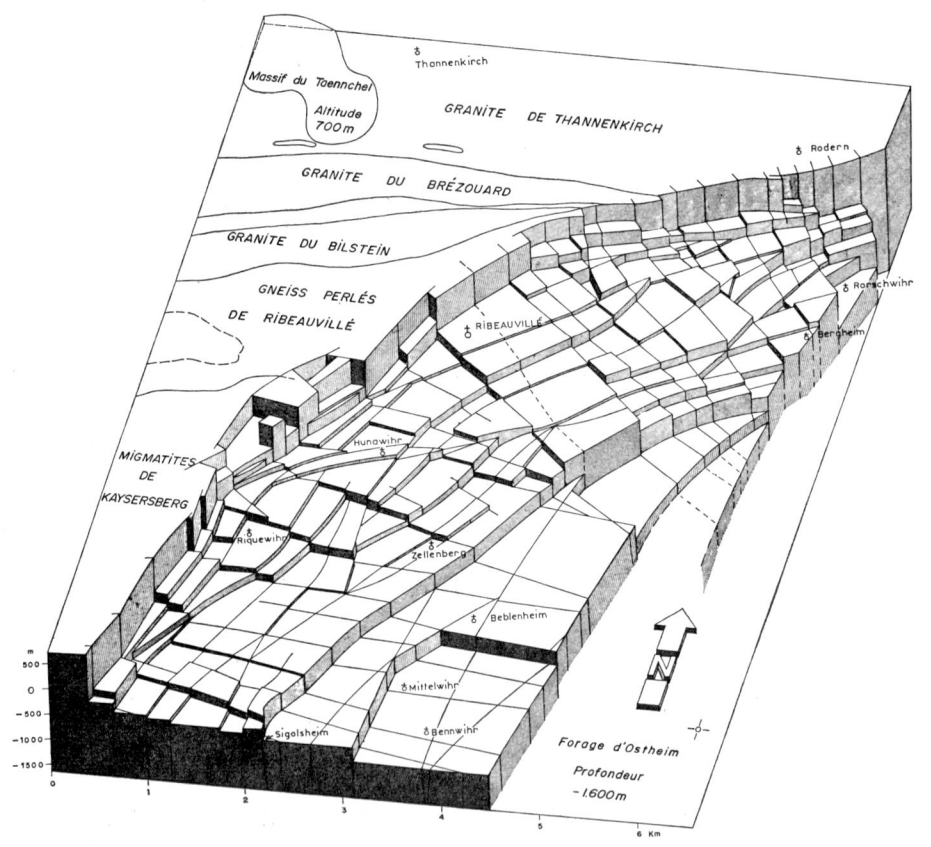

Fig. 2. Bloc-diagramme de la partie Nord du champ de fractures de Ribeauvillé (Haut-Rhin). Le mur du grès vosgien (Buntsandstein moyen) sert de surface de référence; il se trouve à 700 m dans le massif du Taennchel et à — 1 600 m dans le Forage d'Ostheim.

Pour bâtir ce schéma, les failles ont été considérées comme étant verticales, les couches comme étant horizontales. Dans une étape d'analyse ultérieure il faudra tenir compte du pendage réel des failles et des couches afin de mieux appréhender le mode de mise en place du dispositif structural.

Le réseau de failles

Un réseau complexe de failles se développe entre la faille vosgienne et la faille rhénane. L'analyse de leur répartition fait ressortir quatre orientations dominantes associées deux à deux:
— les failles N 10 à N 30 associées à des failles N 90 à N 110;
— les failles N 40 à N 70 associées à des failles N 120 à N 150.

Ces deux couples coexistent; cependant d'une extrémité à l'autre du champ de fractures l'un ou l'autre couple prédomine. Le premier est prépondérant au S de Ribeauvillé, alors que le second est surtout exprimé au N de cette localité. Le passage de l'orientation N 10 à l'orientation N 70 est progressif et se manifeste à l'approche des structures N 70 — N 80 du socle (granite du Bilstein).

Les failles longitudinales (N 10 à N 70) découpent la zone de bordure du Fossé rhénan en lanières juxtaposées, effilées aux extrémités, orientées N 10 à N 45. Ces lanières sont alors fragmentées par des failles transversales de réajustement orientées N 90 à N 150.

Valeur des rejets

Selon l'amplitude du rejet, les failles peuvent être classées en failles majeures au rejet pluri-hectométrique (300 à 1000 mètres), en failles mineures au rejet décamétrique à hectométrique (50 à 200 m) et en microfailles au rejet centimétrique à métrique. Les exemples cités concernent les failles longitudinales majeures. Le rejet de la faille vosgienne, calculé à partir de la base du grès vosgien dans le massif du Taennchel, est de 360 m à l'W de Hunawihr, 420 m au NW de Riquewihr, 690 à 910 m au SW de Riquewihr, 440 à 660 m au N de Ribeauvillé, 590 à 1040 m aux environs de Rodern. Une seule valeur fiable peut être retenue pour la faille rhénane: 600 à 700 m à l'E de Ribeauvillé (résultat d'une prospection par sondages électriques). Une autre faille majeure court de Bergheim à Sigolsheim; elle isole la sous-unité la plus orientale exclusivement couverte de conglomérat oligocène; son rejet moyen est de 300 m, il atteint 700 m à l'E de Ribeauvillé.

Conclusions

Le bloc-diagramme est l'illustration optimale du dispositif structural du champ de fractures de Ribeauvillé.

Généralement orientés E—W, les terrains cristallophylliens déjà structurés, donc anisotropes, sont découpés en nombreuses lanières subméridiennes effondrées en «marches d'escalier»; au niveau des granites le matériel plus isotrope favorise la naissance de fractures moins fréquentes, mais au rejet plus important. Les failles transverses de réajustement, contemporaines ou légèrement postérieures aux failles longitudinales, débitent les lanières en compartiments. Le socle et sa couverture offrent alors l'image d'une structure fortement fragmentée.

A la hauteur de Ribeauvillé le rejet total entre le point culminant du socle (base présumée du Trias au sommet du Brézouard: 1250 à 1300 m) et le fossé (mur calculé du grès vosgien dans le sondage d'Ostheim: —1600 m) est de 2850 à 2900 m. Le long d'une coupe transversale synthétique, tracée du sommet du Brézouard au sondage d'Ostheim, on distingue trois zones structurales ayant chacune une largeur et un rejet cumulé propres:

	largeur	rejet	rapport rejet/largeur
le socle (du Brézouard à la faille vosgienne)	11 km	0,94 km	0,08
le champ de fractures (de la faille vosgienne à la faille rhénane)	4 km	1,56 km	0,40
le fossé (de la faille rhénane au sondage d'Ostheim)	2,4 km	0,4 km	0,16

Dans le champ de fractures de Ribeauvillé et dans toute la bordure sous-vosgienne on relève essentiellement des directions de faille identiques à celles des accidents anciens du socle contigu. Le Fossé rhénan s'effondre en utilisant préférentiellement des fractures latentes inscrites dans le socle. L'orientation générale N 25 qui en résulte, est une adaptation régionale au régime des contraintes qui régnait dans le bouclier rhénan.

Bibliographie

Dellenbach, J., 1960: Etude géologique de la région fracturée entre les vallées du Strengbach et de l'Eckenbach, au NE de Ribeauvillé (Haut-Rhin). — D. E. S., Fac. Sci. Strasbourg, 25 p., ronéo.
Eller, J. P. von; Fluck, P. & Hameurt, J., 1970: Carte géologique des Vosges moyennes, partie centrale et partie orientale (notice explicative, 1 annexe h.-t. en couleur). — Bull. Serv. Carte géol. Als. Lorr., 23, 1, 29—50.
Hirlemann, G., 1970: Contribution à l'étude géologique du champ de fractures de Ribeauvillé (Haut-Rhin). — Thèse 3ème cycle, Fac. Sci. Strasbourg, 109 p., ronéo.

A Landslide on the Rim of a Graben

by

L. Müller and G. Lögters

With 3 figures in the text

A landslide that occurred in April 1971 on the northern coast of the Peleponnes in Greece was investigated with respect to possible connections with the southern border fault of the Korinthian Graben.

Description of the landslide

The actual slide occurred in the upper part of the slope. The large debris pile now covering highways and railroad consists of masses that were shoved across the rim of the 80 m high artificial highway cut (compare Fig. 1). The main direction of sliding was NE, but considerable amounts of rock and soil were transported exactly northward thus producing the debris cone. A second slide took place one week later directly east of the toe of the debris pile. Its direction of sliding was NW.

Geological situation

The area mainly consists of Upper Cretaceous limestones, and is part of the Olonos-Pindos zone. The rock is extremely folded and faulted due to tectonic forces and downslope movements. In the area of the highway cut as well as behind the slide the limestone is comparatively stable. A number of faults striking parallel to the coast certainly are elements of the Korinthian Graben structure (compare Fig. 2 and 3). Similar synthetic faults may be observed in Neogene conglomerates about 50 km west of Panagopoula. Also the terraces close to the town of Korinth most probably are of tectonic origin and consist of synthetic step faults of the Graben. The geological situation and especially the tectonics of the Korinthian Graben were described by RENZ (1912), TRIKALLINOS (1956) and GARAGUNIS (1967).

Influence of graben tectonics on the landslide

Usually the following factors are taken into account when causes of landslides are investigated:
— static and dynamic water pressure
— strength of rock material
— orientation and shear strength of joints resp. other planes of weakness
— time (progressive failure)

Fig. 1. Panagopoula landslide; a: total view; b: close-up of the actual slide.

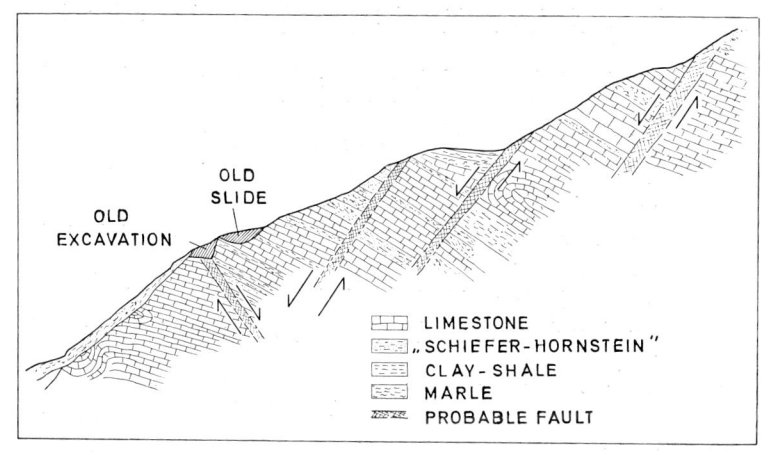

Fig. 2. Geological section of the slope.

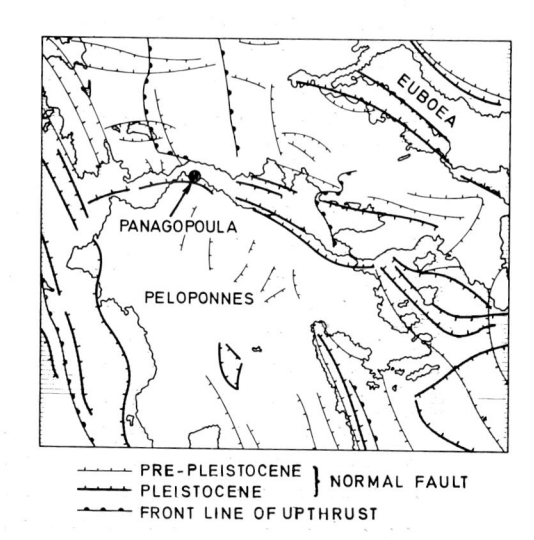

Fig. 3. Fault systems in Greece (BORNOVAS 1971).

Because of lacking scientific background and knowledge the influence of the stress distribution inside the slope is often not included in the investigation. The existence of primary stress fields is doubted in many cases. The Panagopoula landslide in Greece presents an example for the need of new approaches including the stress fields and its variations when attacking landslide problems. Here all the classic solutions were not able to explain the slope failure although neither one of them can be neglected:

Earthquakes certainly had an influence in this seismically active area, where the epicenters of many recent seismic events line up along the two border faults of the Korinthian Graben. But there is the surprising circumstance that the only earthquake in the period of concern happened two days before the slide occurred.

An immediate influence of the **g r o u n d w a t e r** can also not be found. The rainy winter time had recently ended, but rainfall of considerable magnitude was not observed in the period directly preceding the slide.

Also the **a r t i f i c a l   c u t   f o r   t h e   n e w   h i g h w a y** was already completed one year ahead of the event.

The shear **s t r e n g t h   o f   t h e   r o c k   m a s s** certainly is low as in-situ shear tests later have proven; but the results only show that the slopes in this area are in a state of limit equilibrium which is even better demonstrated by their morphology; they do not explain, however, why the slide happened at that and no other time.

The same is true for the **c o a s t - p a r a l l e l   f a u l t s** that have acted as slide surfaces in this case. They were existent before, and they obviously did not trigger this slide.

In this situation it was inevitable to include the influence of the graben tectonics for a more complete explanation of the landslide event.

The recent stress field in a slope along the southern escarpment of the Korinthian Graben is constantly rearranged due to the very small displacements that occur with the upheaval of the Graben shoulders. Thus the slope is brought to a state of limit equilibrium and subsequently to failure in the course of many years.

In Panagopoula this process was probably accelerated by the artificial excavation for the highway in 1969 and 1970. In April of 1971 the state of stress must have been developed very far. The precipitations of the winter time and further earthquake shocks probably have added their portion to increase the high stress field. Only under such conditions the natural process of readjusting equilibrium conditions by rearranging large masses of rock in form of a landslide could start without any extraordinary triggering event in the vicinity.

In the meantime core borings, in-situ shear tests, in-situ strain measurements (after MENCL), deformation measurements by means of chain deflectometers, inclinometers, and geodetically, as well as measurements of water table and permeability, and measurements of rock noise have been executed in many different places on the surface and in boreholes. The records of about one year's time give first indications concerning the tectonic influence.

Conclusion

The transition from primary to secondary state of stress is well known in connection with excavations of underground openings. It would be favourable to introduce this viewpoint into the discussion of slope stability problems, too. Every excavation and any other change of boundary conditions has a comparable effect on the stress field in a slope.

This becomes more pronounced in tectonically active areas. The low velocity of tectonic movements has let us underestimate their effect on civil engineering problems in the past. Not the movement itself but its effect on the stress field in the earth crust causes the actual damage. In most cases it only takes an extremely small impulse to cross the threshold from limit equilibrium to failure.

The correlations between recent tectonic activity and actual civil engineering problems are manifold, but they will rest undetected as long as the geological sciences only deal with the history of tectonics while civil engineering only feels concerned about the superficial symptoms of recent movements in the earth crust. The underestimation of research projects proposing to investigate the connections between tectonics, rock mechanics, and civil engineering will further prevent any progress in the understanding of the very complex, though basic motor of many environmental problems.

References

GALANOPOULOS, A., DELIBASIS, N. & BORNOVAS, I., 1971: Seismotectonic Map of Greece, 1:1 000 000. — Publ. by Inst. for Geol. and Subsurface Res.

GARAGUNIS, C., 1967: Geologie und Tektonik im Bereich des Kanals und der Umgebung von Korinth. — Ann. Géol. Pays Hellén., **18**, Athen.

RENZ, C., 1912: Über den Gebirgsbau Griechenlands. — Z. dt. geol. Ges., **64**.

TRIKKALINOS, J. K., 1956: Die Auswirkung junger, sehr starker diluvialer und rezenter orogener Bewegungen im Gebiete Griechenlands. — Geotekton. Sympos. zu Ehren von HANS STILLE, Stuttgart.

Geothermal anomalies and consequences for diagenesis and thermal waters

Eine geothermische Karte des Rheingrabenuntergrundes*

von

D. Werner und F. Doebl

Mit 6 Abbildungen und 2 Tabellen im Text

A Geothermal Map of the Rhinegraben

Abstract: A geothermal map of the northern Rhinegraben is drawn based on 920 temperature measurements in drill holes for oil prospection. There are considerable local anomalies, whose intensity increases with depth. It is not possible to interprete these results by means of heat conduction. The circulation of water seems to be a more probable explanation.

Allgemeines

Untersuchungen an Grabensystemen und Riftstrukturen haben gezeigt, daß diese Erscheinungen von thermischen Anomalien begleitet sind. Jeder Versuch, die Entstehung der Gräben zu erklären, muß zugleich auch eine Erklärung über deren Wärmehaushalt einschließen. Womöglich fällt hierbei der Geothermik eine Schlüsselrolle zu.

Für das relativ gut erkundete Gebiet des Oberrheingrabens sind bislang nur wenige isolierte geothermische Daten bekannt. Diese Kenntnislücke schließen zu helfen, ist das Ziel der vorliegenden Arbeit. Es wird ein erster Versuch unternommen, eine zusammenhängende geothermische Karte des nördlichen Rheingrabengebietes zu entwerfen. Die Daten hierzu wurden im Laufe vieler Jahre im Zusammenhang mit der Erdölprospektion gesammelt. Sie sind auf die Sedimentfüllung des Grabens beschränkt. Das kartographische Resultat ist daher im eingangs skizzierten Sinne nicht oder doch nur sehr bedingt aussagekräftig. Die Ursache für die starken lokalen Anomalien, wie sie auf den hier vorgestellten Temperaturkarten erkennbar sind, ist offenbar nicht in größeren Tiefen zu suchen, sondern am Ort der Anomalien selbst. Es zeigt sich, daß ein Interpretationsversuch mit Hilfe der Wärmeleitungsgleichung vergebens ist.

Wenn hier von „Anomalien" gesprochen wird — ein Terminus, der ebenso nützlich wie anfechtbar ist —, dann müssen zwei Sachverhalte auseinandergehalten werden:

---

* Beitrag Nr. 66 aus dem Instiut für Geophysik der ETH Zürich.

1. Die großräumige Anomalie (erhöhte Wärmestromdichte), die über das eigentliche Grabengebiet hinausgreift und als thermisches Abbild des gesamten Riftsystems interpretiert werden kann. Sie drückt sich — zumindest qualitativ — deutlich aus auf einer Darstellung von HAENEL (1971), der die bislang bekannten Wärmestromdichtemeßwerte Deutschlands in einer Karte zusammengefaßt hat.
2. Die lokalen Temperaturanomalien im Bereich der Sedimentfüllung des nördlichen Rheingrabens. Auf sie beschränkt sich diese Darstellung.

Messungen

Die Temperaturen wurden während der Bohrlochmessungen (z. B. elektr. Widerstandsmessungen) entnommen.

An der Meßsonde wurde außen, etwa 3 m über der Basis, ein Maximalthermometer angebracht, dessen Meßgenauigkeit ± 0,5° C beträgt.

Die Messungen, bei denen unsere Temperaturwerte entnommen wurden, sind:
1. Das Electric-Log (früher Elektr. Kern.) = ES. Diese Widerstandsmessung beginnt ungefähr 3—4 Stunden nach Beendigung der Spülungszirkulation am Boden des Bohrlochs.
2. Die Microlog-Messung (ebenfalls eine Widerstandsmessung) = ML wurde meist im Anschluß an die ES-Messung vorgenommen. Die Sonde dürfte meist 4—6 Stunden nach der Zirkulation den tiefsten Meßpunkt erreicht haben.
3. Die Gamma-Ray-Messung = GR. Diese Messungen werden oft später, zum Teil sogar Jahre nach Beendigung der Bohrtätigkeit gemacht.

Durch die Spülung wird das Bohrloch stark ausgekühlt. Die Erwärmung auf die ursprüngliche Temperatur dauert lang, jedoch ist nach 3—4 Tagen die ursprüngliche Temperatur fast erreicht. Beispiele dieser Erwärmung zeigt Tab. 1.

Aus der Tabelle ist zu ersehen, daß bereits nach 5 Tagen der stärkste Temperaturanstieg stattgefunden hat und daß die Erwärmung nach über 8 Jahren keine großen Unterschiede zeigt. Außerdem ist zu sehen, daß während der ersten Messungen direkt nach dem Bohren, die normalerweise im Abstand von ungefähr 2—3 Stunden stattfinden, die Erwärmung bereits begonnen hat.

Datenauswertung

Die rohen Temperaturanzeigen, die meist nicht identisch sind mit der wahren Gesteinstemperatur, sind zu korrigieren. Detaillierte Angaben über den zeitlichen Abstand zwischen verschiedenen Messungen am gleichen Ort, wie in Tab. 1 gezeigt, liegen nur für wenige Bohrlöcher vor. Zur Ermittlung eines repräsentativen Korrekturwertes δT wurden die Daten in zwei Klassen eingeteilt: Klasse 1 besteht aus den weitgehend verläßlichen Temperaturwerten, die im Zusammenhang mit der GR-Messung erhalten wurden, Klasse 2 umfaßt alle übrigen (ES, ML, SL).

Insgesamt wurden 920 Wertepaare (Temperatur T, zugehörige Teufe z) verarbeitet. Hiervon entfallen 97 (10,5%) auf die Datenklasse 1. Die Korrektur δT

Tab. 1. Beispiele der Erwärmung im Bohrloch nach Einstellung der Spülungszirkulation.

Bohrung	erste Meßart	Zeit nach Spülungszirkulation	Teufe z [m]	Temperatur T [°C]	zweite Meßart	Zeit nach Spülungszirkulation	Teufe z [m]	Temperatur T [°C]	δ T [°C]	Zeitdifferenz zwischen den Messungen
Rheindürkheim 1	ES	2—4 Std.	2085	81	ML	4—6 Std.	2085	86	5	2 Std.
Leopoldshafen 15	ES	2—4 Std.	1260	56	GR	5 Tage	1245	70 / 70.5 *	14 / 14.5 *	5 Tage
Leopoldshafen 15	ML	4—6 Std.	1259	60	GR	5 Tage	1245	70 / 70.5 *	10 / 10.5 *	5 Tage
Stockstadt 40	ES	2—4 Std.	1735	86	SL[1]	5 Mon.	1719	121 / 122 *	35 / 36 *	5 Mon.
Stockstadt 22	ES	2—4 Std.	1678	88	GR	8 ½ Jahre	1650	110 / 110 *	22 / 23 *	8 ½ Jahre

* Korrektur auf die entsprechende Teufe der ersten Meßart.

[1] Sonic-Log

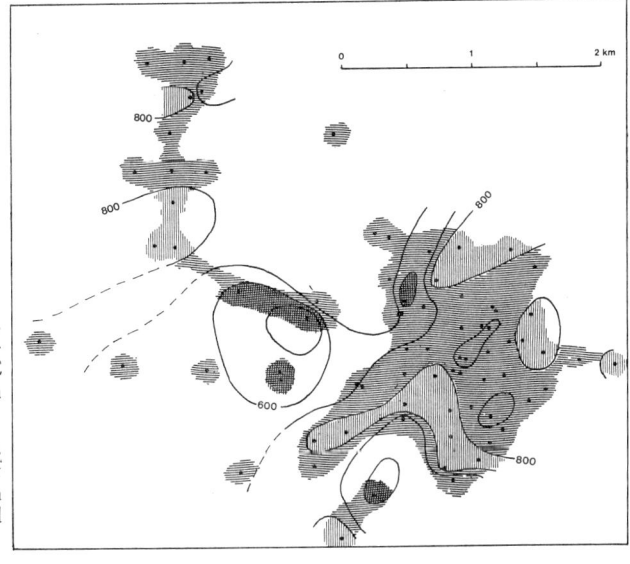

Abb. 1. Feld Stockstadt. Tiefenlinienplan der Geoisothermenfläche für 50°C (Bohrlöcher sind durch Punkte markiert).

Fig. 1. Field Stockstadt. Depth-contour map of the 50°C geoisotherm (drill holes are marked by dots).

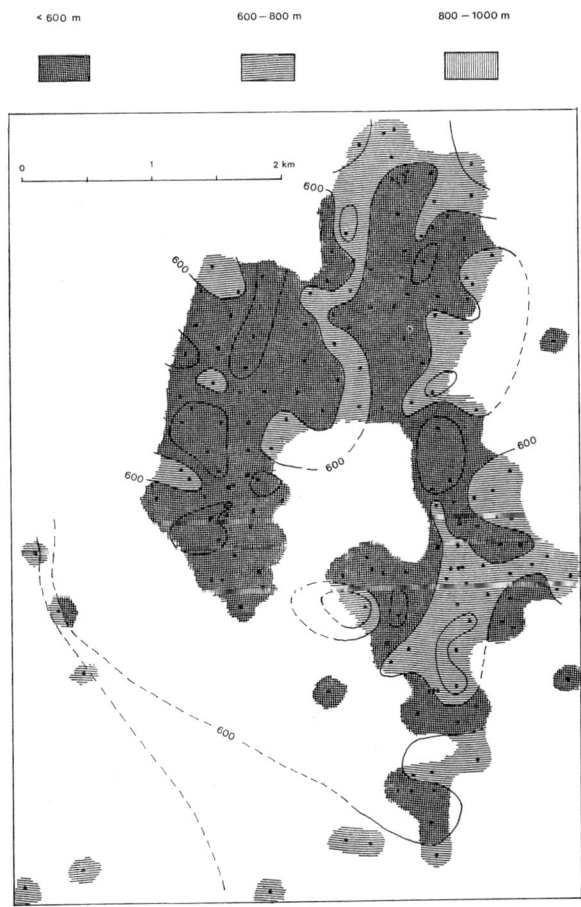

< 600 m   600 – 800 m   800 – 1000 m

Abb. 2. Feld Landau. Tiefenlinienplan der Geoisothermenfläche für 50°C.

Fig. 2. Field Landau. Depth-contour map of the 50°C geoisotherm.

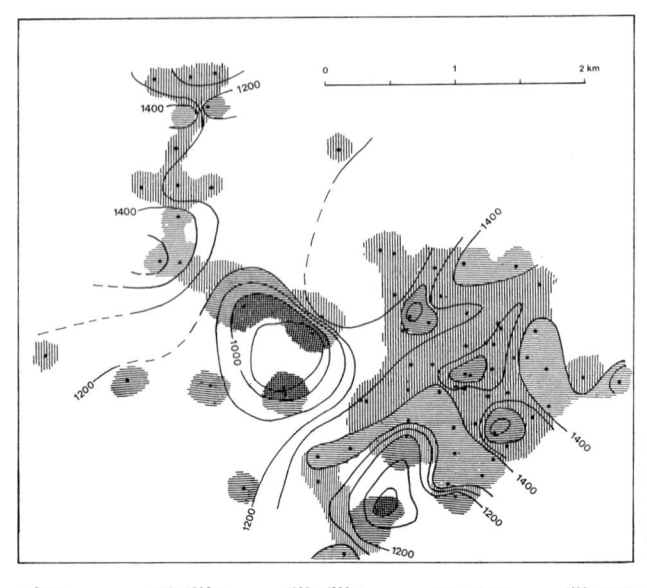

Abb. 3. Feld Stockstadt. Tiefenlinienplan der Geoisothermenfläche f. 80°C.

Fig. 3. Field Stockstadt. Depth-contour map of the 80°C geoisotherm.

< 800 m   800 – 1000 m   1000 – 1200 m   1200 – 1400 m   1400 – 1600 m

Abb. 4. Feld Landau. Tiefenlinienplan der Geoisothermenfläche für 80°C.

Fig. 4. Field Landau. Depth-contour map of the 80°C geoisotherm.

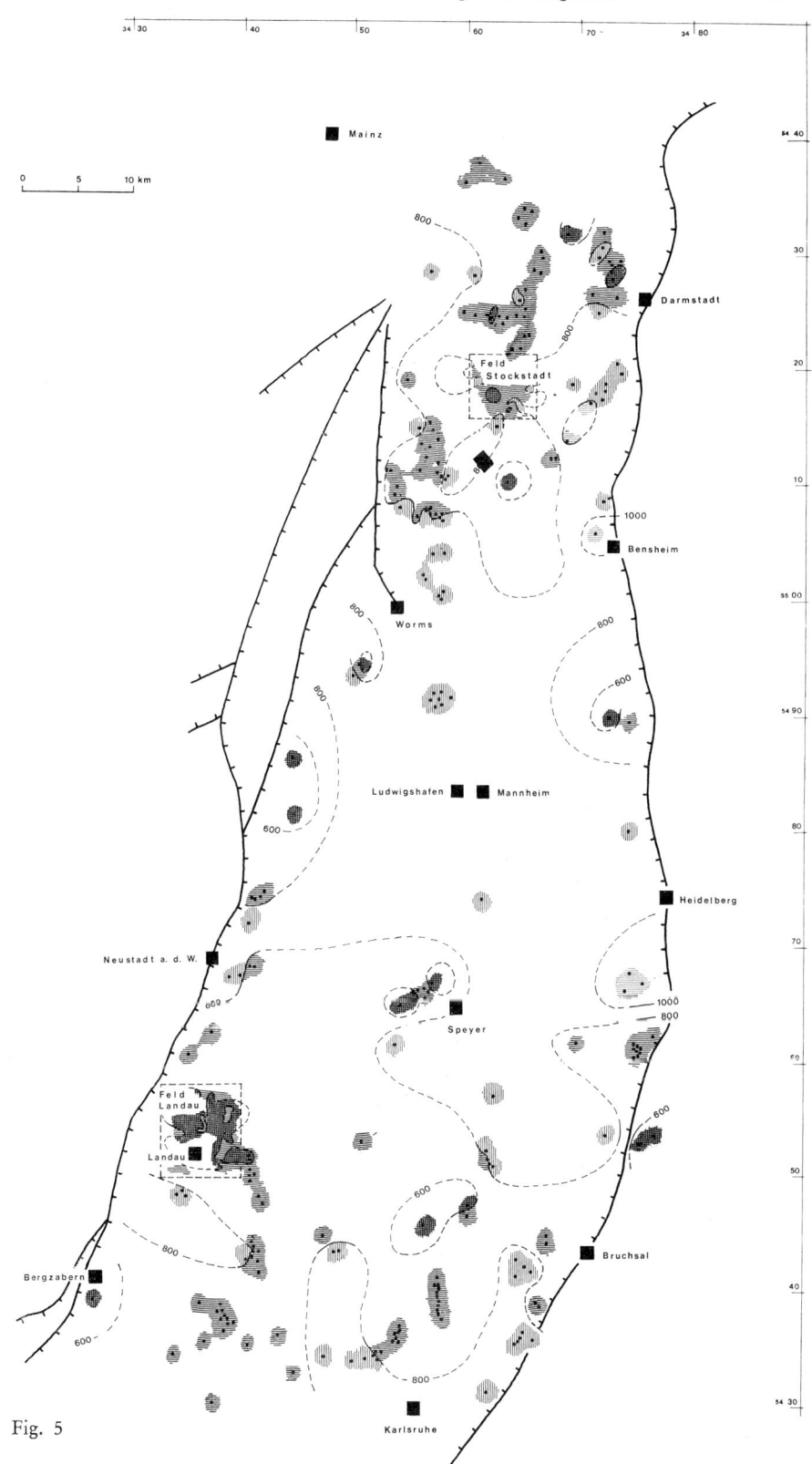

Geothermische Karte des Rheingrabenuntergrundes 187

Fig. 5

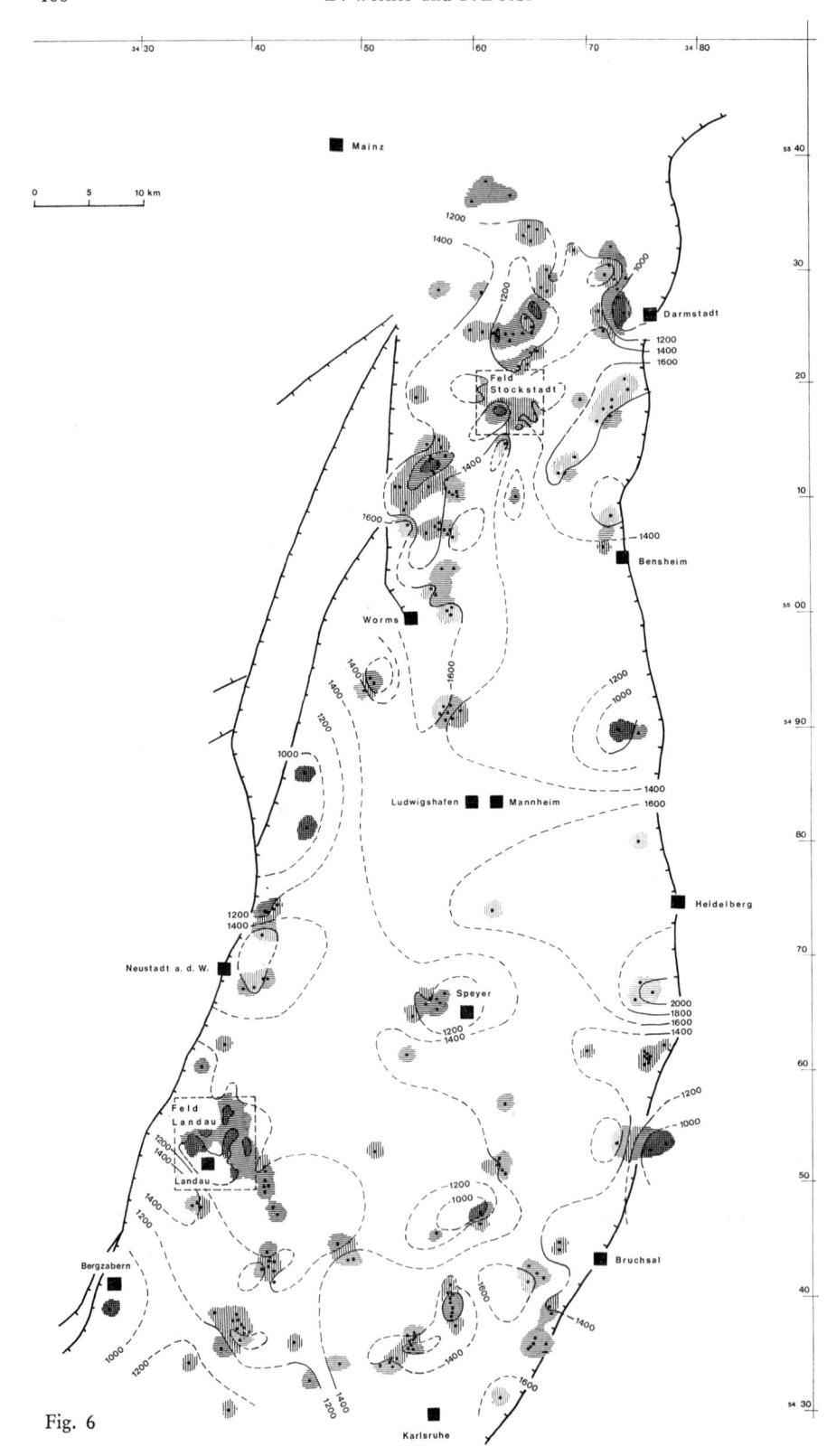

Fig. 6

ergab sich durch Vergleich: Abgefragt wurden solche Meßwerte der Klasse 2, die innerhalb eines Zylinders (Radius 250 m, Höhe 50 m) um einen Meßpunkt der Klasse 1 liegen. Aus 75 derartigen Vergleichsfällen errechnete sich die Korrektur zu $\delta T = +14.6°C$ bei einer mittleren quadratischen Schwankung von $\delta\delta T = \pm 13.5°C$ (vgl. hierzu die Beispiele Leopoldshafen 15 und Stockstadt 22 in Tab. 1).

Die Schwankung $\delta\delta T$ wäre als Maß für die Datenungenauigkeit anzusehen, wenn man voraussetzt, daß das Volumen des Zylinders in jedem Falle klein ist gegenüber den räumlichen Dimensionen der tatsächlichen lokalen Temperaturschwankungen. Diese Voraussetzung ist jedoch nicht ohne weiteres gegeben. Vermutlich ist der Betrag von $\delta\delta T$ zu einem Teil auch durch echte lokale Temperaturunterschiede zustande gekommen. Eine Verkleinerung des Zylindervolumens führt bei geringerer Treffzahl allerdings nicht zu wesentlich anderen Ergebnissen für δT und $\delta\delta T$.

Ein anderes Problem ist die Verteilung der Meßpunkte. Die Anordnung der Bohrlöcher folgte wirtschaftlichen Gesichtspunkten, nicht geothermischen. Daher ist die räumliche Dichte der Meßpunkte sehr unterschiedlich. Den Ölfeldern von Landau und Stockstadt, die temperaturmäßig gut sondiert sind, stehen Gebiete gegenüber, für die nur vereinzelte oder keine Beobachtungen vorliegen.

Zur anschaulichen Darstellung des Temperaturfeldes wurden Tiefenlinienpläne der Isothermenflächen herangezogen (T_{is} bleibt konstant, z_{is} variiert). Die Erdoberfläche, im betrachteten Gebiet ohne nennenswerte Topographie, bildet eine solche Isotherme mit $z_{is} = \text{const.} = 0$ und $T_{is} = 10°C$ (Jahresmittel). Die Berechnung der Tiefenlagen z_{is} gewünschter Isothermenflächen erfolgt in den einzelnen Bohrlöchern zunächst durch lineare Interpolation. Um der Datenunsicherheit ausgleichend entgegenzuwirken, ist bei der Ermittlung der Werte z_{is} außerdem ein gewichteter Einfluß entsprechender Werte der Umgebung berücksichtigt. Die Gewichte sind abhängig von der räumlichen Entfernung und folgen einer GAUSS-Verteilung. Die Umgebung mit dem Radius ϱ soll durch die Halbwertsbreite dieser räumlichen GAUSS-Glocke definiert sein (für die Darstellung der Temperaturfelder von Landau und Stockstadt — Abb. 1—4 — wurde $\varrho = 0,1$ km gewählt, für die Gesamtdarstellung — Abb. 5, 6 — $\varrho = 1,0$ km).

Ein eigens zu Kartierungszwecken entwickeltes Rechenprogramm kombiniert

---

Legende zu Abb. 5 und 6:

Abb. 5. Nördlicher Rheingraben. Tiefenlinienplan der Geoisothermenfläche für 50° C.
Fig. 5. Northern Rhinegraben. Depth-contour map of the 50° C geoisotherm.

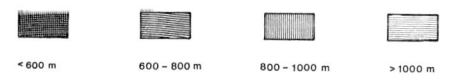

< 600 m   600 - 800 m   800 - 1000 m   > 1000 m

Abb. 6. Nördlicher Rheingraben. Tiefenlinienplan der Geoisothermenfläche für 80° C.
Fig. 6. Northern Rhinegraben. Depth-contour map of the 80° C geoisotherm.

< 1000 m   1000 - 1200 m   1200 - 1400 m   1400 - 1600 m   1600 - 1800 m   > 1800 m

die beiden angegebenen Prinzipien: 1. lineare Interpolation entlang der z-Achse und 2. räumliche Glättung nach dem oben skizzierten Verfahren. Durch zahlreiche Abfragen wird eine Anpassung an die jeweilige Konfiguration der Meßpunkte erreicht — einfachster Fall: isoliertes Bohrloch mit nur einem Meßwert, allgemeinster Fall: Bohrloch mit mehreren Meßwerten und benachbarten Bohrlöchern mit jeweils mehreren Meßwerten.

Die schließlich erhaltenen Werte z_{is} — Stützpunkte der Isothermenflächen — werden einer Probe unterworfen, indem aus ihnen durch lineare Interpolation an den ursprünglichen Meßteufen z die zugehörige Temperatur T_1 zurückberechnet wird. Die Differenz aus T_1 und der Ausgangstemperatur T überschreiten nur selten das Fehlermaß $\delta\delta T$, ihr quadratisches Mittel liegt weit darunter.

Das Rechenprogramm liefert Drucker-Plots, die als unmittelbare Vorlage zum Zeichnen der Tiefenlinien dienen. Der Verlauf dieser Iso-Linien ist oft nicht eindeutig auszumachen, insbesondere in Gebieten mit geringer Datendichte. Indessen sind willkürliche Lösungen weitgehend ausgeschlossen, da die Linienführung topologischen Zwängen unterworfen ist. Die vorgelegten Karten verstehen sich ausdrücklich als Entwurf.

Resultate, Interpretationsversuch

Dargestellt werden die Isothermenflächen für 50° C und 80° C, weil in diesem Temperaturbereich die Mehrzahl der Meßdaten liegt und dadurch gewährleistet ist, daß die Stützpunktwerte z_{is} sich fast ausschließlich durch Interpolation (plus Glättung) errechnet haben, selten durch Extrapolation. Die gut vermessenen Gebiete von Landau und Stockstadt (Abb. 1—4) werden im vergrößerten Maßstab gezeigt, um die lokale Struktur der Temperaturschwankungen zu verdeutlichen. Das Landauer Feld ist in einer Darstellung der geothermischen Tiefenstufe bereits bekannt geworden (DOEBL 1970). Als Fortsetzung unserer Karte nach Süden liegt eine geothermische Karte des französischen Grabengebietes vor (DELATTRE, HENTINGER & LAUER 1970), die erkennen läßt, daß dort offenbar die gleichen Erscheinungen wie im nördlichen Rheingraben angetroffen werden.

Ein Vergleich der 50° C- mit der 80° C-Fläche (Abb. 1—6) läßt unschwer erkennen, daß die thermische Unruhe mit der Tiefe zunimmt, zumindest im hier betrachteten Bereich, der Tiefen bis zu ca. 2300 m repräsentativ umfaßt (tiefster Meßpunkt: 3150 m). Während die Tiefenschwankung der 50° C-Fläche mit etwa 600 m angegeben werden kann, beträgt dieser Wert für die 80° C-Fläche bereits rund das Doppelte.

Wie bereits früher angedeutet (WERNER 1970), ist die Wärmeleitungsgleichung kein geeignetes Instrument, die lokale Schwankungsintensität zu interpretieren. Einleuchtend dagegen ist die Annahme eines umfassenden, verzweigten Systems zirkulierenden Wassers. Aufsteigenden Strömen könnten lokal erwärmte Gesteinsregionen zugeordnet werden, während abfließende Ströme erniedrigte Temperaturen hinterlassen. Ein solcher Kreislauf bedarf eines beständigen Antriebs. Dieser könnte darin erblickt werden, daß innerhalb eines tiefreichenden Hohlraumsystems wasserführende Bereiche auf wärmeren, wasserdampfführenden Bereichen aufsitzen, wodurch eine stark instabile Lagerung gegeben ist.

Neben den starken lokalen Temperaturschwankungen ist das Rheingrabengebiet noch durch eine weitere thermische Besonderheit ausgezeichnet, die in den Darstellungen nicht zum Ausdruck kommt. Das Gebiet weist i n s g e s a m t einen beträchtlich höheren Temperaturgradienten dT/dz auf gegenüber normal zu erwartenden Werten um etwa 30° C/km. In Tab. 2 sind die mittleren Gradienten zusammengestellt. Diese Zahlen stehen qualitativ im Einklang mit den bislang bekannten Werten für die Wärmestromdichte im Rheingrabengebiet (CREUTZBURG 1964; KAPPELMEYER 1967; HAENEL 1971).

Tab. 2. Mittlere Temperaturgradienten [° C/km] im nördlichen Rheingraben.

	Temperaturbereich		
	50—60° C	60—70° C	70—80° C
Feld Landau	68.8 ± 8.0	70.7 ± 16.8	81.1 ± 19.1
übriges Gebiet	56.5 ± 13.5	56.2 ± 14.5	58.5 ± 20.0

Als Hauptergebnis dieser Untersuchung wäre dies festzuhalten: Der tiefere Einblick in den Wärmehaushalt des Grabensystems ist zunächst einmal durch starke Störungen verstellt. Deren Ursache und Mechanismen müssen studiert werden, da diese Erscheinung offenbar beträchtlich am Energietransport beteiligt ist. Erst nach Kenntnis dieses Phänomens können Kriterien für die Anwendbarkeit der Wärmeleitungsgleichung im betrachteten Gebiet aufgestellt werden.

Für die Überlassung der Temperaturdaten danken die Autoren den Firmen: Gewerkschaft Brigitta, DEA (jetzt Texaco), Deilmann, Elwerath, Itag, Mobil Oil, Preussag, Wintershall. Hinweise erhielten wir durch Herrn Dr. HELING, Heidelberg.

Programmentwicklung und Rechenarbeiten wurden ausgeführt an den Rechenzentren des Kernforschungszentrums Karlsruhe (IBM) und der Eidgenössischen Technischen Hochschule Zürich (CDC).

Literaturverzeichnis

CREUTZBURG, H., 1964: Untersuchungen über den Wärmestrom der Erde in West-Deutschland. — Kali u. Steinsalz, 4, H. 3, 1—108.

DELATTRE, J. N., HENTINGER, R. & LAUER, J. P., 1970: A Provisional Geothermal Map of the Rhinegraben (Alsation Part). — In: ILLIES, J. H. & MUELLER, ST. (eds.), Graben Problems, 107—110, Schweizerbart, Stuttgart.

DOEBL, F., 1970: Die geothermischen Verhältnisse des Ölfeldes Landau/Pfalz. — In: ILLIES, J. H. & MUELLER, ST. (eds.), Graben Problems, 110—116, Schweizerbart, Stuttgart.

HAENEL, R., 1971: Heat Flow Measurements and a First Heat Flow Map of Germany. — Z. Geophysik, 37, 975—992.

KAPPELMEYER, O., 1967: The Geothermal Field of the Upper Rhinegraben. — In: ROTHE, J. P. & SAUER, K. (eds.), The Rhinegraben Progress Report 1967, 101—103, Abh. geol. L.-Amt Baden-Württ., Freiburg.

WERNER, D., 1970: Geothermal Anomalies of the Rhinegraben. — In: ILLIES, J. H. & MUELLER, ST. (eds.), Graben Problems, 121—124, Schweizerbart, Stuttgart.

Diagenesis of Tertiary Clayey Sediments and Included Dispersed Organic Matter in Relationship to Geothermics in the Upper Rhine Graben

by

F. Doebl, D. Heling, W. Homann, J. Karweil, M. Teichmüller and D. Welte

With 4 figures in the text and 1 table as a folder (see at the end of the volume)

Abstract: A special feature of the Upper Rhine Graben is its high heat flow. Certain regions within the graben are warmer than others. The diagenesis of Tertiary sediments was studied in a "warm" borehole Landau 2 (7.69° C/100 m) and a relatively "cold" borehole Sandhausen 1 (4.16° C/100 m). The porosity of the clayey sediments decreases more rapidly with depth in the warm borehole than in the cold one. The transformation of clay minerals (smectite toward mixed layered minerals) occurs also in a lower depth in Landau 2. The same feature has been observed in other boreholes indicating a relatively high geothermic gradient.

Most striking relationships exist between the transformation of dispersed organic matter and geothermal anomalies (see Tab. 1): the rank of bituminous coal is reached in Landau 2 at a depth of 1500 m, in Sandhausen 1 at 2600 m. The first formation of petroleum hydrocarbons was observed in Landau 2 at a depth of 700 m and in Sandhausen 1 at 2600 m. Since both, coalification and bituminization depend on the temperature and on the time of its exposure, certain conclusions are possible in respect to the geothermic history. Calculations on the basis of reaction kinetics suggest that the exceptionally high heat flow which is observed in Landau 2 at present, possibly started not earlier than 15 million years ago. In both boreholes the high heat flow, which is typical for the entire Rhine Graben, is suggested to have started at the same time with Rhine Graben taphrogenesis during the Eocene.

1. The problem

(M. Teichmüller)

First subsidence of the Upper Rhine graben began early in the Eocene. Contemporaneously synsedimentary faults became active (Veit 1962). At the same time volcanism reached its climax (Horn et al. 1972). Heat flow measurements which have been investigated during the Upper Mantle Project, have revealed that the complete Upper Rhine Graben area represents a region of positive heat flow anomalies (Hänel 1970). Besides there are local differences of the geothermic gradient (Doebl 1970, Delattre et al. 1970, Werner 1972). Thus "cold" and "warm" areas may be distinguished within the graben.

The question arose whether the diagenesis of the clayey sediments and especially the alterations of the organic matter in these sediments are different in boreholes with different geothermic gradients, and whether coalification and bituminization could be a tool for the geothermic history of the Graben resp. of different regions within it. It is well known that the degree of coalification

Diagenesis of Tertiary Clayey Sediments 193

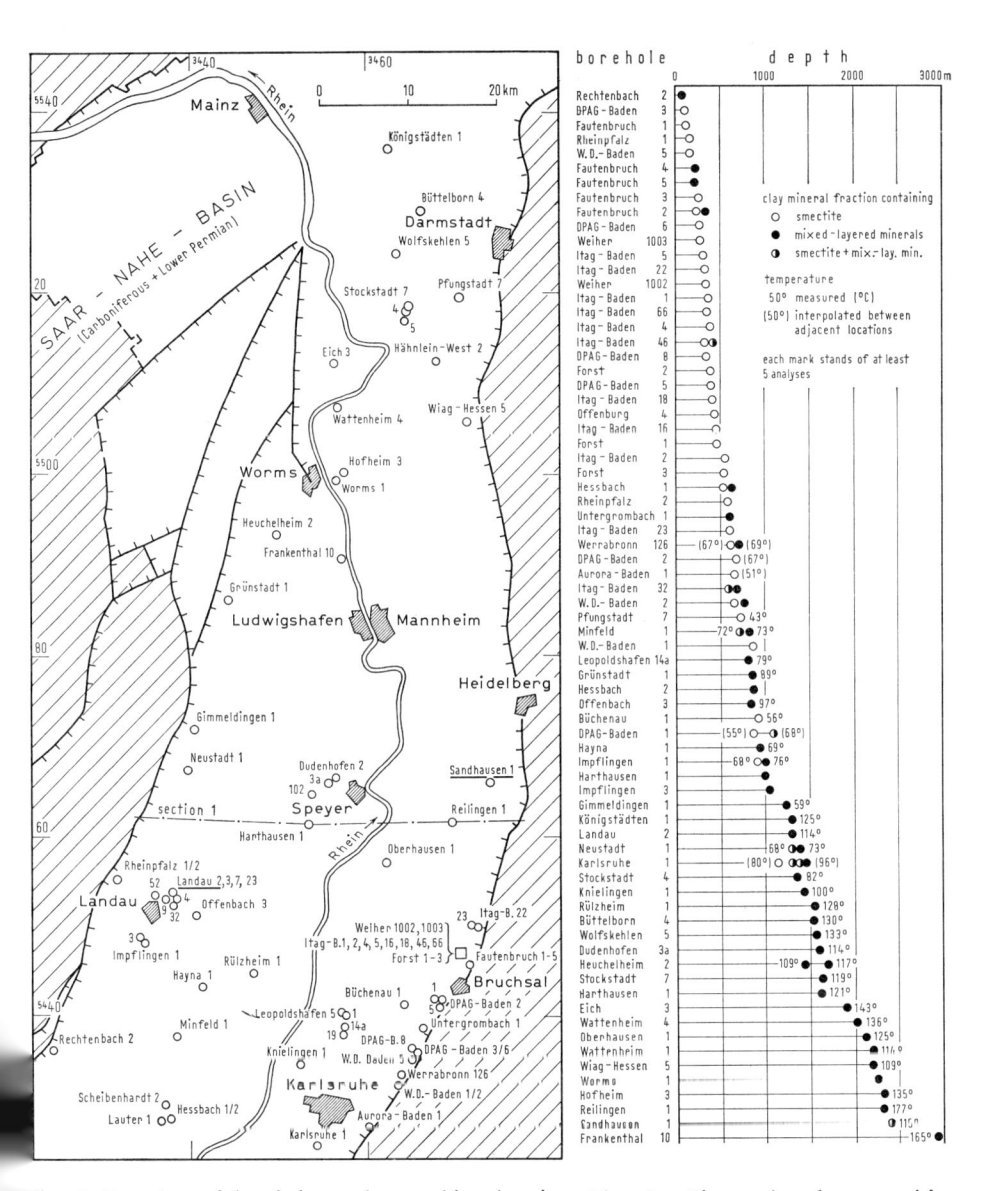

Fig. 1. Location of boreholes under consideration in the Rhine Graben.

Fig. 2. Clay mineral composition (fraction < 2 µ) in the Graue Schichtenfolge in relation to depth and temperature.

13 Approaches to Taphrogenesis

depends mainly on the temperature a coal was exposed to during its geologic history. Another factor for the rank of coal, although of minor importance, is the time of heat exposure. Not only the coal, resp. the coaly inclusions in the sediments but also the bituminous matter is very sensitive to heat exposure. The formation, alteration and destruction of crude oil depends largely on geothermal heat and its duration. The degradation of kerogen, i. e. the solid mother substance of petroleum, is also a temperature governed process.

Hardly anywhere in Central Europe the relation between geothermics on one hand and coalification and bituminization on the other can be investigated better than in the Upper Rhine Graben. It represents a complete sedimentary basin with young sediments of an almost continuous sedimentation, relatively rich in organic matter and with varying degrees of overburden. In the Graben many wells were drilled by oil industries from which samples are still available. Moreover geothermal measurements have been made in many wells. All this was the reason why a team of different specialists — sponsored by the Deutsche Forschungsgemeinschaft — started to study the „Diagenese und Metamorphose organischer Substanzen in Abhängigkeit von der Geothermik" in the Upper Rhine Graben in 1971.

The first aim was to compare two deep boreholes with different geothermal gradients: Landau 2 (2063 m deep) at the western rim of the Graben as an example of a "warm borehole" with a gradient of approximately 8° C/100 m and Sandhausen 1 (2905 m deep) at the eastern rim as an example of a "cold borehole" with a gradient of about 4° C/100 m. From both boreholes which are approximately 40 km apart from each other, stratigraphically identical core-samples out of depths ranging from 330 to 2060 m resp. 500 to 2900 m[1] were studied by the different specialists (see Tab. 1, fig. 1 and 2).

Other boreholes are still under investigation, partly with a vertical series of samples from total sections, partly with single samples belonging to certain stratigraphic horizons in order to determine regional changes of diagenesis. As to the regional studies only certain mineralogical results from a horizon at the Rupelian/Chattian boundary could be obtained so far. Textfig. 1 shows the location of the wells from which samples were taken. Only cores where studied, except for the microscopic reflectance measurements of coaly inclusions which were done on cuttings as well.

2. Geological and geothermic fundamentals
(F. DOEBL)

Tab. 1, fig. 1a and 2a show the stratigraphy, the depth, the position of cores obtained during drilling, the lithology of the sediments, the electriclog (spontaneous potential, resistivity), the rock temperature and the environment of the sediments (salinity) in the boreholes Landau 2 and Sandhausen 1. Because the

---

[1] The uppermost series were excluded because core samples were not available.

facies influences the chemical and microscopical composition of the organic matter, determinations of the facies were made on the basis of lithology, of electrical well logs, and of detailed microfaunistic studies. Relationships which were found between the salinity and the character of bituminous matter are discussed elsewhere (WAPLES et al. in press).

Landau 2 is situated within an eastern marginal block of the Nußdorf horst which began to develop in the Upper Eocene (SCHAD 1962) (see Tab. 1, fig. 4); Sandhausen 1 is located in a block being subject to relatively strong subsidence since the early Tertiary (Tab. 1, fig. 3). Consequently the thickness of the Tertiary and especially of the Aquitanian and younger sediments is considerably smaller in Landau 2 than in Sandhausen 1: Landau 2 reached at total depth of 2063 m the base of the Tertiary (here represented by Upper Eocene: Lymnäenmergel) or perhaps even the Mesozoic (from 1995 m downward), whereas Sandhausen 1 was drilled down to 2905 m where only the Lower Oligocene (Middle Pechelbronner Schichten) was reached. — The facies of the Tertiary commonly is more sandy in Sandhausen than in Landau though certain stratigraphic units are very similar. — In both cases the underlying beds are of Mesozoic age.

The temperature data in Tab. 1 are based on temperature measurements and interpolations described in detail by DOEBL in 1970. It must be considered that in reality the temperatures in the uppermost 500—700 m vary considerably as a result of ground water streams of different temperatures. But as the temperatures are relatively low in these depths (mostly $< 40°$ C, rarely $60°$ C) they cannot have great influence on the diagenesis of the organic matter. Where Pleistocene and Pliocene sediments of little compaction and, consequently, of high porosity and high water content reach a great thickness, as for instance in the „Heidelberger Loch", conditions may be different. The low geothermal gradient in the borehole Sandhausen 1 may be caused partly by the great thickness of these young sandy and water filled sediments.

The highest geothermal gradients have been observed in the Nußdorf horst near Landau, a block in which the Pre-Tertiary has been upliftet to a maximum. The consolidated strata has a greater heat conductivity than the less consolidated sediments of the Tertiary which — moreover — are predominantly clayey (and therefore additionally less conductive). For these reasons the possibility has been discussed that below the base of the Tertiary the heat flow is dammed up which — by the time — may influence the Tertiary sediments too ("socle effect"). But this effect, as well as the different heat conductivities of clayey and sandy layers within the Tertiary, cannot explain the great regional variations of the geothermic gradients in the Upper Rhine Graben and in the two boreholes studied. Landau 2 has a gradient of $7.69°$ C/100 m whereas in Sandhausen 1 a gradient of only $4.16°$ C/100 m was calculated. Although the rock temperatures are considerably higher in Landau 2 at corresponding depths, layers of same stratigraphic age are at higher temperatures in Sandhausen 1, — due to a deeper burial.

3. The diagenesis of the clayey sediments

(D. HELING)

Evidence for geothermal influence during diagenesis was to be expected from compaction data as well as from clay mineral anlyses.

a) Compaction

Compaction, being the most significant postdepositional alteration of clayey sediments, implies reduction of pore space with increasing overburden. In an early stage of burial water which can move freely is squeezed out. The flow rate of this water is controlled principally by permeability. If the overburden increases more rapidly than water can escape, it will require some time until an equilibrium between overburden pressure and porosity is established. In the subsequent state of deep burial, the remaining portion of pore water which is adsorbed on clay mineral surfaces, may be removed by diffusion. This process is ruled essentially by temperature.

The porosity reduction is shown in Tab. 1, fig. 1b and 2b by plots of the void ratio (pore volume/volume of the solid phase) of the samples versus the depth. The scattering in the porosity plots is caused by natural variations in grain size and carbonate cementation (Textfig. 3). By averaging the plots the shallow part of the resulting curves is ± a straight line. In Landau 2 the void ratio curve is bent at about 1000 m depth in the Bunte Niederröderner Schichten, in Sandhausen 1 at about 1500 m in the *Corbicula*-Schichten, indicating that an increasing increment of overburden is needed for a definite amount of porosity reduction. This is charactric for the deep-burial stage. In this stage adjacent particles are in touch with each other. Later, with a further increase in overburden, there is no more considerable compaction. The remaining pore water is mostly in an adsorbed state. It cannot be removed by pressure but by thermal energy. Thus, in this stage of burial, temperature is more important for further compaction than overburden pressure.

The transition zone from shallow to deep burial stage is approximately 500 m deeper in Sandhausen than in Landau. However, in both positions, the mean pore radii run to 140 Å (Tab. 1, fig. 1c and 2c), i. e. in both sections compaction has approximately reached the same state. This may be the consequence of the rate of subsidence of the Sandhausen area having been greater than in Landau. Therefore in Sandhausen the state of equilibrium between overburden load and compaction could not be established completely, in particular, since the maximum of subsidence occurred lately in the geologic history (see Textfig. 4).

Surprisingly in the transition zone the porosity in Sandhausen 1 ($E = 0.12$) is a little smaller than in Landau 2 ($E = 0.14$), although the median pore radii are equal. The smaller porosity is considered to be due to the coarser grain size in the silty shales of the Sandhausen section (Textfig. 3), whereas the fine grained pure shales of the Landau section have a higher specific surface area (and consequently adsorb more water). The specific surface area for the transition zone in Sand-

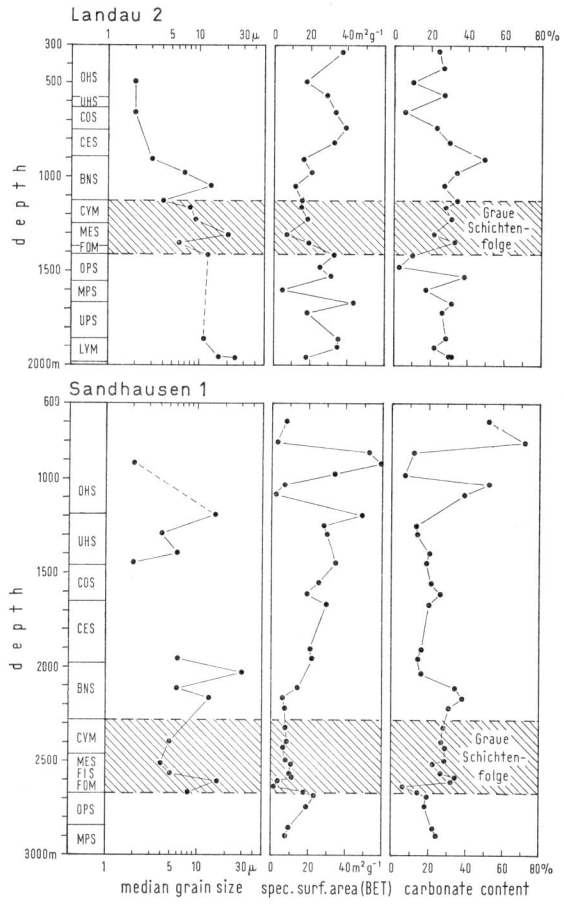

Fig. 3. Median grain size (settling tube method), specific surface area (BET) and carbonate content in samples of the boreholes Sandhausen 1 and Landau 2. OHS Upper Hydrobienschichten, UHS Lower Hydrobienschichten, COS *Corbicula*-Schichten, CES Cerithienschichten, BNS Bunte Niederröderner Schichten, CYM Cyrenenmergel, MES Melettaschichten, FIS Fischschiefer, FOM Foraminiferenmergel, OPS Upper Pechelbronner Schichten, MPS Middle Pechelbronner Schichten, UPS Lower Pechelbronner Schichten, LYM Lymnäenmergel.

hausen is 7 m²/g whereas in Landau it is 15 m²/g. This difference obviously is a consequence of the different grain size.

b) Disappearance of smectites

When analysing the clay minerals particular attention was given to the zonal and regional distribution of smectites which are known to be sensitive against diagenetic changes.

For X-ray analysis, the rocks were crushed, treated with H_2O_2 for removal of organic material, and the minus 2 micron fraction was separated by settling tube techniques. Three pipette slides of each sample were prepared for qualitative determination of the clays and three slides for semiquantitative estimation of the relative abundances. The first slide was X-rayed in air-dry state, the second after treatment with ethylenglycol and the third after heat-treatment up to 350° C for several hours.

The X-ray analysis (Table 1, fig. 1d and 2d) showed that the minus 2 micron fraction is composed of illite, kaolinite and chlorite, small amounts of quartz and feldspar. There is up to 20% calcite, sometimes some dolomite and seldom gypsum, and very sporadically halite. This assemblage is supplemented by a larger portion of expandable clay minerals which is composed of mixed-layered minerals of randomly interstratified smectite and illite (ML-minerals). In certain samples, the portion of expandable minerals is composed of both mixed-layered minerals and pure smectite (S). The smectite is found mainly in the Upper Rupel and Lower Chatt formations or „Graue Schichtenfolge". The same mineral composition was reported by SITTLER (1965) in boreholes of the central part of the Rhinegraben. WEAVER (1959) and BURST (1969) describe the transformation of smectite into mixed-layered smectite-illite and finally illite at slightly elevated temperatures, sufficient potassium supply provided.

In the Graue Schichtenfolge of Sandhausen pure smectite was found. None was detected in Landau 2. Although the thickness of overburden at Sandhausen 1 exceeds that at Landau 2 by 1250 m, there is almost the same temperature of approximately 100—120° C in both sections. In other words: smectite is found in the deeper location, but it is lacking in the shallower one, although the temperature is equal. Provided that smectite was also deposited in Landau it must have been transformed.

WIRTH (1953) and SCHAD (1962, 1964) agree that the Graue Schichtenfolge represents a marine to partly brackish deposition of great lateral uniformity throughout the graben. Therefore regional differences of the detrital smectite distribution are improbable. At the time of deposition, smectite very probably was present all over the area under investigation. Thus, differences in the distribution of the smectite as found today are due to diagenetical changes.

In order to detect possible facies-changes in respect to smectite distribution of the Graue Schichtenfolge core sample from some 60 boreholes of the northern part of the Rhinegraben (Textfig. 1) were analysed. The data are compiled in Textfig. 2, p. 193.

Smectite generally is confined to shallow depths. Below 1000—1500 m there is no smectite but only mixed-layered clay-minerals. In some boreholes smectite is included in the upper section of the Graue Schichtenfolge while the lower sections contains ML. In these sections the transformation depths and corresponding temperatures are covered by a sequence of samples. These locations are:

Werrabronn 126	(67°) ... (69°) C
Minfeld 1	72° ... 73°
DPAG — Baden 1	(55°) ... (68°)
Impflingen 1	68° ... 76°
Neustadt 1	68° ... 73°
Karlsruhe 1	(80°) ... (96°).

Average temperatures of alteration range between 68 and 76° C. The scattering is surprisingly small considering that thermal discharge (temperature × time) rather than temperature is the ruling factor. In addition to that also the potassium supply is of importance. The alteration of smectite into mixed-layered minerals as found in the Rhinegraben boreholes is completed at a mean temperature of 72° C. In the case of Sandhausen 1, smectite is found besides ML at temperatures as high as 115° C. According to the young and rapid subsidence of the Sandhausen area it is suggested that high temperatures did not prevail long enough to complete the transformation of smectite.

Of particular interest are the samples from positions close to the main faults bordering the graben. In the Fautenbruch 2, 4, 5 samples — located less than 1.5 km off the eastern main graben fault near the town of Bruchsal — the smectite has been replaced by ML-minerals at the relatively shallow depth of less than 500 m. Tab. 1, fig. 5 gives a section through the Fautenbruch area. The Graue Schichtenfolge near the eastern fault was never buried deeper than today. Therefore geothermal temperatures in this area must have been high. Numerous hot springs in the area confirm this conclusion.

The samples from Itag-Baden 4, Itag-Baden 1 and Itag-Baden 46 lead to similar conclusions. These boreholes are located 1 km, 0.9 km and 0.8 km respectively off the eastern main fault, and the samples of the Graue Schichtenfolge were taken from nearly the same depth of around 350 m. Clay analyses show that the more distant samples Itag-Baden 4 and 1 contain smectite, whereas there are both smectite and ML-minerals in sample Itag-Baden 46 which is closest to the main fault. This is further evidence that temperature rises when approaching the eastern main fault. — Wether there is a temperature increase when approaching the western graben edge has yet to be shown. It seems conspicous, however, that a sample from Rechtenbach 2, contains ML-minerals although it is only buried 73 m.

The reported data on the Rhinegraben and the smectite transformation are in line with replacement temperatures reported from other basins:

Salton Sea Basin (California), Pliocene, Quarternary	93° C	(MUFFLER & WHITE 1969)
Gulf Coast, Eocene	92	(BURST 1959, 1969)
Logbaba (Cameroun), Cretaceous	69	(DUNOYER DE SEGONZAC 1969)
Pierrefeu (Provence), Oligocene	79	(DUNOYER DE SEGONZAC 1969)

The replacement of smectite by ML-minerals under shallow burial in locations close to the graben border faults — as observed by the present clay analyses — indicate that the disappearance of smectite is related to temperature rather than overburden pressure. ILLIES (1965) reported extremely high thermal gradients at the fault zones bordering the graben. He attributed this to the high permeability of these zones allowing hot waters to ascend from abyssal depth. This heat probably caused the transformation of the smectite.

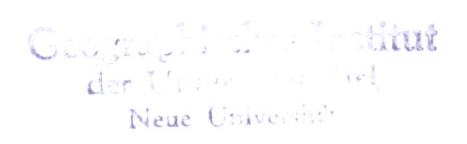

DUNOYER DE SEGONZAC (1970) considered pressure to be essential for the disappearance of smectite. Since we observed smectite to disappear under less than 400 m of overburden, temperature is regarded to be the stronger controlling factor.

4. Alterations of the bituminous matter
(D. WELTE)

The transformation of kerogen and of high molecular weight bitumen into relatively low molecular weight petroleum hydrocarbons is a chemical process, called bituminization, which is mainly based on the rupture of C—C bonds. This process is temperature dependent. This means that the generation (and migration) of crude oil from kerogen requires a certain minimum rock temperature. During bituminization hydrocarbons like n-alkanes, iso-alkanes and cyclo-alkanes are formed. Thus the amount of hydrocarbons (HC) in the sediments as compared to the total amount of insoluble organic carbon present (HC/org. C) is commonly used as a measure of the degree of bituminization resp. of oil formation. The distribution pattern of the different n-alkanes found in the extracts give further information on the "maturity" of the bitumen (see on Tab. 1, fig. 1i and 2i the small diagrams). A well known measure is the so called carbon preference index (CPI) which represents a ratio of the amount of n-alkanes with an odd number of C-atoms versus the amount of n-alkanes with an even number of C-atoms. During the diagenesis of bituminous matter this ratio normally decreases from numbers between 2—4 and approaches 1. CPI-values of less than 1.5 and a systematic trend towards lower numbers indicate that the critical temperature for oil formation has been reached[2].

Tab. 1, fig. 1h and 2h demonstrate that in corresponding stratigraphic units the amount of hydrocarbons (ratio HC/org. C $\times$ 10³) is smaller in Sandhausen 1 than in Landau 2 although — due to the greater depth — the rock temperatures for a given layer are today higher in Sandhausen 1.

The CPI-values are smaller in Landau 2 than in Sandhausen 1 (Tab. 1, fig. 1i and 2i) indicating, likewise, a higher degree of maturity in Landau. An example of the different maturity is given from the *Corbicula*-Schichten[3]:

	Landau 2	Sandhausen 1
depth m	656—736	1552—1613
temperature °C	~ 60	~ 75
ppm hydrocarbons	235—587	9—146
HC/org. C $\times$ 10³	24—45	9—13
CPI	0.8—1.4	1.3—2.5

---

[2] Exceptions from this rule and further characteristics of the bituminization process as well as the methods of bitumen geochemistry are described elsewhere (WELTE 1972; WELTE & WAPLES, in press).

[3] Data from 2 samples out of each well. It should be mentioned that the very low CPI values are partly caused by the saline facies (WAPLES et al., in press; WELTE & WAPLES, in press).

The Meletta-Schichten in Landau 2 were encountered at a depth of 1245—1370 m corresponding to a temperature of about 105° C. In Sandhausen 1 they lie 2470—2630 m deep at a rock temperature of about 115° C. Nevertheless only a small amount of hydrocarbons was formed anew in Sandhausen. Furthermore these hydrocarbons bear the characteristics of immaturity, like a high CPI-value and the dominance of certain components in the HC-spectrum, whereas in Landau 2 the Meletta-Schichten contain already mature, petroleum-like hydrocarbons.

The beginning of the formation of hydrocarbons in the subsurface was observed in Landau 2 in the *Corbicula*-Schichten (Aquitanian, ∼ 700 m, ∼ 60° C), whereas in Sandhausen 1 this phenomenon was not observed before reaching the Fischschiefer (Rupelian, ∼ 2630 m, ∼ 120° C). These differences may be the consequence of the different rate of subsidence since the Upper Tertiary (see Text-fig. 4): in Sandhausen 1 the equilibrium between temperature and formation of hydrocarbons may not yet have been reached. Another reason may be looked for in facies differences (see Tab. 1, fig. 1a and 2a).

5. Alterations of the coaly inclusions
(W. Homann & M. Teichmüller)

The degree of coalification or rank of coal is dependent on temperature and the time of heat exposure. It can be determined under the microscope by reflectance[4] measurements of the humic substances (called "huminites" in lignites and "vitrinites" in hard coals). Most sediments contain more or less tiny inclusions of huminites or vitrinites which can be used for such reflectance measurements (M. Teichmüller 1970). Here, besides reflectance measurements also other microscopic methods were applied: microhardness determinations of humic matter and fluorescence spectra measurements of sporinites (pollen, spores). The latter method (originally developed by Van Gijzel (1961) for fluorescence measurements on isolated sporomorphs in transmitted light) has been applied for the first time for polished sections in incident light. The method will be described in a seperate paper by Homann.

The results of the different microscopical rank measurements are shown in Tab. 1, fig. 1e—g and 2e—g. In the borehole Landau 2 over a depth interval of 1730 m (330—2060 m) the r e f l e c t a n c e of huminites/vitrinites increases with depth from about 0.20 % to about 0.86 % Rm_{oil} (according to the average curve). In Sandhausen 1 the increase is less significant: from about 0.20 % to about 0.73 % Rm_{oil} over a depth interval of 2260 m (640—2900 m). In other words the gradient is higher in Landau 2 (0.038 % Rm_{oil}/100 m) than in Sandhausen 1 (0.023 % Rm_{oil}/100 m). — The transition from lignite to hard coal (Braunkohle/Steinkohle) which corresponds to a reflectance value of about 0.6 % Rm_{oil}, is reached in Landau at a depth of 1500 m (Upper Pechelbronner Schichten), whereas in Sandhausen

---
[4] The reflectance depends on the degree of aromatization and condensation of the humic matter. It increases with the rank of coal.

it was found at a depth of 2600 m (Lower Meletta-Schichten). In Landau 2 the lowermost sample (Lymnäenmergel, 2061 m) near the bottom of the borehole contains coaly inclusions in the stage of high volatile bituminous coal (Gasflammkohle) with about 37% volatile matter (vitrite), whereas in Sandhausen 1 the lowermost sample (Middle Pechelbronner Schichten, 2897 m), although at a greater depth, merely reached the rank of flame coals with about 40% volatile matter.

It should be mentioned that the two curves represent a first example of reflectance increase with depth for the rank range from soft brown coals to high volatile bituminous coals. The reflectance values were obtained from rock sections in which the organic inclusions could be identified as autochthonous humic matter. Organic matter which has been isolated by sink and flow seperations contains coal particles with considerably higher reflectance values (max. 1.3% $R_{m\ oil}$). Those particles are especially abundant in the lower part of the borehole Landau 2. In the lowermost sample (2061 m) Dr. GREBE (Krefeld) identified spores of Permian and Carboniferous age. Perhaps the high rank particles derive from the Permo-Carboniferous of the Saar—Nahe-Basin (see textfig. 1). Pollen of Triassic age (Keuper) were identified in Landau 2 at depths of 1645 m and 2012 m.

The microhardness-values are strongly scattering in both boreholes (Tab. 1, fig. 1g and 2g). Nevertheless, there is a tendency for an increase in microhardness with depth, in Landau 2 to a higher degree (from $\sim$ 19 to $\sim$ 34 kp/mm²) than in Sandhausen 1 (from $\sim$ 16 to $\sim$ 25 kp/mm²).

The fluorescence spectra of the sporinites change their half-peak-width[5] with increasing depth in Landau 2 from about 109 nm to about 153 nm, in Sandhausen 1 from about 108 nm to about 138 nm. This results from a steady shift of the long wavelength half of the spectral curves to greater wavelengths (red) as the degree of coalification increases. Although this change is relatively small the diagrams in Tab. 1, fig. 1f and 2f demonstrate that in Landau 2, again, the gradient (nm/depth) is somewhat higher than in Sandhausen 1.

To sum up: The increase of coalification with depth, measured by means of different microscopical methods, is greater in Landau 2 than in Sandhausen 1. These findings are in agreement with the higher geothermic gradient in Landau 2. In both boreholes down from 330 m resp. 500 m the increase is more or less steady and without breaks. This result indicates a constant geothermic history during Tertiary times, at least up to the Upper Aquitanian, for either of the two boreholes considered.

6. Considerations to the geothermic history on the basis of reaction kinetics

(J. KARWEIL & D. WELTE)

As stated by HÄNEL (1970), the heat flow in the Upper Rhine Graben is markedly higher than it is in the adjacent regions. In fact, in the surroundings of the graben average values of 1.7 µcal. cm^{-2} · sec^{-1} have been found as against

---

[5] The width of the spectral curve in the half height of the maximum peak is called half-peak-width. For further information see HOMANN, separate paper.

2 to 4 μcal. cm^{-2} · sec^{-1} in the graben itself. The distribution of the heat flow is far from being uniform. On the contrary, there are very narrow local peaks. In the Landau area e. g. values up to 4 μcal. cm^{-2} · sec^{-1} have been measured, corresponding to a thermal gradient of 7—8° C/100 m.

It is to be supposed that the increased heat flow is connected with the taphrogenesis. The age of the graben is known from geological investigations, the duration of the heat flow can be derived from the degree of metamorphism of the organic substances (bituminous and humic).

a) Bituminous matter

The amount of hydrocarbons formed anew in a sedimentary unit is dependent on quantity and type of organic matter and its geothermal history during burial. Theoretically the influence of temperature, i. e. paleogeothermal conditions can be reconstructed with the help of the ARRHENIUS equation.

However, this is only possible when the activation energy E and the frequency factor s are known, or, when comparing the same stratigraphic unit in two different boreholes where E and s are identical for both boreholes. As a measure for the reaction rate k the extractable amount of hydrocarbons as compared to the amount of insoluble organic matter $\left(\frac{HC}{OC}\right)$ can be taken. The ratio $\frac{HC}{OC}$ for samples from the borehole Landau 2 may be called Q_L and the corresponding ratio for Sandhausen Q_S. The temperatures (T) in both wells are expressed as T_L (Landau) and T_S (Sandhausen).

With these data and with the gas constant R the following equation can be formulated:

$$(1) \qquad \ln Q_L + \frac{E}{RT_L} = \ln Q_S + \frac{E}{RT_S}$$

$$(2) \qquad \frac{1}{T_S} - \frac{1}{T_L} = \frac{R}{E} \cdot \ln \frac{Q_S}{Q_L}$$

Out of the second equation a formal temperature difference can be calculated for a certain stratigraphic unit between the two boreholes. This temperature difference, however, cannot be considered to be a real paleogeothermal difference. This is so because these temperatures are calculated on the basis of the amount of hydrocarbons generated under the influence of heat. They are, therefore, average temperatures which are related to the total heat flow a particular stratigraphic unit has suffered during geologic history.

It should be kept in mind that a calculation of this type is only realistic when there is no difference in facies between comparable stratigraphic units in both boreholes. There are, however, some facies differences between Landau and Sandhausen, i. e. the sediments in Sandhausen are more sandy and consequently may contain a different type of organic matter.

Nevertheless, the temperature values which have been calculated by means of equation (2) clearly show that the Landau area was subjected to a higher integral heat flow during the geologic past than Sandhausen. The differences observed cannot be due to facies differences only.

b) Humic (coaly) matter

The coalification process is a sequence of chemical reactions. Their rate depends on the temperature and the duration of its action or, expressed in geological terms, on the depth of subsidence (burial), the thermal gradient and the duration of heat exposure. At the beginning of the coalification process, water and CO_2 are formed by dissociation reactions. Later on, when the state of bituminous coal has been reached, methane is formed by condensation reactions. It is impossible, of course, to characterize the kinetics of these different types of reactions by one single frequency factor and one and the same activation energy. We must confine ourselves to describe the development of the reactions by mean values of these two magnitudes. In this way, it becomes possible to get an insight into the kinetics of reaction within the range of bituminous coals. Thus certain conclusions for the paleogeothermal gradient are possible, if the geological history and the rank of coal are known (KARWEIL 1955).

The reflectance measurements on samples taken from the borings Landau 2 and Sandhausen 1 can be used to check the geothermal history of the two boreholes; with this aim in view, the graphically determined mean reflectance values (average curves on Tab. 1, fig. 1e and 2e) were converted to the corresponding volatile matter contents on the basis of a diagram (reflectance versus volatile matter) published by M. TEICHMÜLLER in 1970. To determine the duration of the different temperature steps during the subsidence of a certain layer its geological history must be known. It is indicated for the lowermost units (Lymnäenmergel in Landau, Middle Pechelbronner Schichten in Sandhausen 1) on Textfig. 4 based on data supplied by Dr. KÖWING. The subsidence curves have been subdivided schematically into 4 successive steps in order to make KARWEIL's diagram (1955) applicable. The temperatures corresponding to the different steps have been calculated on the basis of the present temperature gradients.

The reflectance of the coaly inclusions in the Middle Pechelbronner Schichten of the S a n d h a u s e n 1 borehole has reached a mean value of 0.73% Rm_{oil} in the course of 37 million years. A calculation based on the present temperature gradient, viz 4.16° C/100 m leads to a reflectance of 0.76% Rm_{oil}. The result is different, if the calculation is based on a mean temperature gradient of 3.3° C/ 100 m as determined by HEDEMANN (1963) for the Münsterland 1 borehole, and allowance is made for the magma cushion assumed by HÄNEL (1970) which would have the effect of raising the temperature gradient for the last 2 million years to the present values. The reflectance value calculated in this way (0.65% Rm_{oil}) is lower than the measured one (0.73% Rm_{oil}).

In the same way the data of the L a n d a u 2 borehole can be used. Assuming a subsidence time of 44 million years and a geothermal gradient of 7.69° C/100 m, corresponding to the value measured in the borehole, a reflectance of 1.3% Rm_{oil}

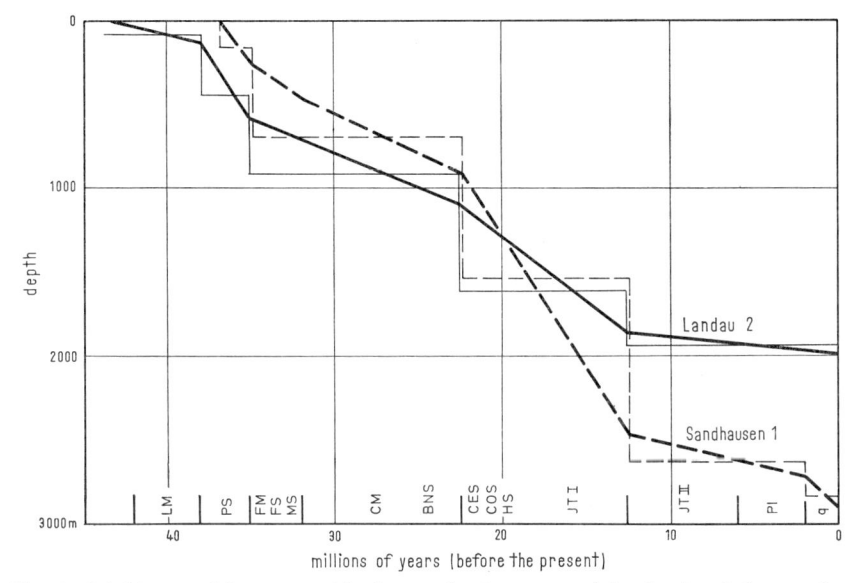

Fig. 4. Subsidence of lowermost Tertiary units (encountered in the boreholes Landau 2 and Sandhausen 1) during geological history. LM Lymnäenmergel, PS Pechelbronner Schichten, FM Foraminiferenmergel, FS Fischschiefer, MS Melettaschichten, CM Cyrenenmergel, BNS Bunte Niederröderner Schichten, CES Cerithienschichten, COS *Corbicula*-Schichten, HS Hydrobienschichten, JT I Jungtertiär I, JT II Jungtertiär II, PL Pliocene, Q Quaternary.

is calculated for the carbonaceous substance included in the Lymnäenmergel at a depth of 2060. The measured reflectance was 0.86%. From this lower value it can be concluded, that in the past the temperature gradient was lower than it is at present. Because the coalification curve (Tab. 1, fig. 1e and 2e) is continuous and without distinct steps it is assumed that either the temperature gradient increased continuously since the Eocene, or that the gradient was constant until the Lower Miocene followed by a sudden strong increase of heat flow in Post Lower Miocene times (for which coalification data could not be evaluated). If, however, the calculation is based on the "normal" temperature gradient (Central Europe: 3.3° C/100 m) making allowance for Hänel's magma cushion, then, the same as in Sandhausen, the resulting reflectance, namely 0.65% Rm_{oil}, is lower than the determined value (0.86% Rm_{oil}).

Thus for both boreholes agreement between the calculated and the measured values is only reached if it is assumed that a temperature gradient higher than normal in Central Europe has been effective during the whole subsidence period unter consideration. This suggests that the increased heat flow began at the same time when the subsidence of the Tertiary sediments started. H e a t f l o w a n d t a p h r o g e n e s i s , t h e r e f o r e , a r e o f t h e s a m e a g e .

The mechanism of the genesis of the graben and reasons for the high heat flow will be the subject of another publication (KARWEIL 1973).

Acknowledgements

This work was supported by DFG-grant Te 53.

We thank the Deutsche Forschungsgemeinschaft for the generous financial help. For valuable discussions we are grateful to Dr. A. HOLLERBACH (Göttingen), Dr. K. KÖWING (Krefeld), Prof. Dr. SCHAD (Kassel), Dr. C. SITTLER (Strassbourg), Dr. R. TEICHMÜLLER (Krefeld), Dr. E. WIRTH (Bruchsal). — For providing samples and geological information we thank the following oil companies: C. DEILMANN AG, Deutsche Erdöl AG (now Deutsche Texaco AG), Gewerkschaften BRIGITTA-ELWERATH Betriebsführungsgesellschaft mbH, ITAG, MOBIL OIL AG in Deutschland, PREUSSAG AG, and especially WINTERSHALL AG. — Many samples were obtained from the collection of borehole samples in the Geological Institute of the University Heidelberg (director: Prof. SIMON).

References

BURST, J. F., 1959: Postdiagenetic clay-mineral environmental relationsship in the Gulf Coast Eocene. — Proc. Ntl. Conf. Clays Clay Minerals, 6th Ntl. Acad. Sci. Ntl. Res. Council, Publ. 1957, 327—341.
— 1969: Diagenesis of Gulf Coast clayey sediments and its possible relation to petroleum migration. — Bull. Amer. Assoc. Petrol. Geol., 53, 73—93.
DELATTRE, J. N., HENTINGER, R. & LAUER, J. P., 1970: A provisional geothermal map of the Rhine-Graben (Alsatian part.). — In: ILLIES & MUELLER, Graben Problems, 107—110, Schweizerbart, Stuttgart.
DOEBL, F., 1970: Die geothermischen Verhältnisse des Ölfeldes Landau/Pfalz. — In: ILLIES & MUELLER, Graben Problems, 110—116, Schweizerbart, Stuttgart.
DOEBL, F. & BADER, M., 1971: Alter und Verhalten einiger Störungen im Ölfeld Landau/Pfalz. — Oberrhein. geol. Abh., 20, 1—14.
DUNOYER DE SEGONZAC, G., 1969: Les minéraux argileux dans la diagenèse passage au métamorphisme. — Mem. Serv. Carte Geol. Alsace Lorraine, 29, 320 pp.
— 1970: The transformation of clay minerals during diagenesis and low grade metamorphism: A review. — Sedimentology, 15, 281—346.
GIJZEL, P. VAN, 1961: Autofluorescence and age of some fossil pollen and spores. — Kon. Nederl. Ak. Wetensch. B 64, 56—63.
HÄNEL, R., 1970: Interpretation of the terrestrial heat flow in the Rhinegraben. — In: ILLIES & MUELLER, Graben Problems, 116—120, Schweizerbart, Stuttgart.
HEDEMANN, H. A., 1963: Die Gebirgstemperatur in der Bohrung Münsterland und die geothermische Tiefenstufe. — Fortschr. Geol. Rheinld. u. Westf. 11, 403—418.
HOMANN, W. (Manuskript): Zum spektralen Fluoreszenzverhalten des Sporinits in Kohlen-Anschliffen und seine Bedeutung für Inkohlungs-Untersuchungen.
HORN, P., LIPPOLT, H. J. & TODT, WO., 1972: Kalium-Argon-Altersbestimmungen an tertiären Vulkaniten des Oberrheingrabens. — Eclogae geol. Helv., 65, 131—150.
ILLIES, H., 1965: Bauplan und Baugeschichte des Oberrheingrabens. — Oberrhein. geol. Abh., 14, 1—54.
KARWEIL, J., 1955: Die Metamorphose der Kohlen vom Standpunkt der physikalischen Chemie. — Z. dt. geol. Ges. 107, 132—139.
— 1973: Kosmogonie und Krustendynamik. — Essen.
MUFFLER, L. J. P. & WHITE, D. E., 1969: Active metamorphism of Upper Cenozoic sediments in the Salton sea geothermal field and the Salton trough, Southeastern California. — Bull. Geol. Soc. Amer., 80, 29—40.
SCHAD, A., 1962: Voraussetzungen für die Bildung von Erdöllagerstätten im Rheingraben. — Abh. Geol. L.-Amt Bad.-Württ., 4, 29—40.
— 1962: Das Erdölfeld Landau. — Abh. Geol. L.-Amt Bad.-Württ., 4, 81—102.
— 1964: Feingliederung des Miozäns und die Deutung der nacholigozänen Bewegungen im mittleren Rheingraben. — Abh. Geol. L.-Amt Bad.-Württ., 5, 1—56.

SITTLER, C., 1965: Le Paleogène des fossés Rhénan et Rhodanien. Etudes sedimentologiques et paleoclimatiques. — Mem. Serv. Carte Géol. Alsace Lorraine, 24, 392 pp.
TEICHMÜLLER, M., 1970: Bestimmung des Inkohlungsgrades von kohligen Einschlüssen in Sedimenten des Oberrheingrabens — ein Hilfsmittel bei der Klärung geothermischer Fragen. — In: ILLIES & MUELLER, Graben Problems, 124—142, Schweizerbart, Stuttgart.
TISSOT, B. & PELET, R., 1971: Nouvelles données sur les mecanismes de genèse et de migration du pétrole; simulation mathématique et application à la prospection. — Proc. World Petrol. Congr. (Moscow, June 13—19, 1971), 1, 1—23.
VEIT, E., 1962: Die Öl- und Gasvorkommen der Südpfalz außerhalb von Landau (mit Beiträgen von A. SCHAD). — Abh. geol. L.-Amt Bad.-Württ., 4, 103—122.
WAPLES, D. W., HAUG, PAT & WELTE, D., in press: Occurrence of a regular C^{25} isoprenoid hydrocarbon in Tertiary sediments representing a lagoonal type, saline environment. — Geochim. Cosmochim. Acta.
WEAVER, C. E., 1959: The clay petrology of sediments. — In: Clays and clay minerals. Proc. Nat. Conf. Clays Clay Minerals, 6, 154—187.
WELTE, D. H., 1972: Petroleum exploration and organic geochemistry. — J. Geochem. Explor., 1, 117—136.
WELTE, D. H. & WAPLES, D. W., in press: Über die Bevorzugung geradzahliger n-Alkane in Sedimentgesteinen. — Naturwissenschaften.
WERNER, D., 1970: Geothermal anomalies of the Rhinegraben. — In: ILLIES & MUELLER, Graben Problems, 121—123, Schweizerbart, Stuttgart.
WIRTH, E., 1953: Grundlagen und Aussichten der Erdölsuche im Rheintalgraben. — Z. dt. geol. Ges., 105, 32—46.
— 1962: Die Erdöllagerstätten Badens. — Abh. Geol. L.-Amt Bad.-Württ., 4, 63—80.

Die Wärme-Anomalie der mittleren Schwäbischen Alb (Baden-Württemberg)

von

W. Carlé

Mit 1 Abbildung als Beilage

A Hot-Spot Area in the Swabian Alb, Western Germany

Abstract: In central Wurtemberg an area of anomalous high heat flow has been observed. The geothermal anomaly reveals an oval extension following south and parallel adjacent to the 140 km long Bebenhausen Fracture Zone. The anomaly comprises 335 Upper Miocene basaltic necks of the Urach volcanic area and numerous wells of acidulous water. The maxima of the anomaly are situated near Boll (geothermic degree 9,1 m/° C) and Neuffen (g. d. 11,1 m/° C). Actually there are several drillings for thermal water in progress. Related deep structures as are quoted to explain magmatic, carbon dioxide and thermal ascensions have not been found neither by magnetic nor gravimetric investigations. It is assumed that the Miocene volcanics as well as the recent thermal and acidulous waters have penetrated the crust using the fissure systems which accompany the Bebenhausen Fracture Zone. The cushion of mantle-derived material underneath the graben and its shoulders are considered to have caused both, volcanic action and increased heat production. There are some geophysical indications for a local updoming of the cushion surface underneath the hot-spot proper.

1. Geophysikalische Erkenntnisse

Vor 15 Jahren wurden die 17 damals bekannten Meßpunkte der geothermischen Tiefenstufe in Mittel-Württemberg zusammengestellt; diese wenigen Punkte ermöglichten aber immerhin die Konstruktion von Kurven gleicher geothermischer Tiefenstufe rund um das Uracher Vulkangebiet (CARLÉ 1958). Schon damals zeigte sich sehr deutlich, daß die thermische Anomalie keineswegs auf die an der Oberfläche sichtbare Verbreitung der obermiozänen Vulkanschlote beschränkt ist, sondern das Ausbruchsgebiet in Form einer ostnordöstlich erstreckten Ellipse umschreibt. Hieraus wurden seinerzeit Aussagen über die Existenz und relative Tiefenlage wärmespendender Herde abgeleitet. Seither erhöhte sich die Zahl der Meßpunkte auf 45. Die ursprüngliche Konstruktion der Anomalie erwies sich als im Prinzip richtig, doch nahm sie jetzt eine viel detailliertere Gestalt an.

Als Ergebnis sehr kleinräumig angeordneter erdmagnetischer Messungen erhöhte sich die Zahl der ursprünglich 125 Schlote auf 335 Ausbruchsstellen (MÄUSSNEST 1969). Durch entsprechend angesetzte Messungen suchte man vergebens Anhaltspunkte für eine große, überregionale magnetische Anomalie, in welcher sich ein Tiefenherd unter dem Vulkangebiet abzeichnen könnte. Man fand jedoch in ENE-Richtung angeordnete Teilanomalien, sowie eine seither unbekannte, sehr starke positive magnetische Anomalie bei Boll, die später noch kurz zu erwähnen ist (MÄUSSNEST 1970). Bei diesen im Prinzip schon von NEUMANN (1939) ermittelten Anomalien scheint es sich jedoch mit großer Wahrscheinlichkeit um Störkörper innerhalb des variszischen Grundgebirges zu handeln.

Sprengseismische Untersuchungen haben im Bereich der „Schwäbischen Wärme-Anomalie" ein anomales Verhalten seismischer Fortpflanzungs-Geschwindigkeiten im Grenzbereich zwischen Kruste und Mantel ergeben. Eine eindeutige Erklärung lassen die vorläufigen seismischen Profile jedoch nicht zu (EMTER 1971).

2. Kurze Beschreibung der Wärme-Anomalie

Von Schramberg bis östlich von Steinheim am Albuch beträgt die Längserstreckung der Anomalie 150 km. Ihre Achse ist sanft gebogen und nach NW konkav. Der Wärmeabfall erfolgt nach S zum Molasse-Trog hin erheblich rascher als nach N in Richtung auf Stuttgart und Heilbronn. Außerordentlich stark ist das Gefälle nach W zum Schwarzwald und, wenn auch nur auf kurze Erstreckung, in der Umgebung von Ulm. Die 18 m/°C-Linie bezeichnet den Kern der Anomalie. Wo sie aufgrund zahlreicher Meßpunkte genauer zu erfassen ist, werden Details offenbar. Das ein Jahrhundert lang als Maximum betrachtete, durch zwei Meßpunkte nachgewiesene Gebiet um Neuffen einerseits, sowie das 1972 durch eine Thermalwasser-Bohrung entdeckte absolute Maximum von Boll andererseits liegen auf der Achse der Anomalie. Das Maximum von Boll mit 9,1 m/°C deckt sich weitgehend mit der ebenfalls neu gefundenen magnetischen Anomalie von Boll. Am äußersten SW-Ende des Phänomens, ebenfalls auf der Achse, liegt hart am Rand des Schwarzwälder Wärmetiefs das Wärmehoch von Schramberg.

Die Häufung von Meßpunkten am oberen Neckar erbringt keine Detaillierung im Kurvenverlauf, sondern erhärtet nur den raschen Temperatur-Abfall in Richtung auf den Schwarzwald-Schild. Dagegen zwingen die zahlreichen Meßpunkte im Fils-Gebiet und auf der östlichen Alb zu besonderer Linienführung. In merkwürdiger Weise sind positive und negative Wärmeanomalien einander zugeordnet. Dem Maximum von Boll ist das Minimum von Albershausen eng benachbart, ebenso das Minimum von Westerstetten dem Maximum von Urspring. Sehr auffällig ist das an der Donau gelegene, hier eigentlich kaum zu erwartende Maximum von Donaustetten; dieses ist die einzige Detaileinheit, die den großen harmonischen Verlauf der Anomalie etwas stört.

Dem weiteren Ausschwingen der zentralen Aufheizungszone nach Sulz und Schramberg in südwestlicher Richtung entspricht die ostwärts ziehende Ausbuchtung nach Steinheim am Albuch. Diese zielt genau auf das weiter östlich befindliche nächste Wärmehoch, das in Nördlingen mit 12,3 m/°C ermittelt wurde (NATHAN 1957).

3. Wärme-Anomalie und Bruchtektonik

Schon früher wurde festgestellt, daß sowohl die obermiozänen Alb-Vulkane als auch das damals als Kern der Anomalie angesehene Wärmehoch-Gebiet von Neuffen durch die ENE-streichende Bebenhäuser Zone tangiert wird, während gleichzeitig das NW-streichende Filder-Bruchsystem etwa die Breite des Vulkan-Gebietes besitzt und auf dieses hinzielt. Auch das NNE-gerichtete Lauchert-Bruchsystem zieht von S her an das Vulkan-Gebiet heran, ohne es zu erreichen (CARLÉ 1955, 1958).

Da die langgestreckte Anomalie und ihre mutmaßliche Fortsetzung bei Nördlingen auf die gesamte Erstreckung der längs ihrer Nordflanke verlaufenden Bebenhäuser Zone parallel zieht, müssen Wechselbeziehungen zwischen dieser Störungszone und der Wärme-Anomalie angenommen werden. Daher ist eine kurze Charakteristik dieser Zone erforderlich (CARLÉ 1950, 1955, SEIBOLD 1950, WIEDEMANN 1966). Es ist eine allein im württembergischen Gebiet über 140 km lange Bruchzone, an der generell die angrenzende Süd-Scholle um Beträge bis zu 25 m absinkt. Streckenweise bewirken gegenfallende Brüche schmale Graben-Späne, oder Störungs-Schwärme bedingen verwickelte Bruchsysteme. Da südwärtige Abschiebungen überwiegen, fällt der Hauptbruch nach S ein; er dürfte also in etwa 15 km Tiefe dort zu erwarten sein, wo sich an der Oberfläche das Wärme-Maximum befindet. Die so auffallende Diskrepanz zwischen der enormen Länge der Störung und ihrer geringen Sprunghöhe hat zu Überlegungen geführt, die Störung entweder als durchgepaustes Lineament aus dem varistischen Untergrund oder als Blattverschiebung zu interpretieren. Letztere Vorstellung wird durch die Tatsache unterstützt, daß die nördlich der Zone stark ausgeprägte NW-Bruchtektonik in der Regel von der Bebenhäuser Zone abgefangen wird und hierbei mit gebogenen Aufsplitterungs-Zonen aus der NW- in die ENE-Richtung einlenkt. Auch die stellenweise beobachteten horizontalen Harnische auf Verwerfungsflächen lassen auf horizontale Schollen-Bewegungen schließen. Offenbar haben sowohl vertikale Schollenbewegungen als auch linkshändige Blattverschiebungen stattgefunden (WIEDEMANN 1966).

Stellt man eine Rangfolge der Störungen hinsichtlich ihrer Beziehung zur Wärme-Anomalie auf, so rangiert die Bebenhäuser Bruchzone an erster Stelle. Die Lauchert-Zone reicht von S her bis nahe an das Vulkangebiet heran, während die NW-Störungen des Filder-Systems mehr oder weniger von der ENE-Zone abgefangen werden.

Auf der nunmehr vorgelegten Karte fällt auf, daß die bedeutendsten Teilanomalien in Fortsetzung der beiden maßgeblichen Randbrüche des engeren Filder-Grabens aufgereiht sind, so Neuffen und Donaustetten im Fortstreichen der Vaihinger Verwerfung. Boll und Ursprung, aber auch die ihnen benachbarten negativen Flecken liegen in der Fortsetzung der Cannstatter Verwerfung. Daß letztere sich im Fortstreichen auch tektonisch andeuten könnte, wurde durch eine ins Detail gehende Aufnahme des ehemaligen Eisenerz-Bergwerks Grube Karl bei Geislingen als möglich angesehen (CARLÉ in CARLÉ & GROSCHOPF 1967); auf diese tektonisch stark beanspruchte Zone ist auch das bedeutende Kohlendioxid-Vorkommen von Bad Überkingen konzentriert (CARLÉ 1972). Freilich können bei Verdichtung des Beobachtungsnetzes weitere Teilanomalien gefunden werden, die sich dieser Regel nicht fügen.

4. Wärme-Anomalie und Kohlendioxid-Exhalation

Es ist auffallend, daß das Kerngebiet der Wärme-Anomalie von Kohlendioxid-Austritten bedeutenden Ausmaßes besetzt ist. Das größte, auch industriell ausgenutzte Gasfeld befindet sich am oberen Neckar und seinen rechten Nebenflüssen zwischen Rottenburg und Haigerloch. Sehr bedeutend ist auch das große Vorkommen des Filstales mit Schwerpunkten in Göppingen und Bad Überkingen; jüngst wurde in Boll ein Thermal-Säuerling aus dem Muschelkalk erbohrt. Erstaunlicherweise sind Säuerlinge im Herzen des Vulkan-Gebietes selten. Ursprünglich gab es nur einen freien Austritt in Kleinengstingen (CARLÉ 1954); in den vergangenen zwei Jahrzehnten wurden jedoch ein Säuerling in Nürtingen, ein Thermal-Säuerling in Urach und eine Mofette in Buttenhausen erschlossen. Es liegt nahe, sowohl die anomale Erdwärme als auch die ausströmende Kohlensäure und das in Ober-Miozän extrudierte Magma auf die gleiche Ursache zurückzuführen (CARLÉ 1958).

Leider wird das plausible Bild dadurch gestört, daß sich bedeutende Kohlendioxid-Vorkommen auch außerhalb der Anomalie befinden, so in Bad Cannstatt, aber auch im Schwarzwald, wo im Renchtal zahlreiche Säuerlinge sprudeln. In Bad Teinach harmoniert die geothermische Tiefenstufe von 41 m/°C nicht mit dem reichen Gas-Vorkommen.

5. Diskussion

Zunächst ist zu fragen, ob die geothermische Anomalie eine ursächliche Beziehung zum obermiozänen Vulkanismus besitzt, ob es sich also um Restwärme eines erloschenen Vulkanherdes handelt. Dann müßten in der Tat Zwischenherde in verhältnismäßig geringen Tiefen vorausgesetzt werden, denn das geförderte olivinisch-nephelinitische Magma entstammt dem oberen Erdmantel. Eine Existenz von derartigen Zwischenherden erscheint aufgrund der Assimilation von Grund- und Deckgebirgs-Bestandteilen in den Tuffen prinzipiell möglich (WEISKIRCHNER, Vortrag 1972), ist aber weder magnetisch noch gravimetrisch eindeutig nachweisbar.

Der aszendente Wärmestrom kann seit der vulkanischen Ausbruchsphase, also seit etwa 12 Millionen Jahren andauern. Auf den gleichen Wegen, auf denen heute die Wärme, wohl transportiert durch Wasser, aufsteigt, kann auch das Magma emporgedrungen sein. Die geothermische Anomalie wäre in diesem Falle in Anlehnung an MORGAN (1972) als hot-spot aufzufassen. Angesichts der räumlichen Beziehungen darf es als sehr wahrscheinlich gelten, daß beide Vorgänge die Bebenhäuser Zone und das ihr zugehörige Spalten-Inventar als Aufstiegsweg benutzt haben; auch das Kohlendioxid scheint gleichen Wegen zu folgen. Zu örtlichen Steigerungen des Wärme-Aufstieges scheint es dort zu kommen, wo die zweitrangigen NW-Bruchstörungen in das Gebiet der Anomalie einmünden.

Schließlich ist die Frage zu stellen, woher Magmen, Wärme und Kohlendioxid herzuleiten sind. Die Zusammenschau geologischer und geophysikalischer Daten hat zur Erkenntnis einer kissenförmigen Struktur aus Mantel-Materie unterhalb des Oberrhein-Grabens und seiner Schultern geführt (ILLIES 1970). Dieses Kissen steigt unterhalb des Grabens bis zu einer Tiefe von 24 km auf und taucht gegen die Flanken allmählich ab. Die aktive Grabentektonik und damit verbunden die Vorgänge im Grenzbereich zwischen Erdkruste und Erdmantel setzen sich bis in die Gegenwart fort. Die Interpretation sprengseismischer Daten durch EMTER (1971) läßt die Möglichkeit offen, daß die Oberfläche des Kissens unterhalb der Wärme-Anomalie eine nochmalige Aufwölbung erfährt. Nimmt man mit SEIBOLD (1951) an, daß sich die Bebenhäuser Störungszone über den mittleren Schwarzwald unmittelbar bis zum Oberrhein-Graben fortsetzt, so würde sich ein Zusammenhang zwischen dem württembergischen hot-spot und der Tiefentektonik des Oberrhein-Grabens zwanglos ergeben.

Literaturverzeichnis

CARLÉ, W.: Neue Beobachtungen zur Deckgebirgs-Tektonik von Südwest-Deutschland. — Jber. Mitt. oberrhein. geol. Ver. **32**, 10—33, Stuttgart 1950.
— Der Säuerling von Kleinengstingen — die einzige Mineralquelle der Albhochfläche. — Z. dt. geol. Ges. **105**, 252—267, Hannover 1954.
— Bau und Entwicklung der Südwestdeutschen Großscholle. — Beih. geol. Jb. **16**, 272 S., Hannover 1955.
— Kohlensäure, Erdwärme und Herdlage im Uracher Vulkangebiet und seiner weiteren Umgebung. — Z. dt. geol. Ges. **110**, 71—110, Hannover 1958.

CARLÉ, W.: Geologie und Hydrogeologie der Mineral- und Thermalwässer von Bad Überkingen, Landkreis Göppingen, Baden-Württemberg. — Jh. geol. L.-Amt Baden-Württ. **14**, 69—143, Freiburg 1972.

CARLÉ, W. & GROSCHOPF, P.: Zur Stratigraphie, Genese und Tektonik des Dogger-Eisenerz-Vorkommens von Geislingen an der Steige. — Jh. Ver. vaterländ. Naturkd. Württ. **122**, 67—91, Stuttgart 1967.

EMTER, D.: Ergebnisse seismischer Untersuchungen der Erdkruste und des obersten Erdmantels in Südwestdeutschland. — Diss. Univ. Stuttgart, 108 S., Stuttgart 1971.

ILLIES, H.: Graben Tectonics as Related to Crust-Mantle Interaction. — Graben Problems. Internat. Upper Mantle Proj. **27**, 4—26, Stuttgart 1970.

MÄUSSNEST, O.: Die Ergebnisse der magnetischen Bearbeitung des Schwäbischen Vulkans. — Jber. Mitt. oberrhein. geol. Ver. **51**, 159—167, Stuttgart 1969.

— Regionalmagnetische Vermessung Mittelwürttembergs. — Geol. Jb. **88**, 567—586, Hannover 1970.

MORGAN, W. J.: Deep mantle convection plumes and plate motions. — Amer. Assoc. Petrol. Geol. Bull. **56**, 203—213, Tulsa 1972.

NATHAN, H.: Wasserbohrungen im Ries. — Geol. Jb. **74**, 135—146, Hannover 1957.

NEUMANN, G.: Regionale magnetische Variometermessungen in Südwestdeutschland 1932. — Beitr. angew. Geophysik **8**, 19—44, Leipzig 1939.

SEIBOLD, E.: Der Bau des Deckgebirges im Oberen Rems-Kocher-Jagst-Gebiet. — N. Jb. Geol. Paläont. Abh. **92**, 243—366, Stuttgart 1950.

— Das Schwäbische Lineament zwischen Filder-Graben und Ries. — N. Jb. Geol. Paläont. Abh. **93**, 285—324, Stuttgart 1951.

WEISKIRCHNER, W.: Vorläufige Ergebnisse der Untersuchung an vulkanischen Produkten der Schwäbischen Alb. — Vortrag Karlsruhe 1972.

WIEDEMANN, H. U.: Die Geologie der Blätter Göppingen (7223) und Lorch (7224) in Württemberg mit Nachträgen zu Blatt Weilheim (7323) 1:25 000. — Arb. geol.-paläont. Inst. Techn. Hochschule Stuttgart **53**, 226 S., Stuttgart 1966.

Volcanism of the Rhinegraben: potassium-argon ages, local setting, petrology, and gravity anomalies

Apparent Potassium-Argon Ages of Lower Tertiary Rhine Graben Volcanics

by

H. J. Lippolt, W. Todt and P. Horn

With 2 figures and 3 tables in the text

Abstract: Potassium-argon age determinations of 56 volcanic rocks of mainly alkalibasaltic composition from the Upper Rhine Graben area yield a range of Senonian to Pliocene ages with a frequency maximum of Eocene ages. Upper Cretaceous and early Tertiary ages occur in nearly all areas bordering the Rhine Graben.

A group of basalts having Upper Oligocene to Lower Miocene ages lies in the Otzberg area. Kaiserstuhl, Hegau and Vogelsberg volcanism occurred in the Miocene.

There appears to be a time span of about 20 Ma between the onset of volcanism in the Upper Cretaceous and Rhine Graben sedimentation in the Eocene. This indicates that tensional rupturing accompanied by volcanic eruptions preceded step faulting in the graben area.

Introduction

The Upper Rhine Graben region, which had been dormant since Permian times, saw renewed volcanic activity in the Tertiary period with the eruption of various types of basic to intermediate lavas, which are typical of continental graben structures.

The volcanics, which occur as pipes, dykes and occasionally as flow sheets, irrupted almost exclusively along faults in the highlands bordering the Rhine Graben. It is noteworthy that the main graben fault is free of basalts with the exception of the Kaiserstuhl which lies on the intersection of this fault and the Bonndorf Graben Fault System.

The oldest indication that tensional forces were operable in the Upper Rhine Graben region is the occurrence of granite porphyry and lamprophyre dykes in the Schwarzwald basement. This magmatic activity was followed by the emplacement of N—S trending hydrothermal veins. In addition, the distribution of Permian rhyolites demonstrates the previous existence of E—W tensional forces along the present margins of the graben, long before the graben formation.

The graben began to form in Middle Eocene times according to stratigraphical evidence (WAGNER 1953). What began as down-warping took on the character of step-faulting in Chattian to Aquitanian times. The zone of greatest subsidence migrated from SW to NE during Tertiary times, as can be seen in the paleotectonic maps published by ILLIES (1965).

Stratigraphic evidence for the time of eruption

The sequence and time of eruption at the various volcanic centers is important because of the probable close relationship between the Rhine Graben tectonism and the graben volcanism.

LEPSIUS, as early as 1892, proposed that the nepheline basalts in the northern Odenwald and in the Mainz basin are considerably older than the Miocene eruptions of the Vogelsberg and the Lower Main Valley, as they are nowhere in contact with Tertiary sediments. RÜGER (1933) inferred an early Tertiary age for the Katzenbuckel on geomorphological grounds. On the western edge of the Rhine Graben, boulders of basalts occur in Middle Oligocene sediments, indicating a minimum age for the corresponding eruptions (DOEBL 1964, STELLRECHT 1964).

Paleontological evidence indicates the onset of the Kaiserstuhl volcanism in Middle Miocene times (TOBIEN 1958). From stratigraphical evidence the volcanic activity in the Hegau lasted from the Upper Miocene until Lower Pliocene times.

New evidence has shown that the trap basalts of the Lower Main area overlie Aquitanian sediments, so that one has to assume that the basalt itself was erupted in Middle to Upper Miocene times (NÖRING 1955).

First age determinations

The majority of the Upper Rhine Graben volcanics occur in the highland areas devoid of Tertiary cover: Hence, only radiometric methods can be used for dating.

The first age determinations on these rocks, or mineral separations from them, using the K-Ar method, were carried out by LIPPOLT, GENTNER & WIMMENAUER (1963). The age of 66 Ma for a biotite sample from the Katzenbuckel supported RÜGER's conclusion that the Katzenbuckel eruption occurred in Late Cretaceous or early Tertiary times. K-Ar ages of the rocks of the Kaiserstuhl showed that this volcano was active between 18 and 16 Ma ago. In the Hegau the highest measured age was about 14 and the lowest 7 Ma.

The Katzenbuckel analysis disproved the opinion common at that time that the volcanism in the Upper Rhine Graben region was exclusively of late Tertiary age.

Whole rock dating of basalts

Potassium argon age determinations were carried out on 56 whole rock samples in order to obtain a sharper resolution of the evolutionary sequence of the Upper Rhine Graben volcanism than was achieved in the earlier work. Whole rock samples were analysed because it is very difficult and often impossible to seperate the required minerals from the fine-grained basalts. Moreover, it has been shown in recent years that potassium-argon whole rock analyses on basic volcanics can yield realistic data, provided model-age conditions prevail.

Although only the freshest possible material was sampled, some specimens nevertheless proved to have suffered secondary alteration. The degree of alteration appears to have little influence on the measured age, as can be seen from two examples given in Tab. 1.

Tab. 1. Age results as a function of freshness.

Sample	Potassium (%)	Age (Ma)
Neckarbischofsheim-West, not fresh	2,31	64,8
Neckarbischofsheim-East, fresher than NBH-W	3,15	65,4
Hochstädten, E of Schlossberg, fresh	1,13	46,5
Hochstädten, S of Bangertshöhe, unfresh	1,52	47,1

Because the alteration of the other basalt samples of this study is quantitatively less than, but qualitatively similar to that exhibited by the examples above, it may be assumed that they likewise exhibit no change in their apparent ages.

Whole rock ages on fresh basalts can be lower than the ages on the reliable potassium-argon minerals of the same rocks, owing to argon diffusion losses from the groundmass and the less retentive constituents (LIPPOLT et al. 1963). Unpublished data have shown that this reduction in age can be up to 20 %. Even if one assumes that such losses have occurred in the whole rock samples of this study, the general conclusions remain unchanged.

Another possible source of error could be extraneous argon from wall rock inclusions. As one can assume that such contamination would vary among cogenetic samples from a single outcrop, pairs of samples from specific outcrops were analysed. The results are given in Tab. 2.

Tab. 2. Apparent ages of different, but cogenetic samples.

Sample		Potassium (%)	Age (Ma)
Rossberg	I	0,998	42,5
	II	0,996	42,5
Katzenbuckel-Shonkinite		4,59	55,2
-Nephelinite		1,58	53,1
Hinter-Hauenstein	I	0,963	34,4
	II	1,18	34,8
Strütt	I	0,395	44,3
	II	0,195	43,0
Vogelsberg-Trapp:			
Klein-Steinheim		0,892	13,9
Teufelskaute		0,872	13,1
Frankfurt		0,921	13,5
Hohl-Hörstein		1,725	70,3
Hohl-Rückersbach		1,825	59,8

The two Rossberg samples were taken from two different levels of the quarry. They show neither petrographic nor geochronological differences. The two major rock types of the Katzenbuckel (shonkinite, nephelinite) show an age difference of about 4 %, which may be due to differences in argon retentivity. The olivine-nephelinites from Hauenstein near Hornberg yield identical ages. A further example is the pair of olivine-nephelinite samples from Strütt near Aschaffenburg. The samples Klein-Steinheim, Teufelskaute and Frankfurt belong to the same lava flow of the Vogelsberg volcano. Their potassium contents fall in a very narrow range and the difference in the apparent ages is well within the limits of analytical error. The nepheline-basanites from Hohl-Rückersbach and from Hohl-Hörstein are neighbouring dykes in pre-Variscan gneisses. They show an age difference just outside the error limits. It cannot be decided whether this difference is real or whether it is the effect of argon contamination or argon loss.

From the first five examples it is concluded that a single sample from any one outcrop gives a close approximation to the potassium-argon age of the rock.

Duplicate analyses for argon and potassium were carried out on all samples. From the statistics of the potassium and argon measurements it is concluded that for a total amount of argon of about $2 \cdot 10^{-6}$ cm^3, of which 30 % is atmospheric argon, the error on the ages is about 4 %.

Geographical and temporal distribution of the volcanic events

The potassium-argon ages published by HORN, LIPPOLT & TODT (1972) are presented in Tab. 3 together with new data from TODT, LIPPOLT & BARANYI (in preparation). Given are the sample names, the laboratory numbers, the percentage content of atmospheric argon in the samples (to which the analytical errors are sensitive) and the apparent ages.

The geographical distribution of the localites sampled by the various authors mentioned above and for the work of LIPPOLT et al. (1963) is illustrated in Fig. 1, which consists of a sequence of four sketch maps of the Rhine Graben region. Map 4 contains the locations of the Upper Cretaceous and Paleocene volcanics, Map 3 and 2 those of the Eocene and Oligocene volcanics, respectively, Map 1 combines the data for the Miocene and Pliocene times.

From Fig. 1 the volcanic history of the Upper Rhine Graben region can be seen to have developed as follows:

Map 4: Upper Cretaceous and Paleocene ages are found in the Taunus and Vor-Spessart areas in the north, near the Kraichgau-Zabern depression and on both sides of the Graben at the latitude of Freiburg. This early volcanism occurred in restricted areas over the whole length of the Graben.

Map 3: In the Eocene, which marks the time of greatest intensity of volcanism, the majority of the occurrences is concentrated in the extreme north of the region, where, about 20 Ma later, the greatest thickness of graben sediments was deposited. This means that the most intensive volcanism preceded the greatest amount of sedimentation in the same general area.

K-Ar-
Alter

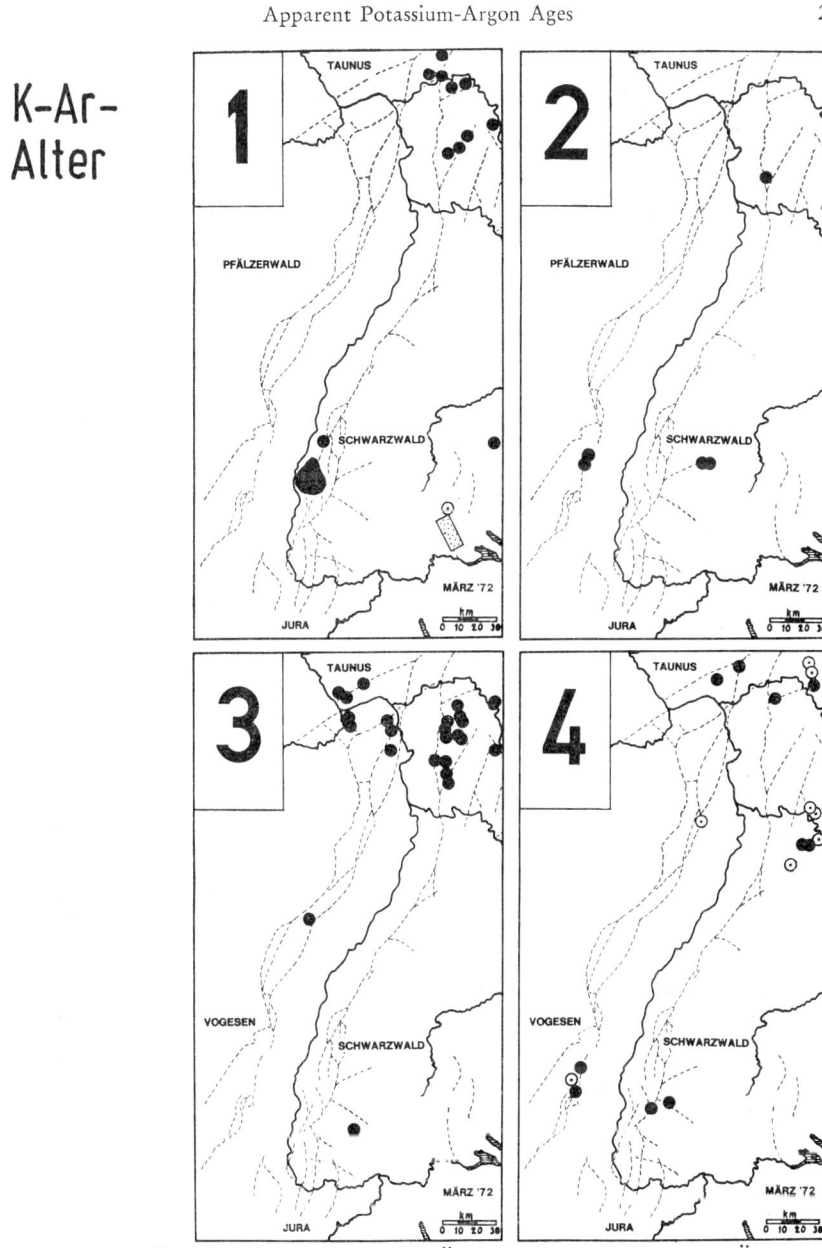

1 ⊙ PLIOZÄN   2 OLIGOZÄN   4 ⊙ PALEOZÄN
  (T < 10 Ma)   (26 Ma<T<38 Ma)   (52 Ma<T< 65 Ma)

● MIOZÄN   3 EOZÄN   ● OBERKREIDE
  (10 Ma<T<26 Ma)   (38 Ma<T< 52 Ma)   (T > 65 Ma)

Fig. 1. K-Ar ages of the different outcrops of Upper Cretaceous to Pliocene volcanics in the Rhinegraben area.

Map 2: In Oligocene times this northern area became virtually dormant, whereas the activity in the south continued as before.

Map 1: In Miocene times there was a recrudescence in the northern part of the region with the emplacement of volcanic rocks in close association with the Otzberg Zone tectonism and the production of trap basalts belonging to the Vogelsberg Volcanic Suite. One of the earliest members of this suite is the phonolite from Grebenhainer Schutzhütte with an age of 19 Ma. Two ages of Vogelsberg lava flows can be distinguished, one of 13.5 Ma and one near 16.5 Ma.

Tab. 3. Apparent K-Ar ages of 56 samples from volcanic rocks of the Upper Rhine Graben area.

Sample	Sample Number	Ar (atm) %	Age (Ma)
Odenwald:			
Götzenhain	3	75	40
Dieburg	13	24	40
Kühruh	38	34	41
Katzenschneise	1	33	44
Stetteritz	4	33	44
Roßberg	7	40	43
Traisa	9	23	35
Forstberg	11	31	45
Otzberg	6	38	22
Spitalfeld	22	28	28
Webern	23	67	25
Hochstädten	2	25	47
Mitlechtern	10	9	45
Oberlaudenbach	37	38	38
Zotzenbach	49	36	47
Hoher Berg	43	8	65
Farenberg	52	41	20
Eisenbach	50	13	49
Katzenbuckel (Na-shonkinite)	44	12	55
Katzenbuckel (Sa-nephelinite)	55	15	53
Vor-Spessart:			
Strütt	29	65	44
Hohl-Hörstein	53	64	70
Hohl-Rückersbach	54	31	60
Rückersbacherschlucht	SP	16	53
Kraichgau:			
Neckarelz-Diedesheim	42	23	60
Neckarbischofsheim	21	20	65
Steinsberg	14	18	55
Schwarzwald:			
Hinter-Hauenstein	15	40	35
Attental	61	19	81
St. Georgen	35	29	70
Hochkopf	62	77	42
Vogesen:			
Reichshofen	8	36	44
Le Valtin	59	49	61
Reichenweier	68	96	32

Tab. 3, continued

Sample	Sample Number	Ar (atm) %	Age (Ma)
Taennchel	67	70	29
Vorder-Marbach	69	57	83
Trois Epis	70	68	93
Taunus:			
Hörkopf	20	30	41
Waldburghöhe	19	15	41
Bossenheim	18	22	52
Eppstein	17	18	76
Oberberg	48	46	68
Mainzer Becken:			
Rochusberg	45	49	51
Sarmsheim	40	22	49
Olm	41	34	58
Nierstein	5	27	48
Hillesheim	16	50	44
Forst	12	28	53
Lower Main area:			
Galgenberg	47	41	16
Frankfurt (Bohrloch N)	28	63	17
Klein-Steinheim	25	66	13,5
Teufelskaute	26	58	13,5
Frankfurt	30	46	13,5
Vogelsberg:			
Grebenhainer Schutzhütte	V 8	30	19
Kaiserstuhl area:			
Mahlberg	60	60	16
Hegau:			
Wartenberg	33	72	8,5

In the southern part of the region, volcanism reached its peak intensity with the eruption of the Kaiserstuhl volcanics and continued through the Upper Miocene and Lower Pliocene with the development of the new volcanic center of the Hegau area.

Ages and rock-types

The ages of the post-Permian volcanic rocks of the Upper Rhine Graben region show a certain correlation with rock type (Fig. 2).

Olivine-nephelinitic magma was the first to be intruded, which supports the view of WIMMENAUER (1967) that it is primary in character. This rock type is represented throughout the whole age range from 93 to 16 Ma. The basanites lie in the interval 70 to 35 Ma. The emplacement of both these rock types appears to have culminated at the same time, 40 to 50 Ma ago. In the interval from 16 to 8 Ma the tholeiites from the Vogelsberg, the melilite-ankaratrites and the phonolites predominate (Kaiserstuhl, Hegau). These rock types are represented only sporadically among the older volcanic rocks (Reichshofen, Le Valtin,

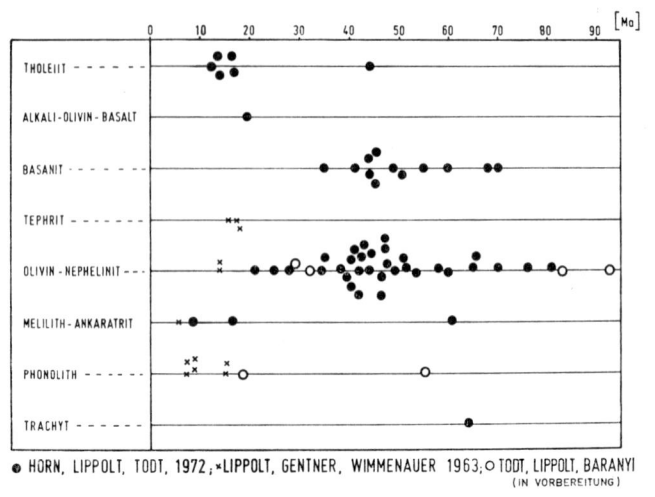

Fig. 2. Age distribution of the various rock types.

Rückersbacher Schlucht). Tephrites are known only from the Kaiserstuhl, having ages of about 18 Ma. The single alkali-olivine-basalt from Farenberg near Aschaffenburg is of doubtful provenance as the sample was taken from a boulder. The only trachyte dated up to now has an age corresponding to the Cretaceous to Tertiary boundary. It follows that with few exceptions the early Tertiary volcanic rocks of the Rhine Graben region are either nephelinites or basanites.

Summary

Upper Cretaceous and Early Tertiary alkali-basaltic rocks occur in the marginal zones of the Rhine Graben.

There are authors who hold to the hypothesis that there is interplay between the Alpine orogenic activity and taphrogenic tension zones (TRUNKO 1970). Several geochronologists recognize an early Alpine phase at about 80 Ma. This date corresponds approximately to the initiation of volcanism in the Rhine Graben region.

The maximum intensity of the volcanism in this region, as measured by the frequency of occurrences, occurred in Middle and Upper Eocene times. In the Odenwald a few basalts within the NE — trending Otzberg Zone have Upper Oligocene to Lower Miocenes ages. The tholeiitic trap basalts of the Vogelsberg suite were extruded during Middle or Upper Miocene times or both. In the south the Kaiserstuhl volcanics were erupted in Middle Miocene times and were followed by the Hegau volcanism which lasted until Lower Pliocene times.

A correlation between the chemistry of the rocks and the age exists in so far as olivine-nephelinites and basanites were produced first and predominate until about 40 Ma whereas the remaining rock types occur episodically, reaching their greatest frequency between 20 and 8 Ma ago. The tectonism which permitted eruption of basaltic lavas began in the Upper Cretaceous; however, since the oldest graben sediments are of Middle Eocene age, the major phase of subsidence began some 20 to 30 Ma later. Similarly, the time of maximum sedimentation followed the time of maximum volcanism by a comparable time span.

Acknowledgement: The authors thank their colleague Dr. M. S. BREWER for reading the manuscript and offering helpful criticism.

References

DOEBL, F., 1964: Ein Beitrag zur Frage der Altersstellung des Basaltes von Hillesheim. — Oberrhein. geol. Abh. 13, 123—129.

HORN, P.; LIPPOLT, H. J. & TODT, W., 1972: K-Ar-Altersbestimmungen an tertiären Vulkaniten des Oberrheingrabens. — Eclogae Geol. Helv. 65/1, 131—156.

ILLIES, H., 1965: Bauplan und Baugeschichte des Oberrheingrabens. — Oberrhein. geol. Abh. 14, 1—54.

LIPPOLT, H. J.; GENTNER, W. & WIMMENAUER, W., 1963: Altersbestimmungen nach der K-Ar-Methode an tertiären Eruptivgesteinen Südwestdeutschlands. — Jh. geol. L.-Amt Baden-Württ. 6, 507—538.

NÖRING, F., 1955: Diskussionsbemerkung zu: TOBIEN, H., Eine miozäne Säugerfauna aus vulkanischen Tuffen des Vogelsbergs. — Z. dt. geol. Ges. 105, 588.

RÜGER, L., 1933: Paläomorphologische Probleme aus dem Odenwald und das Alter der Katzenbuckeleruption. — Zbl. Miner. etc. Abt. B, 542—552.

STELLRECHT, R., 1964: Der tertiäre Vulkanismus bei Forst am Pfälzer Rand des Oberrheingrabens. — Jber. u. Mitt. oberrh. geol. Ver. NF. 46, 97—128.

TOBIEN, H., 1958: Das Alter der Eruptionen am Limberg (Kaiserstuhl). — Z. dt. geol. Ges. 110, 4—5.

TODT, W.; LIPPOLT, H. J. & BARANYI, I., 1973: The age distribution of the volcanic rocks of the southern Rhine Graben (in preparation).

TRUNKO, L., 1970: Rheingraben, Karpatenbecken und die Interferenz von Orogenese und Taphrogenese. — Oberrhein. geol. Abh. 19, 61—71.

WAGNER, W., 1953: Das Gebiet des Rheintalgrabens während des Mesozoikums und zu Beginn seiner tertiären Ablagerungen. — Notizbl. hess. L.-Amt Bodenforsch. 81, 164—194.

WIMMENAUER, W., 1967: Igneous rocks of the Rhinegraben. — Abh. geol. L.-Amt Baden-Württ., 6, 144—148.

Der paleozäne Basalt-Vulkanismus im Raum des Unteren Neckars

von

O. Mäussnest

mit einem Beitrag von E. BECKSMANN

Mit 2 Abbildungen im Text

Paleocene Basaltic Volcanism in the Area of the Lower Neckar River

Abstract: About 60 million years ago volcanic activity occurred in the area of the lower Neckar river. This basaltic volcanism, however, was limited to only a few centers. The volcanics are connected to NNE—SSW running fissures, which appear to be as old as the volcanism itself. The opening of the fissures may be referred to tensional forces in the crust as the result of the formation of the Rhinegraben during the Eocene. These fissures are supposed to be rejuvenations of old lineaments within the Hercynian basement. The observed faults in the northern Kraichgau as well as those in the southern Odenwald have not been volcanically active. This observation lets us assume that these faults have not dissected the entire crust as it may be expected from this portion of the rift shoulder.

Vor etwa 60 Millionen Jahren (HORN, LIPPOLT & TODT 1972) und damit ungefähr an der Wende Kreide-Tertiär waren vulkanische Kräfte in der als Kraichgau bezeichneten Landschaft zwischen Odenwald im Norden und Schwarzwald im Süden aktiv. Es handelt sich um einen auf wenige Punkte beschränkten basaltischen Vulkanismus. In den geologischen Karten und im einschlägigen Schrifttum sind folgende Vorkommen von Westen nach Osten aufgeführt: der aus Basalten und ihren Tuffen aufgebaute Schlot Steinsberg bei Weiler südlich Sinsheim, zwei kleine Basaltgängchen bei Neckarbischofsheim, der schon im Odenwald gelegene große Katzenbuckelschlot, ein Basaltgang bei Strümpfelbrunn SE Katzenbuckel, ein 800 m langer Basaltgang NE Neckarelz sowie drei kleine Basaltgängchen SW Neckarelz.

Auf Anregung und in enger Zusammenarbeit mit Prof. Dr. E. BECKSMANN, Freiburg i. Br., erfolgten in dieser unbedeutenden Vulkanprovinz insbesondere magnetische Feldmessungen in ΔZ. Einer genauen geologischen Kartierung der dortigen vulkanischen Vorkommen standen erhebliche Schwierigkeiten infolge mangelnder Aufschlüsse und sehr mächtiger Lößlehmüberdeckung entgegen. Alle Vorkommen dieser Vulkanprovinz zeigen nur positive Anomalien in ΔZ.

Die maximale Anomalie des Steinsberges beträgt + 5990 gamma, der zugehörige Randeffekt erreicht − 1040 gamma in ΔZ. Der Schlot zeigt eine gewisse Längserstreckung in nordöstlicher Richtung.

400 m vom Mittelpunkt der Steinsberganomalie entfernt in Richtung N 25° E befindet man sich auf einem bei den Feldmessungen neu gefundenen Basalttuffgang, der selbst in der Richtung N 35° E streicht, 200 m lang ist und eine maxi-

male ΔZ-Anomalie von + 120 gamma zeigt. Dieser Gang befindet sich in der Flur Kaiser der Markung Weiler.

Nach den Feldmessungen befinden sich bei Neckarbischofsheim nicht nur zwei kurze, im Gelände erfaßbare Basaltgängchen. Es liegt vielmehr ein ganzes Basaltgangsystem vor, das sich über 5,5 km erstreckt und N 25° E streicht.

Die Entfernung zwischen dem Nordende des Kaiserganges und dem Südende des Neckarbischofsheimer Gangsystems beträgt 8 km; die Verbindungslinie streicht ebenso wie die Neckarbischofsheimer Gänge etwa N 25° E.

Das Neckarbischofsheimer Gangsystem läßt sich in zwei große Gänge unterteilen; der Südgang ist 1,9 km lang, der Nordgang 4,3 km, wobei auf 700 m beide Gänge in 200 m Entfernung nebeneinander laufen. Der Südgang besteht aus einem Stück und zieht sich etwas geschlängelt durch das Gelände. Der Nordgang dagegen setzt sich aus insgesamt 10 kleineren Gangstücken zusammen, wobei das die Fortsetzung nach N übernehmende Gangstück jeweils etwas nach E verschoben ist. Im Übergangsgebiet zwischen zwei Gangstücken laufen oft die beiden Gangstücke noch einige Meter nebeneinander her.

Ein ΔZ-Profil quer zum Gangstreichen zeigt die Abb. 1; links ist der auskeilende südliche Teilgang zu sehen, rechts der neu einsetzende nördliche Teilgang. Die maximale ΔZ-Anomalie des Gangsystems beträgt + 2100 gamma. Das ganze Gangsystem steht weitgehend senkrecht.

Geht man vom Nordende des Neckarbischofsheimer Gangsystems ungefähr in der Richtung N 25° E weiter, so gelangt man nach 17 km zum Katzenbuckel. Auf dem Gipfel des Katzenbuckels befindet sich Blitzschlagmagnetisierung an drei Stellen. Der Schlot ist mit Basalten und deren Tuffen sowie mit Shonkiniten, einer

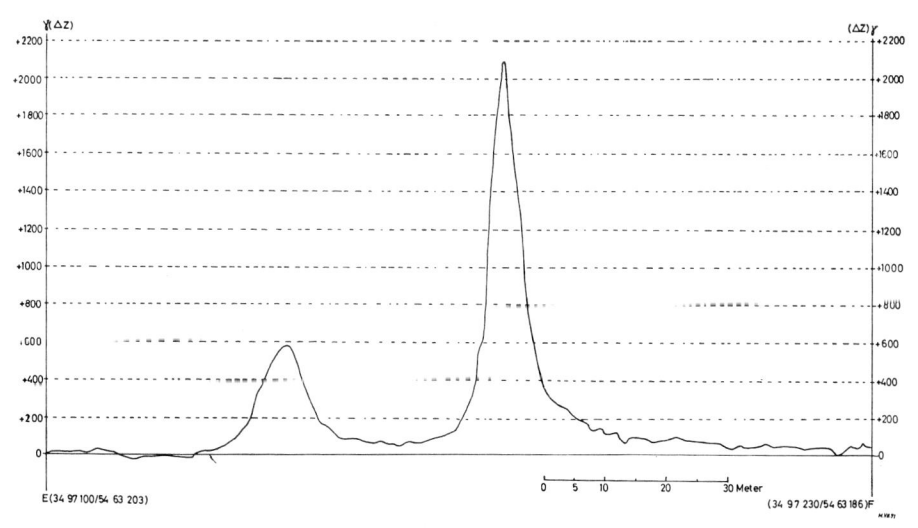

Abb. 1. ΔZ-Profil, gemessen senkrecht zum Streichen des Neckarbischofsheimer Gangsystems. Links ist der auskeilende südliche Gangteil, rechts der neu einsetzende nördliche Gangteil zu sehen.

Tiefenfazies des Basaltes, gefüllt. Die maximale ΔZ-Anomalie beträgt außerhalb des Gebietes der Blitzschlagmagnetisierung knapp $+ 8000$ gamma. Die Umgrenzung des Schlotes ist insbesondere im Norden nach Feldmessungen zu berichtigen; die mächtigen eiszeitlichen Basaltblockmeere machten eine einwandfreie geologische Kartierung der Schlotgrenze unmöglich.

Etwa 2,5 km SE des Katzenbuckels wurde kürzlich im Schrifttum ein Basaltgang erwähnt, der bei einer Wasserbohrung angetroffen worden sei (CARLÉ 1970). In dieser Bohrung wurden tatsächlich einige Vulkanitbrocken gefunden, jedoch bestanden diese wenigen Brocken nicht nur aus Basalt, sondern auch aus Shonkinit. Der Shonkinit ist aber hier eine absolute Unmöglichkeit, da das Gebiet nie aufgeheizt war und damit für die Ausbildung des grobkristallinen Shonkinites keine Voraussetzung gegeben war. Damit ist klar erwiesen, daß hier kein gangförmiges Vulkanitvorkommen angefahren wurde, sondern daß es sich um einige vermutlich von Spaziergängern in die Bohrung hineingeworfene Steine handelt. Da im ganzen Gebiet rings um den Katzenbuckel Wege und Straßen mit Basalten und Shonkiniten beschottert sind, wurden natürlich diese Steine hineingeworfen. Obwohl uns nach unserer Kenntnis dieses Gebietes ein vulkanischer Gang und insbesondere Shonkinit hier eine Unmöglichkeit erschien, wurden zur Sicherheit sehr engmaschige Feldmessungen in ΔZ durchgeführt, die erwartungsgemäß keine magnetische Anomalie und damit keinerlei Bestätigung eines vulkanischen Ganges bei Strümpfelbrunn ergeben haben.

Zehn km östlich der N 25° E streichenden Linie Steinsberg—Basalttuffgang Kaiser—Basaltgangsystem Neckarbischofsheim—Katzenbuckel befindet sich bei Neckarelz-Obrigheim ein weiteres Basaltgangsystem. Nach unseren Untersuchungen ist dieses Basaltgangsystem insgesamt 4 km lang und zeigt ebenfalls die mittlere Streichrichtung N 25° E. Zwischen Neckarelz und dem Neckar befinden sich zwei parallel verlaufende Basaltgänge. Zum Westgang gehört ein bisher unbekannter Schlot mit etwa 100 m Durchmesser und einer maximalen ΔZ-Anomalie von $+ 3740$ gamma. Die maximale Anomalie in ΔZ über den Basaltgängen von Neckarelz-Obrigheim betrug $+ 625$ gamma. Der SW von Neckarelz gelegene Teil des Basaltgangsystems besteht aus durchziehenden Gängen, der NE von Neckarelz gelegene Teil des Gangsystems ganz wie der Nordgang des Neckarbischofsheimer Gangsystems aus einzelnen Gangstücken. Zum Unterschied ist jedoch das die Fortsetzung nach Norden übernehmende Gangstück etwas nach W versetzt.

Neuerdings durchgeführte weitere Feldmessungen ergaben weder noch unbekannte vulkanische Vorkommen in Fortsetzung des Basaltgangsystems Neckarelz-Obrigheim bzw. des Systems Katzenbuckel—Steinsberg noch zwischen den einzelnen bereits bekannten Vorkommen auf der Linie Steinsberg—Katzenbuckel. Die Abb. 2 zeigt die von uns neugefundenen bzw. bestätigten vulkanischen Vorkommen im Gebiet des unteren Neckars sowie die wichtigsten tektonischen Linien dieses Raumes. — Die Felduntersuchungen waren nur dank der Unterstützung durch die DFG möglich.

Weder die Basalt-Gänge noch die Schlote sind an Zerrungsstörungen (Verwer-

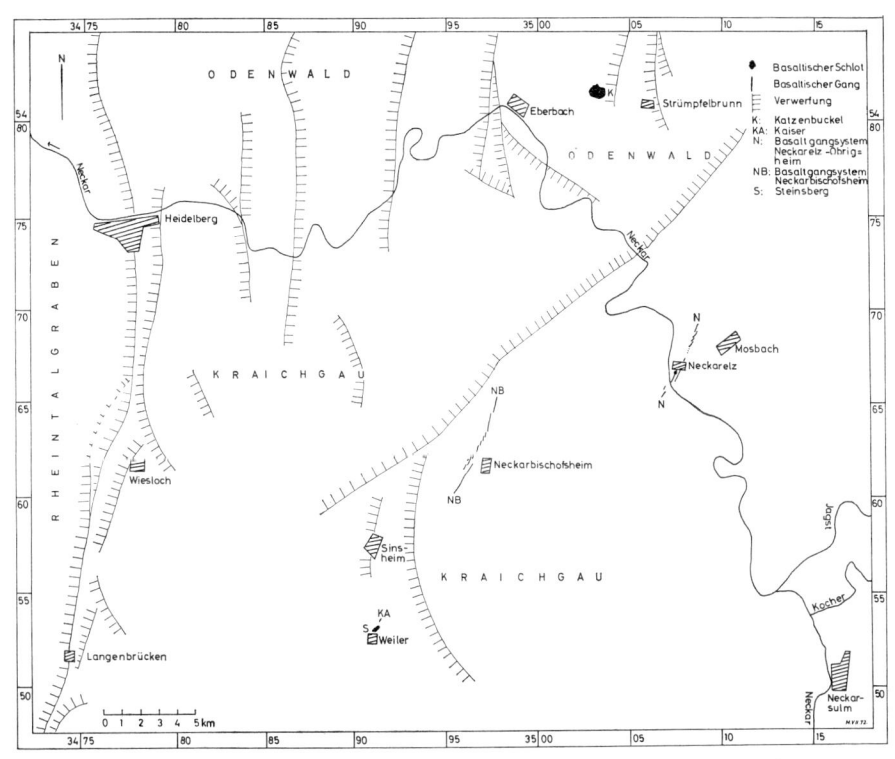

Abb. 2. Übersichtskarte des paleozänen Vulkangebietes am unteren Neckar.

fungen), die Schollen gegeneinander versetzen, sondern an NNE—SSW (= rheinisch) streichende Spalten gebunden. Diese Trennfugen müssen so tief aufgerissen sein, daß sie dem basaltischen Magma Aufstiegswege bieten konnten. Sie sind damit gleichen Alters wie der Basalt, also bereits im Paleozän entstanden. Da sie parallel zum Hauptteil des Oberrhein-Grabens das östliche Gebiet bis über den Neckar hinaus nach Osten durchsetzen, dürften sie auf dieselbe Krustenspannung zurückzuführen sein, die schließlich im Laufe des Eozäns dazu führte, daß das Rheinische Lineament in Form eines Grabens wieder auflebte. Der physikalischen Datierung der Basalte (HORN, LIPPOLT & TODT 1972) ist es zu verdanken, daß die rheinischen Spalten wie der an sie gebundene basaltische Vulkanismus nunmehr als Begleiterscheinungen eines frühen Embryonalstadiums dieser Taphrogenese verständlich werden.

In seinem nördlichen Drittel folgt der Oberrhein-Graben, dessen Einsenkung, wie die ältesten Teile seiner Füllung zeigen, im Laufe des Eozäns einsetzt, nicht mehr der rheinischen Richtung, sondern schwenkt in die NS-Richtung ein. Annähernd parallel dazu verlaufen die lang gestreckten, meist schwach gegen E gekippten Schollen im Bereich des Odenwaldes, dessen Aufbeulung offenbar jünger als paleozän ist. Die Brüche sind im Laufe des Tertiärs mehrfach bewegt worden (BECKSMANN 1970). Nur im südöstlichen Randbereich des Odenwaldes erfolgen die Bewegungen allein während des Pliozäns. In der Kraichgau-Senke, wo die Lößüberdeckung bis über 25 m mächtig wird, begegnet eine solche zeitlich-regionale Aufgliederung der tertiärzeitlichen Bruchtektonik noch erheblichen Schwierigkeiten. Weder im südlichen Odenwald noch im nördlichen Kraichgau konnte aber bislang ein an Brüchen aufgestiegenes Basalt-Vorkommen gefunden werden. Das läßt auf einen geringeren Tiefgang dieser Verwerfungen schließen und deutet auf mechanische Beziehungen zur Grabentektonik.

Literatur

BECKSMANN, E., 1970: Die zeitliche Aufgliederung der Bruchtektonik im Odenwald und Kraichgau. — Z. dt. geol. Ges., **121**, 119—123.

CARLÉ, W., 1970: Neue bemerkenswerte Aufschlüsse im Neckarland von Mittel-Württemberg und Nord-Baden. — Jh. Ges. Naturkde. Württemberg, **125**, 49—60.

HORN, P.; LIPPOLT, H. J. & TODT, W., 1972: Kalium-Argon-Altersbestimmungen an tertiären Vulkaniten des Oberrheingrabens I. Gesamtgesteinsalter. — Eclogae geol. Helv., **65**, 129—154.

The Subvolcanic Breccias of the Kaiserstuhl Volcano (SW-Germany)

by

I. Baranyi

With 1 figure and 2 tables in the text

The volcanic hills of the Kaiserstuhl lie in the southern quarter of the 300 km long, tectonically formed Rhine Graben. The central part of the volcano is deeply eroded; it forms today a truncated cone only. In the center of this cone, the subvolcanic intrusives are exposed, whereas on the outer slopes, volcanic tuffs and lava flows can be seen (WIMMENAUER 1957, 1959, 1962, 1963). The volcanic formations rest upon Middle Jurassic, Oligocene and, to a lesser extent, upon Miocene sedimentary rocks. The sedimentary base is mostly covered by loess.

The occurrences of subvolcanic breccias are confined to a region of approximately 6 km² within the center of the volcano and along the eastern slopes. They consist of xenoliths and xenocrysts, mostly rounded but occasionally sharp-edged in shape, embedded in a fine-grained volcanic matrix (BARANYI 1971).

Various processes of brecciation produced four types of rocks:

1. Breccias of the vent filling:
   a) rounded constituents predominant; mostly small rock fragments among fine rock and mineral detritus (probably formed by fluidization).

b) angular and rounded rock fragments of strongly varying size; disordered texture (probably formed by more violent explosions).
2. Brecciated country rock with carbonatite cement:
   a) ± in situ breccias; country rock fragments in carbonatite matrix.
   b) breccias with transport phenomena; rounded country rock fragments in carbonatite matrix with flow structure.

The most abundant type are the intrusive tuff breccias (1. a), arising from fluidization, i. e. formed by gas transport within the vent.

Interpenetration relationships of the breccias and the adjoining carbonatite are often observed (occurrence of xenoliths of breccia in carbonatite and xenoliths of carbonatite in breccia); they suggest a rapid alternation in the sequence of the carbonatite and breccia.

The components making up the breccias can be divided into the following groups:

1. Fragments of those rocks of the Kaiserstuhl, which crop out at the present surface; these are rocks of the plagioclase-bearing essexitic-theralitic family, rocks of the alkali feldspar-bearing phonolitic family, and feldspathoidic subtypes of the above-named two rock families, and carbonatites.
2. Fragments of basic and ultrabasic rocks, such as hornblendite, pyroxenite, tawite as well as xenoliths and xenocrysts of peridotitic rocks. The hornblendites consist mainly of common and titanian hornblende, accompanied by more or less abundant titanian augite and biotite. The jacupirangitic pyroxenites are composed predominantly of aegirine-augite, titanian augite, apatite and ore minerals. Closely related to the jacupirangites are the so-called tawites, bearing haüyne as leucocratic constituent. Frequently, the haüynes are altered to a mixture of calcite, zeolites, chlorite and clay minerals. The amphibole peridotite and the fragments of the peridotitic minerals (olivine picotite and chromian diopside) occur exclusively in the breccias of the central part of the Kaiserstuhl volcano.
3. As a result of the influence of the heat from the carbonatitic magma and by interaction with its alkali-rich volatiles, the gneisses and granites of the basement have been transformed into fenites. Fragments of fenites also occur in the breccias; they are essentially composed of potassic feldspar and aegirine-augite (syenitic fenites in the series described by Hsiao 1971).
4. Non-fenitized fragments of the crystalline basement of the Kaiserstuhl.
5. Fragments of the sedimentary graben filling.
6. Fragments of phlogopite, diopside and basaltic hornblende, often nearly idiomorphic, up to one centimeter in size; presumably originating from the "juvenile" magmatic source of the breccias.
7. Fine to medium grained matrix; consisting of mineral detritus and in-situ-formations: biotite, chlorite, calcite, zeolites and apatite.

Only fragments of hornblendites, pyroxenites, tawites and essexite porphyries were found in quantities sufficient for detailed petrographic and geochemical investigations. In order to obtain useful information about the chemistry of the mineral fragments, which make up 67—86 % by volume of the breccias, the trace element content of the whole breccia was also determined.

Tab. 1. Chemical compositions of the breccias and of rock fragments in the breccias (in weight %).

1. Average of 3 hornblendites
2. Average of 4 jacupirangites
3. Average of 6 tawites
4. Pyroxene fragment in the breccia
5. Hornblende fragment in the breccia
6. Phlogopite fragment in the breccia
7. Melanite in a melanite tawite fragment
8. Average of 7 whole breccias

	1	2	3	4	5	6	7	8
SiO_2	34,0	39,2	39,1	45,2	39,0	37,4	37,1	36,4
TiO_2	3,13	3,28	2,57	1,84	2,87	4,21	2,72	1,58
Al_2O_3	10,65	7,53	10,08	8,46	14,17	16,64	6,53	10,91
Fe_2O_3	11,10	7,30	7,01	3,42	4,20	2,39	17,76	6,09
FeO	6,51	6,06	5,48	3,73	6,61	6,50	2,86	2,26
MnO	0,32	0,28	0,27	0,05	0,07	0,00	0,58	0,45
CaO	14,75	20,33	18,78	21,14	11,46	0,26	29,01	15,93
MgO	9,38	10,78	7,76	13,73	14,51	19,76	0,98	4,69
Na_2O	1,47	0,84	0,80	0,76	2,19	0,53	0,01	1,91
K_2O	1,51	0,46	0,60	0,14	2,31	9,15	0,11	4,28
H_2O^-	0,43	0,27	1,15	0,16	0,17	0,33	0,00	2,02
$CO_2 + H_2O^+$	4,48	2,23	3,30	0,50	1,92	3,27	1,45	12,02
P_2O_5	1,53	1,38	2,48	0,66	0,26			0,61
Total	99,26	99,94	99,38	99,79	99,74	99,54	99,11	99,15

The compositions of the whole breccias and the rock fragments of type 2 (see p. 227) are generally basic to ultrabasic (Tab. 1). The contents of the elements Sr, Rb, Nb, Cr and Ni are considerably higher in the whole breccias than in the particular rock fragments named above (Tab. 2). Due to the metasomatic influence of carbonatitic fluids — recognizable by carbonatization and biotitization of the breccias in the neighbourhood of the carbonatite body — the matrices of the breccias were enriched in Ba, Sr and Rb. Owing to the presence of carbonatitic minerals, such as pyrochlore and dysanalyte, the breccias obtain their high niobium values. The abundance of Cr and Ni in the breccias is presumably caused by the mineral fragments of olivine, picotite and chromian diopside occurring in the matrix. A good covariant relationship between Cr and Ni is shown by the whole breccia samples, whereas the rock fragments show low Cr- and very low Ni values.

Tab. 2. Trace element contents of the breccias and of the fragmentes in the breccias (in ppm).

	1	2	3	4
Cr	108	136	60	179
Ni	20	45	38	172
Ba	4253			3409
Sr	533	310	1253	1979
Rb	31	18	37	168
Nb	200	17	100	254

1. Average of 3 hornblendites
2. Average of 5 jacupirangites
3. Average of 6 tawites
4. Average of 38 whole breccias
   (but average only of 8 Ba values of whole breccias)

The question of the origin of the pyroxenites and hornblendites in the breccias shall be briefly discussed. WIMMENAUER (1963) suggested that all of the Kaiserstuhl volcanic rocks could have arisen from a single olivine nephelinitic parent magma. Indeed, experiments of BULTITUDE & GREEN (1967) indicate the possible formation of hornblendite and biotite hornblendite from such an olivine nephelinitic melt at moderate depth (30—40 km) and high water vapour pressure. However, evidence is given by FRÖHLICH (1960) that amphiboles are able to take up large amounts of chromium from a melt. The hornblendite fragments of the breccias, however, are characterized by low amounts of Cr and Ni, opposed to the high content of these elements in the olivine nephelinites of the Kaiserstuhl. Therefore, one would not expect a direct relationship between olivine nephelinite and the hornblendites of the breccias.

Rather, the trace element distributions of the hornblendites, jacupirangites and tawites are comparable to those of the phonolitic rocks of the Kaiserstuhl. It is therefore proposed, that the mafitic xenoliths of the breccias just named represent melanocratic differentiates of the phonolitic magma of the Kaiserstuhl (Fig. 1). In agreement with this proposal is the fact that WIMMENAUER (1962) found many similar jacupirangite-, tawite- and yamaskite-like xenoliths in rocks of the phonolitic family.

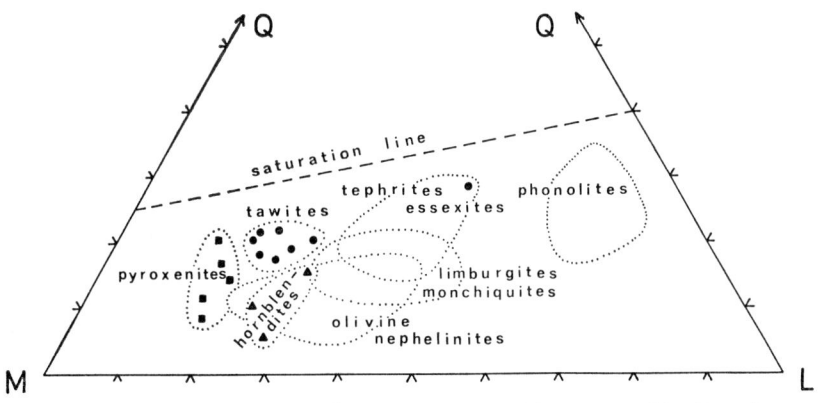

Fig. 1. QLM-diagram of igneous rocks of the Kaiserstuhl. The hornblendites ▲, pyroxenites ■ and tawites ● are fragments in the breccias (BARANYI 1971; WIMMENAUER 1970).

The distribution of minor elements in the breccias allows to define characteristic local differences. The breccias of the center have the highest Cr- and Ni contents; they decrease towards the periphery. Nb, Ba, Sr and Rb are most abundant near to the contacts with carbonatite. Nb is also found in remarkable amounts in some breccias of the eastern Kaiserstuhl. It probably indicates a geochemical influence from a carbonatite body in the depth, also witnessed by numerous fragments of fenites in these breccias.

In conclusion, the following model is offered for the origin of the subvolcanic breccias and their components:

Due to slow tension and foundering in the southern Rhine Graben region, numerous fractures were formed. Sudden decrease of pressure led to the escape of volatiles from the magma and to the formation of breccia-filled vents and fissures. The fractures, created at that time, reached to various depths. The vents of the central Kaiserstuhl yielded sparse fragments of rocks and minerals from the level of the Upper Mantle: amphibole peridotite and a "peridotitic" mineral assemblage olivine, chromian diopside and picotite. The same vents also yielded jacupirangite, tawite and hornblendite; rocks which probably formed in shallower magma chambers. On the other hand, the peripheral vents of the Kaiserstuhl transported mineral and rock fragments exclusively from shallow magma chambers, from the crystalline basement, the fenitic envelope of the carbonatite intrusion and from the sedimentary graben filling.

After the emplacement of the breccias, their components were subject to pneumatolytic and hydrothermal alteration, which obscured many of the primary features.

I am greatly indebted to Prof. Dr. W. WIMMENAUER for critically reading the manuscript. Special thanks are due to Dr. R. BECKER, Chicago, for help in translation.

References

BARANYI, I., 1971: Petrographie und Geochemie der subvulkanischen Breccien des Kaiserstuhls und seiner Umgebung. — Inaug. Diss. Univ. Freiburg i. Br.

BULTITUDE, R. J. & GREEN, D. H., 1967: Experimental study at high pressures on the origin of olivine nephelinite and olivine melilite nephelinite magmas. — Earth Planet. Sci. Lett. **3**, 325—337.

FRÖHLICH, F., 1960: Beitrag zur Geochemie des Chroms. — Geochim. Cosmochim. Acta, **20**, 215—240.

HSIAO, C., 1971: Petrologische und geochemische Untersuchung der Auswürflinge vom Limberg und von Eichstetten am Kaiserstuhl. — Inaug. Diss. Univ. Freiburg i. Br.

WIMMENAUER, W., 1957: Beiträge zur Petrographie des Kaiserstuhls. I. — N. Jb. Miner. Abh. **91**, 131—150.; I. II. III. 1959. — N. Jb. Miner. Abh. **93**, 133—173.; IV. V. 1962. — N. Jb. Miner. Abh. **98**, 367—415.; VI. VII. 1963. — N. Jb. Miner. Abh. **99**, 231—276.

— 1970: Zur Petrologie der Magmatite des Oberrheingrabens. — Fortschr. Miner. **47**, 242—262.

Bouguer-Anomalienkarte für den Kaiserstuhl

von

P. Thiele

Mit 3 Abbildungen im Text

Abstract: In the Kaiserstuhl paleovolcano (southern Rhinegraben) gravity measurements along profiles have been carried out by geo-students of the Karlsruhe University in the years 1967 until 1971.

Field technique and computational procedures are described. The results of approximately 230 points are presented as a map of Bouguer anomalies.

The accuracy of a single Bouguer anomaly is evaluted, mainly affected by the uncertainty of the terrain correction values.

The presented map indicates a significant maximum in the centre of the Kaiserstuhl and a rather smooth but unsymmetrical decrease of the anomalies with a minimal gradient in south-western direction.

To the west of the maximum an evident border of a plateau indicates a corresponding N—S-disturbance. In the utmost south-west of the Kaiserstuhl a further decrease of the anomalies is found in south-western direction, already published by REICH, CLOSS & SCHOENE (1940). This decrease certainly corresponds with the high values of Quarternary sedimentary fill of the Rhinegraben area in the south-west of the Kaiserstuhl volcanics.

1. Einleitung

Der Kaiserstuhl, im südlichen Teil des Oberrheingrabens gelegen, ist seit einigen Jahren das Übungsgebiet für das geophysikalische Feldpraktikum der Geo-Studenten der Universität Karlsruhe. In der vorliegenden Arbeit sind dabei die Schweremessungen der Jahre 1967 bis 1971 verarbeitet.

2. Feldmessungen

Die flächenhafte Aufnahme (ca. 230 Meßpunkte) setzt sich aus mehreren jahresweise bearbeiteten Einzelprofilen zusammen, der mittlere Punktabstand im Profil beträgt je nach Gelände 0,2 bis 0,5 km (Abb. 1).

Für die Schweremessungen standen als Gravimeter zur Verfügung: Worden Master Nr. 666 (ab 1967), LaCoste & Romberg Nr. 156 (ab 1969), Nr. 249 (ab 1971). Die Anlage der Gravimetermessungen erfolgte in Profilabschnitten, ab 1969 konnten mit zwei Gravimetern unabhängige Parallelmessungen vorgenommen werden. Die Wahl der Meßpunkte erfolgte so weit wie möglich nach dem Gesichtspunkt minimaler topographischer Reduktionen für den Nahbereich; zur lagemäßigen Festlegung standen amtliche Karten im Maßstab 1 : 5 000 zur Verfügung. Die Punkthöhen wurden durch tachymetrisch gemessene Höhenzüge oder — in flachen Bereichen — durch Nivellements bestimmt.

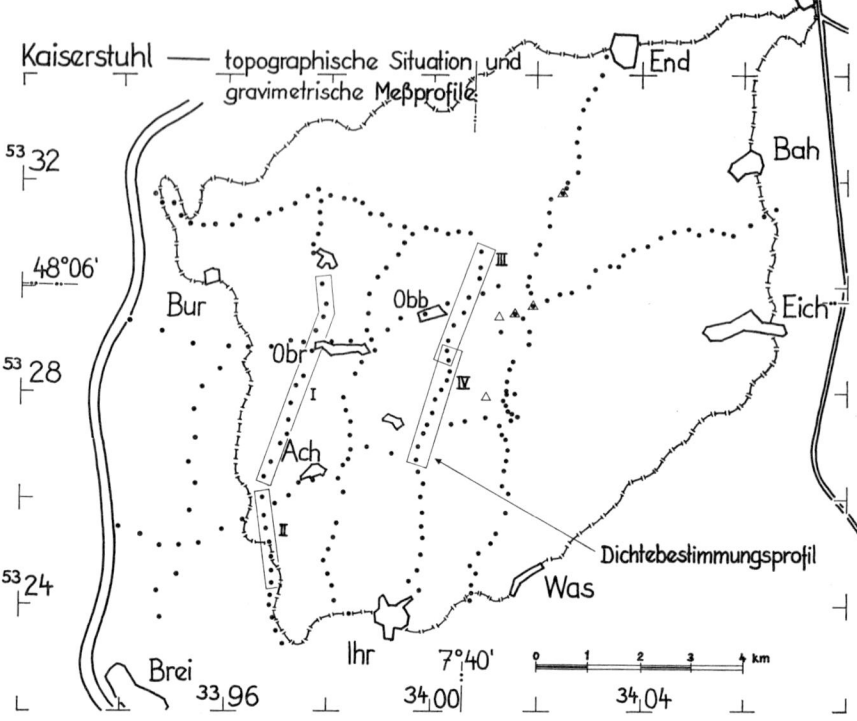

Abb. 1. Kaiserstuhl — topographische Situation, gravimetrische Meßprofile und Dichtebestimmungsprofile.

3. Bearbeitung der Feldmessungen

3.1 Bouguer-Anomalie

Ziel der Messungen war für alle Punkte die Berechnung der Bouguer-Anomalie
$$\Delta g_B = g - \gamma + \delta g_F - \delta g_B + \delta g_t.$$
Hierin bedeuten:

g   gemessene Schwere, bezogen auf den Punkt Freiburg (7913/5 A) des Deutschen Schweregrundnetzes: $g_{7913/5\,A} = 980\,840{,}92$ [mgal];

γ Normalschwere, berechnet nach der Internationalen Schwereformel, Stockholm 1930;

δg_F Freiluftreduktion, $\delta g_F = 0{,}3086 \cdot H$ [mgal]
    mit H[m]: Stationshöhe;

δg_B Bouguer-Reduktion, $\delta g_B = 0{,}0419 \cdot \sigma \cdot H$ [mgal]
    mit H[m]: Stationshöhe und σ [g cm^{-3}]: Gesteinsdichte;

δg_t topographische Reduktion [mgal], berechnet nach dem graphisch aufbereiteten Verfahren von Bursa (1965) für einen Bereich von 0 bis 5,24 km. Die Informationen für die Nahzone wurden hierbei über die Deutsche Grundkarte

(1 : 5 000) oder über Feldmessungen mit dem Koinzidenztelemeter WILD TM 10 gewonnen.

3.2 Gesteinsdichte

Der für die Berechnung von δg_t und δg_B benötigte Gesteinsdichtewert wurde mit 2,4 g cm^{-3} zugrunde gelegt. Gesteinsdichtebestimmungen nach JUNG (1943) und SIEGERT (1942) ergaben für vier Profilabschnitte (Abb. 1) einen Mittelwert von 2,39 g cm^{-3}; die Einzelergebnisse [g cm^{-3}] weichen vom Mittelwert nur um geringe Beträge ab.

Profilstück	I	II	III	IV
Verfahren JUNG	—	2,40	—	2,43
Verfahren SIEGERT	2,40	2,34	2,39	2,37

3.3 Genauigkeiten

Die Schwerewerte (g) können mit einer Genauigkeit von wenigen 0,01 mgal angegeben werden, die exakt berechenbare Normalschwere (γ) ist nur mit vernachlässigbar kleinen Fehlern von $\leq \pm 0,01$ mgal behaftet, die aus Kartierungs- und Punktentnahmefehlern in den Kartenunterlagen resultieren. Eine für tachymetrische Höhenzüge geländeabhängige Höhenungenauigkeit von $\pm 1 \ldots 3$ dm bewirkt in der Bouguer-Anomalie einen Fehler von $\pm 0,02 \ldots 0,06$ mgal. Bei nivellitischer Höhenbestimmung sind die Fehler vernachlässigbar klein. Die größten Ungenauigkeiten treten bei der topographischen Reduktion δg_t auf. Die verfahrensbedingte δg_t-Ungenauigkeit von einigen % ergibt für $\delta g_t = 1 \ldots 5$ mgal Fehler von 0,1 bis 0,2 mgal, in ungünstigen Fällen muß mit noch größeren Beträgen gerechnet werden. Die Genauigkeit von δg_t stellt aufgrund des vergleichsweise großen Betrages somit gleichzeitig nahezu die Genauigkeit der Bouguer-Anomalie für einen Meßpunkt dar. Zusätzlich müssen — wenigstens im Bereich außerhalb der Dichtebestimmungsprofile — auch Ungenauigkeiten in der Reduktionsdichte σ angenommen werden; diese führen dann in der Bouguer-Anomalienkarte zu überlagernden höhenabhängigen Scheinanomalien.

4. Meßergebnisse

4.1 Anomalienkarte

Die vorliegende Karte (Abb. 2) läßt ein sehr deutliches Schwerehoch im Zentrum des Kaiserstuhls und eine relativ ruhige, aber ungleichmäßig abklingende und nach SW stärker auskeilende Scharung der Isanomalen erkennen. Die Lage des Maximums fällt mit dem Sattel zwischen den Orten Vogtsburg und Schelingen nahezu zusammen.

Etwa 1,5 km westlich des Maximums weist ein von N nach S verlaufender Plateaurand auf eine entsprechende N—S-Störung hin. Inwieweit das Schwerehoch mit den Karbonatit-Intrusivmassen in Verbindung gebracht werden kann, müßten Modellrechnungen zeigen. Vom Maximum aus treten — über größere Strecken — nach Norden etwas größere Gradienten als nach Süden auf. Ob aber in dieser

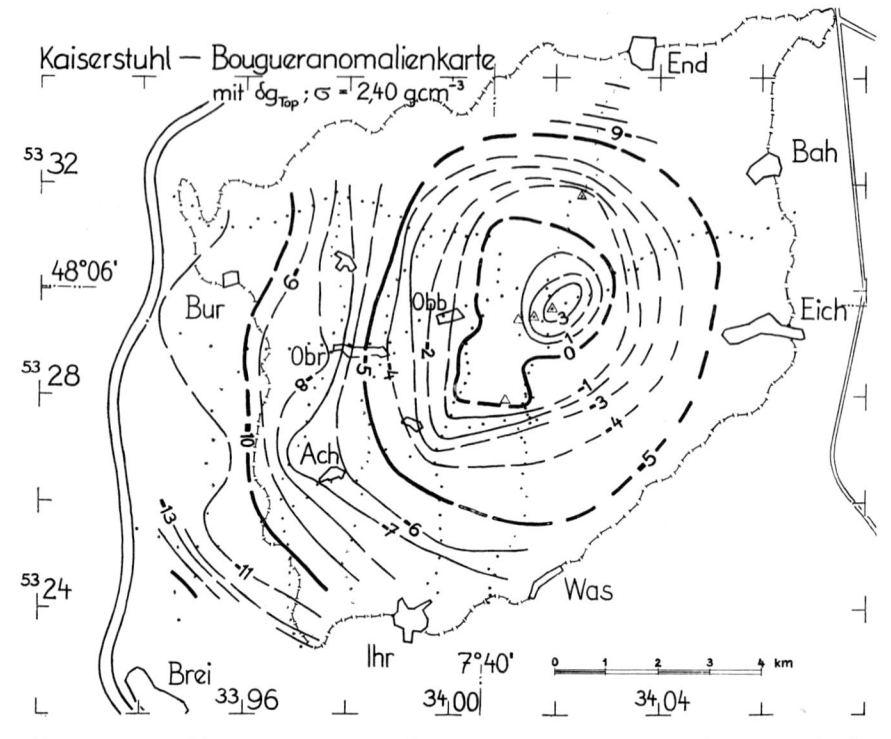

Abb. 2. Kaiserstuhl — Bouguer-Anomalienkarte ($\sigma = 2{,}40 \, \text{g cm}^{-3}$) mit topographischer Reduktion.

Gradientendifferenz ein regionaler Anteil enthalten ist, kann aus den bisherigen Messungen für die N—S-Richtung nicht nachgewiesen werden; in O—W-Richtung zeigt sich ein regionaler Anteil recht deutlich.

Am SW-Zipfel des bisherigen Untersuchungsgebietes tritt bei Breisach ein südwestlich gerichteter starker Steilabfall auf, der aber aufgrund noch fehlender Messungen bislang nicht genauer erfaßt werden konnte.

4.2 Vergleich mit Ergebnissen früherer Jahre

Die Anomalienkarte der SEISMOS (1951) enthält trotz ihres relativ großen Punktabstandes das ausgeprägte Schwerehoch (Abb. 3), die Verwendung der Normalschwereformel von Helmert (1901) führt aber zu abweichenden Anomalienwerten: Diff $\triangle g_{B \, \text{Kaiserstuhl}} = 12{,}7$ mgal.

Der Steilabfall im NO von Breisach — in der Seismoskarte aufgrund der zu geringen Punktdichte nicht enthalten — wurde jedoch schon von REICH, CLOSS & SCHOENE (1940) über Drehwaagemessungen festgestellt und für das gesamte Vorland im SW des Kaiserstuhls detailliert angegeben.

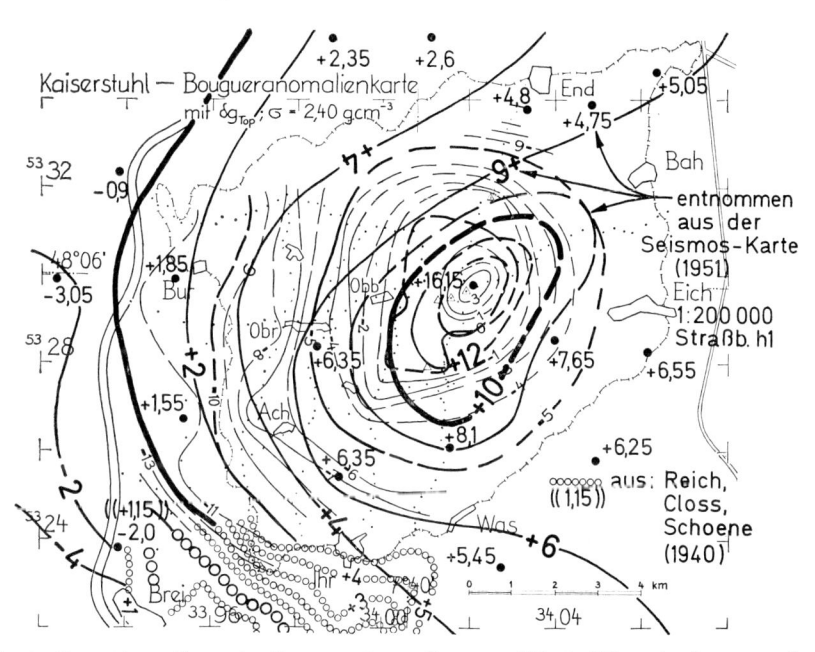

Abb. 3. Gegenüberstellung der Bouguer-Anomalien aus Abb. 2 [dünn durchgezogen bzw. gestrichelt] aus SEISMOS (1951) [dicker durchgezogen bzw. gestrichelt] und aus REICH et al. (1940) [gepunktet].

4.3 Ausblick auf weitere Untersuchungen

Diese Informationen über den Steilabfall zum Vorland hin, wobei sich nach REICH et al. (1940) am westlichen Tunibergrand Gradienten bis zu 7 mgal/km ergeben, dürften in enger Verbindung mit dem Maximum quartärer Grabenfüllung südlich des Kaiserstuhls und den rezenten Erdkrustenbewegungen in diesem Gebiet stehen.

Verdichtungsmessungen im bisher bearbeiteten Meßgebiet und eine Erweiterung des Meßgebiets — speziell im SW des Kaiserstuhls — lassen für die nächsten Jahre weitere aufschlußreiche Ergebnisse erwarten.

Literatur

BURSA, M., 1965: Consideration of Influence of Near Topographic Masses in Determining Terrain Corrections (in russ.). — Trav. Inst. Géophys. Acad. Tchécosl. Sci., No 217, Geofysikální sborník XIII, 43—69, Prag 1966.
JUNG, K., 1943: Über die Bestimmung der Bodendichte aus den Schweremessungen. — Beitr. angew. Geophys., 10, 154—164.
REICH, H., CLOSS, H. & SCHOENE, H., 1940: Über magnetische und gravimetrische Untersuchungen am Kaiserstuhl. — Beitr. angew. Geophys., 8, 45—77.
SEISMOS-GmbH, 1951: Gravimetermessungen für das Amt für Bodenforschung. — Bouguer-Anomalienkarte 1:200 000, Blatt Straßburg h 1.
SIEGERT, A. J. F., 1942: Determination of the Bouguer Correction Constant. — Geophysics, 7, 29—34.

The northern end of the Rhinegraben: mantle rise, structure, and recent tectonic activity

The Northern End of the Rhinegraben due to some Geophysical Measurements

by

R. Meissner and U. Vetter

With 5 textfigures

Abstract: Results from five refraction profiles obtained by quarry blasts, from a wide angle profile obtained by bore hole shots, and from some near vertical reflection measurements at the northern end of the Rhinegraben are discussed and compiled. Velocity-depth profiles from different geologic units are rather uniform, especially in the deeper parts of the crust. A dip of the Moho on both flanks of the graben reveals a deep-seated uplifting process.

1. Introduction

Seismic refraction measurements in the northern part of the Rhinegraben were performed in 1966 and 1967 and were reported by MEISSNER et al. (1967, 1970). Additional measurements along a wide angle profile with a common reflection element carried out by means of special bore holes with 4 shots on each side of the Rhine were observed in 1968 and reported by MEISSNER et al. (1972). Fig. 1 shows the situation of the profiles which cross different geologic units. By comparing results of these profiles together with some information from gravity surveys (GERKE & WATERMANN, 1960) an attempt is made to derive some meaningful geophysical data for the northern extension of the graben which seems to follow the Hesse Depression in a NNE direction according to many geologic features. Data from earthquake activity, on the other hand, indicate a more northwesterly extension with a connection to the Kölner Bucht along the present course of the Rhine river (AHORNER, 1970).

2. Results from refraction observations

Lines I, III, IV, V and VI in Fig. 1 are refraction profiles from quarry blasts, line II was observed as a wide angle profile but contains refracted arrivals, too. The profiles show 3 to 5 correlatable travel time branches. There are

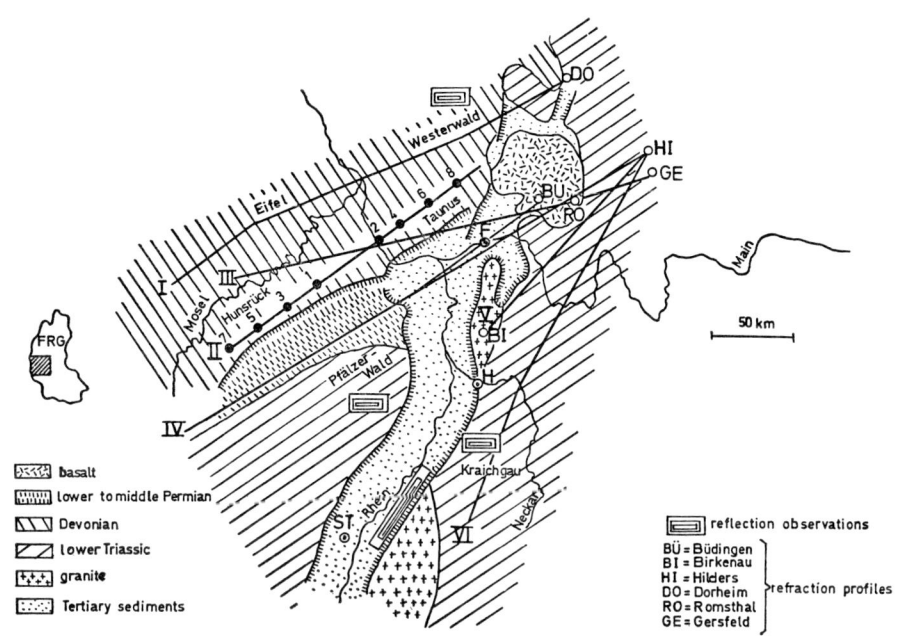

Fig. 1. Situation of refraction lines and reflection observations; names of quarry blasts, profile numbers, and some geologic features indicated.

i) the P_g (or P_1)-wave with a maximum observation distance of 120 km. This branch shows a slightly concave curvature indicating a velocity gradient in the uppermost part of the crystalline basement;

ii) the P_2-wave, mostly with a slightly convex, hyperbolic curvature, probably in connection with the Conrad Discontinuity (= CD);

iii) the P_3-wave, often also hyperbola-like and in connection with a discontinuity between CD and the Mohorovičić Discontinuity (= MD);

iv) the P_m (= P_4)-wave, mostly the strongest wave group with a convex hyperbola-like curvature, indicating true reflections from the MD between 40 and 70 km observation distance and diving waves from a strong velocity gradient on top of the MD at observation distances larger than about 100 km;

v) the P_n (= P_5)-wave, the refracted wave from the uppermost part of the mantle, observed at distances larger than 120—150 km, sometimes showing slight deviations from a straight line indicating undulations of the MD.

L i n e I is not yet completed. P_g, P_M-, and P_n-waves are clearly defined. The P_n-wave comes from a greater depth than does P_M as indicated by an apparent intersection of both branches when extrapolating P_n to shorter distances. As mentioned by MEISSNER (1967) lamellation of the deeper crust and/or a small velocity gradient in the mantle are responsible for such an apparent intersection. MD-depths are rather small and Bouguer anomalies are positive although the average elevation is about 400 m above sea level.

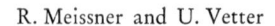

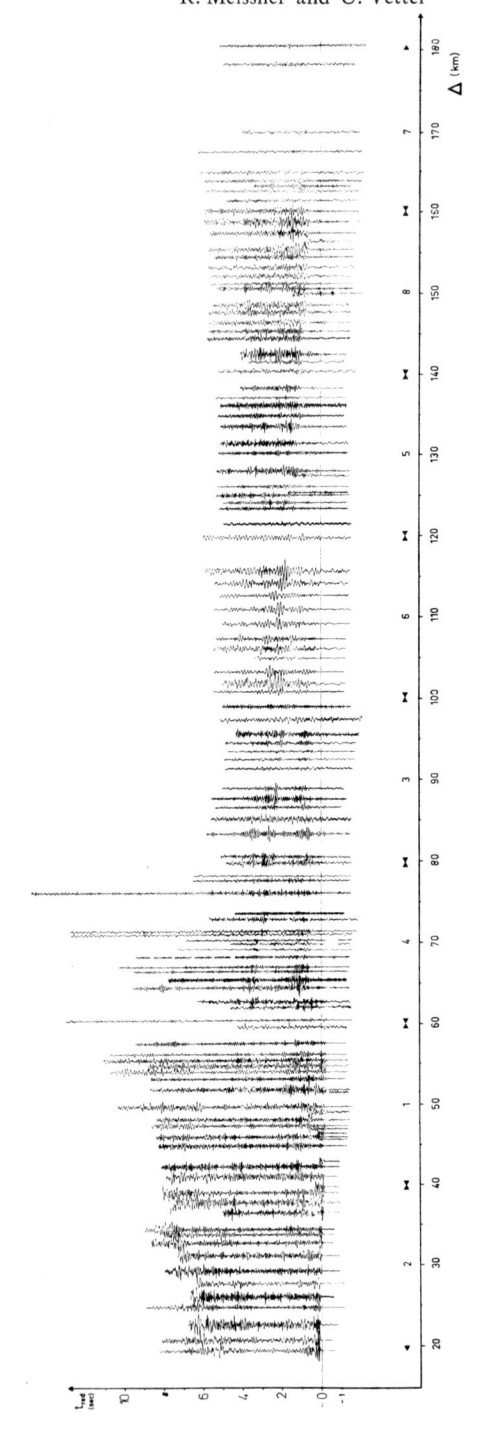

Fig. 2. Record section of profile II; reduced time scale $t_{red} = t - \dfrac{\Delta}{6}$.

Line II contains P_g-, P_2-, P_3-, and P_M-wave branches. P_n is very poor. This profile, extensively commented by MEISSNER et al. (1972), also shows shallow MD-depth with the highest point slightly less than 28 km at shotpoint 8 in the direct extension of the Rhinegraben axis to the north. This information was obtained by a near vertical reflection survey performed at each of the 8 shotpoints, as will be mentioned in the next paragraph. The seismogram section of line II is presented in Fig. 2. As this line was observed with the common depth point method the velocity-depth curve does not contain possible distortions from undulations of the boundaries, so most accurate velocity functions may be obtained though only for the center of the profile (Fig. 3, curve II).

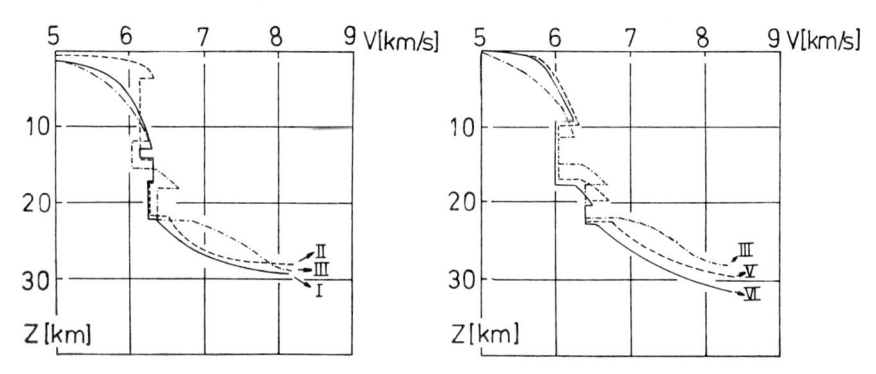

Fig. 3. Velocity-depth functions as derived from refraction observations, mostly from P_g-, P_2-, P_3-, and P_M-waves; numbers correspond to profile numbers of Fig. 1.

Line III crossing the Hanau-Seligenstadt depression shows only poor P_2-waves. P_M-waves are partly strong but not well correlatable as probably different MD-depths and structures are present along the profile. P_n-waves are clearly defined, a backward extension leads to an intersection with the P_M-branch as already found on line I.

Line IV, discussed by MEISSNER et al. (1967 and 1970) had revealed a pillow shaped structure in the northern part of the Rhinegraben. Already at depths of 24 to 25 km a velocity of 7.2 to 7.4 km/s is clearly defined. These velocity values are not found on refracted arrivals of the other profiles in the surrounding of the graben. They indicate a specific and limited structure near the crust-mantle boundary. No clear P_M-arrivals are found along the whole profile and no P_n-waves are observed inside the graben zone and in the Hanau-Seligenstadt depression. P_g-waves clearly indicate a lifting of both flanks in the upper part of the crystalline basement. The top of the pillow on the western flank corresponds to the gravity high in the Mainz area.

Line V, interpreted by STROBACH (1969) and MEISSNER et al. (1974) shows clear P_2- and P_M-waves indicating the existence of a CD and MD. Correlation is similar to the adjacent profile VI.

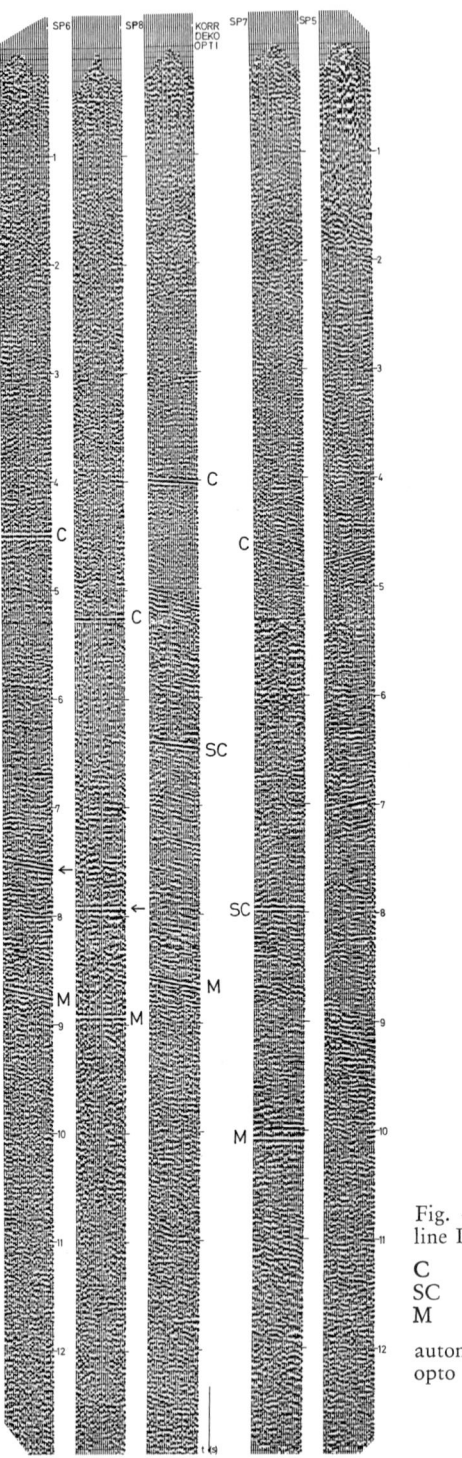

Fig. 4. Near vertical reflection observations on line II;

C = Conrad
SC = Sub-Conrad } Discontinuity
M = Moho

automatic correction, spike deconvolution, and opto filter applied.

L i n e V I, discussed by WILDE (1969) and MEISSNER et al. (1970) runs along the eastern flank of the Rhinegraben into the Black Forest. The P_2- and P_M-waves are rather strong, P_n also shows clear arrivals although some deviations indicate an undulation of the MD corresponding to depths of 29 ± 2 km which is not in agreement with observations of ANSORGE et al. (1970). On this line also an apparent intersection between P_n and P_M takes place showing that reflections from the MD certainly come from depths smaller than 29 km.

Velocity- depth functions from profiles I to VI are presented in Fig. 3. The penetration depth of the P_g-wave is abnormally small on profile II where only a thin layer of Devonian slates lies on top of the basement. The velocity gradient zones on top of the MD are similar to each other on lines I and II and on V and VI. Line III crossing the Hesse Depression shows velocity values of 7.2 to 7.4 km/s already at a depth of 24 to 25 km, while at the other profiles depths of 27 to 28 km correspond to these velocities. This might be an indication for a similar, though definitely weaker, process of rifting in the Hesse Depression compared to the Rhinegraben.

3. Results from near vertical reflection measurements

As already indicated by Fig. 1 reflection observations were carried out in various areas around the northern part of the graben (see SCHULZ, 1957; DEMNATI & DOHR, 1965; DÜRBAUM, 1966; and DOHR, 1972). The near vertical reflection observations of the 8 shotpoints of profile II carried out by using modern equipment and digital analysing techniques showed clear reflections from the CD and MD. An example is seen in Fig. 4. A group of reflections at about 5 sec travel time corresponding to the CD also appears in the Kraichgau and Pfälzerwald area but not in the Rhinegraben. In all areas groups of reflections between 7.0 and 7.8 and between 8.5 and 9.0 sec are found. They all belong to the velocity gradient zone on top of the MD obtained from the refraction profiles. The reflections in this zone indicate a strong lamellation. No reflections appear beyond a traveltime of 10 sec and above 4.5 sec indicating seismically homogeneous material at the corresponding depths.

4. Conclusions

Based on the velocity-depth functions and reflection observations the MD has been mapped around the northern part of the Rhinegraben (see Fig. 5). Whereas in the inner part of the graben the MD could not be identified the lifting of the flanks at both sides of the graben is clearly defined. So, an extension of the crust in the rift valley seems to be associated with some vertical movements of the mantle which must have their origin in deep-seated processes. High heat flow values, volcanism, and the forming of the graben pillow are further indications for such rather long time processes; shallow epicenters indicate the recent tectonic activity. These features are well known from other rift systems where sometimes the

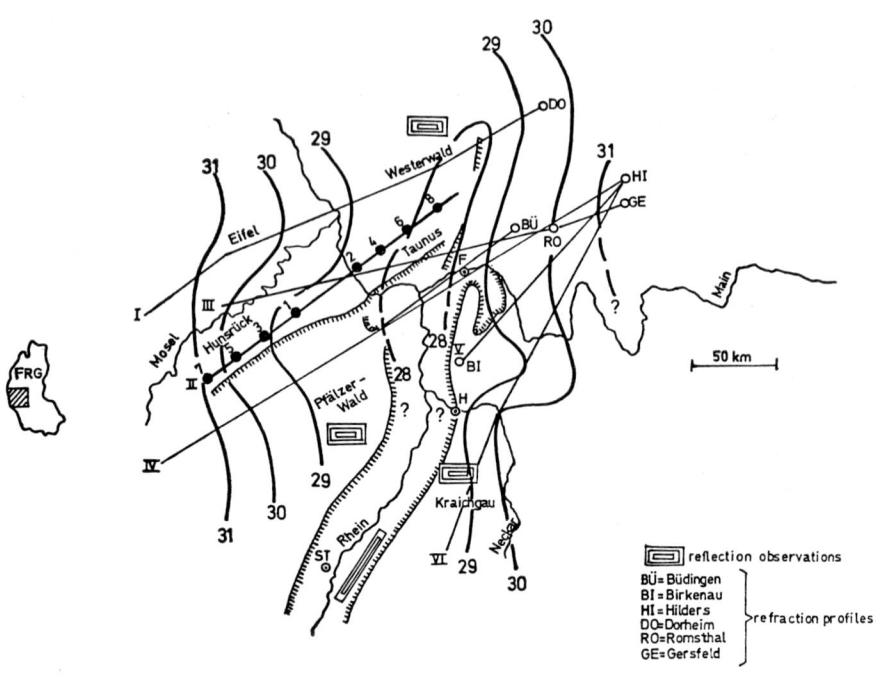

Fig. 5. Depth contour map of the Moho; in the Rhinegraben lines have been omitted because of doubtful correlation.

direction of the main activity changes, some branches of a system become inactiv while others come into life. A similar process may happen to the northern extension of the Rhinegraben where the formerly active Hesse Depression is at present seismically dead while the seismic activity runs in a NNW direction. The deep structures, on the other hand, like that of the MD depth, run along the straight extension of the graben to the north, much longer than the surface features blocked by the Rhenish Massif. This seems to be another indication for long time and deep-seated processes along the Rhinegraben rift structure.

References

AHORNER, L., 1970: Seismo-tectonic relations between the graben zones of the upper and lower Rhine valley. — Graben Problems, Internat. Upp. Mantle Proj. Rep. No. 27, 155—166.

ANSORGE, J., EMTER, D., FUCHS, K., LAUER, J. P., MUELLER, ST. & PETERSCHMITT, E., 1970: Structure of the crust and upper mantle in the rift system around the Rhinegraben. — Graben Problems, Internat. Upp. Mantle Proj. Rep. No. 27, 190—197.

BARTELSEN, H., 1970: Deutung der seismischen Weitwinkelmessungen in der Rheinischen Masse unter Verwendung neuer Kontroll- und Auswerteverfahren. — Dipl.-Arb., Inst. f. Met. u. Geoph., Univ. Frankfurt/M., unveröff.

DEMNATI, A. & DOHR, G., 1965: Reflexionsseismische Tiefensondierungen im Bereich des Oberrheingrabens und des Kraichgaus. — Z. Geoph., 31, 229—249.

Dohr, G., 1972: Reflexionsseismische Tiefensondierung. — Z. Geoph., **38**, 2, 192—220.
Dürbaum, H. J., 1966: Deep seismic sounding in the E part of the Rhenish Massif. — ESC Kopenhagen.
Gerke, K. & Watermann, H. 1960: Übersichtskarten der Schwere und mittleren Höhen von Westdeutschland. — Dt. Geodät. Komm., Bayr. Akad. Wiss., **46**, Teil III.
Meissner, R., 1967: Zum Aufbau der Erdkruste, Ergebnisse der Weitwinkelmessungen im bayr. Molassebecken, Teil 1 + 2. — Gerl. Beitr. z. Geoph., **76**, 211—254, 295—314.
Meissner, R. & Berckhemer, H., 1967: Seismic refraction measurements in the northern Rhinegraben. — Abhandl. Geol. L.-Amt Baden-Württ. **6**, 105—108.
Meissner, R., Berckhemer, H., Wilde, R. & Poursadeg, M. 1970: Interpretation of seismic refraction measurements in the northern part of the Rhinegraben. — Graben Problems, Internat. Upp. Mantle Proj. Rep. No. 27, 184—190.
Meissner, R., Bartelsen, H., Glocke, A. & Kaminski, H., 1974: An interpretation of the wide angle measurements in the Rhenish Massif. — DFG Monograph, 1974, in press.
Schulz, G., 1957: Reflexionen aus dem kristallinen Untergrund des Pfälzer Berglandes. — Z. Geoph., **23**, 225—235.
Strobach, K., 1969: pers. communication.
Wilde, R., 1969: Refraktionsseismische Untersuchungen auf dem Kraichgauprofil. — Dipl.-Arb., Inst. f. Met. u. Geoph., Univers. Frankfurt/M., unveröff.

Block Tectonic Interrelations between Northern Upper Rhine Graben and Southern Taunus Mountains

by

H.-J. Anderle

With 1 figure on a folder

Abstract: A new structural map for the northern end of the Upper Rhine Graben is presented. It is based on several thousand unpublished well logs, the literature including geologic maps, and field work on structural analysis in the southern Taunus mountains. The structural pattern, composed of faults in the SW, WNW, NW, and NNE directions, was formed by Tertiary to Recent crustal extension. It depends on the Hercynian planar fabric. Regional differences in the structural pattern are to be explained by contrasting structural behaviour of the Rheno-Hercynian and Saxo-Thuringian crustal blocks. The deep-reaching structural disconnection of these blocks by the Saar-Saale Zone produced the northern end of the Upper Rhine Graben.

1.0. Introduction

Although geologic investigation in the area at the northern end of the Upper Rhine graben (URG — Oberrhein-Graben) reaches back more than a century, a structural synthesis has not been made until recently. An extensive cover of Quaternary sediments like fluviatile sand and gravel, loess, and rock debris made it difficult to discover faults and establish a structural concept. Only a large number of bore holes together with geophysical investigations could improve that.

While KINKELIN (1892) had a rather simple tectonic concept, WENZ (1914) already knew some of the more important faults. CLOOS (1939) drew attention to the special structural style at the southern end of the URG comparing it with similar patterns in other graben endings. But he did not deal with the northern end, although it is shown in a sketch map. In the structural synthesis on SW-Germany by CARLÉ (1955) the northern end of the URG stayed to be a marginal area.

An approach towards a structural map of the northern end of the Upper Rhine Graben was made during the International Upper Mantle Project by the author (ANDERLE 1970). In the meantime new material from different sources has been added to this map (Fig. 1). The additional information drawn from the geologic maps 1:25 000 is not equivalent, as the years of issue of the single sheets range from 1891 to 1972. Extensive field work for structural analysis was done by the author in the southern Taunus mountains. The results give a notion about fault blocks in this area and their interrelation with the URG (2.1.) and also about the Hercynian planar fabric which influenced Tertiary to Recent structural development (3.0.).

2.0. Regional features of the structural pattern

2.1. Southern Taunus Mountains

The map (Fig. 1) shows the fault pattern for that part of the Taunus mountains only which has been investigated by the author. The area is characterized by SW striking rock units (not shown in Fig. 1) and a Metamorphic Zone (M. Z.) in the south. These are divided into 16 main blocks by transverse faults. There is a striking difference between the areas on both sides of the Idstein Depression (Idsteiner Senke). W of it there is only slight influence of young block faulting. There is one downthrow fault morphologically visible as an escarpment and one graben at Wiesbaden. E of the Idstein Depression extensive young block faulting is shown by morphological features. The fault blocks are bounded by NW striking transverse faults. Grabens form depressions with rather few outcrops, rocks altered by Tertiary weathering, and much rock debris. Horsts are morphological highs with numerous outcrops of fresh rock and little rock debris. Between Hofheim and Bad Homburg 3 of these structures cross the SW striking southern border fault of the Taunus mountains and continue SE into the area of the URG (see 2.3.).

It seems that the southern border fault of the Taunus mountains is just one of a series of SW striking step faults. That this one is quite easy to recognize depends on the contrasting rocks separated by it. Fault steps are indicated e. g. by several wells which encountered Rheno-Hercynian basement rocks under up to 230 m of Tertiary sediments up to 2 km S of the southern border fault (v. REINACH 1890).

2.2. Block-faulted area of Rheinhessen

The block tectonic style of Rheinhessen is similar to that of the western Taunus mountains. The whole area is slightly block-faulted. The faults are following the

NW and SW directions. The blocks are rising like steps from 2 marginal graben zones (WAGNER 1933) marked by the course of river Rhine. To the E — approaching the URG — NNE trending structures are added. The whole forms kind of a basin of Tertiary rocks framed by structural highs with pre-Tertiary rocks rising to the surface. Extensive loess coverage made it difficult to give the exact extension of Pliocene sediments in Rheinhessen (Dr. V. SONNE, Mainz, gave advise in that question).

2.3. Upper Rhine Graben

The constant NNE direction of the URG, marked by its parallel striking border faults and escarpments, is such an impressive structural and also morphological feature that in German geologic literature the adjective "rheinisch" is used to name that particular structural direction. But this over-all impression cannot conceal the fact that this clear picture is disturbed N of Heidelberg. From here to the N a NNW to NW trend is added to the structural pattern. This trend is represented by faults and also by the distribution of the isopachs of Tertiary plus Quaternary sediments in the graben (compare the isopach map compiled by DOEBL 1967). As the border faults do not run parallel any longer — the eastern one running to the NNW between Wiesloch S of Heidelberg and Darmstadt — the width of the graben is reduced distinctly from up to 40 km S of Heidelberg to about 20 km N of Darmstadt. Although N of Darmstadt and Oppenheim the border faults again strike to the NNE, the internal fault pattern of the graben is almost entirely formed by NNW to NW striking faults (STRAUB 1962). Most important is an oblique graben inside the URG which might be an equivalent of the Idstein Depression further NW in the Taunus mountains.

NE of the Gräfenhausen fault zone (Gräfenhausener Störungszone, VEIT 1954) which forms the northeastern margin of this graben, calcareous Miocene sediments have been encountered beneath Quaternary NW of Langen — a fact that shows the depth reduction of the URG. This continues N of river Main where calcareous Miocene sediments beneath Quaternary and at the surface delineate a NW striking horst. This horst is accompanied by 2 less distinct grabens. All 3 structures continue NW into the Taunus mountains. They form a structural connection of URG and Taunus mountains. NNE striking structures on the other hand do not continue into the Taunus mountains — neither the western border fault of the URG nor the Nidda Graben — although both are distinct structural features N of river Main.

At the present time it is not possible to complete the isopach maps of DOEBL (1967) for the area between the line Darmstadt — Rüsselsheim and the Taunus mountains, because bore holes penetrating the graben sediments are only available from the margins of this area. An isopach map (ANDERLE 1968) shows a considerable reduction of thickness of the fluviatile Quaternary sediments for the area between Gräfenhausen fault zone and river Main if compared with the URG further S. The area N of river Main has already been uplifted during Quaternary.

2.4. Block-faulted area of the western Wetterau

The block field adjoining the Taunus mountains in the E consists of 2 structurally different parts, the structures of which are intersecting in the areas of Butzbach and Bad Nauheim. (1) Between Bad Nauheim and Giessen NNW striking blocks form a transition from the uplifted eastern Taunus mountains to the depressed Vogelsberg volcanic area. The distinct NNW striking blocks are jointed by less distinct NE striking structural lines as at Butzbach. The clastic Tertiary sediments which cover the Rheno-Hercynian basement rocks are given the Miocene signature in Fig. 1 because they are covered in parts by Miocene basaltic rocks, and in one place NE of Butzbach Miocene lignite has been found beneath these basaltic rocks (KEGEL 1929). But as fossils are very rare parts of the Tertiary sediments may also be of Oligocene age. The largest thickness encountered yet is 109 m of Tertiary sediments (SCHOTTLER 1919). (2) Between Frankfurt and the southern Vogelsberg NNE striking faults are most important. Step faults in this direction form kind of a complex graben between Taunus mountains in the W and Sprendlingen Horst axis in the E. But there is no continuous central graben. Small graben structures like Nidda Graben and Horloff Graben are cut off by transverse faults and seem to be shifted along NE striking faults which form a small graben E of Bad Homburg. Although the Horloff Graben cuts into the Vogelsberg volcanics it has not yet been confirmed that NNE striking structures continue beneath and past the Vogelsberg palaeovolcanic area. HUMMEL (1929) has denied a direct connection with NNE striking depressions in northern Hesse.

The Tertiary sediments of the narrow fault step between Bad Homburg and Bad Nauheim are supposed to be of Oligocene age because of their structural position in a marginal uplifted block and one well which encountered rocks containing an Oligocene microfauna (records of the Geological Survey of Hesse).

2.5. Sprendlingen Horst axis

The most conspicuous structure in the whole area is the Sprendlingen Horst (Sprendlinger Horst) whose northern extension can be traced as far as the southern Vogelsberg. It starts narrowing in the Neu-Isenburg area, possibly caused by faulting, and is split in the Offenbach area (SOLLE 1951). Nevertheless its stronger eastern branch contains the structural highs of Palaeozoic greywacke at Erbstadt (sheet 5719 Altenstadt) and of Upper Permian rocks at Rabertshausen (too small for Fig. 1; sheet 5519 Hungen). The eastern border fault of this horst axis continues further N (SCHOTTLER 1925). The seeming absence of faults in the Sprendlingen Horst is due to the uniformity of Lower Permian rocks which does not favour the discovery of faults. Intense fracturing is indicated by the large number of volcanics, the structurally preserved Eocene rocks at Messel (MATTHESS 1966), and the strike of geologic boundaries E of Darmstadt.

2.6. Hanau-Seligenstadt Depression

Very little is known about the structural pattern of the Hanau-Seligenstadt Depression (Hanau-Seligenstädter Senke) although it is a rather shallow secondary structure compared with the URG. The depression shows a marked NNE—SSW extension. Its outlines are strongly influenced by NW striking structures which are most distinct in the southeastern corner. In the N where Miocene rocks rise to the surface, structural lines are visible which are running to the SW, NW and NNE. The southern extension at Dieburg is bordered in the E by the Otzberg Zone, an important NNE striking fracture zone continuing further S into the Odenwald mountains. There is no agreement yet as to its structural character, whether it is a zone of considerable left-lateral displacement originated in the Palaeozoic (v. BUBNOFF 1926, ILLIES 1965) or a fracture zone formed by mere block-faulting in the Tertiary (KLEMM 1925, SCHÄLICKE 1969). In this zone volcanic rocks of Permian and Tertiary age are exposed. The basaltic Tertiary rocks are 20—30 m.y. of age (HORN et al. 1972), a result that agrees well with a SSW trend in Miocene palaeogeography of the Hanau-Seligenstadt Depression which has been stated by GOLWER (1968).

2.7. Spessart Rise

In the Spessart mountains Saxo-Thuringian basement rocks are exposed which are covered by Buntsandstein further E. Although bordered by N striking normal faults in the W, this upwarped block is characterized by NW striking fractures. These fractures are normal faults, generally showing relative downthrow to the SW in the SW and relative downthrow to the NE in the NE. By this they form a NW striking horst-like uplift which continues NW into the Wetterau. There the basement is covered by Lower Permian rocks of unknown thickness. The relative downthrow or downwarp to the NW is caused by the SW striking Kinzig Zone (MURAWSKI 1963). The structural pattern in the area covered by Buntsandstein (for the most part outside of Fig. 1) shows that SW striking faults are present too (STREIT & WEINELT 1971, WEINELT 1962). These are easier to recognize here than in the SW trending basement rocks.

3.0. Causes and events responsible for the structural pattern

In order to explain the causes for both form and development of the structural pattern in a block-faulted area, it is necessary to study the fabric of the basement rocks. Because a rigid crust will react upon any endogenetic motions according to pre-existent fabrics. And the fault pattern originated in an undeformed sedimentary cover will depend on the structural response of the underlying basement. The pre-Tertiary structural elements which influenced Tertiary to Recent structural development at the northern end of the URG are (1) the fabric formed by Hercynian orogeny in the basement rocks, (2) the Rheno-Hercynian and Saxo-Thuringian Zones of the Hercynian basement and their differing structural behaviour, and (3) the Saar-Saale Zone — a narrow mobile zone of the crust which separates these two basement blocks since the Palaeozoic.

3.1. Hercynian structural heritage

Macroscopic structural analysis was carried out by the author in the southern Taunus mountains (in the area for which the fault pattern is given in Fig. 1). The results (which will in detail be published elsewhere) can be outlined as follows. Several rock units of Lower Devonian age striking about 60° (not outlined in Fig. 1) are accompanied in the SE by a Metamorphic Zone of lowest grade greenschist facies (MEISL 1970). The exact age of its rock units — sediments as well as volcanics — is not known, but it must be pre-Hercynian as is shown by its Hercynian fabric. The SW striking vertical to steeply dipping 1st schistosity which is parallel to the bedding in the sedimentary units is the principal s-plane. The generally NW dipping 2nd schistosity is less important and of less interest here. The joint subfabric is composed of tension joints and a conjugate system of shear joints. All joints are in general steeply dipping to vertical except where tilted by later rotation. In the orientation diagrams the tension joints form minima except near transverse faults. Their medium strike is about NW to NNW. The shear joints form maxima in most diagrams. They are intersecting in a steeply dipping to vertical axis — the c-axis of the joint subfabric. The medium strike of the shear joints is about WNW and NNE respectively. It is subject to change according to changing angles of intersection. In horizontal to low dipping strata outside the Metamorphic Zone the orientation of the joint subfabric is generally the same as in the steeply dipping to vertical strata of the Metamorphic Zone. As the joint subfabric has been formed by lateral orogenic compression the strata in the Metamorphic Zone must have been upturned prior to jointing. This seems possible only by rapid subsidence in an adjoining mobile zone further SE: the Saar-Saale Zone.

The Saar-Saale Zone — a SW striking geosuture — showed subsidence at the beginning of Hercynian orogeny (as shown above), as well as during Upper Carboniferous (FALKE 1971), Lower Permian (FALKE 1969, TRUSHEIM 1964), and — between Bad Kreuznach and Hanau (Fig. 1) — again in Tertiary times (GOLWER 1968, SONNE 1970). It separates two crustal blocks of contrasting structural behaviour: the Rheno-Hercynian and Saxo-Thuringian Zones of the Hercynian orogen (KOSSMAT 1927). The Rheno-Hercynian Zone encloses that part of the Hercynian geosyncline which had gathered sediments of great thickness. During Hercynian orogeny they were jointed, thrusted and folded — a planar fabric was formed. After that the orogen was uplifted, block-faulted and eroded. The Rhenish block — the Taunus mountains are part of it — stayed to be uplifted since. The Saxo-Thuringian Zone on the contrary is a deeper, more consolidated range of the Hercynian orogen with high-grade metamorphism and intrusions (BARTH 1972, NICKEL 1965). It was, at least in some parts, under erosion during Devonian (MEYER 1970), Carboniferous (FALKE 1971) and Lower Permian (FALKE 1969). From Triassic on it was covered by sediments. The recent exposures of the Saxo-Thuringian basement are due to horst structures formed by Tertiary to Recent block-faulting.

3.2. Tertiary to Recent block-faulting

When lateral extension affected the crust in the Upper Rhine area the whole Hercynian fabric of joints, s-planes, faults and thrusts was fit to react. The formation of a block field composed of all these structural elements in uniform distribution might have been expected. And there is indeed one phenomenon that shows the amount of fracturing: the great number of basaltic necks and dikes in the vicinity of the URG. In the southern Taunus mountains basaltic dikes have been found running in all directions of the Hercynian planar fabric. All the necks and dikes marked by triangles in Fig. 1 are without exception exposed in pre-Tertiary rocks. K-Ar dating of some of these necks and dikes has shown that most of them are of Eocene age (HORN et al. 1972). That is, magmatic activity found at lot of fissures in the beginning of Tertiary crustal extension. But the overall structural picture is different to-day. Certain structural directions are locally predominant. These singular structural features have been explained before (2.0.). At this place it is important to note the distinct structural difference between the areas belonging to the Rheno-Hercynian and the Saxo-Thuringian basement blocks. Different Hercynian fabrics according to different depth ranges in the Hercynian orogen (3.1.) resulted in different structural patterns during Tertiary to Recent block-faulting (see CLOOS 1939: 462).

In the more consolidated Saxo-Thuringian basement block a few large and distinct NNE striking structures have developed; i. e. URG, Sprendlingen Horst and Hanau-Seligenstadt Depression. That the NNE direction has been favoured might be due to two causes: (1) the NNE striking planes being almost perpendicular to the tensional forces (KNETSCH 1967) and (2) the existence of Hercynian shear or fracture zones (ILLIES 1965). The other directions of the Hercynian fabric are also present but they are of secondary importance.

As soon as the NNE striking structures reach the Saar-Saale Zone (which is marked by the distribution of the signature for Lower Permian rocks in Fig. 1) NW and SW striking structural lines gain importance. At the same time the depth of URG and Hanau-Seligenstadt Depression is reduced as is shown by reduction of sediment thickness (2.3.) and by deeper stratigraphical units rising to the surface.

In the more fractured Rheno-Hercynian basement block a kind of structural splitting takes place. In the Taunus mountains as well as in the Wetterau (HUM MEL 1929) numerous small structures (grabens, basins, horsts) are visible. One thing is particularly striking; the NNE running structures of the URG do not continue into the eastern Taunus mountains although their Hercynian fabric contains a well developed system of NNE striking joints (3.1.). The fault blocks of the eastern Taunus mountains are nevertheless running NW (2.1.). How can this be explained? There is a structural disconnection through deep-reaching vertical s-planes at the boundary of Rheno-Hercynian and Saar-Saale Zone (3.1.). Right-lateral displacement in this SW striking zone is proved by horizontal to low-dipping slickenside striae on SW striking s-planes in the southern Taunus mountains. As these slickensides are not found too often it must be assumed that the main zone of right-lateral displacement is further SE — now covered by post-Hercynian rocks. The remaining amount of lateral extension caused block-faulting along pre-existent transverse faults in the Taunus mountains immediately adjoining the URG, i. e. E of the Idstein Depression. These faults must have been of less cohesion than the NNE striking joints. The Taunus mountains W of the Idstein Depression remained almost unaffected by young block-faulting as they are adjoining the western shoulder of the URG.

Several km further E, however, the Taunus mountains are truncated by a NNE striking graben-like structure between Bad Homburg and Bad Nauheim (2.4.). This structure is distinctly smaller than the URG, and there is no direct structural connection (2.4.). But it might nevertheless be called a branch of the main graben that continues the NNE running fracture zone. That the Taunus mountains were cut off just here must be attributed to an older structural line. The recent eastern margin of the Rhenish block — of which the Taunus mountains are the southeastern part — is older than Tertiary block-faulting. In this area there was a facial boundary during Lower Devonian (NÖRING 1939) and for the first time a coast line in Upper Permian (RICHTER-BERNBURG 1959). This continued during Buntsandstein. The Rhenish block delivered sediments into a NNE trending basin in the E (DIEDERICH 1966, LAEMMLEN 1966), but the coast line is less well defined.

ILLIES (1972) who synthesized the results of the International Rhinegraben Research Group with special reference to deep-seated causes for graben formation, has explained the structural patterns in the zones of lateral offset of the rift structure as being caused by a transform fault mechanism in the basement. Data are particularly impressive for the southern end of the URG. At its northern end slickensides in the southern Taunus mountains are an additional detail favouring the transform fault hypothesis. Their right right-lateral displacement is resulting from differing rates of extension in the URG and the southern Taunus mountains.

Acknowledgements: I wish to thank the Geological Survey of Hesse (Hessisches Landesamt für Bodenforschung) at Wiesbaden, where this work was carried out, and its Director Prof. F. NÖRING, who made accessible the records of the Geological Survey. In particular, I am grateful to H. HICKETHIER, E. KÜMMERLE and O. SCHMITT, Wiesbaden, for providing their own unpublished results, to H.-J. SCHARPFF, Wiesbaden, and V. SONNE, Mainz, for helpful discussions and comments and to E. PAULY and H.-R. VÖLK, Wiesbaden, and P. ROTHE, Heidelberg, for critically reading a first version of the manuscript. This paper was supported by the German Research Association (Deutsche Forschungsgemeinschaft) at Bad Godesberg.

References

AMRHEIN, R., 1971: Geologische und hydrogeologische Untersuchungen im Ostteil von Blatt 6120 Obernburg. — Diplomarbeit, Frankfurt a. M. (unpublished).
ANDERLE, H.-J., 1968: Die Mächtigkeiten der sandig-kiesigen Sedimente des Quartärs im nördlichen Oberrhein-Graben und der östlichen Untermain-Ebene. — Notizbl. hess. L.-Amt Bodenforsch., **96**, 185—196.

ANDERLE, H.-J., 1970: Outlines of the structural development at the northern end of the Upper Rhine Graben. — In: ILLIES, J. H. & MUELLER, ST. (eds.), Graben Problems, 97—102, Schweizerbart, Stuttgart.

BARTH, H., 1972: Petrologische Untersuchungen im Felsberg-Zug (Bergsträßer Odenwald). — Abh. hess. L.-Amt Bodenforsch., **66**.

BENDER, H., 1969: Über ein Tuffvorkommen bei Bermbach (Bl. 5715 Idstein/Ts.). — Jb. nass. Ver. Naturk., **100**, 14—21.

BRAND, E., 1940: Ein neues Basaltvorkommen auf Blatt Eltville im Rheingau. — Senckenbergiana, **22**, 74—76.

BUBNOFF, S. v., 1926: Studien im südwestdeutschen Grundgebirge. II. Die tektonische Stellung des Böllsteiner Odenwaldes und des Vorspessarts. — N. Jb. Miner. Geol. Paläont., Abt. B, Beil.-Bd. **55**, 468—496.

CARLÉ, W., 1955: Bau und Entwicklung der Südwestdeutschen Großscholle. — Beih. Geol. Jb., **16**.

CLOOS, H., 1939: Hebung — Spaltung — Vulkanismus. Elemente einer geometrischen Analyse irdischer Großformen. — Geol. Rdsch., **30**, 401—527 u. 637—640.

DIEDERICH, G., 1966: Fazies, Paläogeographie und Genese des Unteren Buntsandstein norddeutscher Auffassung im südlichen Beckenbereich. — Notizbl. hess. L.-Amt Bodenforsch., **94**, 132—157.

DOEBL, F., 1967: The Tertiary and Pleistocene Sediments of the Northern and Central Part of the Upper Rhinegraben. — Abh. geol. L.-Amt Baden-Württ., **6**, 48—54.

EHRENBERG, K.-H., KUPFAHL, H.-G. & KÜMMERLE, E., 1968: Geol. Kte. Hessen 1:25 000, Bl. 5913 Presberg, 2. Aufl., Wiesbaden.

FALKE, H., 1960: Rheinhessen und die Umgebung von Mainz. — Slg. geol. Führer, **38**, Borntraeger, Berlin.

— 1969: Zur Paläogeographie der Randgebiete des nördlichen Oberrheingrabens zur Zeit des Rotliegenden. — Notizbl. hess. L.-Amt Bodenforsch., **97**, 130—151.

— 1971: Die paläogeographische Entwicklung des Oberkarbons in Süddeutschland. — Fortschr. Geol. Rheinld. u. Westf., **19**, 167—172.

GOLWER, A., 1968: Paläogeographie des Hanauer Beckens im Oligozän und Miozän. — Notizbl. hess. L.-Amt Bodenforsch., **96**, 157—184.

HEINRICHS, T., 1968: Geologische Untersuchungen im Hochtaunus zwischen Falkenstein und Oberems (Bl. 5716 Oberreifenberg und 5816 Königstein i. T., Rheinisches Schiefergebirge). — Diplomarbeit, Frankfurt a. M. (unpublished).

HOFFMANN, CHR., 1966: Geologische Beobachtungen im Taunus zwischen Saalburg und Hohemark. — Diplomarbeit, Frankfurt a. M. (unpublished).

HORN, P., LIPPOLT, H. J. & TODT, W., 1972: Kalium-Argon-Altersbestimmungen an tertiären Vulkaniten des Oberrheingrabens, I. Gesamtgesteinsalter. — Eclogae geol. Helv., **65**, 131—156.

HUMMEL, K., 1929: Die tektonische Entwicklung eines Schollengebirgslandes (Vogelsberg und Rhön). — Fortschr. Geol. Paläontol., (24) **8**.

ILLIES, H., 1965: Bauplan und Baugeschichte des Oberrheingrabens. Ein Beitrag zum „Upper Mantle Project". — Oberrh. geol. Abh., **14**, 1—54.

ILLIES, J. H., 1972: The Rhine Graben Rift System — Plate Tectonics and Transform Faulting. — Geophys. Surv., **1** (1972), 27—60.

KEGEL, W., 1929: Erl. geol. Kte. Preußen 1:25 000, Lfg. 275, Bl. Kleeberg-Kirchgöns, Nr. 3221, 50 S., Berlin.

KLEMM, G., 1925: Bemerkungen über die Tektonik des Odenwaldes. — Notizbl. Ver. Erdk. u. hess. geol. L.-Anstalt, (V) **7**, 8—22.

KNETSCH, G., 1967: Changing Tectonic Rôles of the Upper Rhine Lineament in the Course of Geological Times and Events. — Abh. geol. L.-Amt Baden-Württ., **6**, 13—15.

KOSSMAT, F., 1927: Gliederung des varistischen Gebirgsbaues. — Abh. sächs. geol. L.-Anstalt, **1**.

Kubella, K., 1951: Zum tektonischen Werdegang des südlichen Taunus. — Abh. hess. L.-Amt Bodenforsch., **3**.

Kümmerle, E. & Semmel, A., 1969: Geol. Kte. Hessen 1:25 000, Bl. 5916 Hochheim a. M., 3. Aufl., Wiesbaden.

Laemmlen, M., 1966: Der Mittlere Buntsandstein und die Solling-Folge in Südhessen und in den südlich angrenzenden Nachbargebieten. — Z. dt. geol. Ges., **116**, 908—949.

Lippert, H.-J. & Hentschel, H., 1968: Ein neues Basaltvorkommen bei Wingsbach auf Blatt 5814 Bad Schwalbach (Taunus, Rheinisches Schiefergebirge). — Jb. nass. Ver. Naturk., **99**, 86—91.

Matthess, G., 1966: Zur Geologie des Ölschiefervorkommens von Messel bei Darmstadt. — Abh. hess. L.-Amt Bodenforsch., **51**.

Meisl, St., 1970: Petrologische Studien im Grenzbereich Diagenese—Metamorphose. — Abh. hess. L.-Amt Bodenforsch., **57**.

Meyer, D. E., 1970: Stratigraphie und Fazies des Paläozoikums im Guldenbachtal/SE-Hunsrück am Südrand des Rheinischen Schiefergebirges. — 307 S., Diss. Rhein. Friedrich-Wilhelms-Univ. Bonn.

Moayedpour, E., 1970: Geologische und hydrogeologische Untersuchungen auf dem Blatt Obernburg 6120 W mit einer Neuuntergliederung des Unteren Buntsandsteins. — Diplomarbeit, Frankfurt a. M. (unpublished).

Murawski, H., 1963: Die Bedeutung der „Kinzigtalzone" als Scharnierbereich zwischen der (Spessart-)Schwelle und dem nördlich vorgelagerten (Hessischen) Becken. — Notizbl. hess. L.-Amt Bodenforsch., **91**, 217—230.

Nickel, E., 1965: Das Intrusionsniveau des Odenwaldes (Beiträge zur Tektonik von Fließgefügen III). — N. Jb. Miner. Mh., **1965**, 43—53.

Nöring, F. K., 1939: Das Unterdevon im westlichen Hunsrück. — Abh. preuß. geol. L.-A., N. F. **192**.

Nöring, F., 1951: Neue Nachweise des Untermain-Trapps. — Notizbl. hess. L.-Amt Bodenforsch., (VI) **2**, 41—43.

Okrusch, M., Streit, R. & Weinelt, W., 1967: Erl. geol. Kte. Bayern 1:25 000, Bl. Nr. 5920 Alzenau i. Ufr. — 336 S., München.

Reinach, A. v., 1890: Das Bohrloch im Neuen Wiesbadener Schlachthaus. — Jb. nass. Ver. Naturk., **43**, 33—38.

Richter-Bernburg, G., 1957: Zur Paläogeographie des Zechsteins. — Atti Convegno Milano (1957) Giacimenti gassiferi Europa occident., **1**, 87—99, Roma.

Sandberger, F., 1850: Über die geognostische Zusammensetzung der Umgegend von Wiesbaden. — Jb. Ver. Naturk. Herzogth. Nassau, **6**, 1—27.

Schälicke, W., 1969: Geologische Untersuchungen zur Struktur der Otzberg-Zone im Odenwald. — Notizbl. hess. L.-Amt Bodenforsch., **97**, 296—330.

Fig. 1. Structural map of the northern end of the Upper Rhine Graben based on the maps compiled by Anderle (1970) and Falke (1960), completed from Amrhein (1971), Bender (1969), Brand (1940), Ehrenberg et al. (1968), Heinrichs (1968), Hoffmann (1966), Kegel (1929), Kubella (1951), Kümmerle & Semmel (1969), Lippert & Hentschel (1968), Moayedpour (1970), Murawski (1963), Nöring (1951), Okrusch et al. (1967), Sandberger (1850), Schottler (1925), Sharifzadeh (1970), Sonne (1972), Stenger (1961), Straub (1962), Streit & Weinelt (1971), Werding (in Weyl 1967), and an unpublished structural map of the southern Taunus mountains by the author. Additional information has been drawn from the unpublished sheets 5618 Friedberg (E. Kümmerle, personal communication) and 5620 Ortenberg (H. Hickethier, personal communication) and also from a map prepared by O. Schmitt for this volume.

The main transverse faults of the Taunus mountains are emphasized. Vertical and horizontal numbers at the margin put together in that order give sheet numbers of the topographic maps 1:25 000.

SCHOTTLER, W., 1919: Beiträge zur Geologie der nördlichen Wetterau auf Grund neuer Bohrungen. — Notizbl. Ver. Erdk. u. hess. geol. L.-Anstalt, (V) 4, 57—87.
— 1925: Die Geologie von Salzhausen nebst einem Überblick über den Bau der Wetterau und des Vogelsberges. — Notizbl. Ver. Erdk. u. hess. geol. L.-Anstalt, (V) 7, 23—55.
SHARIFZADEH, Y., 1970: Geologische und hydrogeologische Untersuchungen im SE-Teil des Blattes 6119 Gross-Umstadt. — Diplomarbeit, Frankfurt a. M. (unpublished).
SOLLE, G., 1951: Geologie, Paläomorphologic und Hydrologie der Main-Ebene östlich von Frankfurt am Main. — Abh. senckenb. naturf. Ges., 485.
SONNE, V., 1970: Das nördliche Mainzer Becken im Alttertiär. Betrachtungen zur Paläoorographie, Paläogeographie und Tektonik. — Oberrh. geol. Abh., 19, 1—28.
— 1972: Geol. Kte. Rheinland-Pfalz 1:25 000, Bl. 6015 Undenheim, Mainz.
STENGER, B., 1961: Stratigraphische und gefügetektonische Untersuchungen in der metamorphen Taunus-Südrand-Zone (Rheinisches Schiefergebirge). — Abh. hess. L.-Amt Bodenforsch., 36.
STRAUB, E. W., 1962: Die Erdöl- und Erdgaslagerstätten in Hessen und Rheinhessen. — Abh. geol. L.-Amt Baden-Württ., 4, 123—136.
STREIT, R. & WEINELT, W., 1971: Erl. geol. Kte. Bayern 1:25 000, Bl. Nr. 6020 Aschaffenburg. — 398 S., München.
TRUSHEIM, F., 1964: Über den Untergrund Frankens. Ergebnisse von Tiefbohrungen in Franken und Nachbargebieten 1953—1960. — Geologica Bavarica, 54.
VEIT, E., 1955: Die Tiefbohrungen bei Pfungstadt und der Bau des Rheintalgrabens im Raum um Darmstadt. — Z. dt. geol. Ges., 105, 150—151.
WAGNER, W., 1933: Die Schollentektonik des nordwestlichen Rheinhessens. — Notizbl. Ver. Erdk. u. hess. geol. L.-Anstalt, (V) 14, 31—45.
WEINELT, W., 1962: Erl. geol. Kte. Bayern 1:25 000, Bl. Nr. 6021 Haibach. — 246 S., München.
WENZ, W., 1914: Grundzüge einer Tektonik des östlichen Teiles des Mainzer Beckens. — Abh. senckenb. naturf. Ges., 36, 71—104.
WEYL, R. (ed.), 1967: Geologischer Führer durch die Umgebung von Gießen. — 184 S., Mittelhess. Druck- u. Verlagsges., Gießen.

Zum Verlauf der westlichen Randverwerfung des zentralen Oberrheingrabens zwischen dem Rhein südlich Nackenheim und dem Main bei Rüsselsheim

sowie über eine Grundwasserkaskade und Bauschäden im Bereich dieser Störungszone

von

O. Schmitt

Mit 2 Abbildungen im Text und auf 1 Beilage

About the Extension of the Western Marginal Fault of the Central Upper Rhine Graben between the River Rhine South of Nackenheim and the River Main near Rüsselsheim, a Groundwater Cascade and Damages to Buildings along this Fault-Zone.

Abstract: The above mentioned section of the western marginal fault, mainly mapped by the author by means of drilling results, cuts obliquely (general strike NNE) through the Upper Rhine Valley, which forms a vast plain in this area. The Plio-Pleistocene filling of the Central Graben lies adjacent to Rotliegend and Tertiary sediments of the so-called Mainspitze, which are only covered by Quaternary sediments of minor thickness. Tectonic movements along this fault-zone continuing unto this day caused damages to buildings at Rüsselsheim. A groundwater cascade is specially mentioned, along which groundwater, circulating in the Pleistocene sands and gravels, falls over an extended area to the lower water table developed in the Pleistocene sands and gravels of the Central Graben. The groundwater cascade as well as the demages to buildings (locally) could be utilized for the mapping of the fault.

1. Einleitung

Der nordwestliche Endabschnitt des Oberrheingrabens i. w. S. gliedert sich in zwei Großschollen: die zentrale Grabenzone (Zentralgraben) und das westlich davon liegende Mainzer Becken. Sie werden durch die Westrandverwerfung des Zentralgrabens voneinander getrennt.

Im Zentralgraben wurde bei Königstädten im Plio-Pleistozän eine über 450 m mächtige Füllung aus Kies, Sand, Schluff und Ton aufeinander geschichtet; das Pleistozän allein ist hier über 100 m mächtig. Das Mainzer Becken ist aus marinen und limnischen Tertiärgesteinen aufgebaut, die örtlich von Rotliegend-Aufbrüchen unterbrochen werden. Pliozäne und pleistozäne Rhein- und Mainschotter sind im Mainzer Becken zwar auch in Form von Terrassen vorhanden, kommen hier jedoch nicht in so großer Mächtigkeit und auch nicht zusammenhängend wie im Zentralgraben vor, sondern verteilen sich auf dem unterschiedlich herausgehobenen Schollenmosaik derart, daß ältere Terrassen in der Regel ein höheres Niveau einnehmen als jüngere.

Zwischen Oppenheim und dem Gebiet südlich Nackenheim brechen am linken Rheinufer die plateauartigen Höhen Rheinhessens auf breiter Front gegen Osten zum Zentralgaben hin ab, dessen Westrand hier geologisch und morphologisch augenfällig in Erscheinung tritt. Der weitere Verlauf des südlich Nackenheim auf das rechte (östliche) Rheinufer überwechselnden Grabenabbruches ist dagegen weder morphologisch noch — von der Oberfläche aus gesehen — geologisch gekennzeichnet. Die Randverwerfung verläuft hier schräg (NNE) durch die Oberrheinische Tiefebene auf die Stadt Rüsselsheim zu. Im Bereich dieses Grabenrandabschnittes ist die plio-pleistozäne Grabenfüllung gegen Rotliegend- und Tertiär-Gesteine zum großen Teil synsedimentär verworfen; diese Gesteine bauen die sogenannte Mainspitze auf, wie der tektonisch höher liegende, unter einer geringmächtigen Quartärüberdeckung verborgene Schollenverband zwischen Rhein, Main und Randverwerfung genannt wird. Weiter nördlich findet die Randverwerfung ihre Fortsetzung in der S—N verlaufenden Westrandverwerfung des Hattersheimer Grabens, der als ein NNW-Ast des zentralen Oberrheingrabens anzusehen ist.

2. Zum Verlauf der Randverwerfung

Das auf der Mainspitze nur an einer Stelle zutage tretende Tertiär bei Bauschheim veranlaßte bereits LEPSIUS (1883) und KINKELIN (1885) zu der Annahme, daß die westliche „Rheinspalte" östlich dieses von pleistozänen Mainsanden und -kiesen umrahmten, aquitanen Kalkaufbruches zu suchen ist. STEUER hat aufgrund dieses Aufschlusses und einiger Bohrungen den Verlauf der Randverwerfung kurz beschrieben (1905) und dann (1907) auf einer Karte wiedergegeben. Mit Hilfe weiterer Bohrungen hat BURRE (1952) versucht, den Verlauf der Randverwerfung kartographisch genauer festzulegen. Auch der von ANDERLE (1968) vermutete Verlauf beruht auf der Auswertung von Bohrungen. Bei den übrigen zahlreichen kartographischen Darstellungen ist der Verlauf der Randverwerfung mehr oder weniger schematisiert.

Die vorgelegte Kartierung der Randverwerfung (Abb. 1) stützt sich auf zum großen Teil bisher noch nicht bekannte Brunnen- und Baugrundbohrungen sowie auf eigene, mit dem Schlagbohrgerät des Hessischen Landesamtes für Bodenforschung niedergebrachte Sondierbohrungen. Auf der Mainspitze haben — wenn man von der Hohenauer Tertiär-Teilscholle absieht (s. u.) — diese Bohrungen das Rotliegende und das Tertiär im Mittel bereits bei ca. 6 m unter Flur angetroffen. Es war deshalb der Schluß erlaubt, daß in der Nähe der Randverwerfung abgeteufte Bohrungen, die bei annähernd 10 m unter Gelände das Rotliegende bzw. das Tertiär noch nicht erreicht hatten, d. h. in dieser Teufe immer noch im Pleistozän standen, sich in der Regel bereits im Zentralgraben befanden.

Außerdem konnte zwischen Bauschheim und Rüsselsheim der Verlauf der Randverwerfung im Zuge einer Grundwasserkaskade gut verfolgt werden (Abb. 1 und 2). Auch boten zahlreiche Häuserschäden in Rüsselsheim (Abb. 2) wertvolle Anhaltspunkte für die Kartierung der Verwerfung.

Wo die Randverwerfung auf das östliche Rheinufer übertritt, dürfte oberflächennah die plio-pleistozäne Grabenfüllung nicht gegen die Astheimer Rotliegend-Scholle, die als ein nordöstlicher Ausläufer des Alzey-Niersteiner Horstes anzusehen ist, verworfen sein, sondern von Tertiärschichten (u. a. Rupelton) begrenzt werden. Diese sind auf der rechten Rheinseite südlich des Gutes Hohenau, im Gebiet der sog. Hohenauer Tertiär-Scholle, durch 7,0 bis 11,5 m tiefe Sondierbohrungen nachgewiesen worden. Sie sind hier der Astheimer Rotliegend-Scholle im Osten vorgelagert. Der Abbruch der Hohenauer Tertiär-Scholle zum Zentralgraben, d. h. die Randverwerfung, war allerdings nicht zu ermitteln, weil das auf dem Tertiärsockel liegende Quartär in der Nähe der Abbruchzone bereits so mächtig entwickelt ist, daß es technisch nicht möglich war, die Tertiärunterlage mit dem eingesetzten Schlagbohrgerät zu erreichen.

Nördlich der Hohenauer Tertiär-Scholle bricht auf einer Länge von 3,4 km die Astheimer Rotliegend-Scholle unmittelbar zum Zentralgraben ab; die Randverwerfung weist hier eine generelle Streichrichtung von N 15° E (rheinisch) auf. In diesem Wert sind Abweichungen bis nach W 357° N einerseits und bis nach N 30° E andererseits inbegriffen.

Mit Beginn des nördlich anschließenden, tertiären Schollenverbandes ist eine Richtungsänderung der Randverwerfung auf generell N 27° E mit Abweichungen bis nach N 36° E bzw. N 4° E zu verzeichnen. Auf diese 4,2 km lange Strecke folgt ein 2,2 km langer, nach NE umbiegender Abschnitt, der zunächst N 31° E, dann N 42° E und schließlich N 54° E streicht. Es macht sich hier in der Streichrichtung das erzgebirgische Element des variszischen Grundgebirges stärker bemerkbar. Es folgt nun der 0,8 km lange, N 19° E verlaufende Endabschnitt der Randverwerfung, der ca. 370 m südlich des Maines endet, wo die generell S—N verlaufende Westrandverwerfung des Hattersheimer Grabens beginnt.

3. Grundwasserkaskade

An der entlang der Randverwerfung zwischen Bauschheim und Rüsselsheim ausgebildeten Grundwasserkaskade „stürzt" das Porengrundwasser der Mainspitze, das sich über einer wassersperrenden oberflächennahen Sohlschicht (Tertiärsockel) in den geringmächtigen pleistozänen Sanden und Kiesen angereichert hat und durch einen hochliegenden Grundwasserspiegel ausgezeichnet ist [Bauschheimer Wasserberg, BURRE (1952); Grundwasserkuppe, HOFMANN (1942)], auf breiter Front über die Randverwerfung zum tieferliegenden Porengrundwasser der pleistozänen Grabenfüllung ab (Abb. 1 und 2). Die heute nicht mehr existierende Grundwassermeßstelle 219 (s. Abb. 2) der Stadtwerke Mainz AG, die sich am Fuße der Grundwasserkaskade befand, wies im Jahre 1913 einen mittleren Wasserstand von etwa 84,95 m über NN auf. Der Grundwasserspiegel unmittelbar oberstrom der Kaskade ist im Jahre 1913 im Mittel mit etwa 86,10 m zu veranschlagen, so daß damals an dieser Stelle die Grundwasserkaskade nur ca. 1,15 m hoch und möglicherweise mehr stromschnellenartig ausgebildet war. Der heutige Höhenunterschied zwischen dem ober- und unterstromigen Grundwasser an der Kaskade ist, wie die Wasserstände der zahlreichen, auf Abb. 2 eingezeichneten Baggerseen (Grundwasserblänken) zeigen, mit maximal mehr als 4 m wesentlich größer. Die Ursache hierfür ist die Grundwasserentnahme des seit dem Jahre 1928 in Betrieb befindlichen Wasserwerkes Hof Schönau der Stadtwerke Mainz AG,

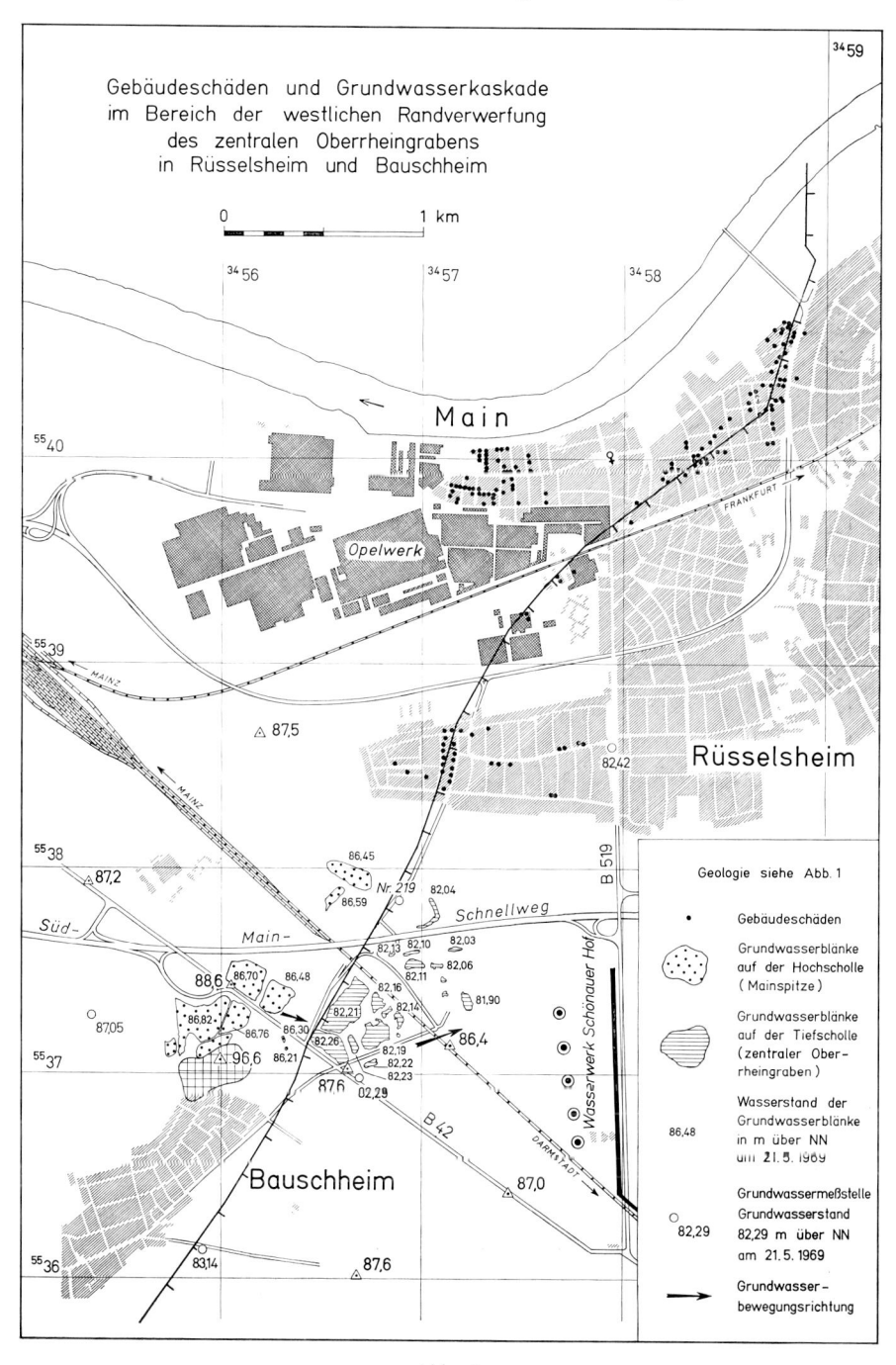

Abb. 2

dessen Entnahmetrichter an der Kaskade abrupt endet, weil hier zwischen dem Porengrundwasser der Mainspitze und dem Porengrundwasser des Zentralgrabens die hydraulische Verbindung abgerissen oder zumindest so stark gestört ist, so daß sich die von den Wasserwerksbrunnen ausgehende Absenkung nur bis an den Fuß der Kaskade und nicht darüber hinaus bis zu dem Porengrundwasser der Mainspitze auswirken kann. Mit der Grundwasserförderung des Wasserwerkes Hof Schönau vergrößerte sich die Grundwasserkaskade an der Randverwerfung zwischen Bauschheim und Rüsselsheim nicht nur höhenmäßig, sondern die Kaskade dehnte sich auch auf die heutige Breite über 5 km aus, und zwar allein dadurch, daß der Grundwasserspiegel durch die genannte Grundwasserentnahme bis zur Randverwerfung stark abgesenkt wurde, während dabei die Lage des Grundwasserspiegels auf der Mainspitze praktisch unverändert blieb.

4. Bauschäden

FAHLBUSCH (1962) sowie MÜLLER & PRINZ (1966) berichteten über Gebäudeschäden, die am Ostrand des Zentralgrabens im Gebiet des Abbruches des kristallinen Odenwaldes (Bergstraße) entstanden sind. Verfasser (1954)[1] wies hier auf Bauschäden an der katholischen Kirche in Jugenheim hin und brachte sie in Beziehung zu einer dort vermuteten Störungszone. Wiederholungsmessungen von Feinnivellements des Hessischen Landesvermessungsamtes haben gezeigt, daß vertikale Bodenbewegungen heute noch an der Bergstraße im Gange sind (HEIL 1957, 1959; SCHMITT 1954[1], 1966; KUTSCHER, PRINZ & SCHWARZ 1968; PRINZ & SCHWARZ 1970).

In der Stadt Rüsselsheim sind im Verlauf der westlichen Randverwerfung zahlreiche Gebäudeschäden zu beachten, deren Lage auf Abb. 2 gekennzeichnet ist. Es handelt sich um mehr oder weniger steile, im Mauerwerk aufsteigende Risse, die zuunterst am stärksten klaffen (bis 3 mm) und nicht nur den Mauerfugen folgen, sondern auch Ziegelsteine und Natursteine gespalten haben. Die Risse treten an Außenwänden, Innenwänden, Kaminen, Türschwellen und Geschoßdecken auf. Dabei sind Tür- und Fensterrahmen verzogen worden. Auffallend ist, daß statische Schwächezonen der Gebäude von den Rissen besonders betroffen sind. In der Regel gelingt es nicht, diese Risse durch bautechnische Maßnahmen auf die Dauer zu schließen.

Die Randverwerfung wird zwischen Bauschheim und Rüsselsheim von der entlang der B 42 (s. Abb. 2) verlaufenden Nivellementlinie I. Ordnung 29 des Hessischen Landesvermessungsamtes gekreuzt. Beim Vergleich der Messung[2] von 1938 und der ersten Wiederholungsmessung aus dem Jahre 1955 an den drei zwischen der Randverwerfung und dem Wasserwerk Hof Schöngau liegenden Meßpunkten,

---

[1] SCHMITT, O.: Gutachten des Hessischen Landesamtes für Bodenforschung, erstattet für das Hessische Landesvermessungsamt, über geologisch-tektonische Ursachen der zwischen den Messungen 1938 und 1953 eingetretenen Höhenunterschiede der Nivellementslinien I. Ordnung: 37 von Groß-Sachsen nach Bensheim, 38 von Bensheim nach Darmstadt, 39 von Bensheim nach Worms, S. 1—38, Wiesbaden 1954.

[2] Freundliche Mitteilung von Herrn Obervermessungsrat E. SCHWARZ, Wiesbaden.

die von der Randverwerfung 0,235 km, 1,23 km und 1,875 km entfernt sind, zeigten sich geringe Absenkungen (—2 mm, —1 mm, —3 mm/17 Jahren). Die im Jahre 1967 durchgeführte zweite Wiederholungsmessung bestätigte die geringe Absenkungstendenz nicht an dem zuerst genannten Meßpunkt, der 9 mm/12 Jahren aufgestiegen war[3]; bei den beiden anderen im Zentralgraben liegenden Meßpunkten war jedoch auch weiterhin eine Absenkung (—3 mm bzw. —5 mm/12 Jahren) zu verzeichnen.

Die erste Wiederholungsmessung auf der Mainspitze zeigte, daß sich hier der 0,470 km von der Randverwerfung entfernte Meßpunkt höhenmäßig nicht verändert hatte; der daran anschließende, von der Randverwerfung 1,33 km entfernte Meßpunkt war in diesen 17 Jahren sogar 2 mm aufgestiegen. Von der zweiten Wiederholungsmessung sind für diese beiden Meßpunkte keine Werte vorhanden.

In der Stadt Rüsselsheim wird die Randverwerfung von der Nivellementlinie III. Ordnung 36/4 zweimal gekreuzt. Die Nivellementlinie verläuft zunächst S—N im Zuge der B 519 (s. Abb. 2); sie knickt dann an der Stadtkirche nach NE um. Südlich der ersten Kreuzung treten im Zentralgraben an 5 Punkten im Bereich einer 3 km langen Strecke in der Zeitspanne 1953 bis 1968 durchweg Absenkungen (—5 mm, —4 mm, —6 mm, —10 mm[4] und —5 mm) auf. Nördlich dieser Kreuzung konnten auf der Mainspitze in den o. g. 15 Jahren nur sehr geringe (—1 mm, —2 mm und —1 mm) oder gar keine (± 0 mm) Senkungen verzeichnet werden. Dagegen zeigte der Meßpunkt, der sich ca. 35 m ö s t l i c h der zweiten Kreuzung im Zentralgraben befindet, wiederum eine Absenkung (—5 mm/ 15 Jahren).

Es stellt sich nun die Frage, ob die Gebäudeschäden in Rüsselsheim dadurch zustande kamen, daß sich an der Randverwerfung oder an quer dazu verlaufenden Störungen steile, zu Senkungen führende Verwürfe bis in die Gebäude hinein fortgesetzt haben. Anzeichen hierfür waren nicht zu erkennen; vielmehr wird angezeigt, daß für die Gebäudeschäden hauptsächlich Zerrungen verantwortlich zu machen sind, die nicht nur unmittelbar an der eigentlichen Verwerfungsfläche, sondern auch in den Randzonen neben der Verwerfung an synthetischen und antithetischen Klüften und Rissen (Spalten) aufgetreten sind. Allgemein darf gesagt werden, daß die Gebäudeschäden in Rüsselsheim durch eine bis in die quartären Schichten sich durchpausende Zerrtektonik (Dehnungsbeanspruchung) an und in dem näheren beiderseitigen Bereich der Randverwerfung entstanden sind. Dabei könnten auch Erdbebenerschütterungen insofern mitgewirkt haben, als durch sie in dem durch Dehnung beanspruchten Untergrund entstehende und sich im Laufe der Zeit verstärkende latente Spannungen ausgelöst werden und zu Riß-

---

[3] Möglicherweise sind für diese auffallende Hebung Wurzeln einer nur 0,20 m vom Meßstein wachsenden Robinie verantwortlich zu machen.

[4] Dieser hohe Absenkungsbetrag ist nicht allein auf tektonische Bewegungen zurückzuführen, sondern dürfte etwa zur Hälfte auf Bodensetzungen beruhen, die durch eine zusätzliche Absenkung des Grundwasserspiegels infolge der Inbetriebnahme von hier im Jahre 1957 gebauten Brunnen hervorgerufen worden sind.

bildungen führen. Ob dies auch für anthropogene Erschütterungen angenommen werden darf, ist eine noch zu beantwortende Frage. Sie würde sich im besonderen für das durch größere Häuserschäden ausgezeichnete, weit nördlich der Randverwerfung liegende Gebiet zwischen dem Opelwerk und dem Main (s. Abb. 2) stellen, falls hier — was nicht unwahrscheinlich ist — Störungen vorhanden sind, in deren Bereich rezente Dehnungserscheinungen den Untergrund beanspruchen. Die Risse in den Häusern werden in diesem Gebiet von der Bevölkerung u. a. auf Erschütterungen zurückgeführt, die noch bis vor wenigen Jahren in verstärktem Maße von der benachbarten Hammerschmiede der Opelwerke ausgingen.

Zerrbewegungen im Bereich der Randverwerfung sind möglicherweise auch als Ursache dafür anzusehen, daß an der Bundesbahnstrecke Mainz—Frankfurt a. M., die bei km 11,93 die Randverwerfung schräg schneidet (s. Abb. 2), beiderseits der Verwerfung zwischen km 11,7 und 12,1 eine unruhige Gleisstrecke in Erscheinung tritt, wo jährlich Unterhaltungsarbeiten erforderlich werden.

Literatur

ANDERLE, H.-J., 1968: Die Mächtigkeit der sandig-kiesigen Sedimente des Quartärs im nördlichen Oberrhein-Graben und der östlichen Untermain-Ebene. — Notizbl. hess. L.-Amt Bodenforsch., 96, 185—196, Wiesbaden.
BURRE, O., 1952: Die Ursachen der Grundwasserentwicklung im nordwestlichen Teile des Kreises Groß-Gerau in Hessen (Mainspitze) in den Jahren 1927—1950. — Notizbl. hess. L.-Amt Bodenforsch., (VI) 3, 199—250, Wiesbaden.
FAHLBUSCH, K., 1962: Bauwerkschäden in Heppenheim (Bergstraße) und ihre Ursache. — Notizbl. hess. L.-Amt Bodenforsch., 90, 393—411, Wiesbaden.
HEIL, R., 1957: Die Vorbergzone bei Heppenheim/Bergstraße und der Abbruch zum Rheintalgraben. — Diss., 103 S., Darmstadt.
— 1959: Die Bergsträsser Diluvialterrasse zwischen Odenwaldquelle (südl. Heppenheim) und Bensheim. — Jber. u. Mitt. oberrh. geol. Ver., NF, 41, 35—45, Stuttgart.
HOFMANN, E. F., 1942: Grundwasserstandsänderungen im Oberrheintal. — Pumpen- u. Brunnenbau, Bohrtechnik, 38, 299—301, 315—320, 340—355, 351—352, Berlin.
KINKELIN, F., 1885: Senkungen im Gebiet des Untermainthales unterhalb Frankfurts und des Unterniedthales. — Ber. senckenb. naturforsch. Ges., 1884—1885, 235—258, Frankfurt a. M.
KUTSCHER, F.; PRINZ, H. & SCHWARZ, E., 1968: Bodenbewegungen in Hessen und ihre geologische Deutung. — Z. Vermessungswesen, 93, 45—54, Stuttgart.
LEPSIUS, G. R., 1883: Das Mainzer Becken. — Festschr. z. 50jähr. Bestehen rhein. naturforsch. Ges., V + 181 S., Darmstadt.
MÜLLER, K.-H. & PRINZ, H., 1966: Zur Frage rezenter tektonischer Bewegungen am Oberrheingrabenabbruch. — Notizbl. hess. L.-Amt Bodenforsch., 94, 390—393, Wiesbaden.
PRINZ, H. & SCHWARZ, E., 1970: Nivellement und rezente tektonische Bewegungen im nördlichen Oberrheingraben. — In: ILLIES, J. H. & MUELLER, ST.: Graben Problems, Internat. Upper Mantle Project, Sc. Rep. 27, 177—183, Schweizerbart, Stuttgart.
SCHMITT, O., 1966: Über pleistozäne Ablagerungen am Rand des Odenwaldes. — Z. dt. geol. Ges., 116, 987—989, Hannover.
STEUER, A., 1905: Erl. geol. Kte. Großherzogt. Hessen 1:25 000, Bl. Groß-Gerau, 27 S., Darmstadt.
— 1907: Bodenwasser und Diluvialablagerungen im hessischen Ried. — Notizbl. Ver. Erdk. u. großh. geol. L.-Anst. Darmstadt, (IV) 28, 49—94, Darmstadt.

Levelling Results at the Northern End of the Rhinegraben

by

Ernst Schwarz

With 8 figures in the text

The Hessian Department of Surveying has undertaken a second relevelling of the traverses of the first order German Height Datum (DHHN) for the northern end of the Rhinegraben and the results are now available.

In order to adjust these levels into geologically supposed fixed level heights of the DHHN, thorough research was undertaken, which in terms of volume (Figs. 1 and 2) goes far beyond the work of PRINZ & SCHWARZ (1970).

The stable levelling points used as reference are listed in Fig. 2. The comparison of the results of the newly measured levelling lines with measurements carried out in the past, starting and ending at the above mentioned points, shows interesting aspects of the changes of movements at the northern end of the upper Rhinegraben. Also a clear picture is obtained of the changes in heights of depressions (Fig. 3) which match well with the tectonic outlines (ANDERLE, sketch Spring 1972) (Fig. 1).

The sliding down of the earth's crust is represented on the eastern Graben fault (in Langen — Fig. 4 and in Bensheim — Fig. 5).

During future relevellings an effort will be made to separate the negative potential parts of the graben depressions from positive potential parts of the uplifting of the flanks at the Graben edges.

With the figures available in 1972 the absolute stability of the points can only be related to certain geological formations with ± 0 (e. g. Triassic).

The sliding down of the heights at the southern foot of the Taunus towards the Mainz basin (Fig. 6) could be compared to the measurements of 1911. The figures for the altitude of Bingen seem to be stable and this is explained by the curvature of the fault.

Particular attention has been paid to the extraction of ground water and other fluid resources (Figs 7 and 8). For those exploration areas lying within the graben area, control levellings were carried out by longer levelling lines starting from stable points situated outside the graben system. The extent of depression is controlled partly by tectonic movement of the earth's crust on the one hand and by human extraction of sub-soil resources on the other.

In future each phase of repeated levelling will be combined with gravimetric surveys. Only by doing this can the height differences be split up into geotectonic parts and parts changing into geopotential surfaces.

The copying of the maps was permitted by the Survey Department of Hesse in Wiesbaden, No. 015/72.

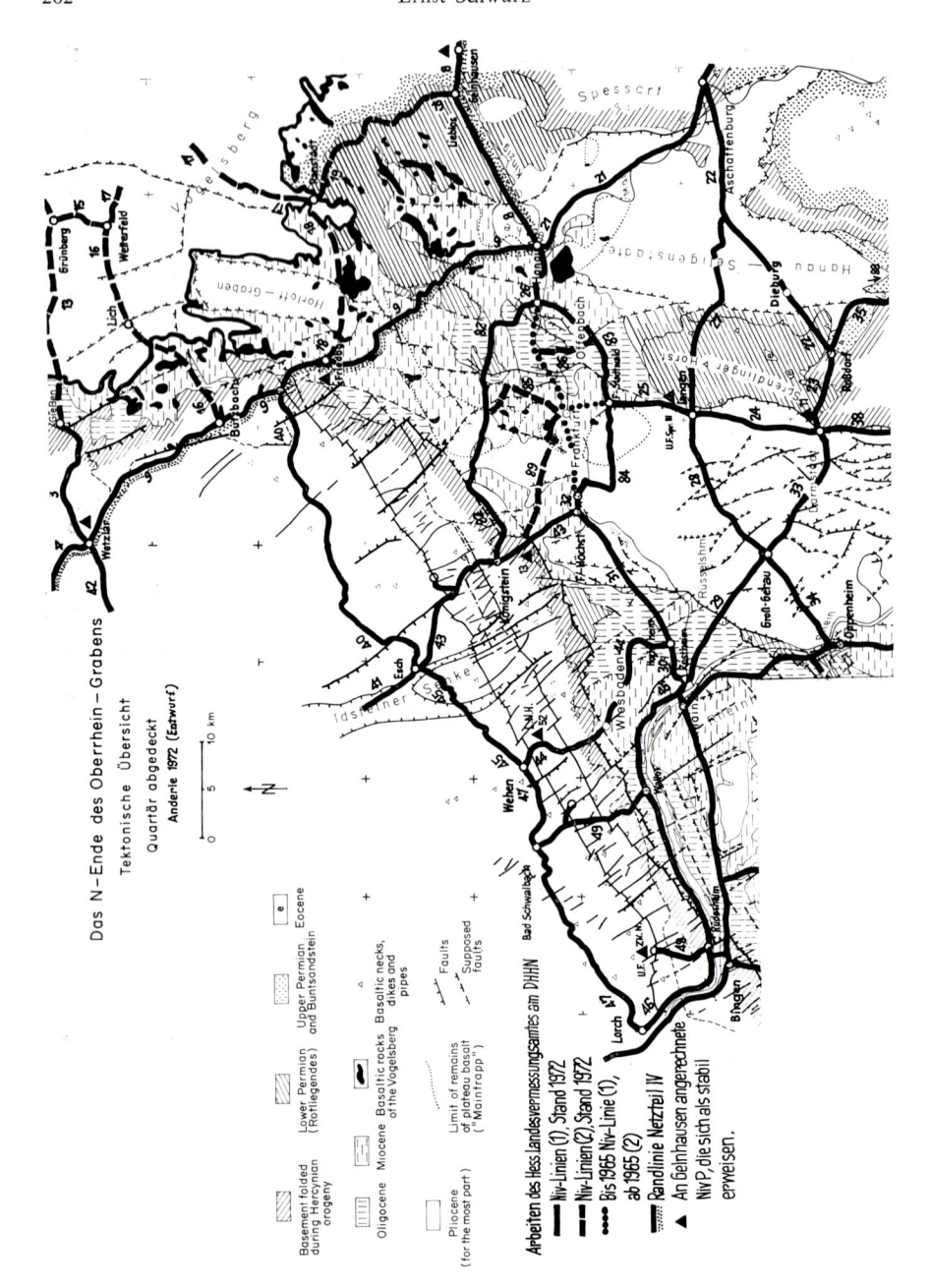

Fig. 1. Levelling lines up to 1972 at the northern end of the Rhinegraben with tectonic outlines from a sketch by ANDERLE, Spring 1972.

Untersuchung der Stabilität aus gemessenen Höhenunterschieden bezogen auf H.M.Gelnhausen,Marienkirche(8/89) (Buntsandstein)					
zum Nivellementpunkt	Meßweg km	1938 mit OK	1938/50 mit OK	1954/57 mit OK	1965/71 mit OK
U.F.Sprendlingen III (Sprendlinger Horst) Widerspruch in mm	58,90	x74,477 87 0	x74,475 68 - 2,19	x74,474 37 - 3,50	x74,476 47 - 1,40
M.B.Darmstadt,Fiedlerweg 7 (23/11) (Granit) Widerspruch in mm	80,92	4,252 06 0	4,250 56 - 1,50	4,248 81 - 3,25	4,251 46 - 0,60
P.B.Würzberg,Landesgrenze (35/88) (Buntsandstein) Widerspruch in mm	136,45		340,164 14 0	340,163 51 - 0,63	
M.B.Bad Soden,Königst.Str.88 (43/13) (Devon) Widerspruch in mm	78,84		x6,116 14 0		x6,118 24 + 2,10
M.B.Wiesbaden,Jagdschloßruine (44/52) (Sandsteine und Quarzite Widerspruch in mm der Hermeskeil-Schichten)	116,04		343,304 04 0		343,305 66 + 1,62
U.F.Geisenheim,Zimmersköpfe Nord (Quarzite) Widerspruch in mm	141,30		264,091 40 0		264,090 79 - 0,61

Fig. 2. Stability of the levelling points used as reference based on H. M. Gelnhausen, Marienkirche (8/89).

References

PRINZ, H. & SCHWARZ, E., 1970: Nivellement und rezente tektonische Bewegungen im nördlichen Oberrheingraben. — In: ILLIES & MUELLER, Graben Problems, Internat. Upper Mantle Project, Sci. Rep. No. 27, 177—183, Schweizerbart, Stuttgart.

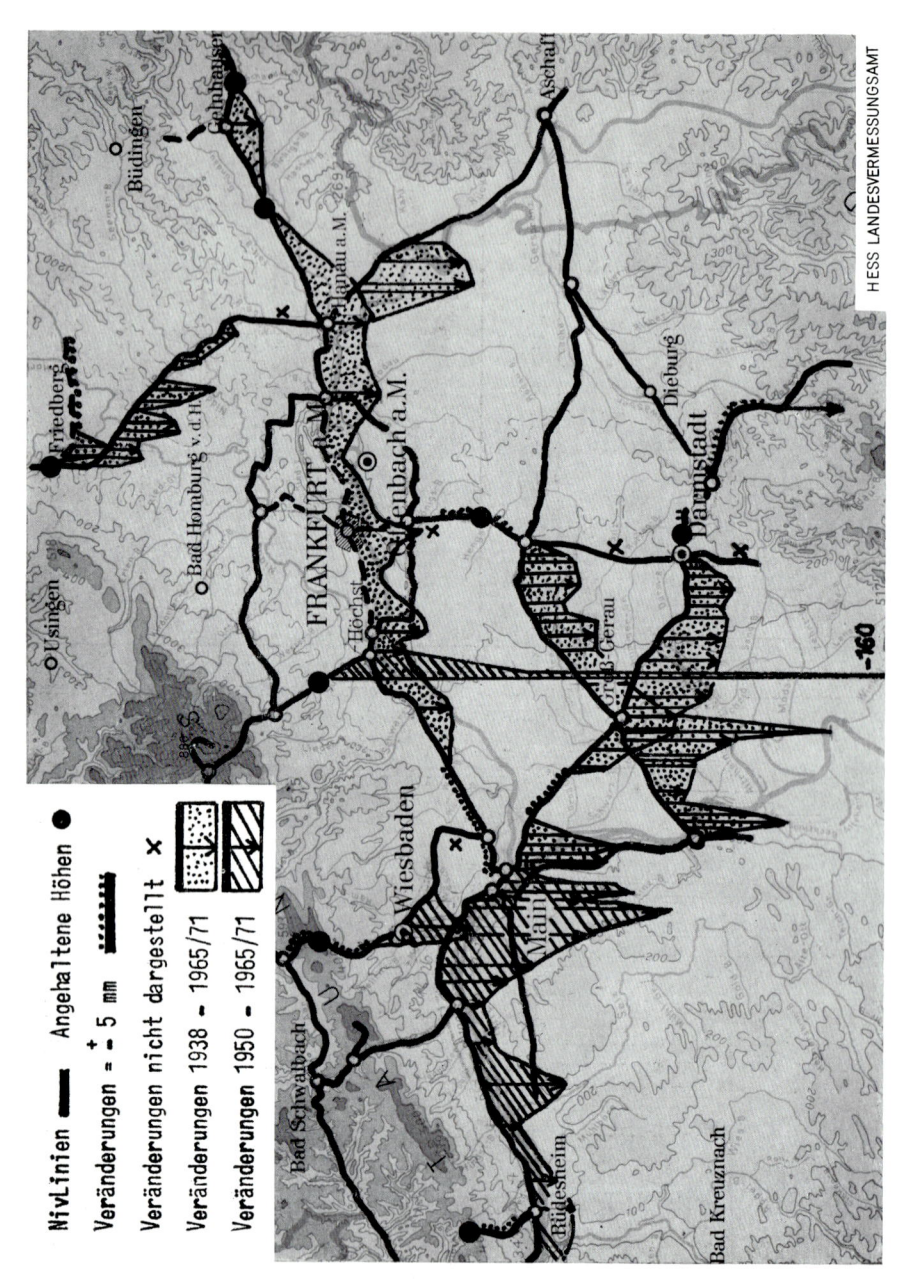

Fig. 3. Changes in heights at the northern end of the Rhinegraben after fitting the relevelling into the stable reference points (Fig. 2). Scales: distance 1:600 000, height variation 1:2.

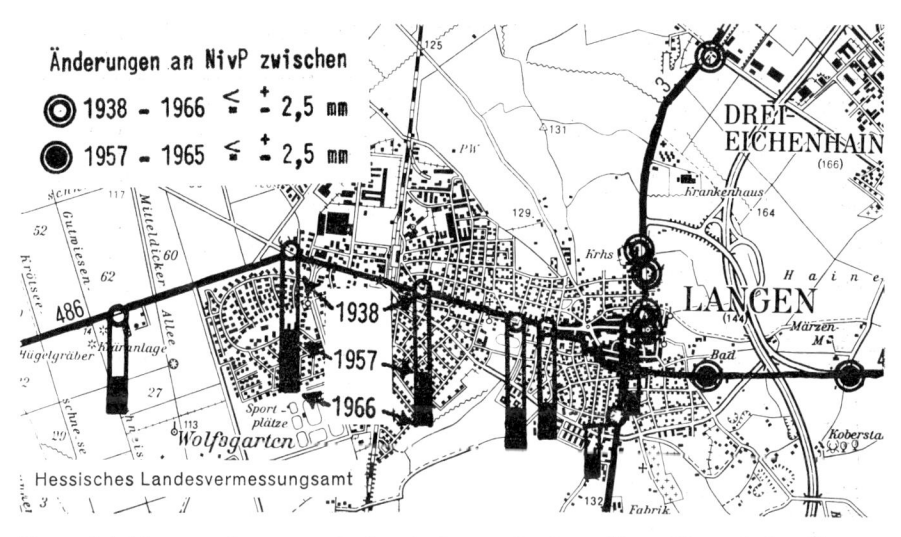

Fig. 4. Subsidence at the eastern Graben fault near the Sprendlinger Horst. Scales: distance 1:50 000, height variation 1:1.

Fig. 5. Subsidence at the eastern Graben fault near the Odenwald. Scales: distance 1:25 000, height variation 1:1.

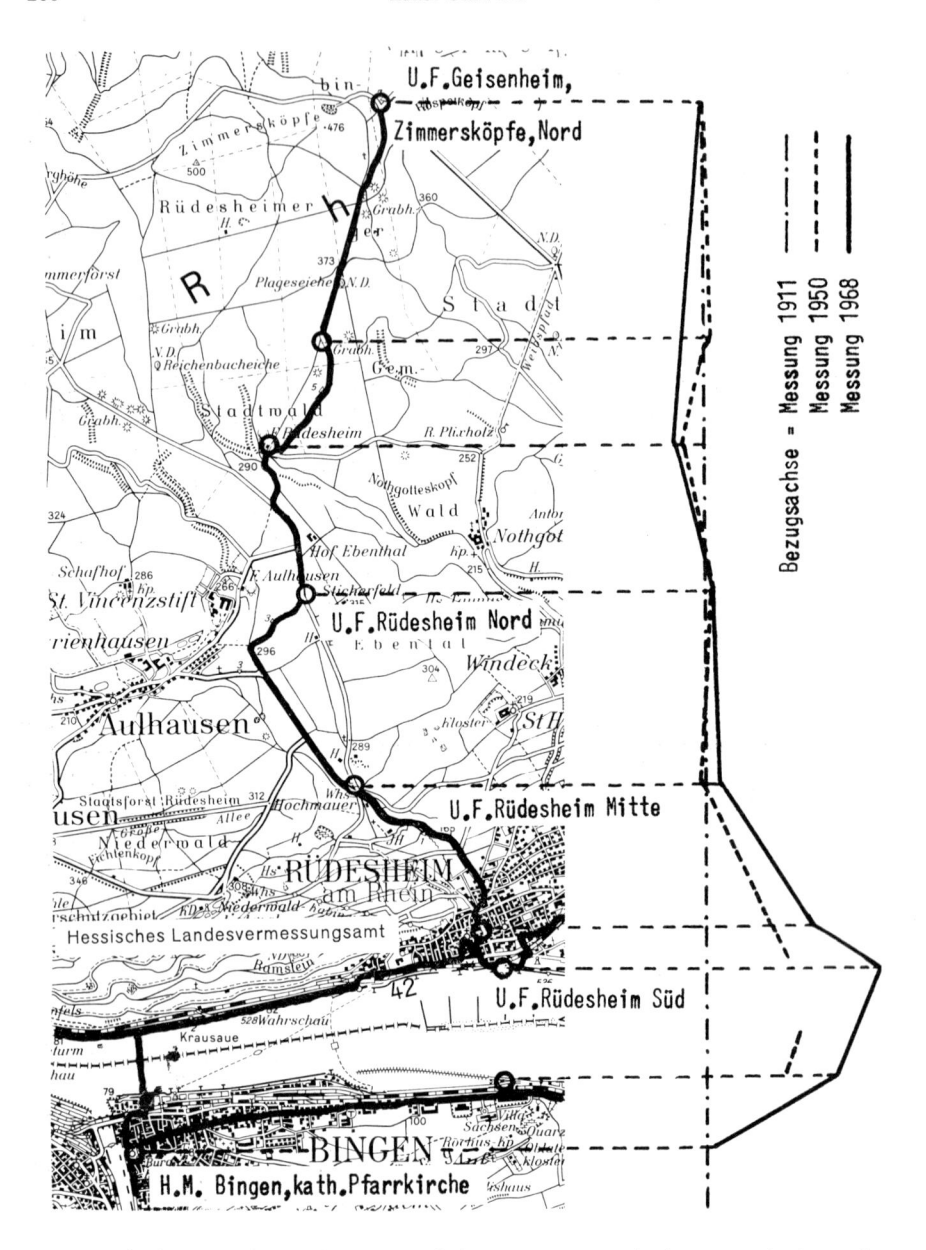

Fig. 6. Subsidence at the southern foot of the Taunus towards the Mainz basin. Scales: distance 1:50 000, height variation 1:1.

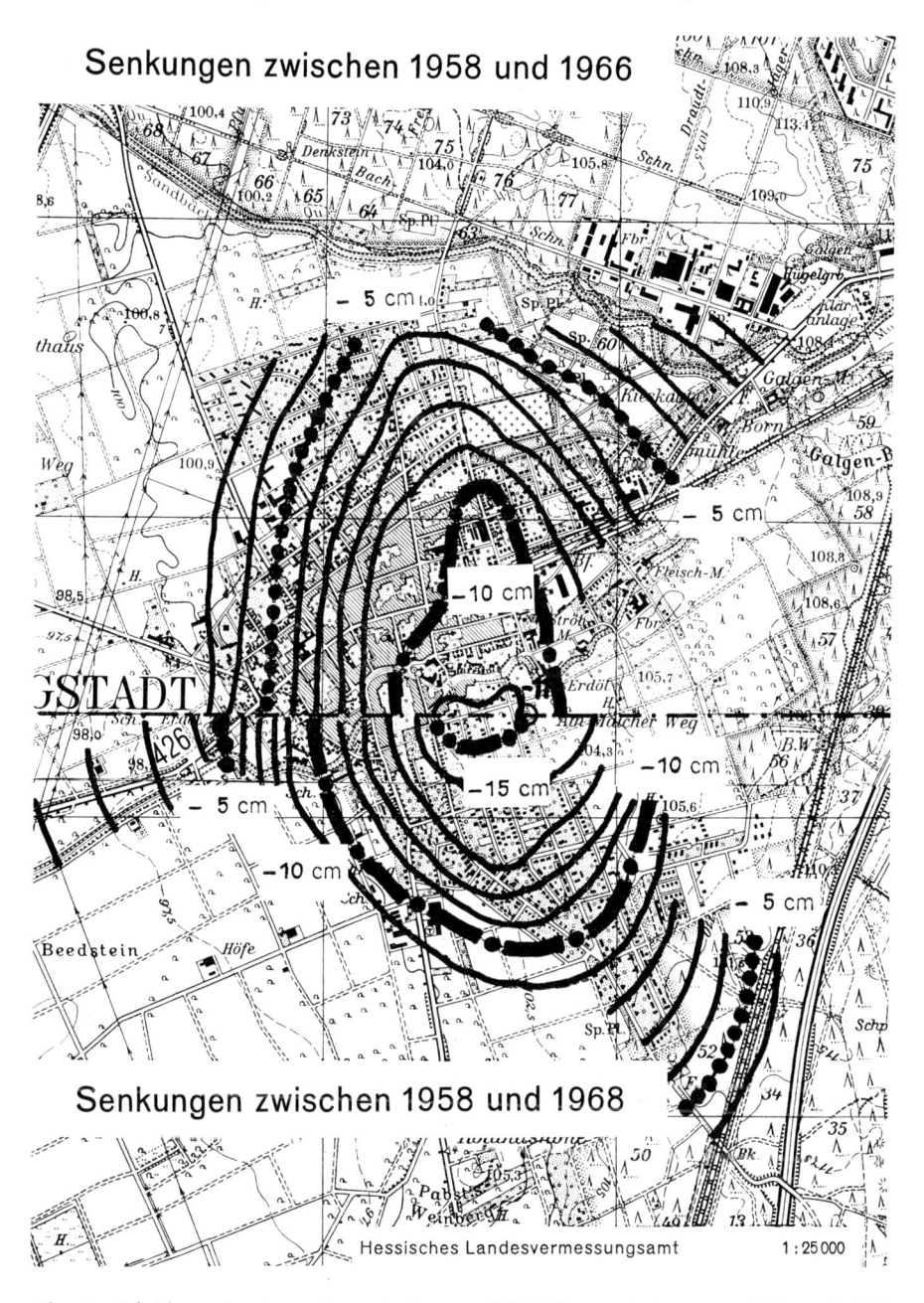

Fig. 7. Subsidence in the region of the gas field Pfungstadt between 1958 and 1966 (upper part), 1958 and 1968 (lower part).

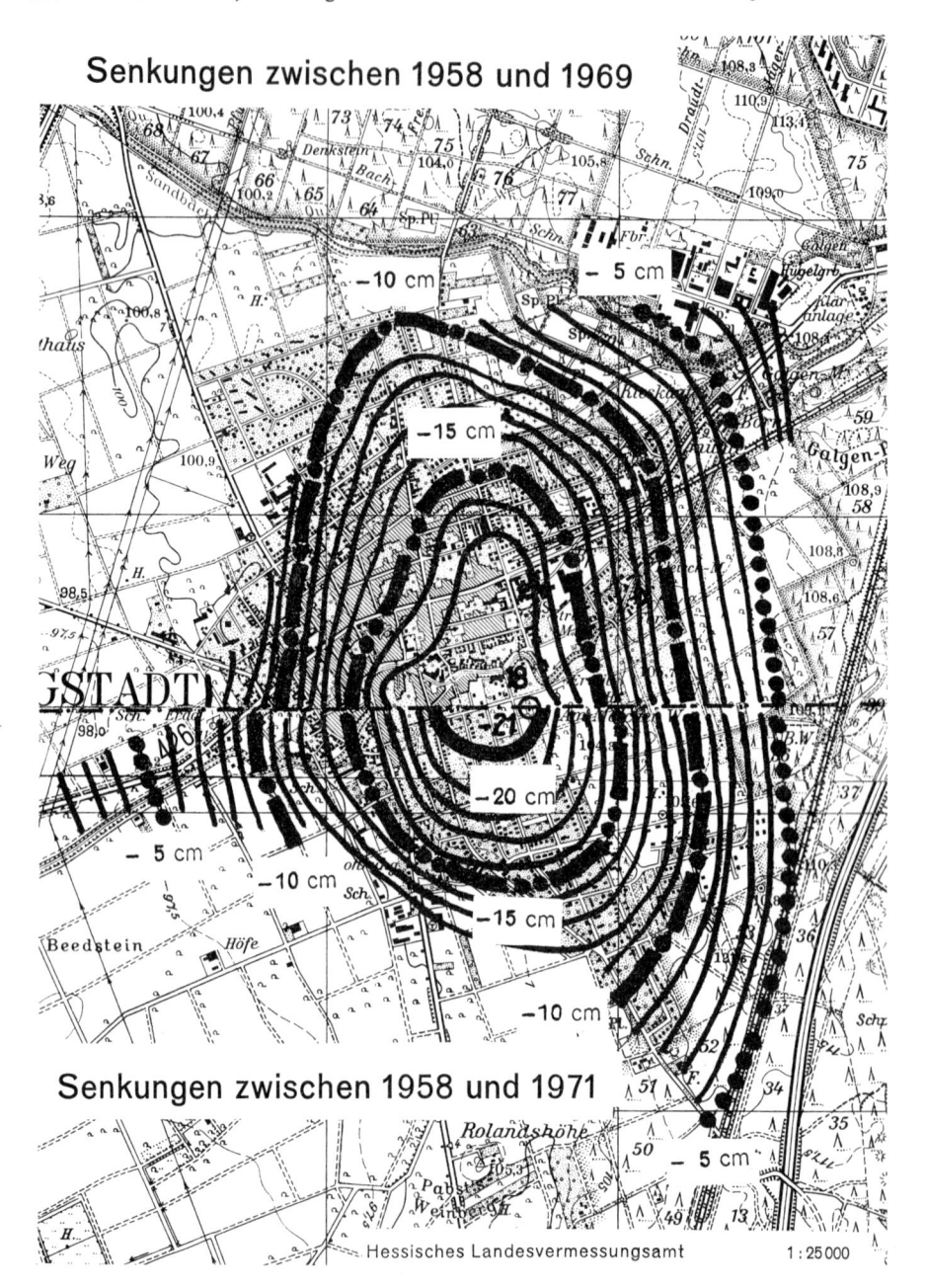

Fig. 8. Subsidence in the region of the gas field Pfungstadt between 1958 and 1969 (upper part), 1958 and 1971 (lower part).

Adjacent fault troughs north of the Rhinegraben

Die tektonische Entwicklung der Niederrheinischen Bucht

von

R. Teichmüller

Mit 12 Figuren im Text

Abstract: In contrast to the deeply subsided Upper Rhine Graben this part of the great Rhenish fault zone — which crosses Europe from the Mediterranean to the North Sea — is characterized by a much lesser thickness of the Tertiary sediments. This facilitates the study of the relationships between the young fracturing and the tectonics of the Hercynian basement. In the same way as in the Upper Rhine Graben, also within the Lower Rhine fault zone the dilatation of the crust caused a tilting of the individual blocks. Also in the Lower Rhine basin the western margin of the fault zone seems to be the more important one. Here, the subsidence started during the Maastrichtian and the Paleocene. It also was here that thermal waters carrying lead solutions ascended. Moreover it was at the western margin of the Lower Rhine fault zone where the surveyors measured the most intense tectonic movements of the present, and where most of the epicenters of earthquakes were registered. — Whereas the crustal movements during the Upper Oligocene and the Lower Miocene are closely connected with the volcanic activity within the Siebengebirge, the "carbon dioxide lines" of the Lower Rhine basin are likely related for the most part with the foiditic volcanism of Pleistocene age.

In Westeuropa läßt sich die große „rheinische" Trennfuge vom Zentralgraben der Nordsee bis zum Mittelmeer verfolgen (Fig. 1). Der Abschnitt am Oberrhein ist in den vergangenen Jahren gründlich mit den verschiedenen Methoden untersucht worden. Beim 4. Kolloquium der Internationalen Rheingraben-Forschungsgruppe tauchte der Wunsch auf, nun auch einmal einen vergleichenden Blick auf die tektonische Entwicklung weiter im Norden zu werfen, d. h. auf die Niederrheinische Bucht.

Beginnen wir mit den Verhältnissen an der Rheinmündung, d. h. in den Niederlanden. Keizer & Letsch (1963) ist hier eine Karte der Tertiärmächtigkeit zu verdanken. Zwei Senkungszentren lassen sich unterscheiden: das eine ist der Zentralgraben der Niederlande, das andere, flachere ist an den Bereich des Jjsselsees gebunden. Die Mächtigkeit des Tertiärs erreicht beiderorts noch nicht 1000 m. Immerhin liegt aber die Basis des Pleistozäns im Bereich des Jjsselsees und des Zentralgrabens mehr als 500 m unter dem Meeresspiegel.

Die jüngsten Bewegungen sind also recht bedeutend. Dementsprechend wird z. B. im Venlograben auch das Pleistozän von starken Störungen zerstückelt

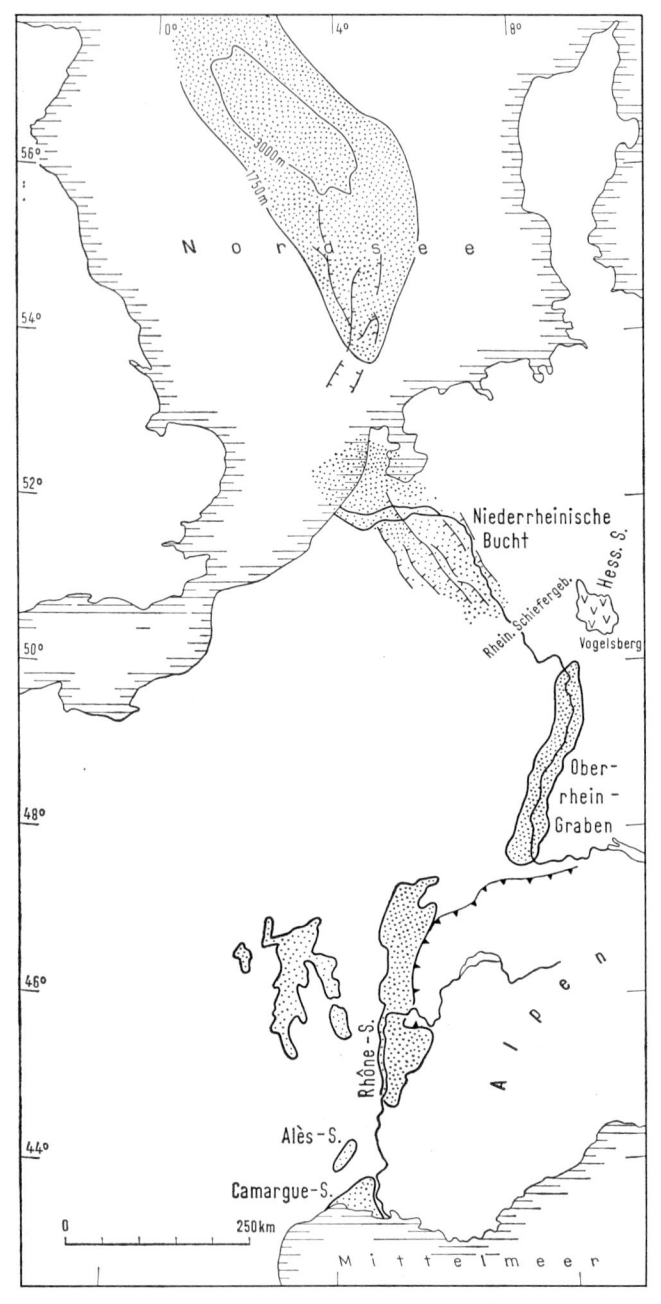

Fig. 1. Die große rheinische Trennfuge Westeuropas. Mächtigere tertiäre Ablagerungen in dieser Zone sind punktiert.

(ZAGWIJN 1963). Aus dem 45 m-Versatz in der sog. Sterkselformation ergibt sich hier eine Absenkung des Grabens von 0,1 mm/Jahr seit der Mindeleiszeit. — Diese **fortlebenden Störungen** können bis in die eigentliche Niederrheinische Bucht verfolgt werden. Auch hier haben viele Brüche noch das Pleistozän verworfen. Aus dem Vergleich der Fein-Nivellements aus den Jahren 1933 und 1952 ermittelten QUITZOW & VAHLENSIEK (1955) das Ausmaß der jüngsten tektonischen Bewegungen (Fig. 2): Der Rurgraben, der die südliche Fortsetzung des Zentralgrabens der Niederlande darstellt, sank in 19 Jahren 5 cm tiefer und die Erftscholle etwa 3 cm, die Kölner Scholle dagegen „nur" 1—2 cm. — Trotzdem sind die Tertiär-Mächtigkeiten auch im Senkungsfeld der Niederrheinischen Bucht gegenüber denen im Oberrheingraben klein. Die Pleistozän-Mächtigkeiten sind in der Niederrheinischen Bucht schon wesentlich geringer als in den Niederlanden. Denn wir nähern uns bereits dem Rheinischen Schiefergebirge, das nach QUITZOW (1962) seit 400 000 Jahren bis zu 200 m aufgestiegen ist mit einer Durchschnitts-Geschwindigkeit von 0,5 mm im Jahr.

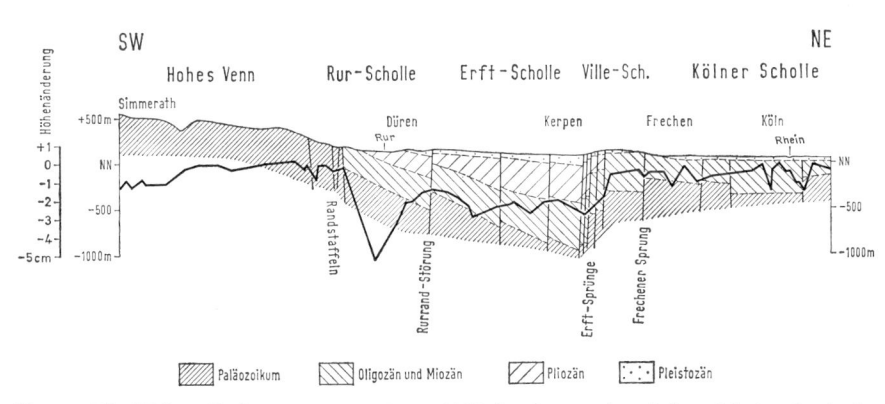

Fig. 2. Die Höhen-Änderungen von 1933—1952 in einem tektonischen Schnitt durch das niederrheinische Senkungsfeld. Nach QUITZOW & VAHLENSIECK 1955.

Das **Senkungszentrum** der Niederrheinischen Bucht im engeren Sinne liegt dort, wo sich zwei große Senken kreuzen (Fig. 3): die Eifeler N—S-Zone und eine NW—SE-streichende Senke. Die Eifeler N—S-Zone wird charakterisiert durch die Eifelkalkmulden, die Zechsteinsalze Na 1 und 2, die Trias von Mechernich, den Lias von Drove und Bislich, das Hauterive vom Niederrhein, das teilweise marine Maastricht von Irnich, das terrestrische Paläozän von Antweiler und Liblar und die ungewöhnlich mächtige Braunkohle des Erftbeckens.

Das Vorkommen von Maastricht bei Irnich und die Paläozän-Funde bei Antweiler und Liblar beweisen, daß die **epirogene Absenkung** im Bereich der späteren Rur- und Erftscholle und damit wohl auch die Dehnung im Senkungsfeld der Niederrheinischen Bucht schon in der jüngsten Kreidezeit begann — zur gleichen Zeit also, als die ersten Vulkanite im Bereich des Oberrheingrabens ihr Dach durchschlugen (HORN et al. 1972). Die Senkung setzte hier also schon

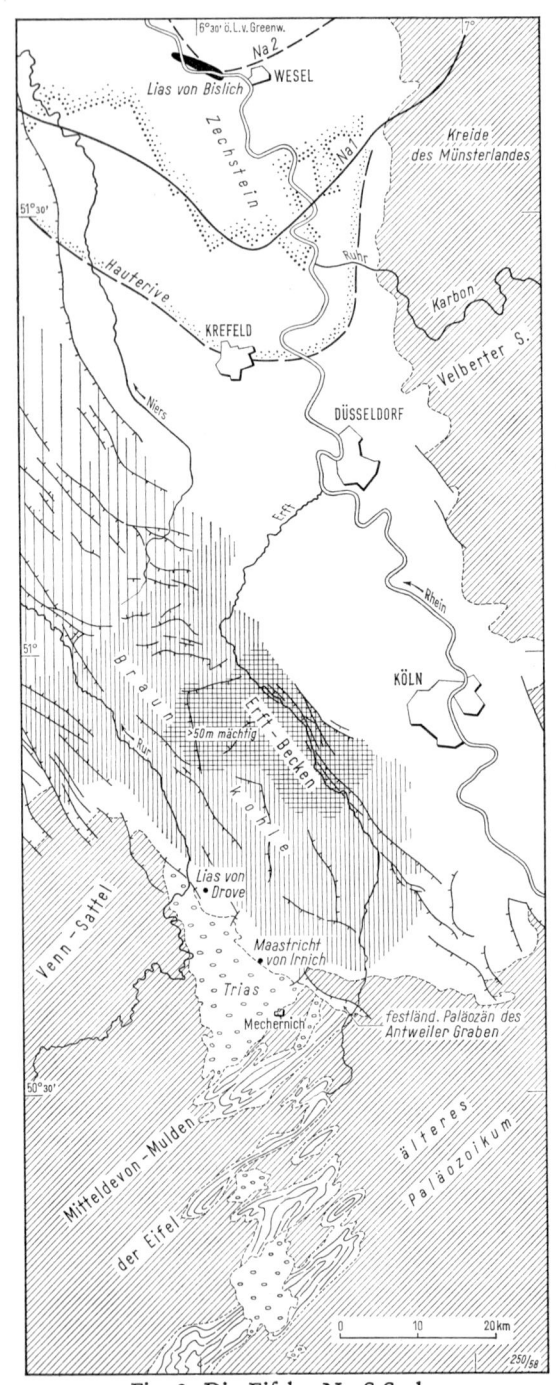

Fig. 3. Die Eifeler N—S-Senke.

etwas früher ein als die Bildung der ersten Becken im Bereich des Oberrheingrabens. — Im Eozän lag das Senkungsfeld der Niederrheinischen Bucht trocken. Erst im Oligozän stieß die Nordsee sukzessive nach Süden vor. Der Sedimentationsraum des Oberoligozäns erstreckte sich damals schon über das gesamte Senkungsfeld (Fig. 4). Im Miozän und Pliozän zog sich das Meer wieder schrittweise zurück (Fig. 5). Bei dieser Regression entstanden die großen Braunkohlenmoore im inneren Teil der Bucht. Sie waren durch Nehrungen gegen das Meer geschützt. Das Alter der marinen Ablagerungen ist durch die Fauna und z. T. auch durch radiometrische Untersuchungen der Glaukonite gesichert (Fig. 6). Die festländischen Ablagerungen sind naturgemäß schwerer einzustufen. Aufgrund der vielen Bohrungen konnten jedoch ANDERSON (1962—1969) und GLIESE 1971 die Verzahnung der einzelnen Flöze und Flözgruppen mit marinen Sedimenten weiter klarstellen und das Alter der Flöze eingabeln. Zum anderen verzahnen sich die von V. D. BRELIE (1968, 1972) palynologisch fixierten Unterflöze der Kölner Schichten landeinwärts mit den mehr als 100 m mächtigen Trachyttuffen des Siebengebirges (Fig. 7), deren Alter TODT et al. (1972) in Bohrungen des Geol. Landesamtes von Nordrhein-Westfalen durch K/Ar-Bestimmungen der Sanidine sozusagen Schicht für Schicht ermittelten (Fig. 8). Dabei ergab sich 1. die Abnahme der Sedimentationsrate der Tuffe mit dem Verklingen der Taphrogenese und der Tufförderung und 2. die überraschende Tatsache, daß die wegen ihres Fossilreichtums berühmten Blätterkohlen von Rott am Siebengebirge nicht dem Chatt angehören, sondern dem Hemmoor[1]. Denn sie entstanden zwischen 18,2 und 20,6 Millionen Jahren, während das Chatt nach BERGGREN (1972) schon bei 22,5 Millionen Jahren endete (typisches Hemmoor datierten KREUZER et al. 1973 bei 19,2—20,4 Mill. J.). — Relikte der Trachyttuffe wurden von PFLUG und FRECHEN (PFLUG 1959) auch noch im Hauptflöz der Ville bei Liblar in Form von Trachyt-Gläsern, Sanidinen, Biotit- und Augit-Körnern nachgewiesen (Fig. 9). Darüber hinaus gelang es PIETZNER & WOLF (1964), im ganzen Bereich der Ville auffallend hohe Titanmaxima in der Kohle des Hauptflözes geochemisch zu verfolgen. Diese Titanmaxima sind — wie die Autoren betonen — nicht mit höheren Aschengehalten verbunden. Auch ist die Vergelung der Kohle hier besonders gering. Deshalb können die Titanmaxima wohl nicht auf subaquatische Bildungen, d. h. Einschwemmungen, zurückgeführt werden. Der auffallend große Fusitgehalt läßt vielmehr an vulkanische Aschenregen im Kohlenmoor denken. In diesem Zusammenhang sei an die starke Beimengung von vulkanischem Material im Tertiär des Venusberges bei Bonn erinnert: SINDOWSKI (1939) stellte hier in einer Schwermineralanalyse 6,0 % Hornblende, 10,0 % Augit und 17,4 % Titanit fest. Aus solchen Mineralen sind wohl die Titanmaxima im Hauptflöz abzuleiten.

---

[1] Nach dem K/Ar-Alter der Vulkanite im Westerwald (s. u.) sind auch dort die *Microbunodon-* und *Anthracotherium*-führenden Schichten ins Untermiozän zu stellen. Da THENIUS (1951) *Anthracotherium* auch noch aus sicherem Untermiozän in der Steiermark beschrieben hat, besteht zwischen den K/Ar-Datierungen und den paläontologischen Befunden kein unüberbrückbarer Gegensatz.

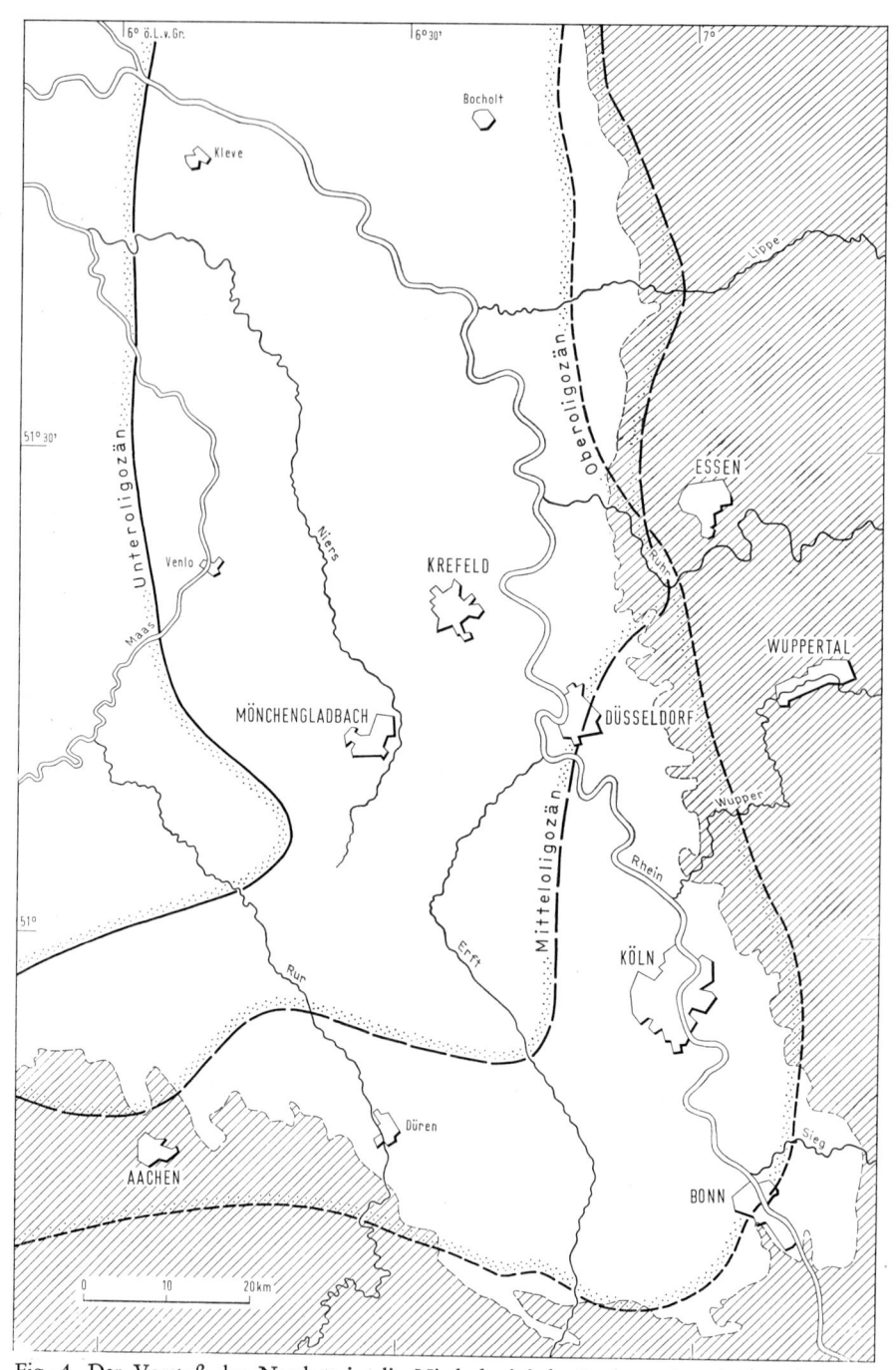

Fig. 4. Der Vorstoß der Nordsee in die Niederrheinische Bucht während des Oligozäns. Nach QUITZOW 1971.

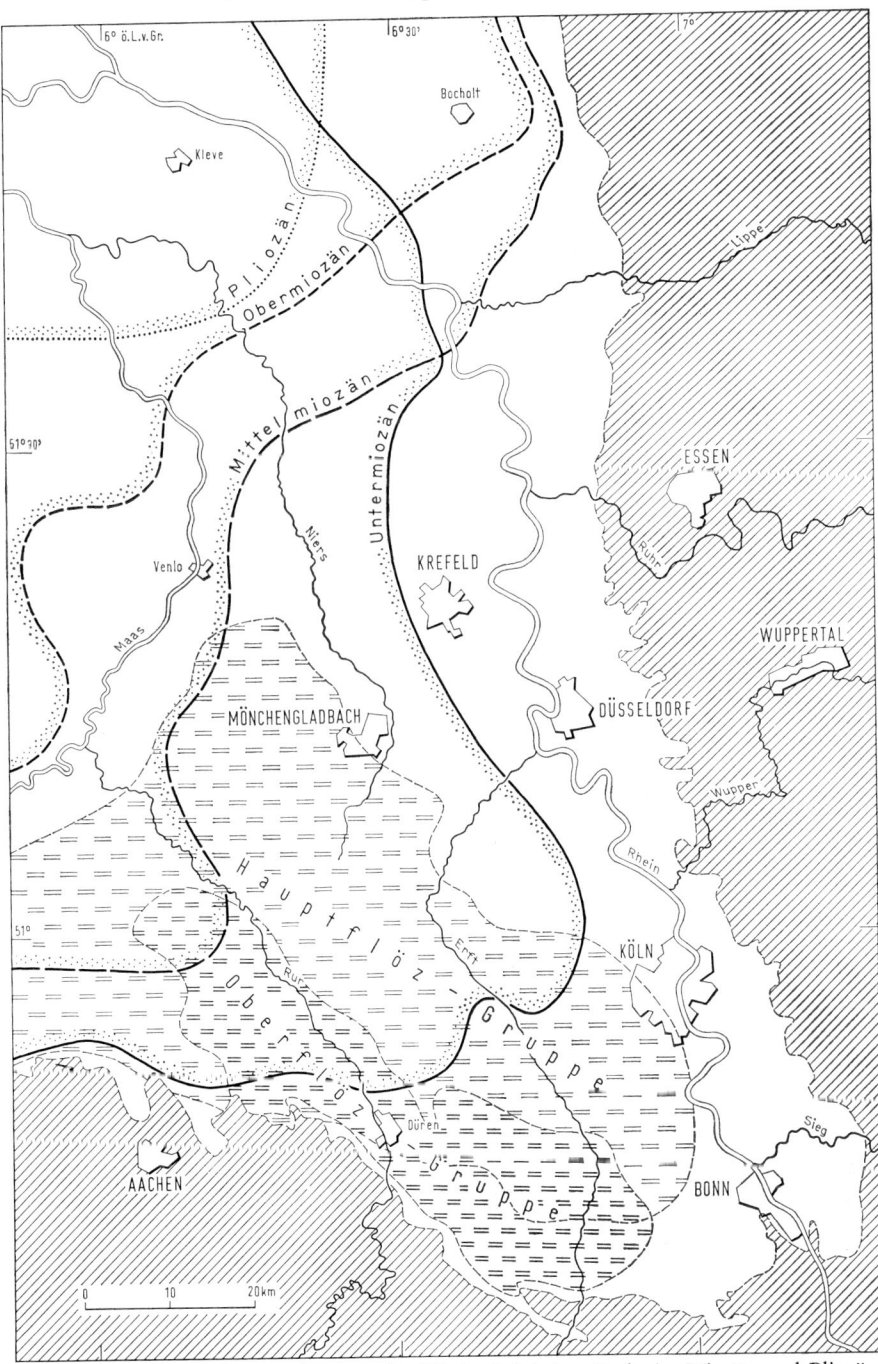

Fig. 5. Der Rückzug der Nordsee aus der Niederrheinischen Bucht im Miozän und Pliozän und die Ausdehnung der miozänen Braunkohlenmoore. Nach QUITZOW 1971.

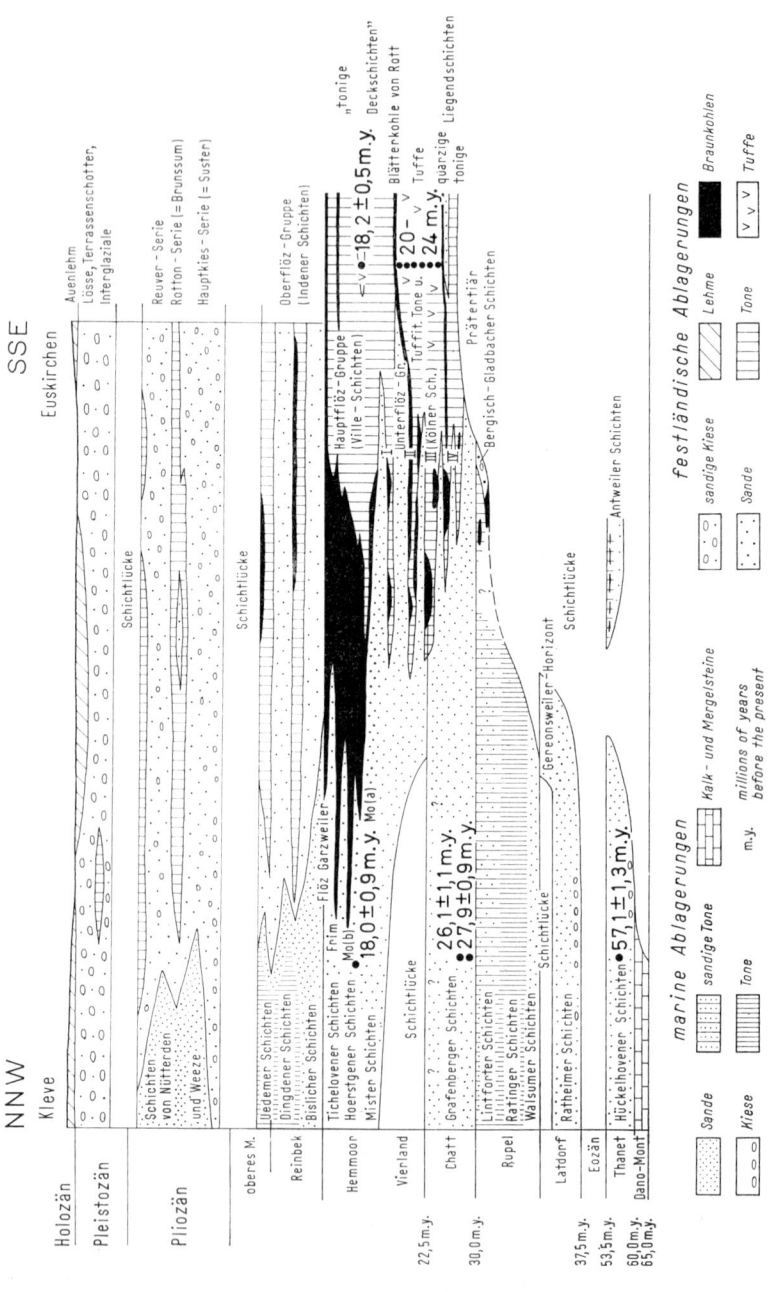

Fig. 6. Die Verzahnung der marinen und festländischen Sedimente im Tertiär der Niederrheinischen Bucht (Schema). Nach ANDERSON 1969 und QUITZOW 1971, etwas abgeändert. Die K/Ar-Datierungen der Glaukonite und Sanidine nach TODT et al. 1972. Zahlen am linken Rand nach BERGGREN 1972. Mo: Flöz Morken. Fri: Flöz Frimmersdorf.

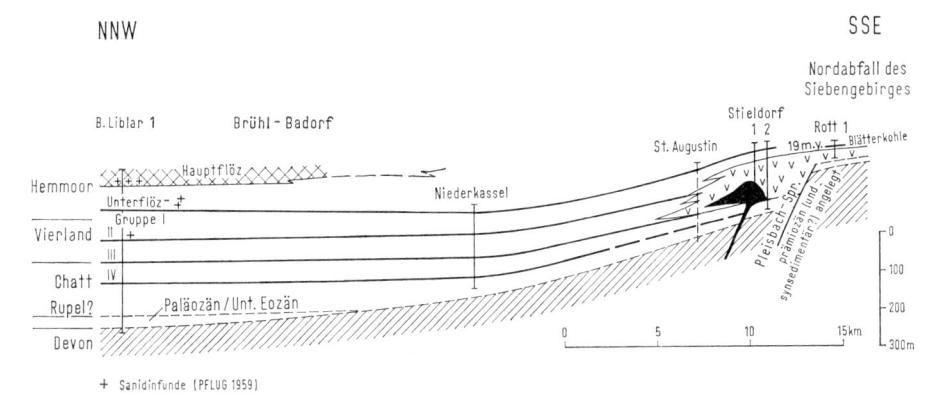

Fig. 7. Die Verzahnung der sogenannten Unterflöze (Kölner Schichten) mit den Trachyttuffen des Siebengebirges.

Die Ablagerungen der Tertiärzeit wurden in der Niederrheinischen Bucht von Brüchen zerstückelt, die wie gesagt bis in die Gegenwart fortleben. Zum **Alter der Bruchtektonik**: Die ersten Brüche rissen gegen Ende der Triaszeit auf. Neben diesen altkimmerischen sind auch jungkimmerische Bruchschollenbewegungen nachzuweisen. Tiefreichende Spalten im Steinkohlengebirge des Niederrheins wurden dabei mit marinem Hauterive gefüllt (SCHAUB 1955). Auch an subherzynen Bewegungen hat es nicht gefehlt: Santon und Maastricht greifen, wie u. a. SCHAUB (1955), ARNOLD (1971) und HOYER (1971) gezeigt haben, auf das ältere Mesozoikum und das Karbon des Krefelder Gewölbes über.

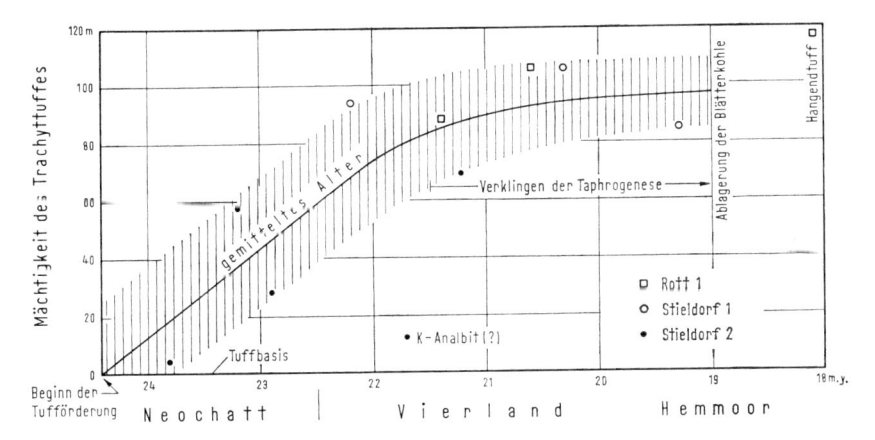

Fig. 8. Das K/Ar-Alter der Sanidine in den einzelnen Schichtpaketen des Trachyttuffes im Siebengebirgsgraben. Rott 1 sowie Stieldorf 1 und 2 sind Kernbohrungen des Geologischen Landesamtes Nordrhein-Westfalen. Nach den Datierungen von TODT 1971, Rott 1 ergänzt nach TODT et al. 1972.

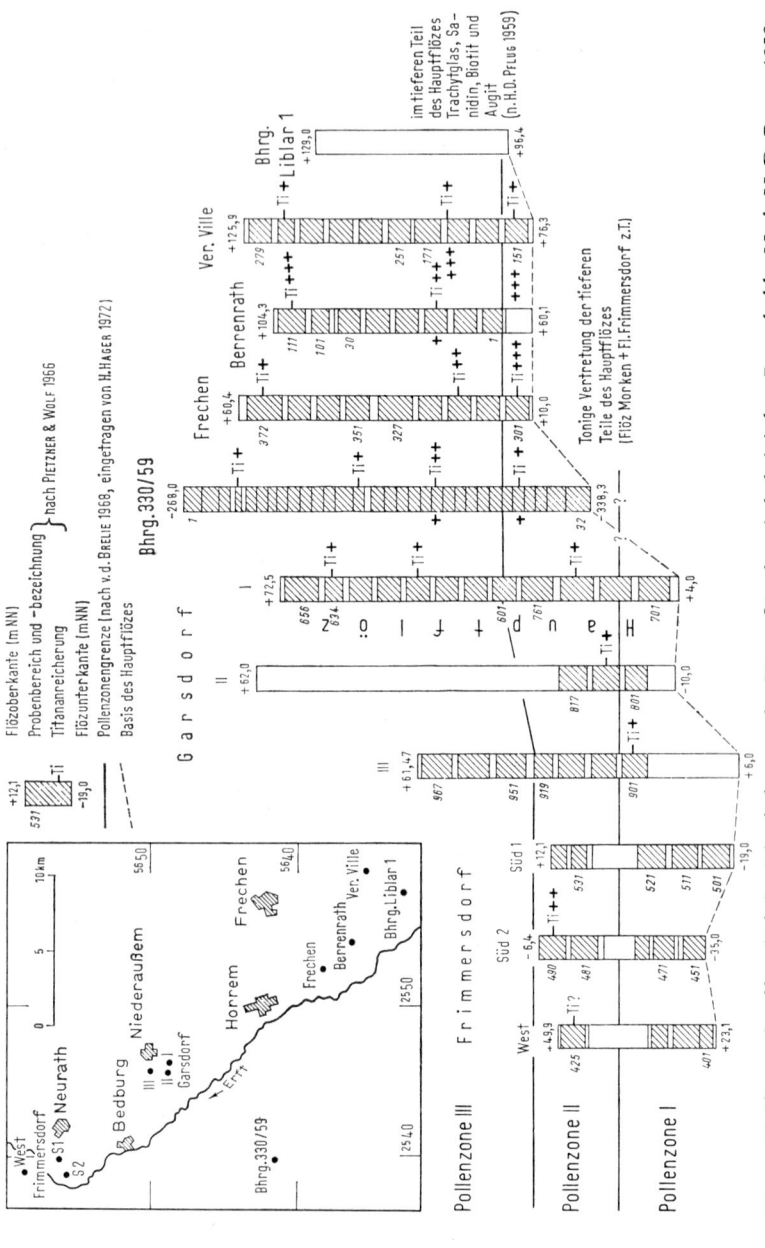

Fig. 9. Tuffite und tuffverdächtige Einschaltungen im Hauptflöz der niederrheinischen Braunkohle. Nach H. D. PFLUG 1959 und PIETZNER & WOLF 1966.

Die tektonische Entwicklung der Niederrheinischen Bucht 279

In engem Zusammenhang mit den Krustenbewegungen an der Wende Oligozän/Miozän steht der Vulkanismus des Siebengebirges (und übrigens auch des Westerwaldes). Er begann, wie die K/Ar-Bestimmungen von TODT (1971) bewiesen, im Chatt und endete im Hemmoor, abgesehen von wenigen jüngeren Alkalibasalten (Fig. 10). Eine eindeutige Differentiationsfolge ist nach diesen Untersuchungen im Siebengebirge und Westerwald nicht mehr festzustellen. Spalteneruptionen im Streichen NW—SE und NE—SW machen es im Siebengebirge wahrscheinlich, daß viele Fugen während des Vierlands und Neochatts aufrissen. — Ähnlich wie bei Wiesloch im Oberrheingraben kam es auch bei Maubach und Mechernich in der Niederrheinischen Bucht zu einem Aufstieg von bleireichen Thermalwässern an den jungen Brüchen und zu ausgedehnten Vererzungen in der Trias. Es ist übrigens nicht ganz ausgeschlossen, daß auch die

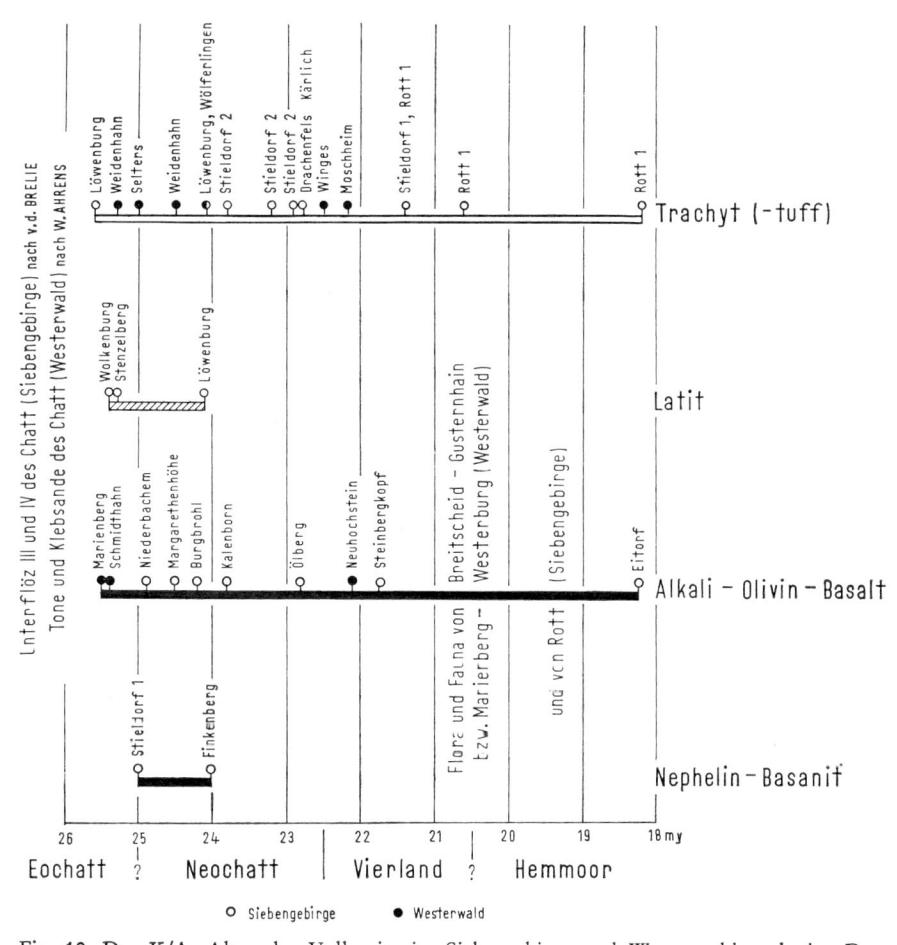

Fig. 10. Das K/Ar-Alter der Vulkanite im Siebengebirge und Westerwald nach den Datierungen von TODT 1971. Rott 1 ergänzt nach TODT et al. 1972.

tektonisch völlig unbeanspruchten Blei-Zink-Erze im Karbon von Lintorf und Selbeck nördlich von Düsseldorf erst im Miozän entstanden. Denn der sog. Ruhrwehrsprung, auf dem sie sitzen, verwirft nach Mitteilung von Dr. BACHMANN das Oligozän noch um 20 m. — Messungen der Wärmestromdichte sind in der Niederrheinischen Bucht noch nicht durchgeführt worden. Wohl aber stellte BALKE (1973) am Rurrand und Erftsprung einen Anstieg der Isothermen auf etwa 20° C im Niveau 100 m unter NN fest. Die Mineralisation der Wässer nimmt mit der Tiefe beträchtlich zu. Der Gehalt an Natrium-Hydrogenkarbonat wächst dabei auf Kosten des Calcium-Hydrogenkarbonates. Dementsprechend wird auch der Chlorid-Gehalt mit der Tiefe größer. In der Kölner Scholle fällt in einigen schmalen Zonen die besonders hohe Konzentration an Kohlendioxyd, Natrium und Chlorid im tieferen Grundwasser auf. Sie ist wohl als Folge-Erscheinung des jungen foiditischen Vulkanismus (Beispiel: der Rodderberg-Krater südöstlich von Bad Godesberg) zu deuten.

Infolge der verhältnismäßig kleinen Tertiär-Mächtigkeit von noch nicht 1000 m haben viele Bohrungen in der Niederrheinischen Bucht den prätertiären Untergrund, d. h. das Devon und Karbon, erreicht. Deshalb konnten HERBST & THOME (1971) eine relativ gut belegte Karte vom variszischen Unterbau der Niederrheinischen Bucht entwerfen. Wenn man diesen alten Unterbau mit der jungen Bruchtektonik vergleicht, so zeigt sich, daß die junge Bruchtektonik im wesentlichen auf den Westflügel des Krefelder Gewölbes beschränkt ist (Fig. 11). Das gilt vor allem für die fortlebenden Verwerfungen, wie HOYER (1971) dargetan hat. Auch die weitaus meisten Erdbebenherde sind auf die Randbrüche der Kippschollen beschränkt, denen Rur und Erft folgen (AHORNER 1970). Eine Beziehung zwischen der jüngen Bruchtektonik und dem variszischen Unterbau ist also unverkennbar. Die Querstörungen im variszischen Faltengebirge waren die vorgezeichneten Bahnen, denen die meisten Brüche, die in tertiärer Zeit im Senkungsfeld der Niederrheinischen Bucht aufrissen, folgten. So kam es hier zu dem für eine rheinische Zerrungszone ungewöhnlichem NW—SE-Streichen.

Schließlich wird auch noch der Lauf des Niederrheins unterhalb von Duisburg offensichtlich vom aufsteigenden Krefelder Gewölbe bestimmt. Der Niederrhein begleitet damit ein uraltes Lineament, das sich übrigens schon im Karbon als Grenze von zwei Faziesbereichen bemerkbar macht:

Im Unterkarbon: Im Westen Kohlenkalk, im Osten Kulm;
Im Namur: Im Westen kleine Mächtigkeit-, im Osten große Mächtigkeit des Flözleeren;
Im Westfal: Auskeilen vieler im Ruhrkohlenbecken bauwürdiger Flöze nach Westen.

Bemerkenswerterweise springt auch der gefaltete Bereich des Subvariszikums an diesem Lineament im Osten viel weiter nach Norden vor. Einer relativ flachgründigen Tafel im Westen entspricht ein tiefgründiges subvariszisches Faltenland im Osten. — Leider wissen wir über das Liegende des Mitteldevons am Linken Niederrhein noch nichts Genaueres. Von besonderem Interesse in dieser Hinsicht

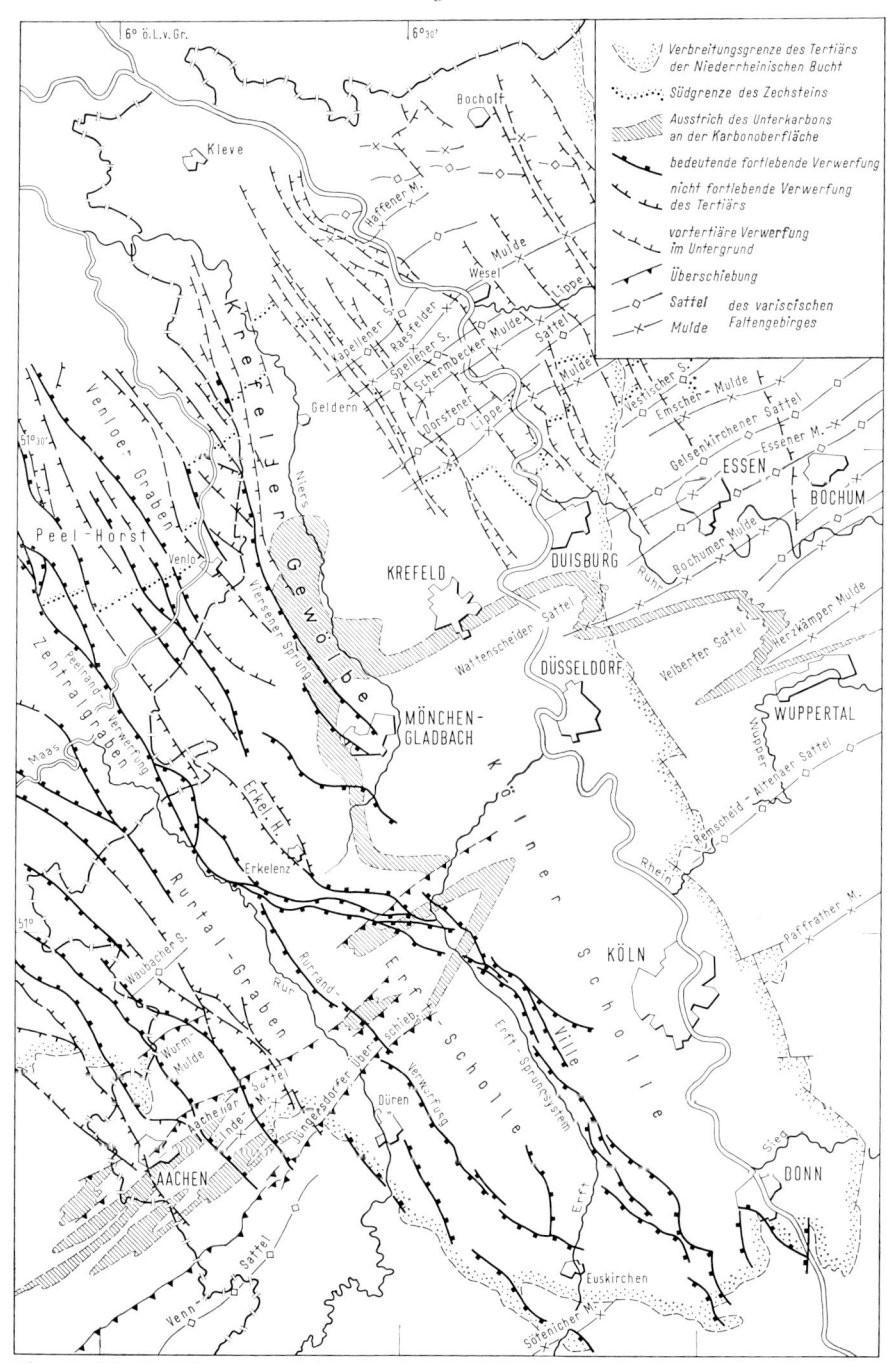

Fig. 11. Die Beziehungen zwischen der jungen Tektonik und dem variszischen Unterbau. Nach HOYER 1971.

Das Tertiär zwischen Köln und Mainz

NW in der Niederrheinischen Bucht südlicher Teil
nach ANDERSON 1969–72, GLIESE 1971, QUITZOW 1971 u.a.

im Westerwald und Neuwieder Becken
nach W. AHRENS 1941 u. 1957, H.D. PFLUG 1959, WIESNER 1970 u.a.

SE im Mainzer Becken
n. FALKE 1960, SONNE 1969–1972, STEPHAN 1971, WIESNER 1970

Cenozoic time-scale nach BERGGREN 1972	NW (Niederrheinische Bucht)	Westerwald / Neuwieder Becken	SE (Mainzer Becken)
Pliozän – 5,0 m.y.	Reuverton-Serie / Rotton-Serie / Hauptkies-Serie	Floren von Dernbach und Berggarten	Sande mit Mastodon arvernensis / Dinotheriensande
Obermiozän	Indener Schichten (Oberflözgruppe)		Flöz Ginnheim
Miozän – Reinbek – Hemmoor 18,0	Ville-Schichten (Hauptflözgruppe) / "Hangendschichten" / Blätterkohle von Rott 20 / 18,2 / bis / 23	jüngere Basalte, Tuffite, Dysodile und Braunkohlentone des Westerwaldes	Landschneckenmergel / oberste Hydrobien-Schichten / Hydrobien-Schichten mit Hastigerinen-Horizont an der Basis / Corbicula-Schichten / Cerithien-Schichten
Vierland – 22,5 m.y.	Obere Kölner Schichten (z.T. marin) mit den Unterflözgruppen I, II u. III z.T.	Floren und Faunen in den Blätterkohlen von Marienberg-Westerburg und Breitscheid – Bustenhain über u. zwischen älteren Vulkaniten des Westerw. / "Knubb" von Kärlich 22,1 / 22,5 / 25,5 Klebsande und Cyrenenmergel	Landschneckenkalk u. Süßwasserschichten / lokale Bildungen
Oligozän – Chatt 26,1 / 27,9 – 30,0 m.y.	Untere Kölner Schichten (z.T. marin) mit den Unterflözgruppen III z.T. u. IV	25,3 / 22,8	Cyrenenmergel z.T. mit Braunkohlenlagen / Verbindung Hess. Senke – Oberrhein-Graben
Rupel	Schichten von Bergisch-Gladbach		Schleichsand / Meeressand / ob. mittl. Rupelton unt.
Latdorf – 37,5 m.y.		Tone mit Braunkohlen, randlich Konglomerate (Mitteleozän – Unteroligozän)	Mittlere Pechelbronner Schichten (brackisch)
Eozän – 53,5 m.y. 57,1	Schichten von Antweiler		Eozäner Basiston ?
Paläozän – 65,0 m.y.			?

Glaukonit-Daten im nördl. Teil der Niederrh. Bucht 26,1 / 27,9

v, V vulkanische Produkte
○ K/Ar-Datierungen nach LIPPOLT 1963, TODT 1971 und TODT et al. 1973
● K/Ar-Datierungen
Alter in m.y. (millions of years before the present)
? K/Ar-Datierung möglich

ist deshalb das unvermittelte Erscheinen von Blockbrekzien im Oberen Mitteldevon des Velberter Sattels nordöstlich von Düsseldorf. Die Blockbrekzien bestehen aus granatführenden Quarziten und gebänderten Phylliten, die an Gesteine des Devillien im Brabanter Massiv erinnern. Sie können nur aus der nächsten Nachbarschaft stammen. Wie seismische Untersuchungen am Linken Niederrhein zeigten, schaltet sich hier zwischen dem flach einfallenden Massenkalk und dem Top des Kristallins noch ein > 3000 m mächtiges Schichtenpaket konkordant ein, das vornehmlich ungefaltetem Kambro-Silur entsprechen dürfte. Die granatführenden Quarzitblöcke stammen demnach wohl aus dem präkambrischen Sockel des Tafellandes am Linken Niederrhein.

Aus dem seismotektonischen Diagramm, das AHORNER 1970 gab, folgt, daß die SW—NE streichenden Falten des Schiefergebirges auch heute noch in der Richtung größter Pressung liegen. Wenn junge Zerrbrüche vorhanden sind, sollten sie sich vor allem im Gewölbescheitel bemerkbar machen. Vielleicht ist der Einbruch des Neuwieder Beckens auch in diesem Sinne zu deuten. H. CLOOS (1939) dachte hier vor allem an eine Dehnung in W—E-Richtung.

Wir kommen damit zur **Frage des Zusammenhangs zwischen dem Oberrheingraben und der Niederrheinischen Bucht und zur Entstehung des Neuwieder Beckens.**

Die Sedimentation begann im Neuwieder Becken, wie H. D. PFLUG (1959) gezeigt hat, erst mit Tonen und Braunkohlen des Obereozäns und Unteroligozäns. Darüber folgen Klebsande und Cyrenenmergel des Chatts, die z. T. von Trachyttuff überlagert werden. Aufgrund des K/Ar-Verhältnisses (LIPPOLT et al. 1963) ist er an die Grenze Oligozän/Miozän zu stellen. An der Basis der übergreifenden Hydrobienschichten des Aquitans fand WIESNER (1970) im Neuwieder Becken *Hastigerina*, eine planktonische Foraminifere, die er bereits im Mainzer Becken nachgewiesen hatte. Aus dem Vierland der Niederrheinischen Bucht ist *Hastigerina* bisher nicht bekannt. Die Ingression der Hydrobienschichten in das Neuwieder Becken ist demnach wohl vom Mainzer Becken her erfolgt[2]. Eine zeitweise Verbindung des Mainzer Beckens mit dem Sedimentationsraum der Kölner Schichten (Chatt bis tiefes Hemmoor) und dem Vierland von Houthalen (HOOBERGS & DE MEUTER 1972) ist aber m. E. nicht ganz von der Hand zu weisen. Es wäre die erste — wenn auch sehr seichte — Verbindung zwischen der Niederrheinischen Bucht und dem Oberrheingraben, eine Verbindung, die bekanntlich im jüngeren Miozän und im Oberpliozän und Pleistozän wieder auflebte (QUITZOW 1962) und sich auch in der Gegenwart durch seine seismische Aktivität auszeichnet (AHORNER 1970). Die fortschreitende Verlagerung des Senkungszentrums im Oberrheingraben nach Norden steht vielleicht mit diesen jungen tektonischen Bewegungen am Mittelrhein in Zusammenhang.

---

[2] Vgl. auch SONNE 1972.

Fig. 12. Übersicht über die stratigraphische Gliederung des Tertiärs zwischen Köln und Mainz. Die K/Ar-Datierungen nach TODT 1971, LIPPOLT et al. 1963 und TODT et al. 1972. Die Möglichkeiten weiterer K/Ar-Datierungen sind angedeutet.

Literatur

AHORNER, L.: Seismo-tectonic relations between the graben-zones of the Upper and Lower Rhine. — In: Graben Problems. Proceed. of an internat. Rift Sympos. held in Karlsruhe Oct. 10—12, 1968, 155—166, hrsg. v. J. H. ILLIES & ST. MUELLER. Schweizerbart'sche Verlagsbuchh., Stuttgart 1970.

AHRENS, W.: Überblick über den Aufbau des Westerwälder Tertiärs mit besonderer Berücksichtigung der stratigraphischen Stellung der vulkanischen Gesteine. — Fortschr. Miner. **35**, 109—116, Stuttgart 1957.

ANDERSON, H. J.: Paläontologische Bemerkungen zur Stratigraphie des Oligo-Miozän in der Niederrheinischen Bucht. — Fortschr. Geol. Rheinld. u. Westf. **6**, 1—18, Krefeld 1962.

— Die Schichtenfolge des Tertiärs und Quartärs. — In: Geologische und bergbauliche Übersicht des rheinischen Braunkohlenreviers. Hrsg. v. d. Dt. Geol. Ges. zur Frühjahrstagung in Köln, S. 2—5, Krefeld 1966.

— Die Molluskenfaunen der marinen Oligocän-Ablagerungen in Nordwestdeutschland. — In: Führer zur Oligocän-Excursion 1969, 28—33, Marburg 1969.

— Das Oligocän in der Niederrheinischen Bucht. — In: Führer zur Oligocän-Excursion 1969, 34—40, Marburg 1969.

ARNOLD, H.: Kreide. — Z. Der Niederrhein, 38. Jg., 97—100, Krefeld 1971.

BALKE, K. D.: Geothermische und hydrogeologische Untersuchungen in der südlichen Niederrheinischen Bucht. — Geol. Jb. Reihe C, H. 5, 62 S., Hannover 1973.

BERGGREN, W. A.: A Cenozoic time scale — some implications for regional geology and paleobiogeography. — Lethaia **5**, 195—215, Oslo 1972.

V. D. BRELIE, G.: Zur mikrofloristischen Schichtengliederung im rheinischen Braunkohlenrevier. — Fortschr. Geol. Rheinld. u. Westf. **16**, 85—102, Krefeld 1968.

CLOOS, H.: Hebung—Spaltung—Vulkanismus. — Geol. Rdsch. **30**, 401—527, Stuttgart 1939.

GLIESE, J.: Fazies und Genese der Kölner Schichten (Tertiär) in der südlichen Niederrheinischen Bucht. — Sonderveröff. d. Geol. Inst. d. Univ. Köln **19**, 91 S., Köln 1971.

HERBST, G. & THOME, K. T.: Der gefaltete Untergrund. — Z. Der Niederrhein, 38. Jg., 89—91, Krefeld 1971.

HOOBERGHS, H. J. F. & DE MEUTER, F. J. C.: Biostratigraphy and interregional correlation of the "Miocene" deposits of Northern Belgium based on planctonic Foraminifera, the Oligocene Miocene boundary on the southern edge of the Northern Sea Basin. — Mededel. Koninkl. Acad. Wetensch. etc. Kl. Wetensch. **34**, 3, 47 S., Brüssel 1972.

HORN, P.; LIPPOLT, H. J. & TODT, WO.: Kalium-Argon-Altersbestimmungen an tertiären Vulkaniten des Oberrheingrabens. — Eclog. geol. Helv. **65**, 131—156, Basel 1972.

HOYER, P.: Der Bau des Niederrheingebietes und seine Entwicklung. — Z. Der Niederrhein, 38. Jg., 108—110, Krefeld 1971.

KEIZER, J. & LETSCH, W. J.: Geology of the Tertiary in the Netherlands. — Verh. Kon. Ned. geol. mijnbouwkundig genootsch. Geol. serie 21—2, 147—172, s'Gravenhage 1963.

KREUZER, H.; DANIELS, C. H. V.; GRAMANN, F.; HARRE, W. & MATTIAT, B.: K/Ar-Dates of some Glauconites of the Northwest German Tertiary Basin. — Fortschr. Miner. **50**, Beih. 3, 94 f., Stuttgart 1973.

LIPPOLT, H. J.; GENTNER, W. & WIMMENAUER, W.: Altersbestimmungen nach der Kalium-Argon-Methode an tertiären Eruptivgesteinen Süddeutschlands. — Jh. geol. Landesamt Baden-Württ. **6**, 507—538, Freiburg/Br. 1963.

PFLUG, H. D.: Die Deformationsbilder im Tertiär des rheinisch-saxonischen Feldes. — Freiberg. Forschungsh. C 71, 1—110, Berlin 1959.

PIETZNER, H. & WOLF, M.: Geochemische Untersuchungen an Braunkohlenaschen und kohlenpetrographische Untersuchungen an Braunkohlen aus dem Niederrheinischen Braunkohlenrevier. — Fortschr. Geol. Rheinld. u. Westf. **12**, 517—550, Krefeld 1964.

QUITZOW, H. W.: Mittel- und Niederrhein. — In: Die Entstehung des Rheintales vom Austritt des Flusses aus dem Bodensee bis zur Mündung. Beitr. z. Rheinkunde H. 14, 9—47, Koblenz 1962, hrsg. v. Rhein-Museum e. V., Koblenz.
— Tertiär. — Z. Der Niederrhein, 38. Jg., 101—103, Krefeld 1971.
QUITZOW, H. W. & VAHLENSIECK, O.: Über pleistozäne Gebirgsbildung und rezente Krustenbewegungen in der Niederrheinischen Bucht. — Geol. Rdsch. **43**, 56—67, Stuttgart 1955.
SCHAUB, H.: Kreidesedimente in Spalten des linksrheinischen Steinkohlengebirges. — Geol. Jb. **69**, 249—254, Hannover 1954.
SINDOWSKI, K. H.: Studien zur Stratigraphie und Paläogeographie des Tertiärs der südlichen niederrheinischen Bucht. — N. Jb. Miner., Beil.-Bd. **82**, 415—484, Stuttgart 1939.
SONNE, V.: Jungtertiäre Ablagerungen („Aquitan") am Nordwestrand des Mainzer Bekkens. — Mainzer geowiss. Mitt. **1**, 137—142, Mainz 1972.
STEHLIN, H. G.: Über die Säugetierfauna der Westerwälder Braunkohlen. — Eclog. geol. Helv. **25**, Basel 1932.
THENIUS, E.: *Anthracotherium* aus dem Untermiozän der Steiermark. — Sitzber. Österr. Akad. Wiss. Math. naturw. Kl. Abt. I, **160**, 3. u. 4. H., 217—226, Wien 1951.
TODT, WO.: Kalium-Argon-Altersbestimmungen an mitteleuropäischen miozänen Vulkaniten bekannter paläomagnetischer Feldrichtung. — Diss. Nat. wiss. Fak. d. Ruprecht Karl-Univ., Heidelberg 1970.
TODT, WO.; LIPPOLT, H. J.; GRÜNHAGEN, H.; KÖWING, KL. & TEICHMÜLLER, R.: Kalium/Argon-Datierungen im Tertiär der Niederrheinischen Bucht. — Vortrag am 12. 9. 1972 in Krefeld.
WIESNER, E.: Das Miozän im östlichen Mainzer Becken unter besonderer Berücksichtigung der Mikrofauna. — Diss. d. Naturw. Fak. d. Johann-Wolfgang-Goethe-Univ. zu Frankfurt a. M., Frankfurt 1970.
ZAGWIJN, W. H.: Pleistocene stratigraphy in the Netherlands based on changes in vegetation and climate. — Verh. Kon. Ned. geol. mijnbouwkundig genootsch. geol. serie 21-2, 173—196, s'Gravenhage 1963.

Nachtrag:

Nach freundlicher brieflicher Mitteilung von Herrn Dr. TODT vom 10. 1. 1974 werden zur Zeit die Datierungen der Sanidine aus den Tuffen im Siebengebirgsgraben noch einmal überprüft.

Die Fortsetzung des Rheingrabens durch Hessen

Ein Beitrag zur tektonischen Analyse der Riftsysteme

von

E. Schenk

Mit 7 Abbildungen im Text

Abstract: The Rhine Graben and its continuations to the north and to the south consist of a fault system which strikes significantly in N—S direction. These faults confine all sections of grabens and horsts of first and minor degrees and even of main parts of the rift system along the Rhône and the Rhine and between Taunus and Harz. They are coordinated to the great dominating Rhenish geofracture system which strikes N 25—30° E from the Mediterranean to the Baltic Sea, while transform faulting apparently cuts this zone e. g. in the area between the Saône and the Rhine and between Main and Rhine.

The Rhenish geofracture in Hessen is accompanied by axes of positive and negative gravity anomalies between Pfälzer Wald and Harz, while the graben axis runs from Worms traversing the area of Frankfurt and the Vogelsberg volcano and continues through the Knüll Mountains into the Göttingen Graben. Magnetic anomalies as well as the graben systems in northern Hessen and Niedersachsen indicate the margins of the geofracture.

Statistical evaluation of magnetic anomalies have revealed feather fissures intruded by basalts. From this result and by means of the analysis of the tectonic pattern strike-slip movements resp. rotations along the Rhenish geofracture in Hessen may be derived.

1. Die Grabenrichtung und Grabenversetzung

Wenn auch von einer Mittelmeer-Mjösen-Zone gesprochen wird, so enden nichtsdestoweniger die Vorstellungen vom Rheingraben, wie das die Karten zu lehren scheinen, an den südlichen Randbrüchen des Rheinischen Schiefergebirges (Abb. 1). Diese Brüche folgen dem Streichen der varistischen Falten und ziehen als breites Bündel am Fuße von Hunsrück und Taunus entlang, queren die südliche Wetterau bis an den Rand der Basaltdecken des Paläovulkans Vogelsberg und begrenzen hier zugleich das jüngere Paläozoikum gegen die tertiäre Sedimentfüllung der hessischen Senke. Im südlichen Bereich des Basaltrandes findet man im Rotliegenden und Buntsandstein entsprechende tektonische Linien längs der Kinzig, der sie nach NE bis in den hessischen Landrücken und in die Rhön folgen. An dieser Rhein—Main—Kinzig-Zone, der Odenwald mit Sprendlinger Horst und Spessart südlich anliegen, erscheint der auf fast 60 km Breite nach Westen ausgeweitete und dabei das Mainzer Becken einfassende Rheingraben bajonettartig versetzt und zudem aufgesplittert in NW zum Niederrhein ziehende Bruchsysteme und eine nach NE bis Nordhessen und Niedersachsen fortsetzende Abzweigung.

In dieser vom ausgesprochen N—S verlaufenden Grabenabschnitt am Oberrhein abzweigenden Fortsetzung scheint die Versetzung sich zu wiederholen zwischen Kellerwald und Vogelsberg bzw. Knüll sowie zwischen Habichtswald

Die Fortsetzung des Rheingrabens durch Hessen 287

Abb. 1. Im Reliefbild von Hessen (Wenschow-Verlag) sind einige morphologische Züge nachgezeichnet, um die Systeme der Bruchtektonik zu markieren. Bei genauer Betrachtung (mit der Lupe) und beim Vergleich mit der geologischen Übersichtskarte kann man erkennen, daß im Bereich der Rheinischen Geofraktur — selbst in der Basaltdecke des Vogelsbergs — Brüche in ungefähr N—S-Richtung sowie im varistischen Streichen diesem N 25—30° E streichenden System zugeordnet sind. Man vergleiche die Kluft-Diagramme in Abb. 5.

und Solling, so daß das niederhessische, an Basaltkuppen so reiche Tertiärbecken von Ziegenhain bis Kassel als Zwischenstück zwischen dem Leinetalgraben und der mittelhessischen Tertiärsenke mit dem Horloffgraben und seiner nördlichen Verlängerung bis zur Amöneburg erscheint.

Solche Versetzung von ziemlich genau N—S verlaufenden Grabenabschnitten nach Osten an NE streichenden Bruchsystemen ist auch in der südlichen Fortsetzung des Rheingrabens mehrfach zwischen Mittelmeerküste und Vogesen im Zuge der Rhône, Saône und Bresse und im kleineren Maßstab schließlich im Rheingraben selbst zu beobachten. Sein gewundener um Nordsüd und Nordnordost pendelnder Verlauf dürfte hierauf zurückzuführen sein. Auch an den anderen großen Gräben der Erde, insbesondere in den ozeanischen Böden (MENARD 1964, HESS 1962), ist die prinzipiell gleiche Strukturierung der Kruste festzustellen. Somit gehören solche Versetzungen zweifelsohne zu den fundamentalen Problemen der Riftbildungen der Erde. Die Strukturen in der Fortsetzung des Rheingrabens durch Hessen haben deshalb mehr als nur lokale Bedeutung.

2. Die morphologische und strukturelle Grabenfortsetzung

Während am Oberrhein die Morphologie und Geometrie des Grabens eindrucksvoll korrelieren, erscheinen die morphologischen und strukturellen Züge in der Grabenfortsetzung kompliziert und unsicher, nicht zuletzt deswegen, weil prägnante morphologische Züge sowie tektonische Linien von ihr wegzuführen scheinen.

Anders als am Oberrhein bilden hier in der Grabenfortsetzung durch über hundert Millionen Jahre getrennte Formationen die Flanken des von tertiären, wenn auch relativ geringmächtigen Sedimenten erfüllten Grabens: Im Westen ist es das aus Devon und Karbon aufgebaute Rheinische Schiefergebirge, im Osten hauptsächlich der von Basalt weithin bedeckte Buntsandstein. Nur ganz im Norden an der Werra (Richelsdorfer Geb. und Werragrauwacken Geb.) und im Harz tritt wieder Paläozoikum auf, dessen Schollengrenze als Leitlinien der Grabentektonik aufgefaßt werden könnten.

Die Basaltdecken verhüllen darüber hinaus weithin die tertiären Sedimentfüllungen und die sie begrenzenden Linien der Schollentektonik, die wiederum engstens verknüpft erscheint einerseits mit den Furchen der schmalen Gräben, Sättel und Mulden Niederhessens und der Egge in der gerade durch sie vielfach zerlegten Buntsandsteintafel, andererseits mit den genau in der Fortsetzung der Odenwald-Randverwerfung in N—S-Richtung bis zur Egge verlaufenden Abbrüchen des Rheinischen Schiefergebirges. Aus diesem N—S-Rand bricht nur die Kellerwald-Scholle aus. So ist denn bis heute Lage und Verlauf der Grabenachse und der eigentlichen Grabenränder unbekannt geblieben. Dies führte letztlich zur Vorstellung von einem senkenartig auslaufenden und verflachenden Graben, der außerdem mit reichem Vulkanismus verknüpft ist, mehr als jeder andere Abschnitt in der unseren Kontinent durchschneidenden Bruchzone.

Allenfalls zwischen dem 880 m hohen Feldberg im Taunus und dem fast 780 m hohen Hoherodskopf und Taufstein im Vogelsberg tritt der Graben auch morphologisch noch klar als Einsenkung hervor. Ähnlich der Breite des Oberrheingrabens ist auch hier eine Breite von 20—30 km zwischen dem devonischen Taunusrand und der Buntsandsteinkante bei Ranstadt und Ortenberg zu verzeichnen. Im paläozoischen Fundament zwischen Bad Homburg und der bei Bad Vilbel hochgekippten Rotliegendscholle, d. h. in der genauen Fortsetzung der Grabenachse vom Oberrhein in den Horloff- und Nidda-Graben beträgt die Breite nur 10 km. Die von hier an nach Norden hin um hunderte von Millionen Jahre verschieden alten Formationen an den Grabenflanken — das Paläozoikum auf der einen, das Mesozoikum auf der anderen Seite — weisen aber eher auf eine von dem großen Vulkan Vogelsberg und den nordhessischen Vulkanreihen verdeckte Verwerfung hin als auf eine Grabenstruktur, zumal der Raum im Osten des Schiefergebirges eine schon während und seit der varistischen Faltung ununterbrochen andauernde Senkung, das Gebirge im Westen dagegen eine das Achsengefälle der Falten übersteigernde Heraushebung erfahren hat. Dafür spricht nicht zuletzt auch der Anstieg der Basisfläche des Tertiärs aus einigen hundert Metern Tiefe unter NN im Rhein-Main-Gebiet bis auf die in 300 und 400 m üb. NN hochliegenden Oberflächen der Buntsandsteintafel nördlich des Vulkangebietes. In ihr geben die vielfältigen Bruchstrukturen der Niederhessischen Gräben und Senken und die Aufreihung von vulkanischen Förderschloten in den an Basaltkuppen und -deckenresten reichen Landschaften keine wirklich prägnante, den Graben kennzeichnende Gliederung. Auch in der Fortsetzung der Senke bis zum Harzrand mit seiner — wie am Taunusrand! — nach Westen bis zum Egge-Gebirge reichenden Ausweitung auf 60 km ist die morphologische Form des Grabens gegenüber dem Oberrheingraben nur kümmerlich entwickelt, auch wenn man den Solling — wie den Odenwald — als Mittelhorst der insgesamt breiteren zugehörigen Graben-Horst-Zone, d. h. dem ganzen über 100 km breiten Riftsystem zuordnet.

3. Grabenfüllung und -begrenzung

Daß aber in Hessen ein echter Graben, wie am Oberrhein, und seine direkte Fortsetzung mit geologisch hohen Ost- und Westflanken vorliegt, haben neuerdings Bohrungen bei Friedberg, Wohnbach und Gießen, im Niddatal bei Rainrod, Nidda und Ranstadt und im Niddertal bei Hirzenhain und Gedern erwiesen. Der bei Ranstadt und Hirzenhain in 100 und 200 m über NN angetroffene Buntsandstein unter rd. 100 m Basalt- und Tuffdecken wurde bei Rainrod in rd. 260 m unter NN erbohrt (Abb. 2). Im zentralen Vogelsberg bei Bermutshain liegt seine Oberfläche in fast 600 m üb. NN, so daß sich hieraus eine gestaffelte Verwerfung mit rd. 800 m Sprunghöhe ergibt. Die vermutlich auch im zentralen Vogelsberg an der Basis der Vulkanitdecke liegenden phonolithischen und trachytischen Gesteine in fast 700 m üb. NN deuten sogar eine Sprunghöhe von nahezu 1000 m an. Nach den seismischen Befunden (German Research Group 1964) liegt die Ober-

Abb. 2. Geologisches Profil durch die Rheinische Geofraktur-Zone in Mittelhessen.

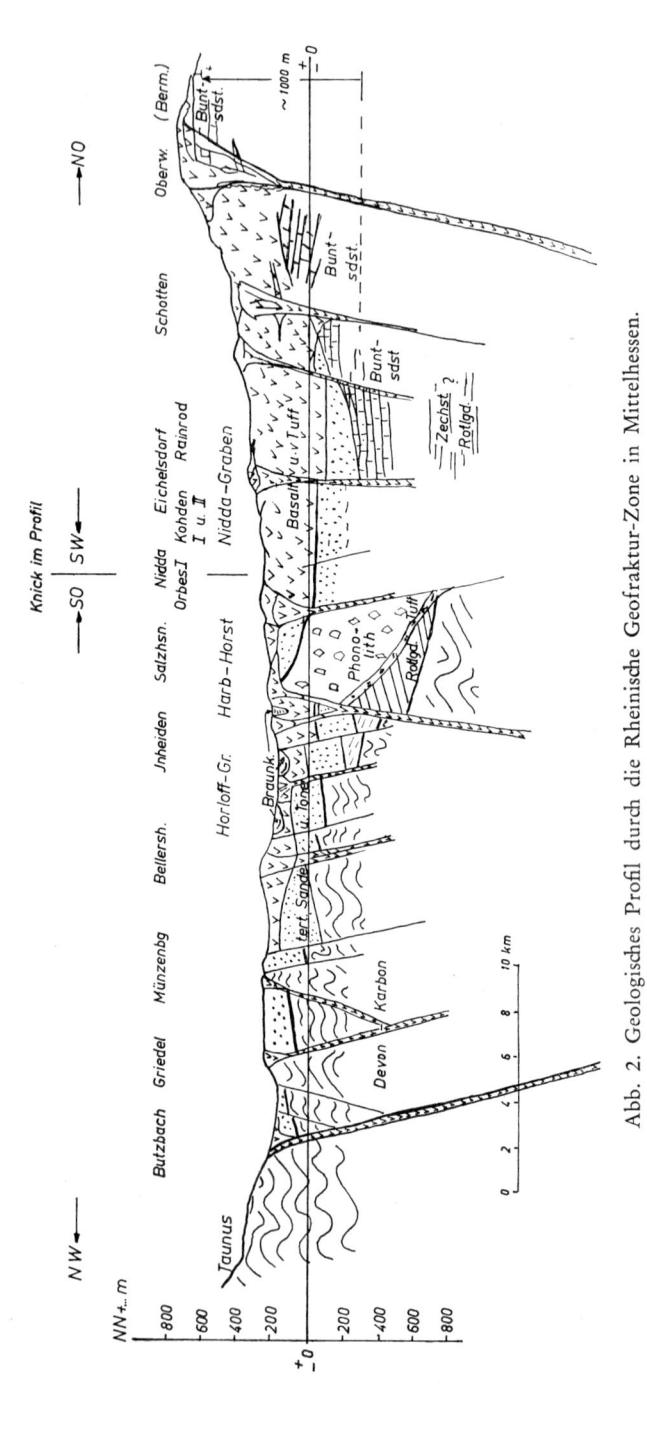

fläche des varistischen Gebirgskörpers dort in etwa 600 m unter NN. Gegenüber der Lage im Taunus ist sie somit rd. 1500 m abgesenkt. In seiner mächtigen Decke wurden auf der hochliegenden Grabenschulter bei Hirzenhain unter rd. 153 m Basalt und Tuff und 52 m tertiären Tonen und Sanden rd. 500 m mittlerer und unterer Buntsandstein erbohrt, bei Ranstadt über 150 m Zechstein und über 260 m Rotliegend. Das darunter liegende und bei Erbstadt an der Naumburg anstehende Karbon wurde in Bad Vilbel bei 115 m Tiefe erbohrt, nordwestlich von Altenstadt in 372 m Tiefe noch nicht erreicht. Die tertiäre Sedimentdecke am westlichen Rand der östlichen Grabenschulter hat nur etwa 30—50 m Mächtigkeit, weiter östlich dagegen über 80—100 m und zudem wieder marinen Charakter, wie die Bohrung bei Gedern ausweist (Chatt/Rupel, nach freundlicher Bestimmung der Mikrofauna von Herrn Doebl). Im Graben selbst wurden zwischen Nidda und Schotten (nach den freundlicherweise durchgeführten Bestimmungen von den Herrn Dr. Doebl und Dr. v. d. Brelie) aquitane und oligozäne Sande, Tone, Braunkohle und Dysodil unter 150 m und 300 m Vulkaniten in über 100 m bis 300 m Mächtigkeit angetroffen, ebenso bei Friedberg. Im Horloffgraben (Bohrungen Wohnbach, Römerstraße und Inheiden; Schenk 1957) waren bei 200 und 300 m Bohrtiefe die Schichten des Aquitans und Chatts noch nicht durchsunken, während unfern am Grabenrand bei Oppershofen und in Münzenberg Devon ansteht.

Die am Westrand des Grabens in Lich am Höhler Berg-Horst anstehenden aquitanen und oligozänen Tone, Braunkohlen und Sande wurden nur wenige Kilometer weiter östlich unter 108 m Basalt und Tuff erbohrt, während sie im Gebiet von Bad Nauheim, Oppershofen und Münzenberg in über 200 m NN dem Devon auflagern. Im Gebiet von Gießen stehen sie in 160—180 m und 200 m üb. NN mit über 150 m Mächtigkeit zweifellos jenseits des westlichen Grabenrandes zutage an (bei Steinbach wurde bereits in rd. 20 m Tiefe Devon erbohrt!). Hier steht man bereits in einer Ausbuchtung der Tertiär-Senke, nämlich auf einer der N—S sich erstreckenden Teilschollen mit ebensolchen noch kleineren Gräben und Horsten, wie z. B. die Aquitan-Vorkommen von Garbenteich, Wieseck und Allertshausen ausweisen.

Vom Rhein her lassen sich also die Grabenränder bis in den Basaltrand (Abb. 1) hinein verfolgen:

Der Westrand verläuft von Bad Dürkheim über Nierstein nach Bad Soden, Bad Homburg, Bad Nauheim, Münzenberg (CO_2- und Salzwasser-Linie!) und östlich Lich. Die hier ansetzende Ausbuchtung schließt das mehrfach in kleine Horste und Gräben gegliederte Becken von Gießen und Amöneburg mit dem Rupelton und marinen Aquitan von Allertshausen (Schenk 1964) und Homberg/ Ohm (Gramann 1958) ein. Hier an der Ohm und Kleen und Wiera kennzeichnen die Grenzen der hochliegenden Schollen von Unterem Buntsandstein mit den kleinen paläozoischen inselartigen Aufbrüchen bei Ruhlkirchen die Abgrenzung des Tertiärbeckens von Neustadt und der Versatz des Momberger Grabens wieder deutlich den westlichen Grabenrand und seine Staffeln, die das Becken von Neustadt mit Melanien- und Rupelton, wie die Bohrungen bei Homberg/Ohm und Wahlen (Schenk 1961) zeigen, dem Graben einbeziehen.

Die östliche Grabenrandverwerfung zieht durch Frankfurt (Rotliegend-scholle im Westhafen) — Bad Vilbel über die Naumburg nach Ranstadt. In der Vulkanitplatte des Vogelsberges selbst ist sie angedeutet durch die bis östlich von Alsfeld reichenden Aufreihungen von Vulkankuppen zwischen den in NNE Richtung verlaufenden Tälern der Nidda und Nidder. Sie quert den Paläovulkan in seinem Zentralgebiet, dem Oberwald, und begrenzt in Staffeln das Alsfelder Becken nach Osten.

Viel deutlicher als die geologischen Karten hat dies die geomagnetische Vermessung der letzten 10 Jahre im Vogelsberg ergeben. Die über 2000 geomagnetischen Anomalien, die dabei festgestellt wurden (Abb. 3), sind nichts anderes als Spalten, d. h. Basaltgänge, Förderschlote, Diatreme usw., deren Aufreihungen die lineare Eruptionstektonik des Vulkans dokumentieren. Auch nach den Befunden ihrer statistischen Untersuchung sind verschiedene Bruchsysteme deutlich unterscheidbar (SCHENK 1972). Hier ist wichtig, daß solch eine Aufreihung mit 3 km

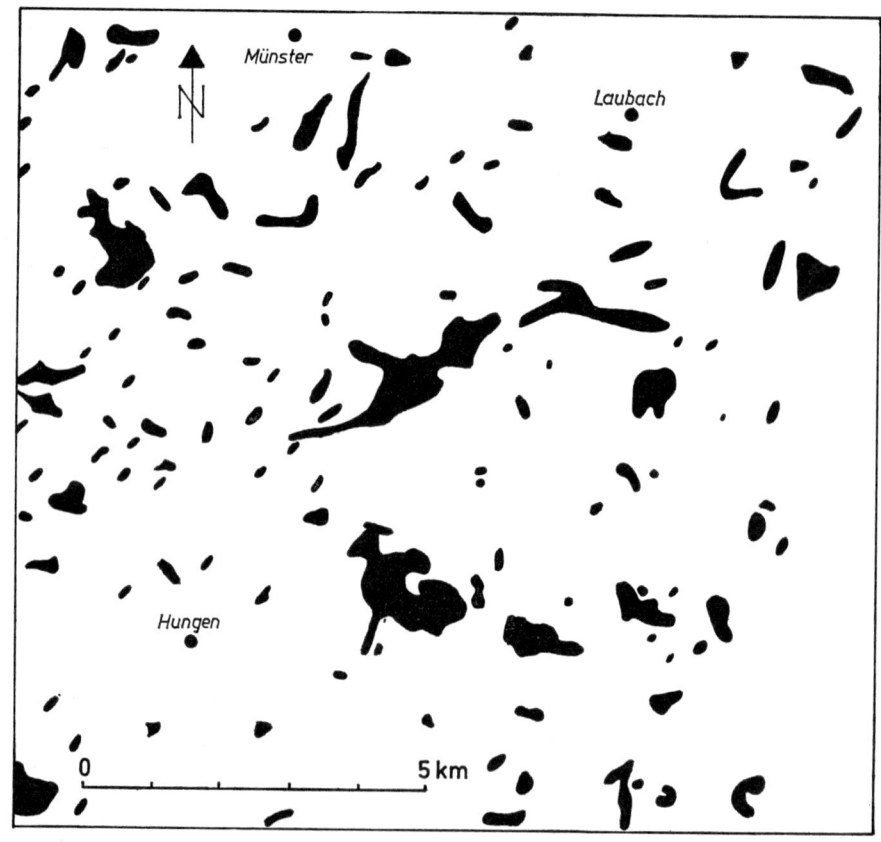

Abb. 3. Die erdmagnetischen Anomalien am westlichen Rand der Rheingraben-Fortsetzung von Münzenberg nach NNE. Eingetragen sind sowohl positiv als auch negativ signifikante △ Z-Anomalien.

Breite zwischen Ranstadt und Ortenberg ansetzt, das Niddatal auf seiner südlichen Seite begleitet, den Oberwald zwischen Taufstein und Ulrichstein quert und östlich von Alsfeld den Basaltrand erreicht, dann auf der Buntsandstein-Hochfläche fortsetzt und genau auf die schmalen Grabenzüge von Ottrau — Weißenborn — Raboldshausen zielt. Damit sind die morphologischen, geologischen und vulkanologischen Befunde ergänzt und bestätigt worden.

In gleicher Weise setzt bei Bad Nauheim — Münzenberg eine über Lich und Reinhardshain bis Homberg/Ohm ziehende 3—5 km breite Aufreihung von Anomalien an, die den westlichen Grabenrand markiert (Abb. 3). Die sich hieraus ergebende Grabenseite im Vulkangebiet selbst beläuft sich auf rd. 20 km und die vom Main aus gerechnete Länge auf fast 100 km.

Die Achse des Grabens (s. Abb. 5) zieht von Höchst am Main über Wollstadt und Hungen nach Romrod bei Alsfeld. Ihre Verlängerung trifft und schneidet das niederhessische Vulkangebiet bis in den Winkel zwischen Eder und Fulda, quert den Kaufunger Wald und läuft dann in den Göttinger Graben ein und westlich am Harz entlang. Damit sind auch hier bereits die Grabenrandverwerfungen angedeutet:

Auf der östlichen Seite sind sie scharf markiert durch die N 30° E streichenden Gräben von Raboldshausen und den Devonaufbruch von Mühlbach, den Grabenzug von Altmorschen bis Großalmerode. Damit ist bereits die Ostflanke des Leinetalgrabens und der Harz erreicht.

Auf der westlichen Seite zieht die Grabenschulter von Homberg/Ohm über Neustadt, Treysa, Wabern, Stellberg und Kaufungen östlich an Kassel vorbei in die Westflanke des Leinetalgrabens. Die Staffel von Fritzlar und die Ausbuchtung nach Norden bis Kassel liegen wie das Mainzer, Gießener und Amöneburger Becken jenseits der Hauptwestrandverwerfung, eben in einer jener problematischen N—S Grabenschollen.

Nördlich des Harzes nehmen die typisch „rheinischen" Aufreihungen der Salzstöcke über Niedersachsen und Schleswig-Holstein (Abb. 4) die Linienführung bis an die Ostsee auf (siehe Geotektonische Karte von Nordwestdeutschland, 1946).

Damit ergibt sich, wenn man die Ausbuchtungen als sekundäre tektonische Erscheinungen auffaßt, eine fast geradlinige Geofrakturzone von der Mittelmeer- bis zur Ostseeküste in N (25—) 30° E von rd. 1500 km Länge quer durch den Kontinent.

4. Das Alter der Geofraktur

Die Entstehungsgeschichte der Geofraktur reicht nach den Befunden in Hessen offensichtlich bis in das Paläozoikum zurück. An ihr tauchen die Achsen der varistischen Falten nach Osten ab. In ihrem Bereich liegen die Schwerpunkte des varistischen Basalt(Diabas)vulkanismus. In ihrer Zone erfolgte seit der varistischen Faltung die Absenkung der mitteldeutschen Zechstein-Triasscholle und die Heraushebung des Rheinischen Schiefergebirges und Harzes, die Entwicklung des speziellen saxonischen Formenschatzes der Falten, Gräben und Horste als echte Vorläufer

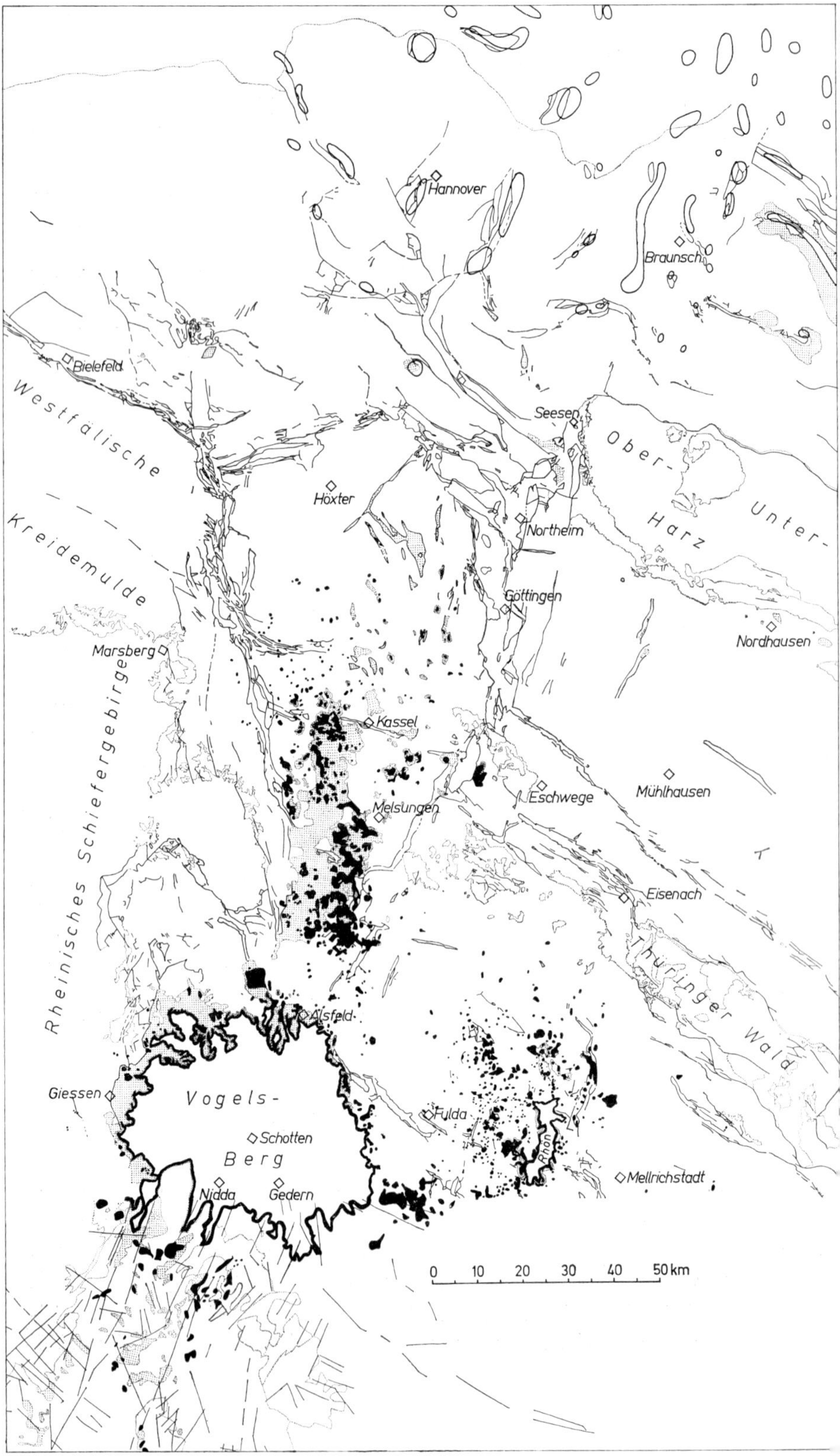

Hannover
Braunsch.
Bielefeld
Westfälische
Kreidemulde
Seesen
Ober-
Unter-
Harz
Höxter
Northeim
Göttingen
Nordhausen
Marsberg
Rheinisches Schiefergebirge
Kassel
Eschwege
Mühlhausen
Melsungen
Eisenach
Thüringer Wald
Alsfeld
Giessen
Vogels-
Berg
Schotten
Nidda
Gedern
Fulda
Rhön
Mellrichstadt
0 10 20 30 40 50 km

Die Fortsetzung des Rheingrabens durch Hessen

Die Bruchsysteme im Buntsandstein – Fundament in der Umrahmung des Vogelsberg – Vulkangebietes

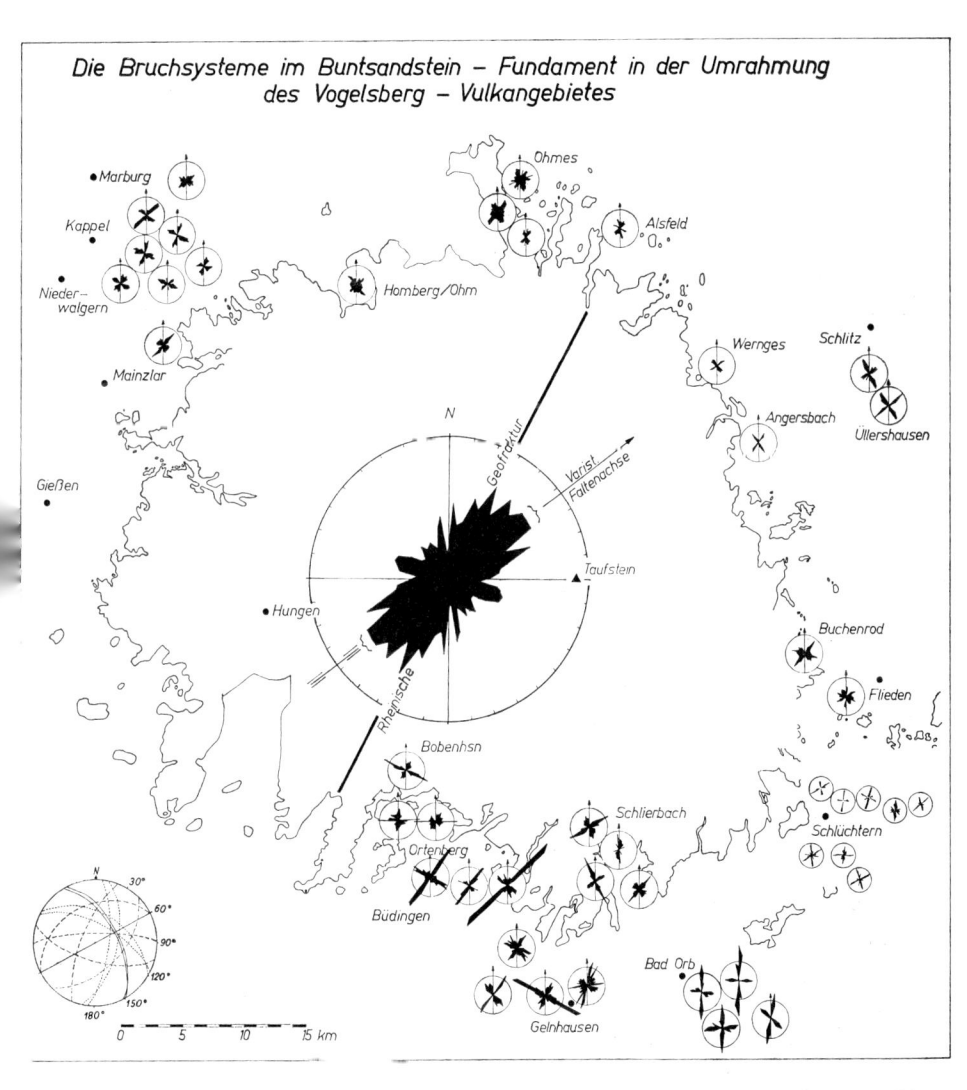

Abb. 5. Die Bruchsysteme (nach Kompaß-Messungen) im Buntsandstein Fundament der Randzonen des Paläovulkans Vogelsberg und nach den Richtungen geomagnetischer Anomalien im Basaltgebiet. Zum Vergleich ist auch die varistische Tektonik (n. SCHENK 1936) eingezeichnet (links unten).

Abb. 4. Die Hauptstrukturen der Grabentektonik in Nordhessen mit den versetzten saxonischen Gräben und Schollenkanten sowie den verbogenen Achsen der saxonischen Falten und der Fortsetzung des Grabens im Bereich der niedersächsischen Salzstöcke.

der tertiären Tektonik und schließlich der Transgressionen des mitteloligozänen und untermiozänen Meeres. Ihre Linienführung bestimmten die saxonischen Schollenkanten als Sättel und Mulden sowie schmale Gräben und breite Becken (Abb. 4). Letztlich führten im Mitteloligozän und Unteren Miozän die gleichbleibenden Bewegungstendenzen zu den deutlichen Konturen des Grabens, die um so schärfer werden — und das erscheint wichtig —, je näher sie dem alpidischen Faltenkörper liegen. Die dabei verstärkte Bruchbildung und Einsenkung der Kruste zum Graben in der Verbiegungszone ermöglichte nicht nur die Transgression des Rupeltons und Aquitan-Meeres in einen schmalen Verbindungskanal zwischen Nord- und Südmeer, sondern schließlich auch den Durchbruch des Magmas, die Entstehung der Vulkane im ausgehenden Aquitan und ihre Aktivität bis ins Obermiozän.

Wie am Oberrhein (Kaiserstuhl) fällt dies auch im Vogelsberg nach den bisherigen Altersbestimmungen (HORN, LIPPOLD & TODT 1972 u. 1973, KREUZER u. a. 1972 u. 1973) in die Zeit vor rd. 12 bis 18 Mill. Jahren. Damit ergibt auch diese Methode eine Zweiphasigkeit der vulkanischen Aktivität. Sie bestätigt die vulkanologischen Befunde, nach denen die Basaltdecken allgemein durch eine mächtige Tuffdecke getrennt sind.

Die Einschaltung von Kalksteinlinsen und -knollen, wie sie in den Bohrungen bei Friedberg, Rodheim, Wohnbach und Römerstraße (SCHENK 1957) in Tuffiten gefunden wurden, ermöglichte andererseits die sichere stratigraphische Datierung der jüngeren, mit dem Beginn des Vulkanismus verknüpften tektonischen Phase der Einsenkung des Grabens zur aquitanen Meeresstraße.

5. Die tektonische Gliederung und Richtung des Grabens und seines Untergrundes

Schon die Gliederung dieses Meeres in Buchten und Verengungen längs der Rhein. Geofraktur (s. Abb. 1 und 4) gibt das Mosaik der Schollenzerlegung zu erkennen. Die Senken vom Main bis zum Harz sind N—S ausgerichtet und selbstverständlich auch die Horste, wobei die Ränder des Schiefergebirges ihrer vorgegebenen varistischen Bruchstruktur entsprechend in ganz charakteristischer Weise im Zickzack vor- und zurückspringen und damit selektiv alte Bewegungsflächen in den von Bruchsystemen zerlegten varistischen Faltenkörpern reaktivieren (Abb. 6). Dessenungeachtet setzt sich die jüngere Tektonik des Grabens mit neuen Brüchen und Bruchrichtungen durch, indem sie nicht nur die Richtung des Grabens und seiner Flanken bestimmt, sondern auch neue Zerlegungen schafft, gewissermaßen als Resultierende in N 45° und 5° und 20° E u. a. zwischen der Generalrichtung des Grabens in N 30° E und den varistischen Bruchsystemen (s. a. Abb. 1 und 7). Damit müssen auch die als Struktur problematischen Ausbuchtungen in N—S mit ihrer Füllung aus tertiären Sedimenten mit der Einsenkung der Grabenscholle entstanden sein. Das ergibt sich auch aus den geologischen Kartierungen bzw. den stratigraphischen Abfolgen mit ihren Lücken.

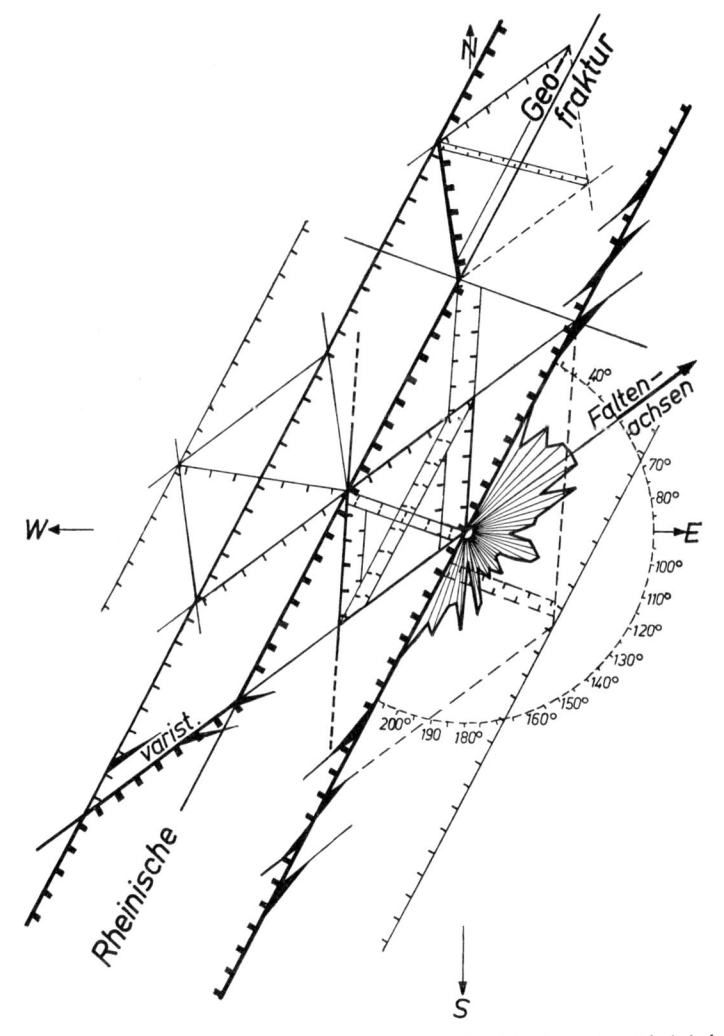

Abb. 6. Schema zur Funktion der Bruchsysteme (gemäß Abb. 5) in der Rheinischen Geofraktur.

Bei der Bildung dieser Brüche bleiben nur die Bruchsysteme im Streichen der varistischen Faltenachsen und die in sehr spitzem Winkel zu ihnen stehenden Scherflächensysteme der Schieferung, wie am Taunussüdrand, für das Strukturbild des Grabens und seine scheinbaren Versetzungen wie im Rhein-Main-Gebiet signifikant (Abb. 5).

Dominant in den geologischen Karten ist wiederum die sogenannte rheinische Richtung, d. h. die N—S Richtung, in der die Grabenabschnitte, Teilgräben und -Horste vom Mittelmeer bis Skandinavien ausgerichtet sind. Die Karten mit den

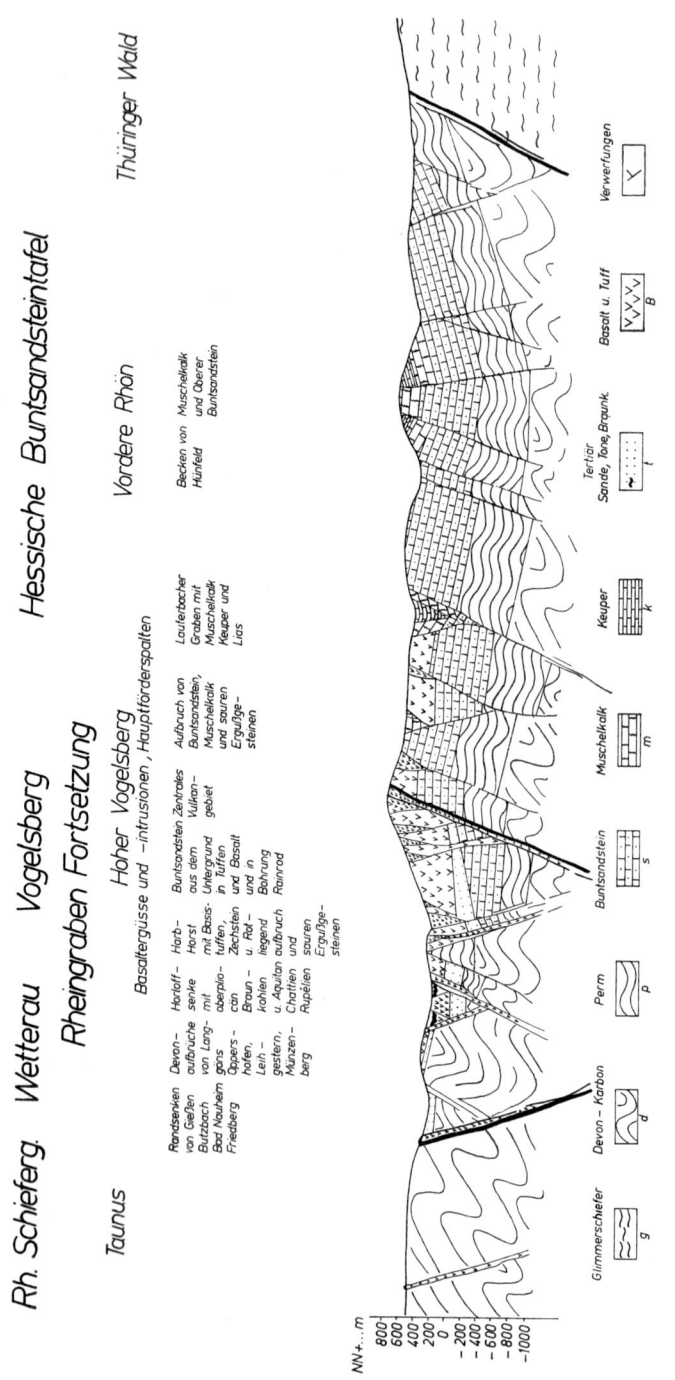

Abb. 7. Profil durch die Rheinische Geofraktur in Hessen.

Befunden gravimetrischer und seismischer Untersuchungen bestätigen und sichern, daß die Geofraktur nicht N—S, sondern N 30° E streicht (Abb. 7).

Genau in der Lage und Richtung der rheinischen Geofraktur, wie sie beschrieben wurde, zieht die Achse der geringsten Schwereanomalien von Kelsterbach-Höchst durch den Nidda-Horloff-Graben quer durch den Vogelsberg bis Alsfeld und dann kaum verbogen am Südostrand des niederhessischen Tertiärbeckens entlang, von wo dann die Isogammen zum Göttinger Graben führen.

Während die westliche Grabenschulter vielfach gegliedert ist, zieht die östliche der Schwereachse parallel vom Sprendlinger Horst über Ranstadt, Ulrichstein und Eifa bei Alsfeld, am Raboldshausener und Altmorschener-Großalmeroder Graben entlang bis Landolfshausen im Osten von Göttingen. Ihr wiederum parallel verläuft mit straff gerafften Isogammen die bekannte negative Schwereanomalie von der Rhön südlich am Spessart und Odenwald (und nicht an der Kinzig!) entlang in N 30° E bis Heidelberg und mündet hier in den Oberrheingraben. Das südöstlich anschließende Schwerehoch ist gekennzeichnet durch den Grenzverlauf des Unteren Muschelkalkes der Süddeutschen Großscholle. Sprengseismische Untersuchungen (German Res. Group 1964, GIESE 1971) zeigen, daß M- und C-Diskontinuität und auch die Oberfläche des varistischen Gebirges die gleiche Gliederung bzw. Linienführung haben und die Strukturen der Oberkruste mit denen des tiefen Untergrundes korrelieren.

In ganz charakteristischer Weise wird der Verlauf dieser N 30°E-Linien des Untergrundes jeweils im nordwestlichen Vorland begleitet, unterbrochen und gegliedert von Ausbuchtungen, die einesteils nach N weisen und anderenteils dem varistischen Streichen in N 55° E sowie quer dazu (N 145° E) folgen. Damit kommen wir auf das eingangs hervorgehobene Problem der Schollengliederung und -bewegung und schließlich der „seitlichen" Versetzung des Grabens und seiner Teilgräben zurück.

6. Die Statistik und Geometrie der Brüche und Spalten

Die an die varistische Faltung gebundene Bruchtektonik ist dadurch ausgezeichnet, daß die Klüfte und Verwerfungen paarweise symmetrisch zur Faltenachse stehen und mit deren Neigung ebenfalls Streichen und Fallen ändern (SCHENK 1936), wie dies in Abb. 5 gezeigt ist. Die Klüftung des triadischen Deckgebirges weicht hiervon trotz mancher Übereinstimmung wesentlich ab, wie die Kluftdiagramme aus dem Buntsandstein Fundament des Vulkans in den Randzonen des Vogelsberges zu erkennen geben. Sie harmonieren vielmehr mit der kartierten Tektonik des Grabens. Die meisten Diagramme lassen als signifikante und dominante Richtung der varistischen Faltenachsen und Schieferung N 55° E erkennen sowie als Richtung der Grabenachse und Schultern N 30° E und ungefähr senkrecht dazu die ESE-Richtung von N 100—120° E. Weniger häufig, doch signifikant treten die um N—S streuenden eigentlichen rheinischen Richtungen auf, wobei die normal zu ihnen liegenden Richtungen sich ebenfalls ändern. Am wenigsten treten die varistischen Querbrüche (145° E) hervor, ohne jedoch zu fehlen.

Örtlich beherrschen sie sogar das Kluftbild, wie im Bereich des Lauterbach-Fuldaer Grabens.

Ein noch viel prägnanteres Bild von den Bruchbildungen geben die Spaltenfüllungen im Vulkangebiet selbst. Während die Monotonie der Basalt- und Tuffgesteine die geologische Auskartierung von tektonischen Störungen fast unmöglich macht, führte die statistische Behandlung der geomagnetischen Anomalien im Basaltgebiet (SCHENK 1972) zu einem überraschend klaren Bild von der Geometrie der mit Basalt gefüllten Förderspalten und -schlote, von ihrer Aufreihung und Anordnung zum Vulkanzentrum (s. Abb. 5, Mitte).

Zu diesem Zweck wurde das Azimut der langen Achsen von über 1100 geomagnetischen Anomalien gemessen, die Häufigkeiten der Richtungen in Klassen von 5° ausgezählt und graphisch dargestellt. Der Vergleich (Abb. 5) dieser Richtungen mit den ebenfalls statistisch ermittelten Kluftsystemen an NE abtauchenden Faltenachsen (SCHENK 1936) zeigt, wie auch die geologischen Karten, daß nur die varistische Streichrichtung, d. h. die Bruch- und Schieferungsflächen in Richtung der Faltenachse, und vielleicht das ihr zunächst gelegene und symmetrisch zugeordnete Kluftflächenpaar für die Spaltenbildungen und den Durchbruch der vulkanischen Schmelze maßgebend waren, aber quantitativ zurücktreten. Alle anderen Richtungen sind neu, d. h. im Zusammenhang mit dem Vulkanismus und infolgedessen auch mit der Grabenbildung genetisch verknüpft.

Signifikant und dominant sind im Diagramm (Abb. 5, Mitte) die in den hessischen Gräben bis zum Leinetal sichtbare Richtung des Grabens mit N (25—) 30° E sowie die dazu in fast rechtem Winkel stehende Querrichtung N 100—120° E. In ihrer Winkelhalbierenden liegen die in allen geologischen Karten hervortretenden und das tektonische Bild des Grabens beherrschenden „rheinischen" und „eggischen" Richtungen mit N 0—5° E und N 170—175° E. Sie sind hoch signifikant. Signifikant ist aber auch die Lücke zwischen ihnen, die nicht übersehen werden darf und einen Streubereich von N 350° über 0 bis 5° E ausschließt. Es müssen also zwei Systeme vorliegen, die der Kinematik der Grabenbildung dienten und möglicherweise auch zeitliche Unterschiede der Deformation fixiert haben.

7. Die Kinematik der Grabentektonik

Um von der Geometrie der Bruchflächen auf ihre Kinematik und die dynamische Beanspruchung schließen zu können, muß die geologische Karte zu Hilfe genommen werden. Alle in W—E verlaufenden Gräben im nördlichen Hessen entsprechen der Signifikanz von N 100—120° E als korrespondierendes System zu den N 25—30° E streichenden Gräben, bzw. den Schollenkanten, die die östliche Grabenrandverwerfung eindeutig und sicher markieren. Sie sind zweifellos schon früh angelegte Brüche, die eine spätere Deformation erlitten haben und erkennen lassen, daß die westlich gelegenen Schollen gegenüber der östlichen nach NNE bewegt worden sind. Am Lauterbacher Graben und seiner Fortsetzung im Momberger Graben (SCHENK 1961) ergibt sich ein relativer Vorschub nach Norden gegenüber den östlichen Schollen in der Größenordnung von 2—4 km (Abb. 4).

In der Borkener Grabenzone ist eine ähnliche Verschiebung meßbar, desgl. am Fritzlar-Homberger Störungssystem, an der Kasseler Grabenzone von Wolfshagen bis Großalmerode, an der Störungszone von Warburg — Hannoversch-Münden — Eichenberg u. a. tektonischen Details der hessischen Buntsandsteinplatte.

In gleicher Weise zeigen die saxonischen Achsen der Sättel und Mulden Verbiegungen, die einem schnelleren oder stärkeren Vorschub der Schollen westlich des Grabens entsprechen, so daß das Bild einer Horizontalflexur deutlich wird (Abb. 4).

Es fällt dazu ferner auf, daß die Braunkohlenlager im Horloffgraben und im niederhessischen Becken und ihre größten Mächtigkeiten nicht zentral im Graben liegen, sondern in den nordwestwärts gelegenen Randzonen von gekippten Schollen (s. SCHENK 1955 und NÖRING, UDLUFT & SCHENK 1957, UDLUFT & LANG 1956). Aus ihren fiederförmigen Staffelungen und möglicherweise als Fiederfalten zu verstehenden Deformationen (SCHENK 1955) und ihrem Abrücken von den tektonischen Störungen ergibt sich wiederum fast eine Gesetzmäßigkeit, die mit der relativ stärkeren Bewegung und Drehung der westlichen gegenüber den östlichen Schollenteilen, wie sie bereits beschrieben wurde, harmoniert.

Schließlich ordnen sich dem kinematischen und dynamischen Diagramm-Schema nicht nur die N—S-Gräben und -Horste, sondern auch die Basaltgänge und Förderspaltensysteme im Streichen der varistischen Achsen ein, und zwar als Fiederspalten, was vielfach schon aus der Form ihrer Füllung und ihrem unmittelbaren Kontakt, d. h. im Schnitt mit den Grabenrandverwerfungen, zu erkennen ist (SCHENK 1972). Dies sichert von sich aus einerseits die Formulierung und Deutung des kinematischen Befundes und nötigt andererseits zur Ableitung des geschilderten dynamischen Planes. Die N—S ausgerichteten Teilgräben und Horste innerhalb des Grabens sind Ausdruck eines von N und S gegeneinander gerichteten Druckes.

Die N—S-Richtungen sind die Resultierenden, an denen sich die Schollen, Teilgräben und -horste, Staffelbrüche und Schollenverschiebungen zwischen den Grabenrändern orientiert haben. Als Gesamtheit bilden sie in Form des Grabens mit seinen Horst-Flanken die Geofrakturzone.

8. Die Folgerungen

Der Graben enthüllt sich demnach als ein System von neben- und übereinander liegenden synthetisch und antithetisch eingesenkten Schollen, die in der Horizontalen zwischen dem System von spitzen Winkeln der Hauptscher- und Bewegungsflächen (N 30° E) und ihren Gegenstücken, den Querbrüchen (N 100—125° E) zugeordnet sind und — bei der dynamisch zwingenden Einfügung und Richtung — vorgegebenen Strukturen — hier dem varistischen Streichen — nach der Seite folgen. Dabei weichen sie bald in dieser, bald in jener Richtung von der Richtung der Geofraktur ab.

Ein besonderes Studium wird zeigen, ob die gesetzmäßigen Ostverschiebungen der Grabenabschnitte in der Geofrakturzone vom Mittelmeer bis Skandinavien (Wettersee und nicht nur Oslo-Graben) ein Gegensystem mit Westverschiebungen südlich des Äquators haben, wie das die ostafrikanischen Gräben anzeigen. Alsdann wäre vielleicht auf Corioliskräfte zu schließen, zumal das Strukturbild der Erdnähte — im Sinne von H. CLOOS (1948) — den experimentellen Befunden von KNETSCH (1965) an der rotierenden Tonkugel nahe kommt.

Man hat es lange als gesicherte Erkenntnis angesehen, daß der Graben die Form und der Ausdruck einer autonomen Dehnung der Erdkruste ist, und in letzter Konsequenz gefolgert, daß die Erde sich ausdehnt. Die Grabentektonik spricht aber vielmehr von Formen einer Kompression, von Zerscherung der Kruste, d. h. einfach von Bewegungen im Bereich von getrennten konsolidierten Großschollen und einem Zerbrechen der oberen Kruste infolge wechselsinniger Verbiegungen und Flexuren im tiefen Untergrund, die sich im Bereich der Moho abbilden. Dies kann nicht Ursache, sondern nur Wirkung sein von Vorgängen im Erdmantel, die dem Vulkanismus sowohl den Stoff als auch die Wege bereiteten, letzteres durch die Schollenbewegungen.

9. Literaturverzeichnis

Geologisches Landesamt, 1945—1946: Geotektonische Karte von Nordwestdeutschland. — Hannover.

German Research Group, 1964: Crustal Structure in Western Germany. — Z. Geophysik, 30, 209—234.

GIESE, P. & STEIN, A., 1971: Versuch einer einheitlichen Auswertung tiefenseismischer Messungen aus dem Bereich zwischen der Nordsee und den Alpen. — Z. Geophysik, 37, 237—272.

GRAMANN, R., 1958: Das Oligozän nördlich des Vogelsbergs, insbesondere im Amöneburger Becken. — Diss. Marburg.

KREUZER, H., BESANG, C., HARRE, W., MÜLLER, P., ULRICH, H. J. & VINKEN, R., 1972: K/Ar-Datierungen an jungtertiären Basalten aus dem Vogelsberg und aus dem Raum zwischen Kassel und Göttingen. — Beiheft Fortschr. Mineral.

LIPPOLT, H. J., HORN, P. & TODT, W., 1972: Kalium-Argon-Altersbestimmungen an tertiären Vulkaniten des Oberrheingrabens. — Eclogae geol. Helv., 65/1, 129—154, 4 Textfig. und 2 Tab., Basel.

MENARD, H. W., 1964: Marine Geology of the Pacific. — McGraw-Hill.

NÖRING, E., UDLUFT, H. & SCHENK, E., 1957: Das Tertiär in der Niederhessischen Senke, im Vogelsberg und in seinen Randgebieten sowie im Untermaingebiet. — „Hydrologische Übersichtskarte 1:500 000 — Erläuterungen zu Bl. Frankfurt" von H. UDLUFT, F. NÖRING & E. SCHENK, Bundesanstalt für Landeskunde, Remagen.

SCHENK, E., 1936: Prinzipielle Bemerkungen zu statistischen Methoden in der Tektonik mit einigen Bespielen aus dem Rheinischen Schiefergebirge. — Zbl. Miner. etc., Abt. B., 4, 129—139, Stuttgart.

— 1961: Ergebnisse einer Bohrung in das ältere Tertiär und den Muschelkalk bei Wahlen und die Tektonik am Nordrand des Vogelsbergs. — Notizbl. Hess. L.-Amt f. Bodenforsch. 98, 310—319, 2 Abb., Wiesbaden.

Das Schollenmosaik des nördlichen Michelstädter Grabens

von

E. Backhaus, A. Rawanpur und M. Zirngast

Mit 4 Abbildungen im Text

The Block-Mosaic of the Northern Michelstadt Graben

Abstract: 20—30 km eastward of the eastern margin of the Rhine Graben the Michelstadt Graben is down-faulted into a platform of Lower Triassic sediments. The northward prolongation of this graben has been doubtful before.

By special mapping of the Bunter under a modern scope the northern extension of the graben segment has been investigated. Bloc diagrams illustrate a general north-southerly trend of the graben. Along the western main fault there is a downthrow of the graben segment of about 350 m and at the eastern margin of 150 m. Confined between the both main faults a mosaic of fault blocks has been observed, most of the single blocks are not larger than 1 km².

The beginning of subsidence is considered to be in connection with the formation of the Upper Rhine Graben in Tertiary times. A continuation of block faulting during the Pleistocene is assumed.

Einleitung

Der Odenwald, die kristalline, östliche Schulter im Nordabschnitt des Oberrheingrabens, wird an einigen Stellen von geringmächtigem Rotliegend, wenn dies fehlt, von wenigen Metern Zechstein und bei dessen stellenweisen Ausfall (z. B. auf der Odenwald-Insel, vgl. BACKHAUS 1961 und 1965) vom Buntsandstein überlagert. Durch die känozoische Herauswölbung des Odenwaldes ist die Schichtenfolge generell um 1—2° nach SE geneigt.

Parallel zur östlichen Rheingraben-Verwerfung (Bergsträßer Rand des Odenwaldes) verläuft im von Süden nach Norden sich vergrößernden Abstand von 10—20 km die Otzberg-Zone, die mit Intrusiva besetzte Trennungslinie zwischen dem variszischen Bergsträßer Odenwald und den variszisch überprägten Böllsteiner Gneisen im Osten (s. Abb. 1).

In einem weiteren Abstand von 5—15 km nach Osten folgt im Buntsandstein-Odenwald der generell rheinisch gerichtete Michelstädter Graben. In ihm sind triassische Schichtenfolgen bis in den Grenzbereich Unterer/Mittlerer Muschelkalk gegen Buntsandstein abgesunken.

Der zentrale Teil des Michelstädter Grabens wurde mit 2—4 km Breite und einer Längserstreckung von 15 km bereits bei den ersten Kartenaufnahmen (KLEMM 1898) erkannt. Nach Süden scheint sich der Graben herauszuheben, an seinem Nordende schien ein Riegel von Schichten des Mittleren Buntsandsteins den Graben zu begrenzen.

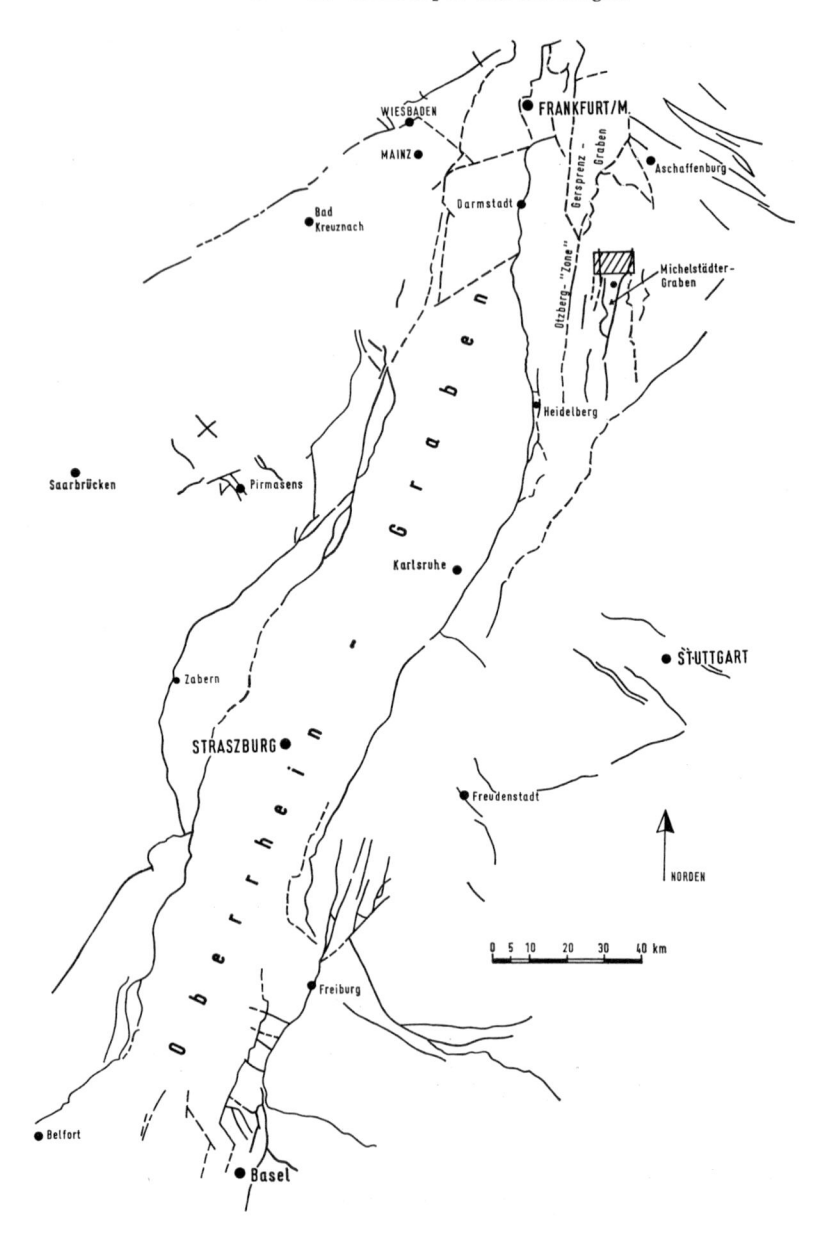

Abb. 1. Die Lage des Arbeitsgebietes (Abb. 2 u. 3) am Nordende des Michelstädter Grabens (Rechteck mit Schraffen) im Bezug zum Oberrheingraben.

Fig. 1. Map of the investigated area (see fig. 2 and 3) at the northern part of the Michelstadt Graben (hatched rectangle) in relation to the Rhine Graben.

Neuaufnahme Nordende des Michelstädter Grabens

Auf der Grundlage einer modernen Gliederung des Buntsandsteins wurde das nach Norden an den Graben anschließende Gebiet, SE-Quadrant der TK 1:25 000, Blatt 6219 Brensbach und SW-Quadrant der TK 6220 Wörth geologisch neu kartiert.

Nach den neu gewonnenen geologischen Karten wurden Schichtlagerungskarten dieser Gebiete bezogen auf die Grenze Rohrbrunn-/Geiersberg-Folge (sm R/G) entworfen und auf der Grundlage dieser Karten die beiden perspektivischen Blockbilder der Abb. 2 und 3 gezeichnet.

Die Grenze Rohrbrunn-/Geiersberg-Folge liegt 80 m über der Basis des 130 m mächtigen Mittleren Buntsandsteins. Sie ist im Kartiergebiet morphologisch, petrographisch und farbmäßig gut faßbar. War dieser Schichtenbereich bereits abgetragen, wurde die Grenze extrapoliert; dabei wurde eine erarbeitete su-Mächtigkeit von insgesamt 270 m zugrunde gelegt.

Das Gebiet ist in ein Mosaik zahlreicher, zumeist nur 1 km² großer Schollen zerbrochen, deren Anordnung ziemlich wahllos anmutet. Zwischen einem herausragenden Komplex im Westen (s. Abb. 2, z. B. Schollen 7, 13 u. 21) — auf der geologischen Karte weitgehend durch Bröckelschiefer, der direkt auf Böllsteiner Gneis aufliegt, erfaßt — und einer gleichfalls relativ hervortretenden Zone im Osten (s. Abb. 3, z. B. Schollen 33, 41 und 42) — auf der geol. Karte schon Oberer Buntsandstein — sind die anderen Schollen unterschiedlich stark abgesunken.

Die extrapolierte Grenze sm R/G liegt im Westen (Scholle 13, Abb. 2) bei ca. 660 m ü. NN, in der ihr nach Osten vorgelagerten Scholle 10 bei lediglich

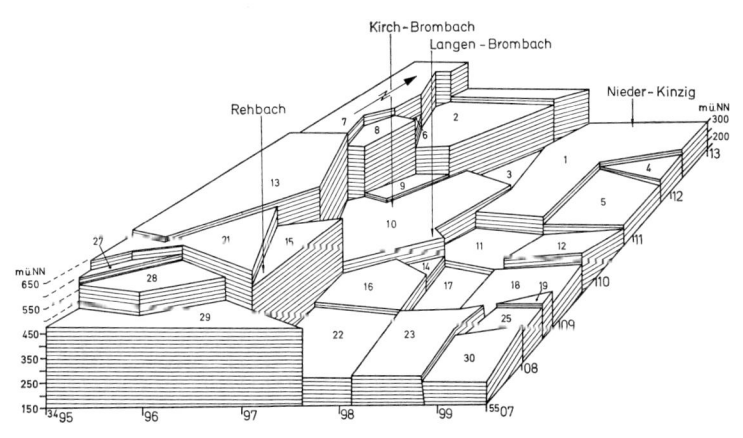

Abb. 2. Perspektivisches Blockdiagramm des Westteiles der Grabenfortsetzung im SE-Quadranten der TK 6219 Brensbach. Gezeichnet auf der Grundlage einer Schichtlagerungskarte der Grenze Rohrbrunn-/Geiersberg-Folge (Mittlerer Buntsandstein).

Fig. 2. Perspective block-diagram of the western part of the northern continuation of the Graben situated in the SE square of official map 6219 Brensbach. Constructed by means of a contour map of the top of the Rohrbrunn-Folge (Middle Bunter).

310 m ü. NN. Dieser Sprung von 350 m ist der höchste Versatz zwischen zwei Schollen im Gebiet. Häufiger ist ein Absinken in einzelnen Staffeln, wie z. B. von Scholle 7 über 8 und 9 nach 10 oder gar 11; hier kommt noch eine Absenkung um weitere 90 m hinzu.

Von der Scholle 7 über 6 und 2 nach 3 ergeben sich Sprunghöhen von 110 + 60 + 200 m und von Scholle 13 über 21 und 15 nach 22 von 40 + 100 + 270 m. Verfolgt man die Absenkung von 10 über 14 zur am tiefsten gelegenen Scholle 17, so wären noch weitere 80 + 40 m zu addieren, so daß im äußersten Falle eine Tieferlegung um 470 m erfolgte, wovon allerdings rund 50 m wegen des Schichtenfallens um 1—2° nach SE abzuziehen wären. Es verbleibt aber eine maximale Absenkung von > 400 m.

Im Osten des Betrachtungsgebietes zeigt sich spiegelbildlich (Abb. 3) das Gleiche nur mit wesentlich geringeren Sprunghöhen. Die Schollen 33, 41 und 42 bilden den relativ stehengebliebenen Rand. Die unmittelbar westlich davorliegenden Schollen sind um 140 m (32 gegen 33), im Bereich der Schollen 42, 41, 39 um 25 + 130 m = 155 m abgesunken. Die im Westen anschließenden Schollen (31, 38) liegen geringfügig (ca. 20 m) höher.

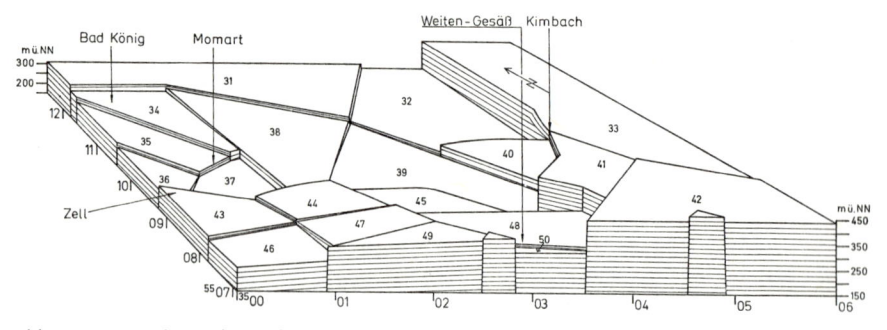

Abb. 3. Perspektivisches Blockdiagramm des Ostteiles der Grabenfortsetzung im SW-Quadranten der TK 6220 Wörth. Darstellungsgrundlage wie Abb. 2.

Fig. 3. Perspective block-diagram of the eastern part of the continuation of the Graben into the SW-square of official map 6220 Wörth. Construction like Fig. 2.

Von der größeren (4 km²) Nordscholle (Scholle 31, Abb. 3) treppen die nach SW folgenden Schollen 34, 35 und 36 um jeweils 10—20 m ab. Durch die besonders tief versenkte Scholle 37 wird zusammen mit 36 und den Schollen 11, 12, 17, westlich der Grabenachse, das Gebiet der stärksten Versenkung von Einzelschollen in der Nähe des Ortes Zell gekennzeichnet. Dadurch erscheint ebenso wie der Nordrand auch die südliche Begrenzung des Kartiergebietes (= Nordende des bisherigen Grabens) etwas herausgehoben.

Diese um 10—20 m abgesetzten Schollenstaffeln zeigen bei der stellenweisen starken Schuttbedeckung der Hänge die Grenze der kartistischen Erfassung der Zerlegung gleichfalls auf. Im inneren Teil des Grabens (evtl. bedingt durch das Mümlingtal) lassen sich hingegen auch zahlreiche gekippte Schollen erkennen

(z. B. 43, 44, 46, 47, 49) mit einer generell westwärtigen Neigung, während die westlich der Grabenachse gelegenen Schollen 5 und 12 entsprechend nach Osten geneigt sind; lediglich Scholle 18 läßt eine südwärtige Neigung erkennen.

Die H a u p t s t ö r u n g s linien verlaufen im Ostteil weitgehend in N—S bis NE—SW-Richtung, demgemäß die zugehörigen Querstörungen W—E bis NW—SE. Am Westrand tritt die erzgebirgische Richtung zugunsten einer geringfügig von der N—S-Richtung abweichenden zurück; die Querstörungen verlaufen hier demgemäß vorherrschend West—Ost.

Die vorgenommenen Kluftmessungen verdeutlichen die starke tektonische Beanspruchung. In der Nähe von Störungen sind sie besonders ausgeprägt und betonen die Richtung der Störungen. In den östlich des Grabens aufgenommenen Kluftrosen herrscht die rheinische Richtung vor neben einer W bis WNW streichenden. Im eigentlichen Grabenbereich sind die Kluftrichtungen NE bis NW besonders markant. Da sie zu den vorherrschenden Störungsrichtungen (N—S bzw. W—E) diagonal liegen, können sie als Hinweis auf eine Blattverschiebung aufgefaßt werden, wie BUBNOFF sie 1922 für die Otzbergspalte vermutete, BECKSMANN sie 1954 im Südodenwald fand und RICHTER-BERNBURG 1968 für das gesamte Bauelement ableitete.

Stellt man das Gesamtgebiet in drei Profilschnitten im Abstand von 2 km dar (Abb. 4), wobei wieder die Schichtgrenze der Geiersberg-/Rohrbrunn-Folge als Bezugsfläche benutzt wird, dann tritt der Grabenbau deutlich hervor.

Unter Berücksichtigung des generellen Einfallens der Schichten des Buntsandstein-Odenwaldes von 1—2° nach Südosten liegen die östliche und westliche Schulter des Grabens in einer geneigten Fläche, dazwischen ist das Schollenmosaik des Grabens mit den randlich abbrechenden Staffeln und den kleineren grabenartigen Tiefschollen abgesunken. Die einzelnen Schollenverkippungen können dabei nicht mehr sichtbar dargestellt werden.

Errechnet man unter der Annahme eines idealen Einfallens der Störungsflächen von ca. 65° aus der Summe der Sprunghöhen bei einem Abstand der Randverwerfungen von 9,9—7,8 km die Z e r r u n g s b e t r ä g e , dann ergeben sich im Norden 300 m, in der Mitte 275 m und für das südliche Profil 270 m. Damit deutet sich eine Verbreiterung des Grabens nach Norden an. Im Vergleich zu dem zentralen bisher von KLEMM schon erkannten Grabenteil ist das mehr als das Doppelte. Es ist aber anzunehmen, daß dieser breite Bereich bei einer nach Süden fortschreitenden Kartierung auch in den Gebieten des Unteren und Mittleren Buntsandsteins erkannt werden wird.

Die östliche Randverwerfung entspricht in ihrem Verlauf einem schon von VOGEL (1898) dargestellten schmalen Grabenfortsatz; eine Fortsetzung in Richtung auf das Maintal will nicht ausgeschlossen erscheinen. Am Westrand hielt KLEMM (1925) einen geringen Versatz des Buntsandsteins gegen das Kristallin für möglich. Unrichtige Ansprache der geröllführenden Sandsteine verhinderte das Erkennen dieser starken westlichen Randverwerfung. Diese westliche Randverwerfung scheint sich nach unseren bisherigen Begehungen im nördlich anschließenden Gebiet fortzusetzen, so daß KLEMMS Auffassung vom Übergang in die östliche Randver-

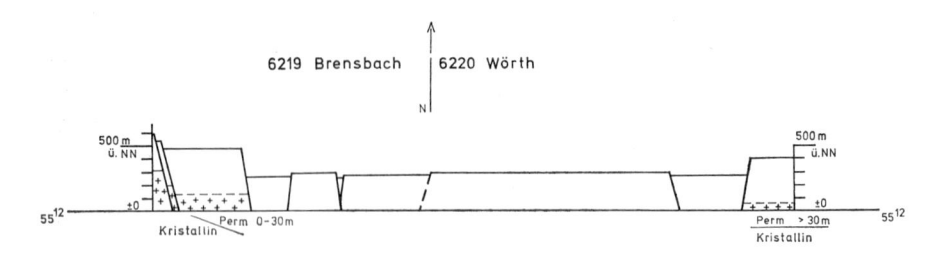

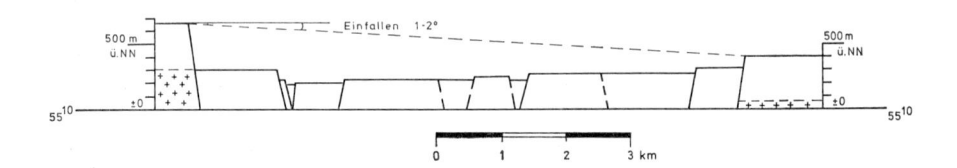

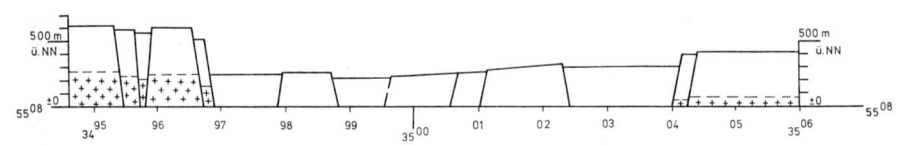

Abb. 4. Drei Profilschnitte durch den Graben entlang den Gitternetzlinien im Abstand von jeweils 2 km durch das Arbeitsgebiet auf der Grundlage der Schichtlagerungskarten. Die Oberkante der Profilschnitte entspricht auch hier der Schichtgrenze Rohrbrunn-/Geiersberg-Folge (Mittlerer Buntsandstein). Das generelle Schichteinfallen nach der Aufwölbung des kristallinen Odenwald-Schildes ist im mittleren Profilschnitt angetragen.

Fig. 4. Three cross-sections cutting the Michelstadt Graben in a distance of 2 km; based on contour maps like Fig. 2 and 3. The general inclination of layers as caused by the uplifting of the Odenwald mountains in Cenozoic times is marked in the middle one of the cross sections.

werfung des Gersprenzgrabens (allerdings Absinken der Westscholle um ca. 200 m) möglich erscheint.

Daß der Michelstädter Graben im ursächlichen Zusammenhang mit der Entstehung des Rheingrabens zu bringen ist, wird von niemand bezweifelt werden können. Durch die Herauswölbung des Kristallins und die Bewegungen der Gesamtscholle entstanden Spannungen, die am Ostrand des Kristallins zum Aufreißen der westlichen Randverwerfung führten. Die starke Beanspruchung bedingt das Zerbrechen in einzelne Schollen. Auf 10 km Distanz sind die stärksten Spannungen abgeklungen, so daß nach einem Randsprung von 150 m die Schichten wieder im gleichen Niveau liegen wie das herausgewölbte Kristallin.

Für das A l t e r dieser Vorgänge lassen sich aus den Sedimenten außer den pleistozänen Füllungen keine Anhaltspunkte gewinnen. BECKSMANN und ZIENERT arbeiten bei der Bestimmung des Zeitablaufs mit tertiären Verebnungsflächen, die sich im wesentlichen im Pliozän ausbilden konnten.

Würden wir diese Verebnungsflächen als tektonisch verwertbar und gleichaltrig in unserem Gebiet annehmen können, dann würden am Ostrand ca. 30 % der Sprunghöhe von 155 m auf die Zeit vor dem Flachrelief (? pontisch) einzustufen sein, während der Rest, ca. 100 m, jünger, also bis ins Pleistozän, zu datieren wäre. Im Westen (westliche Randverwerfung) scheint die Relation nahezu umgekehrt zu sein. Von den 380 m an der Hauptverwerfung wären nur 80—100 m nach der Entstehung der Verebnungsfläche zu datieren, während die übrigen 250 m vorher schon abgesunken sein müßten. Nehmen wir an, wir betrachten in beiden Fällen eine gleichaltrige Verebnungsfläche, dann wären die unterschiedlichen Sprunghöhen vor Ausbildung des Flachreliefs entstanden, während danach an beiden Randverwerfungen nur noch eine gleichstarke Einsenkung von rund 100 m stattgefunden hätte (Groß-Umstadt, Ostrand des Gersprenz-Grabens, 200 m). Wenn BECKSMANN den Beginn der Störungen bis an die Wende Oligozän/Miozän zurückrechnet, dann könnten die auf Miozän bis Pliozän pollenanalytisch datierbaren grabenartig versenkten Tonvorkommen des benachbarten Spessart (BACKHAUS 1967) ihn bestätigen.

Literaturverzeichnis

BACKHAUS, E., 1961: Das fossilführende Zechsteinvorkommen von Forstel-Hummetroth (Nordodenwald) und Bemerkungen zur südwestdeutschen Zechsteingliederung. — Notizbl. hess. L.-Amt Bodenforsch., 89, 187—202.
— 1965: Die randliche „Rotliegend"-Fazies und die Paläogeographie des Zechsteins im Bereich des nördlichen Odenwaldes. — Notizbl. hess. L.-Amt Bodenforsch., 93, 112—140.
— 1967: Die vermeintliche pliozäne Schotterterrasse des Mains von Schippach im Spessart (Bl. 6121, Heimbuchenthal). — Veröff. Geschichts- u. Kunstver. Aschaffenburg, 10, 165—174.
BECKSMANN, E., 1954: Zur jüngeren Baugeschichte der Heidelberger Landschaft. — Verh. naturhist.-med. Ver. Heidelberg, N. F. 20, 51—65.
— 1970: Die zeitliche Aufgliederung der Bruchtektonik im Odenwald und Kraichgau. — Z. dt. geol. Ges., Jg. 1969, 121, 119—123.
BUBNOFF, S. VON, 1922: Die hercynischen Brüche im Schwarzwald, ihre Beziehung zur carbonischen Faltung und ihre Posthumität. — N. Jb. Miner. Geol. Paläont. Beil.-B., 45, 1—120.
KLEMM, G., 1925: Bemerkungen über die Tektonik des Odenwaldes. — Notizbl. V. Erdk. u. Hess. Geol. L.-Anstalt zu Darmstadt, 5. Folge, H. 7, 8—22.
— 1928: Erläuterungen zur Geologischen Karte von Hessen im Maßstab 1:25 000. — Bl. Erbach u. Michelstadt, 2. Aufl., 45 S.
RAWANPUR, A., 1972: Tektonische und sedimentologische Untersuchungen im westlichen Teil des Michelstädter Grabens auf Bl. Brensbach (TK 6219). — Unveröff. Diplomarbeit TH Darmstadt, 140 S.
RICHTER-BERNBURG, G., 1968: Saxonische Tektonik als Indikator erdtiefer Bewegungen. — Geol. Jb., 85, 997—1030.
VOGEL, CHR., 1898: Erläuterungen zur Geologischen Karte des Großherzogthums Hessen im Maßstab von 1:25 000. V. Lfg. — Blatt König, 54 S.
ZIENERT, A., 1957: Die Großformen des Odenwaldes. — Heidelberger Geogr. Arb., 2, 156 S.
ZIRNGAST, M., 1972: Tektonische und sedimentologische Untersuchungen im östlichen Teil des Michelstädter Grabens auf Blatt Wörth am Main (TK 6220). — Unveröff. Diplomarbeit TH Darmstadt, 102 S.

Southern extensions of the Rhinegraben

Relations entre le Fossé rhénan et le Fossé de la Saône
Tectonique des régions sous-vosgiennes et préjurassiennes

par

D. Contini et N. Théobald

Avec 5 figures dans le texte et en supplément

Relations between Rhine and Saône Rift Valleys. Tectonic of "Sous-vosgiennes" and "Prejurassiennes" Areas

Abstract: Since 1954, systematic geological surveys (1/25.000) were made along the southern side of the Vosges and in the "prejurassienne" areas. It is now possible to state the geological structure of the transition zone between Rhine and Saône rift-valleys. A dense system of sub-meridian faults is the major characteristic of the tectonic. Along the southern side of the Vosges, an orthogonal system inherited from hercynian directions may be added to the former. In the sedimentary cover of the "prejurassienne" area, sub-meridian faults act as strike-slip faults and delimit compartments which are undulated by transverse undulations, various anticlines (e. g. diapir) and local shearplanes.

I. Introduction (fig. 1)

Le présent travail expose les résultats nouveaux acquis depuis 1954 par des levers systématiques au 25.000 de toute la région comprise entre le Fossé rhénan à l'E, le plateau de Langres à l'W, la retombée méridionale des Vosges au N et le Jura plissé au S. Cette région est précisément la zone de passage apparente entre le Fossé de Dannemarie, extrémité S du Fossé rhénan et le Fossé de la Saône, extrémité N du Fossé bressan (LAUBSCHER, 1970).

C'est pourquoi il nous a semblé utile de publier des documents montrant la structure géologique de la région.

De la même manière que le Fossé rhénan a servi de modèle à l'analyse des rifts continentaux (J. H. ILLIES, 1967, 1970), la connaissance de la région envisagée ici peut nous permettre de mieux situer les zones de transfert des structures taphrogénétiques appelées failles transformantes.

Nous considérons que notre apport essentiel sera l'exposé des structures levées en surface, constituant un document de travail. Nous présenterons une brève analyse de l'évolution paléogéographique de la région et nous proposerons à titre d'hypothèse une interprétation.

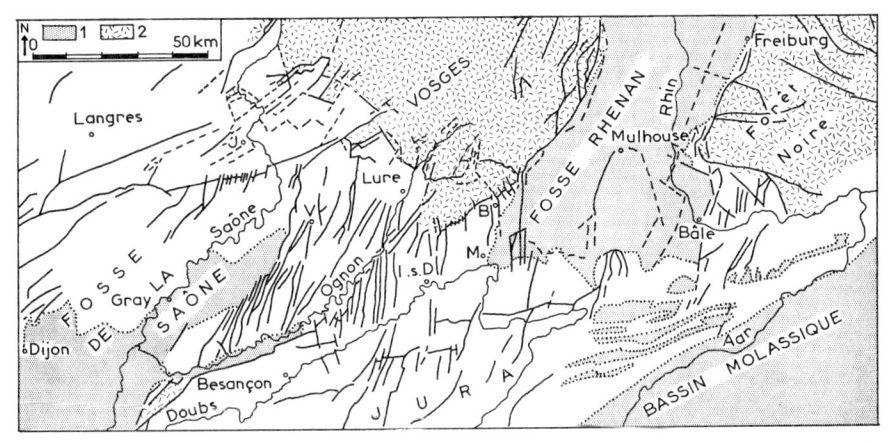

Fig. 1. Schéma structural de la région comprise entre le Fossé rhénan et le Fossé de la Saône. Légende: 1. Extension actuelle des dépôts d'âge éocène, oligocène et miocène. 2. Extension actuelle des dépôts d'âge primaire. B: Belfort, I. s. D: Isle sur le Doubs, J: Jussey, M: Montbéliard, V: Vesoul.

II. Exposé des structures levées en surface (fig. 2)

A. Partie méridionale du Fossé rhénan

En bordure SW du Fossé rhénan, la structure géologique est dominée par une importante faille composite de direction subméridienne: la faille dite de Charmois (voir Laubscher, Theobald & Wittmann, 1967) dont le rejet cumulé dépasse 1800 m au N dans la région de St-Germain-le-Chatelet et atteint encore 200 m au S dans la région de Grandvillars. La faille de l'Allaine se poursuit en direction SE vers Delle et le Jura.

Tandis que la faille de Charmois peut être considérée comme le prolongement de la faille dite faille rhénane, il est plus difficile de préciser le tracé de la faille bordière du massif ancien, dite faille vosgienne. La limite méridionale des Vosges a été précisée par des études récentes (Theobald, 1969).

B. Retombée méridionale des Vosges

Au NE de Belfort, au S du Massif primaire de la Forêt d'Arsot, existe un véritable champ de fractures dit de Roppe, haché de failles de direction SE—NW ou subméridiennes, limitant des compartiments, où les couches d'âge jurassique moyen et supérieur, décollées au niveau des séries marneuses, triasiques et liasiques, redressées jusqu'à la verticale et parfois déversées, soulignent l'importance du relèvement tertiaire du massif vosgien.

Au NW de Belfort et à l'W de la faille de la Savoureuse, faille dédoublée selon des renseignements inédits récents, on peut suivre un relais de failles de direction générale NE—SW, jalonnant le bord S du massif du Salbert où le Jurassique moyen est abaissé contre le Dévono-Dinantien, décroché à plusieurs reprises par des failles subméridiennes en bordure S du massif de Chagey. Ici encore, les compartiments fortement inclinés révèlent souvent la suppression tectonique des marnes bariolées du Trias moyen.

A l'W de Saulnot, apparaissent des failles subméridiennes à important rejet W formant la limite occidentale du massif de Chagey, failles minéralisées au N de Saulnot, limitant les affleurements des terrains dévoniens, ensemble de failles dont il faut signaler celle de Mignavillers et celle de Moffans — La Vergenne formant la limite occidentale d'affleurement des terrains permiens. Ces deux dernières doivent être considérées comme appartenant déjà au système des failles de l'Ognon, importante zone de dislocation, le long de laquelle le rejet cumulé W des terrains de la couverture sédimentaire dépasse 3 à 400 m.

Le massif vosgien se trouve ainsi reporté d'une vingtaine de kilomètres en direction N jusque dans la région de Mélisey.

La dépression de Lure se prolonge en direction NE vers le fossé de Mélisey situé entre le massif du Mont de Vannes et le petit horst du Rocheret. La dépression de Lure s'élargit en direction NW où elle se poursuit par la dépression marginale de la Lanterne jusqu'à Luxeuil-les-Bains. Au NW de cette localité, apparaît le horst de Luxeuil, élément structural important par son extension en direction SW et son soubassement de gneiss et de granite.

Après l'importante faille double de Fontaine-les-Luxeuil, de direction SW—NE, la bordure vosgienne se suit vers le NW de Fougerolles à Bains-les-Bains. Un môle d'importance secondaire s'alignant sur le Bois du Clerjus permet de distinguer le fossé de Fougerolles St-Loup-sur-Semouse au SE, celui de Corre au NW. Dans leur prolongement SW, s'alignent le synclinal de Conflans-sur-Lanterne et plus loin celui de Jussey.

Les structures de Passavant-la-Rochère sont bien connues (THEOBALD, 1960—1963). Le horst de Passavant-la-Rochère—Mont de Paron est séparé par le fossé de la Rochère du môle de Darney. Plus loin, le socle vosgien s'ennoie peu à peu sous la couverture triasique, d'une façon générale en direction W, mais réapparaît encore à la faveur de quelques failles antithétiques. Tel est le cas du pointement cristallin de Chatillon-sur-Saône et de celui de Bussières-les-Belmont, rejeté très loin (50 km environ) en direction SW. En précisant qu'à Passavant-la-Rochère, le socle est recouvert par les grès intermédiaires, à Chatillon-sur-Saône par les grès à Voltzia et à Bussières-les-Belmont par la dolomie-moellon, on souligne le rôle de seuil joué par le plateau de Langres et le détroit morvano-vosgien dès le Trias.

Une vue d'ensemble sur les accidents relevés à l'intérieur du massif vosgien et de sa bordure méridionale conduit à souligner les faits suivants:
— le rôle décrochant de certaines failles tardihercyniennes entrevu par divers auteurs (ILLIES, 1962) et exposé récemment (FOURQUIN, GUERIN, THEOBALD & THIEBAUT, 1971; MATTAUER, 1971),
— l'orientation rayonnante des failles autour du massif vosgien : de WNW—ESE passant à NNW—SSE dans le Bois de Roppe, puis de sub-méridienne à SSW—

NNE dans le massif de Chagey et le long de la vallée de l'Ognon, enfin s'orientant SW—NE au horst de Luxeuil-les-Bains et Passavant-la-Rochère pour atteindre la direction WSW—ENE à l'approche du plateau de Langres (leur ensemble décrivant ainsi un secteur de cercle voisin de 135°),

— leur disposition concentrique nettement illustrée dans la partie de la bordure située à l'W de la faille de l'Ognon où les terrains s'affaissent progressivement vers la région marginale, plus compliquée dans la partie E où le bassin permien de Giromagny s'intercale entre le massif des Ballons et celui de Chagey.

Notons encore que parmi les failles décrites, probablement toutes d'âge alpin, certaines sont incontestablement établies sur des structures d'âge hercynien. Tel semble être le cas d'une partie de la bordure N du bassin permien aux environs de Giromagny, où la faille W—E à rejet actuel conforme se place dans le prolongement d'une faille chevauchante d'âge hercynien (FOURQUIN, GUERIN, THIEBAUT & THEOBALD, 1971).

C. Zone de passage entre le Fossé de la Saône et le Fossé rhénan

Elle sera décrite de l'W vers l'E.

a) Structure de la partie N du fossé de la Saône. On peut considérer comme limite N du Fossé de la Saône l'important accident de direction générale WSW—ENE se suivant de Crancey-le-Château (Côte d'Or) à Faverney (D. CONTINI, 1966) formé d'une suite de failles normales à regard S, à rejet conforme variant de 30 m à plus de 200 m, découpées par de nombreux petits accidents subméridiens, délimitant des compartiments disposés en touches de piano et quelques failles plus importantes de direction SW—NE. Parmi ces dernières, il faut citer les failles allant de Morey à Fouvent-le-Haut, séparant les plateaux de Combeaufontaine à l'W et ceux de Champlitte à l'E.

Au N de cet accident limite, s'étend une vaste zone déprimée, accidentée de quelques horsts et fossés orientés du SW au NE, découpés par des failles plus ou moins orthogonales et où apparaissent quelques brachy-anticlinaux très courts orientés SW—NE et le synclinal de Jussey reconnu de direction WSW—ENE se résolvant en deux dépressions synclinales vers son extrémité orientale.

Ce quadrillage de failles, comparable à celui de la retombée vosgienne, et faisant apparaître des pointements isolés du socle cristallin, montre le peu de profondeur du socle.

La limite E du fossé de la Saône est formée par un ensemble compliqué de failles subméridiennes se suivant de Faverney, en direction S, vers Grandvelle et Gy. Cette partie du fossé appelée «plaines de Gray» est à soubassement crétacé et tertiaire. Le long de ces failles bordières, ont été repérés des décollements limités de certains compartiments subméridiens, légèrement chevauchants en direction N (DREYFUSS & KUNTZ, 1970).

b) Structure géologique des plateaux de Vesoul. La région située entre le Fossé de la Saône et la faille de l'Ognon est une sorte de plateau à sous-sol formé de Jurassique moyen et supérieur, accusant un pendage général vers le S. De nombreuses failles normales, d'orientation générale SSW—NNE, présentant souvent une légère concavité vers l'WNW, découpent l'ensemble en une suite de petits horsts et fossés. Tel est le fossé d'Andelarre (D. CONTINI, 1964).

Un faisceau de failles s'alignant de Chateney, par Noroy-le-Bourg et Valleroy-le-Bois à Dampierre-sur-Linotte, permet de distinguer une partie occidentale marquée par une dépression synclinale se projetant plus au N dans le synclinal liasique de Saulx-les-Vesoul et une partie orientale. Cette portion des plateaux de Vesoul est accidentée par plusieurs ondulations transversales, notamment par la dépression synclinale de la Font de Champdamoy, d'orientation WNW—ESE.

Dans la partie S des plateaux, certaines failles peuvent être en relation avec les accidents de la zone des Avant-Monts (JAVEY, 1966).

La partie orientale des plateaux de Vesoul s'abaisse par paliers successifs vers la dépression synclinale de l'Ognon, dite système des failles de l'Ognon (THEOBALD & CONTINI, 1965; THEOBALD, 1969).

c) Système des failles de l'Ognon. La vallée de l'Ognon suit une importante ligne de dislocations le long de laquelle apparaissent, dans la partie S, des dépôts crétacés et tertiaires, et qui est désignée dans la littérature géologique sous divers noms: synclinaux de l'Ognon ou système des failles de l'Ognon. Cette dernière dénomination semble préférable, car la disposition synclinale ne se retrouve que très localement et dans certains secteurs, tandis que toute la vallée est jalonnée par des failles. Nous l'analyserons en trois secteurs allant du NE au SW.

1° Secteur N: le fossé de Mélisey. Au débouché du secteur vosgien, la vallée de l'Ognon, jalonnée par une faille à rejet W de 240 m entre les plateaux de la Montagne de Fresse et le Grand Sigle, s'ouvre en un fossé subméridien, affaissé de 380 m par rapport au Mont de Vannes, de 70 m par rapport au horst du Rocheret (THEOBALD, 1968).

Ce fossé, large de 3 km au S de Mélisey, s'élargit jusqu'à 6 km sur la transversale Vouhenans-Moffans et se rétrécit en coin vers Villersexel.

2° Secteur médian: de Villersexel à Corcelle-Mieslot. De direction générale NE—SW, la vallée de l'Ognon forme une dépression dominée par les reliefs de la zone préjurassienne le long d'une ligne de dislocation complexe formée par des faisceaux de failles au NE, une flexure au SW. Le système de failles est décroché par les accidents subméridiens des collines préjurassiennes dont les compartiments étroits et allongés chevauchent parfois légèrement la dépression de l'Ognon. C'est le cas notamment entre Rougemontot et Battenans-les-Mines (HEILAMMER, 1956) et vers Montussaint.

De Servigney à Chazelot, la structure est plus continue; elle représente un anticlinal à noyau keupérien et dont le flanc SE est très légèrement incliné alors que le flanc NW est fortement redressé, parfois déversé, et limité par deux failles.

Cet accident ne semble pas pouvoir être interprété comme une pincée (CAILLE-

TEAU, 1966); celle-ci n'existe pas entre Rougemontot et Battenans-les-Mines (BULLE, MARTIN & ROLLET, 1967), ni vers Montussaint et plus au Nord.

Au coeur de cet accident, apparaît par contre le «horst de Chazelot», petit compartiment lenticulaire, large de 250 m, s'allongeant sur 1400 m, à noyau de Muschelkalk. Un sondage (FOURNIER, 1922) a traversé une épaisseur anormale de Muschelkalk supérieur, un Muschelkalk moyen réduit, une épaisseur normale de Muschelkalk inférieur et de Buntsandstein. Ce horst semble pouvoir être interprété comme étant une écaille intracutanée mise en place lors du glissement en direction W de la région préjurassienne sur la dépression de l'Ognon.

3° Secteur SW: de Corcelle-Mieslot à la Serre. D'orientation générale ENE—WSW, cette région montre une dépression synclinale où ont été identifiés localement des dépôts crétacés et tertiaires, dominée au S par les reliefs des Avant-Monts, région diversifiée dans ses structures, où l'on peut distinguer:
— de Mieslot à Chatillon-le-Duc, une zone anticlinale à coeur bajocien ou liasique et à flanc N chevauchant (A. BONTE, 1943; BULLE, MARTIN & ROLLET, 1966—67),
— de Chatillon-le-Duc à Audeux, une série de compartiments plissés chevauchant les uns sur les autres et sur les synclinaux de l'Ognon (JAVEY, 1966),
— d'Audeux à la Serre, un vaste anticlinal dissymétrique rompu au niveau de sa charnière (JAVEY, 1966).

Le massif de la Serre est un horst où réapparaît le socle (gneiss, granites, Permien) recouvert en discordance de terrains triasiques. Le long de son bord SE, la couverture sédimentaire est affectée, par endroits, de failles inverses.

d) Les collines préjurassiennes. Cette région, située au N du Jura plissé, s'étendant de Corcelle-Mieslot à Belfort et à Montbéliard, est hachée de failles subméridiennes. Très nombreuses dans la partie occidentale, elles découpent la couverture en compartiments étroits, à pendage général vers le S, accidentés de quelques ondulations transversales (CAILLETEAU, 1966; CONTINI, BARDET, CAMPY & MASSON, 1967). Dans la partie orientale, les failles sont moins serrées, le pendage général s'accentue progressivement vers le golfe de Montbéliard.

Au N des collines préjurassiennes, on observe un glissement local à Gonvillars et quelques anticlinaux diapyrs au niveau du Lias inférieur (THEOBALD & CONTINI, 1960; CAILLETEAU, 1966).

e) Le golfe de Montbéliard et le dôme de Montbouton. L'approche du Fossé rhénan est annoncée dans la région de Montbéliard par une dépression orientée du SW vers le NE où l'on note la présence de dépôts éocènes (Lutétien lacustre) et oligocènes (système de Bourogne).

Le dôme de Montbouton, horst subméridien accidenté d'un fossé (fossé de Badevel), forme la transition entre le golfe de Montbéliard, le Fossé rhénan et les plateaux de l'Ajoie.

III. Evolution paléogéographique posthercynienne de ces régions

A la suite de l'orogénèse hercynienne, des bassins subsidents à remplissage détritique permien s'installent en bordure méridionale des Vosges. Ils font partie de la cuvette burgonde de H. BOIGK (1970). Ils s'étalent largement vers le S sur le Jura, où l'accumulation des sédiments atteint par endroits 1000 m (WINNOCK, 1967).

Il faut souligner que dans ce bassin apparaissent, dès ce moment, certaines lignes structurales: tel est le décrochement de la faille de l'Ognon, rejetant la sédimentation permienne d'environ 6 km en direction N. Enfin, les manifestations volcaniques du massif de Chagey et du Rocheret soulignent la tectonique active d'âge permien moyen (THEOBALD & THIEBAUT, 1961).

Durant de Trias, la zone subsidente est marquée par une dépression située en bordure S du massif de Chagey, elle est jalonnée par la présence du faciès des calcaires ondulés dans le Muschelkalk inférieur à Saulnot (THEOBALD, 1963) et d'évaporites dans les grès argileux du sommet du Trias inférieur dans le sondage de Buix.

Le schéma précédent se modifie au Jurassique. Après une période de sédimentation relativement homogène, ne montrant que de petites variations de détail au Lias inférieur et moyen, on voit apparaître au Toarcien supérieur des variations de subsidence entre certains compartiments, variations de subsidence dans l'espace, s'accompagnant d'inversions de la subsidence dans le temps.

Le Toarcien supérieur, réduit au S d'une ligne passant par Moirans—Orgelet et vers Fayl-Billot—Langres, est par contre bien développé dans la région intermédiaire.

L'Aalénien inférieur et moyen sont condensés au S du faisceau salinois et du Lomont, alors qu'ils sont bien représentés en Alsace et en Haute-Saône. C'est l'inverse pour l'Aalénien supérieur qui manque ou est réduit au N du faisceau salinois et atteint jusqu'à 70 m au S (D. CONTINI, 1971).

Jusqu'au Bajocien moyen, une zone de «haut-fond» s'étend du Morvan au Jura d'Argovie, séparant les bassins parisien et souabe du bassin jurassien. La subsidence était plus active tantôt au N, tantôt au S.

A partir du Bajocien supérieur, les lignes isopiques se déplacent vers le SE, prenant le Jura en écharpe depuis le S de la Forêt Noire à l'île Crémieux, mais le faisceau salinois joue toujours un rôle important, notamment à l'Oxfordien (R. ENAY, 1966).

Il faut signaler que durant le Trias et le Jurassique, les fossés n'apparaissent pas dans la paléogéographie, mais qu'une zone à subsidence toujours plus active se dessine dans la partie méridionale du Fossé rhénan.

L'extension des mers crétacées est difficile à préciser. Les affleurements actuels les plus septentrionaux sont situés aux environs d'Avilley dans la vallée de l'Ognon, aux environs de Fresnes-St-Mamès dans les plaines de Gray.

Dans le Fossé de la Saône, les dépôts tertiaires lacustres s'étendent à quelques kilomètres plus au Nord (vers Mont le Vernois).

IV. Essai d'interprétation

A. Interprétation géométrique (fig. 3)

Dans une vue d'ensemble sur la structure de la région de passage entre le Fossé rhénan et le Fossé de la Saône, il convient de souligner les points suivants:

1) Certaines structures, nettement orientées selon des directions hercyniennes (SW—NE) peuvent être modelées sur d'anciens accidents primaires.

a) Cette disposition apparaît bien dans la partie méridionale des Vosges (horst de Passavant-la-Rochère, horst de Luxeuil, alignement des massifs de Chagey, du Salbert et de l'Arsot); elle transparaît à travers la couverture triasique dans la région relevée proche du «seuil» de Langres.

b) Le système de failles formant le bord septentrional du Fossé de la Saône se place sensiblement dans le prolongement de la faille de Fontaine-les-Luxeuil, mais son rejet change de sens à l'W de Faverney, à l'intersection avec les failles subméridiennes formant la limite E du Fossé de la Saône. De plus, sa direction devient presque E—W et, de ce fait, elle recoupe les structures de direction hercynienne. Ces dernières ne se font plus sentir que le long de failles NE—SW découpant les plateaux de Champlitte et de Combeaufontaine.

c) Le système des failles de l'Ognon, malgré sa direction générale SW—NE est plus complexe. Une région «anticlinale» pouvait exister le long d'une ligne joignant la Serre au massif de l'Arsot au N de Belfort. Cette structure n'a pu guider le système des failles de l'Ognon qu'entre la Serre et Villersexel. Il convient de souligner l'importance du rejet de la faille tertiaire, qui abaisse le compartiment NW; le rejet cumulé dépasse 1000 m au N de la Serre, il diminue en direction NE jusqu'à environ 300 m, près de Villersexel.

2) Ces structures, de direction hercynienne, sont recoupées par d'autres qui ont permis le relèvement des Vosges méridionales et l'effondrement des fossés. Nous avons déjà souligné la présence, autour des Vosges méridionales, d'un double réseau de failles, les unes rayonnantes, les autres concentriques. Cette disposition a permis le relèvement du socle des Vosges méridionales.

Parmi les failles rayonnantes, il faut souligner l'importance de la faille de Charmois et des failles de la partie orientale du fossé de Mélisey. Elles déterminent un relèvement important du socle, ce qui fait que ce dernier affleure actuellement jusque vers Chagey.

Les failles du fossé de Mélisey appartiennent au système des failles de l'Ognon qui, comme nous l'avons vu, provoquent un enfoncement important du compartiment NW. Le rejet de ces failles dépasse toujours 300 m au S de Mélisey et tombe à 240 m plus au N.

Les positions variables des divers compartiments du socle vont provoquer des réactions tout à fait différentes dans la couverture post-triasique.

— Dans les champs de fractures de Roppe, le relèvement important du môle vosgien a provoqué de décollement de la couverture secondaire au niveau du Keuper et parfois même au niveau du Lias et son glissement (très limité) en direction SE, vers la dépression du golfe de Montbéliard.

— Le long des collines préjurassiennes, au S et au SW du Massif de Chagey, de petits anticlinaux à coeur keupérien peuvent être dûs à la même cause.

— A l'W de Rougemont, le compartiment situé au N des collines préjurassiennes étant effondré, ces dernières vont chevaucher localement vers le N sur la dépression de l'Ognon.

Les collines préjurassiennes sont découpées par de nombreuses failles N—S qui guident les compartiments de couverture. Il faut souligner le fait que ces glissements sont de faible amplitude.

La zone séparant la partie W (secteur Corcelle-Mieslot à Rougemont) qui a tendance à glisser vers le N et la partie E (secteur Saulnot-Roppe) qui, au contraire, a tendance à glisser vers le S, est hachée d'un réseau dense de failles N—S s'étendant de Rougemont à Baume-les-Dames. L'arc plissé bisontin montre un style tout à fait différent de part et d'autre de Baume-les-Dames, les chevauchements vers le N, bien développés à l'W, sont moins nets à l'E où la direction des plis devient E—W.

— Sur le plateau de Vesoul, les failles de la couverture post-triasique ne reflètent plus du tout les structures du socle hercynien, mais semblent plus ou moins parallèles aux failles limitant la bordure E du Fossé de la Saône.

En conclusion: Dans cette région, l'observation des structures permet de distinguer deux étages tectoniques:

— le premier formé par le socle hercynien et sa couverture de grès triasiques,
— le deuxième correspondant à la couverture post-triasique.

Certains grands accidents du premier étage se répercutent dans le deuxième (ex. système des failles de l'Ognon), mais la plupart des accidents de faible amplitude s'amortissent au niveau du Keuper et un système de failles de direction différente affecte les séries jurassiques.

Bien que tous les accidents visibles soient vraisemblablement d'âge tertiaire, le socle laisse apparaître la superposition de deux styles tectoniques: hercynien et tertiaire.

B. Analyse des mouvements et recherche des causes: Interprétation cinématique et dynamique (fig. 4 et 5)

Les fossés continentaux ont des caractères très différents des rifts océaniques et il est difficile de comparer l'ensemble des fossés traversant l'Europe occidentale à un rift océanique en voie de formation. Les fossés continentaux de l'W de l'Europe ne représentent qu'une amorce de dislocation des blocs continentaux. Diverses hypothèses ont été émises à propos de la formation de ces fossés.

ILLIES (1970) estime que l'effondrement des fossés peut être dû à une tension latérale, pouvant elle-même être la conséquence d'un autre processus.

LAUBSCHER (1970) fait cadrer la formation des fossés continentaux d'Europe avec la théorie des plaques et émet l'hypothèse qu'une faille transformante, située entre la faille de l'Ognon et la ligne Mont Terri—Salins, décale le Fossé rhénan par rapport au Fossé de la Saône.

A diverses reprises, il a été proposé d'interpréter l'écartement des deux bords du fossé comme étant la conséquence secondaire d'une compression venant du S; cette interprétation cadre avec l'histoire des Alpes et du Jura. Un schéma faisant intervenir des forces de compression orientées du SE au NW a été proposé par L. Ahorner (1970).

Avant la formation des fossés, il semble que des zones de fractures ou des zones plus fragiles existaient le long du couloir rhodanien et le long du Fossé rhénan. Le long du couloir rhodanien, Mattauer (1971) souligne l'importance de la faille des Cévennes qui aurait joué comme faille transformante au Crétacé. De grands accidents N—S séparant l'ensemble des Alpes occidentales, des Alpes orientales pouvaient, d'après Goguel (1967) se prolonger dans le Fossé rhénan. Certains de ces grands décrochements peuvent être interprétés comme des glissements le long de failles transformantes de la Téthys (J. Dercourt, 1970).

Nous avons souligné dans le paragraphe III, qu'au cours de la paléogéographie post-hercynienne, une zone charnière existait entre le Morvan et la partie S de la Forêt Noire, cette zone s'étendant entre l'actuel seuil morvano-vosgien et le faisceau salinois.

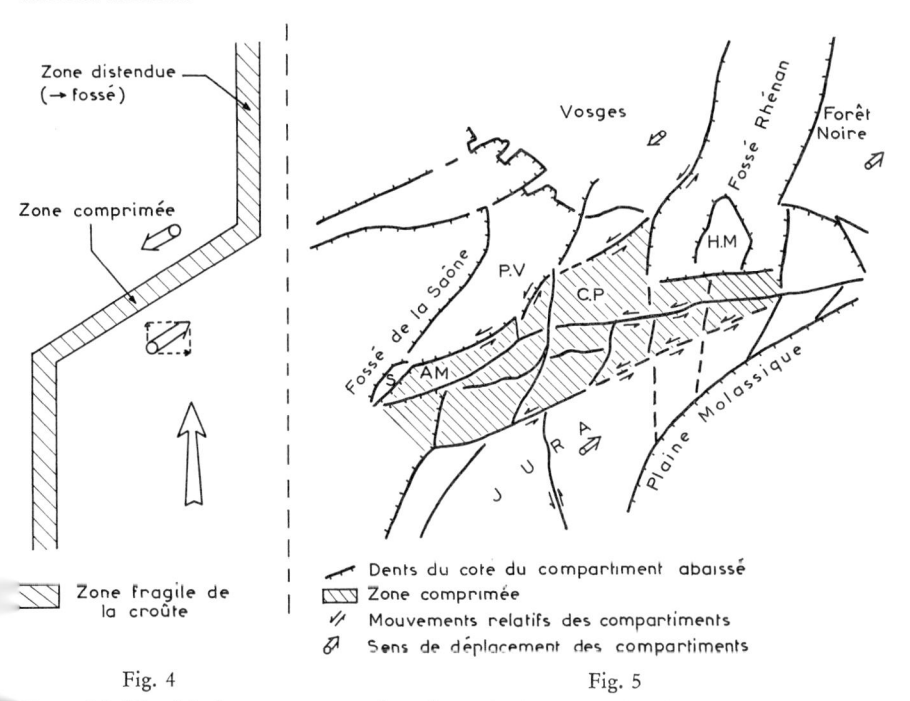

Fig. 4   Fig. 5

Fig. 4. Modèle théorique permettant d'expliquer la formation des fossés sous l'effet d'une poussée venant du S, les blocs glissant les uns par rapport aux autres le long de la zone comprimée.

Fig. 5. Interprétation dynamique de la zone de passage entre le Fossé rhénan et le Fossé de la Saône. PV: Plateau de Vesoul, AV: Avant-Monts, CP: Collines préjurassiennes, HM: Horst de Mulhouse.

Le rapprochement des cratons européens et africano-apuliens, qui a débuté à la fin du Jurassique (DERCOURT, 1970), provoque des modifications dans ces cratons hercyniens: les massifs anciens émergent au Crétacé et à l'Eocène. A l'Oligocène, l'arrêt de l'expansion de l'Atlantique permet au bloc européen de se fracturer sous l'effet de la poussée du bloc africain, le long des zones fragiles.

Le schéma fig. 4 montre deux blocs séparés par une cassure en trois tronçons N—S, NE—SW et N—S. La poussée venant du S provoque un glissement le long de l'accident NE—SW et un écartement des lèvres des cassures N—S. Le mouvement peut se décomposer en une composante E—W, évoquée par LAUBSCHER (1970) qui provoque un déplacement relatif du bloc vosgien par rapport au bloc lédonien et une composante N—S, déjà signalée par de nombreux auteurs en bordure de la Forêt Noire dont WITTMANN, ILLIES, KNETSCH.

Selon WEGMANN (1963), le socle hercynien devait être découpé en une série de compartiments losangiques ou lenticulaires qui ont joué les uns par rapport aux autres. Ainsi, le rejeu du mouvement se répartit sur plusieurs failles et ne provoque pas d'importants décrochements apparents, d'autant plus que la couverture post-triasique s'est désolidarisée du socle et se fracture suivant des directions NNE—SSW sur les plateaux de Vesoul et N—S dans les collines préjurassiennes.

De faibles ondulations E—W ou NW—SE apparaissent sur les plateaux de Vesoul, mais les séries marneuses du Trias et du Lias ne transmettent pas aux calcaires jurassiques les compressions dues aux mouvements du socle.

Conclusion

La géologie de surface de cette région de passage entre le Fossé rhénan et le Fossé de la Saône montre des accidents tectoniques de type rhénan se superposant à des accidents de direction hercynienne.

Une hypothèse est émise pour expliquer la formation des fossés comme la conséquence d'une compression N—S qui aurait provoqué un glissement relatif de blocs de socle de direction NE—SW. La zone de glissement correspondrait à la région comprise entre les failles de l'Ognon et le faisceau salinois (fig. 5).

Les données géophysiques montrent que cette région, particulièrement mobile au cours de la période post-hercynienne, correspond à une zone labile du globe terrestre où la base de l'asthénosphère accuse un fléchissement important (MECHLER & ROCHARD, 1971). Mais les renseignements fournis sont trop incomplets pour donner une interprétation précise des structures profondes.

Aussi, n'avons-nous envisagé l'interprétation dynamique qu'à titre d'hypothèse de travail provisoire. Tandis que nous considérons avoir présenté dans ce travail l'état actuel de nos connaissances sur la structure géologique de la surface.

Bibliographie

AHORNER, L., 1970: Seismo-tectonic relations between the Graben zones of the upper and lower Rhine valley. — Internat. Upper Mantle Project, Sci. Rep. 27, 155—166.

ARTHAUD & MATTAUER, M., 1972: Présentation d'une hypothèse de la virgation pyrénéenne du Languedoc et sur la structure profonde du golfe du Lion. — C. R. Acad. Sc. Paris, 274, 524—527.

BOIGK, H. & SCHÖNEICH, H., 1970: Die Tiefenlage der Permbasis im nördlichen Teil des Oberrheingrabens. — Internat. Upper Mantle Project, Sci. Rep. 27, 45—55.

BULLE, J.; MARTIN, J. & ROLLET, M., 1967: Etude géologique des Avant-Monts entre Venise et Battenans-les-Mines (Doubs). — Ann. Sci. Univ. Besançon, (3), Géol., 3, 51—93.

CONTINI, D., 1970: L'Aalénien et le Bajocien du ‚ura franc-comtois. — Ann. Sci. Univ. Besançon, (3), Géol., 11, 204 p. (y voir bibliographie).

DERCOURT, J., 1970: L'expansion océanique actuelle et fossile: ses implications géotectoniques. — Bull. S. G. F., XII, 261—317.

DREYFUSS, M. & KUNTZ, G., 1970: Carte géologique au 1/50.000, feuille Gy.

FOURQUIN, C.; GUERIN, H.; THEOBALD, N. & THIEBAUT, J., 1971: Données nouvelles sur l'histoire géologique des Vosges méridionales. — C. R. S. S. G. F., 6, 306—308.

GOGUEL, J., 1967: La place du fossé rhénan par rapport aux Alpes occidentales. — The Rhinegraben Progress Rep., 5—6.

ILLIES, J. H., 1970: Graben tectonics as related to crust-mantle interaction. — Graben problems. — Internat. Upper Mantle Project, Sci. Rep., 27, 4—27.

JAVEY, C., 1966: Etude des terrains secondaires de la vallée de l'Ognon entre Voray et Thervay (Jura). — Thèse 3ème cycle, Besançon.

LAUBSCHER, H. P., 1970: Grundsätzliches zur Tektonik des Rheingrabens. — Internat. Upper Mantle Project, Sci. Rep., 27, 79—87.

MATTAUER, M., 1971: Présence de décrochements tardi-hercyniens dans les Vosges méridionales. — C. R. S. S. G. F., 7, 365—367.

MECHLER, P. & ROCARD, Y., 1971: Description d'un relief remarquable présenté sous les Vosges par une couche de discontinuité de vitesse sismique vers 600 km de profondeur. — C. R. Acad. Sc., Paris, 278, 3120—3121.

RUHLAND, M., 1967: Position tectonique du horst de Mulhouse dans le fossé rhénan méridional. — The Rhinegraben Progress Rep., 31—32.

THEOBALD, N. & CONTINI, D., 1960: Présence d'anticlinaux à noyau keupérien dans la dépression au Nord des Collines préjurassiennes et à l'Est de l'Ognon. — C. R. Acad. Sc., Paris, 250, 1092—1093.

— 1965: Structure de la partie orientale du plateau de Vesoul. — Bull. Soc. Hist. Nat. du Doubs, 67, 3, 52—55.

THEOBALD, N., 1969: Rôle des accidents de type rhénan le long de la retombée méridionale des Vosges dans la région de Belfort. — Ber. Naturf. Ges. Freiburg i. Br., 59, 183—191.

— 1972: Géologie et hydrogéologie de la Haute-Saône. — Ann. Sci. Univ. Besançon, (3), Géol., 14, 76 p. (y voir bibliographie).

WINNOCK, E.; BARTHE, A. & GOTTIS, CH., 1967: Résultats des forages pétroliers français effectués dans la région voisine de la frontière suisse. — Schweiz. Petrol. Geol., 33, 83.

WITTMANN, O., 1969: Tektonik des südlichen Oberrheingrabens und seines Rahmens. — Z. dt. geol. Ges., 121, 61—65.

Das Erdbeben von Jeurre vom 21. Juni 1971 und seine Beziehungen zur Tektonik des Faltenjura

von

N. Pavoni und E. Peterschmitt

Mit 4 Abbildungen im Text

The Jeurre Earthquake of June 21, 1971, and its Relationship to the Tectonics of the Jura Mountains

Abstract: The focal mechanisms of the Jeurre earthquake of June 21, 1971, (Fig. 2) as well as of the Clairvaux earthquake of February 5, 1968, (Fig. 3) were investigated. The direction of maximum horizontal pressure as determined by the focal mechanisms of both earthquakes is NW—SE. Whereas the mechanism of the Jeurre earthquake indicates predominantly lateral slip displacement the mechanism of the Clairvaux earthquake indicates thrust movement. Very probably the focus of the Jeurre earthquake was within the sedimentary cover whereas the focus of the Clairvaux earthquake was within the crystalline basement. The NW—SE direction of maximum horizontal pressure corresponds rather well with the direction of maximum shortening as derived from local tectonics (Fig. 4). The contributions of seismological studies to a more thorough knowledge of the tectonics and structural evolution of the Jura Mountains are emphasized.

I. Das Erdbeben von Jeurre vom 21. Juni 1971

Am 21. Juni 1971 um 8.25 Uhr Lokalzeit erfolgte im Gebiet von Jeurre (Tal der Bienne, französischer Jura) ein Erdbeben, welches im Epizentralgebiet eine Intensität von 6 1/2—7 Grad aufwies (Abb. 1). In Jeurre wurden Kamine abgeworfen, und zirka 70% der Häuser, allerdings auf lockerem Untergrund gebaut, zeigten Risse und leichte Schäden im Mauerwerk. Die 3°-Isoseiste ist elliptisch, mit einer etwa 50 km langen Hauptachse in SW—NE-Richtung; die kleine Achse ist 20 km lang. Aus dem makroseismischen Bild läßt sich eine Herdtiefe von 2—3 km abschätzen. Die Magnitude betrug 4 3/4.

Das Beben wurde an zahlreichen mitteleuropäischen Erdbebenstationen registriert. Die Ersteinsätze der P-Wellen ließen vielfach deutlich Kompression oder Dilatation erkennen. Es erschien deshalb angezeigt, eine herdmechanische Analyse des Bebens zu versuchen. Das Resultat dieser Untersuchung ist in Abb. 2 dargestellt. Insgesamt wurden dafür die Aufzeichnungen von 30 Stationen ausgewertet. Wir erkennen für das Beben von Jeurre folgendes Bewegungsbild: Eine erste Verschiebungsebene fällt mit 92°/79° (Azimut/Fallen) steil gegen E. Die Bewegung der hangenden Scholle auf dieser Ebene erfolgte in Richtung 8°/33°. Es handelte sich somit um eine schiefe, sinistrale, laterale Verschiebung. Die zweite Verschiebungsebene fällt mit 188°/58° gegen S. Die Bewegung der hangenden Scholle auf dieser zweiten Ebene erfolgte in Richtung 272°/12° gegen W. Es handelte sich somit um eine dextrale horizontale Verschiebung. Die P-Achse sticht mit 315°/32°, die T-

Das Erdbeben von Jeurre vom 21. Juni 1971 323

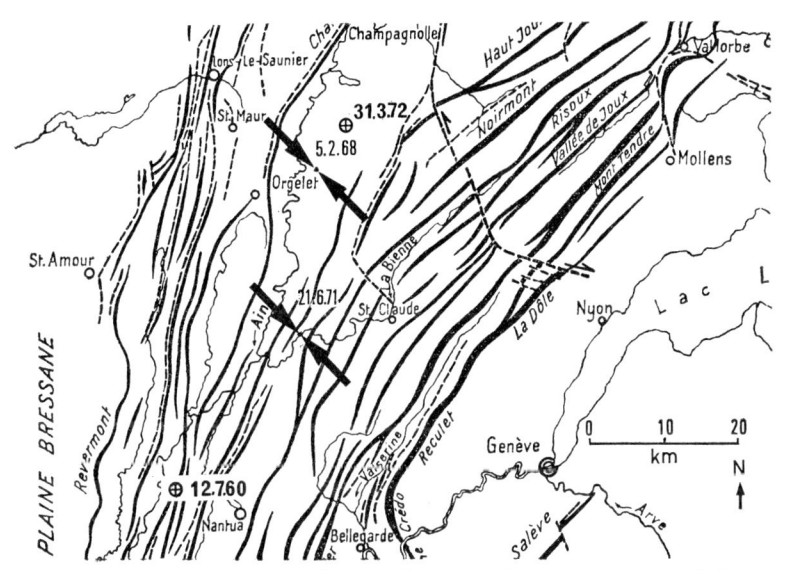

Abb. 1. Lage der Epizentren der Beben in der Gegend von Jeurre. Ausschnitt aus der tektonischen Karte des Jura von A. BERSIER (1934).

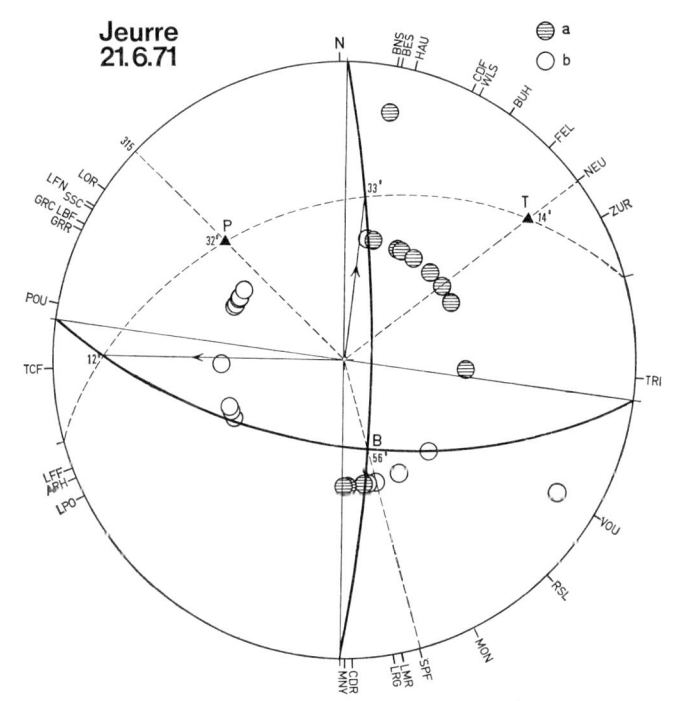

Abb. 2. Herddiagramm des Bebens von Jeurre vom 21. Juni 1971. Untere Halbkugel. a: Kompression, b: Dilatation. P: Durchstichpunkt der P-Achse. T: Durchstichpunkt der T-Achse. B: Durchstichpunkt der B-Achse.

Achse mit 54°/14° und die B-Achse mit 166°/56° durch die untere Halbkugel. Die Horizontalprojektion der P-Achse verläuft somit genau NW—SE. Daß offenbar diese Richtung der P-Achsenprojektion für das in Frage stehende Gebiet nicht zufällig ist, zeigt die Untersuchung eines weiteren Bebens rund 20 km nördlich von Jeurre, des Bebens von Clairvaux vom 5. Februar 1968 (vgl. Abb. 1).

II. Das Beben von Clairvaux vom 5. Februar 1968

In Abb. 3 ist das Ergebnis der herdkinematischen Untersuchung des Bebens von Clairvaux vom 5. Februar 1968 beruhend auf den Daten von 11 Stationen dargestellt. Das Beben besitzt eine Magnitude von 3 1/2. Aufgrund der Intensitätsverteilung ist für dieses Beben mit einer Herdtiefe von 5—6 km zu rechnen. Die Analyse der Herdkinematik ergibt folgendes Bild: Eine erste Verschiebungsebene fällt mit 314°/38° gegen NW, die zweite Verschiebungsebene mit 134°/52° gegen SE. Die P-Achse sticht sehr flach mit 134°/7°, die T-Achse sehr steil mit 314°/83° durch die untere Halbkugel. Die B-Achse liegt horizontal. Beim Bewegungsvorgang handelt es sich für beide in Frage kommenden Ebenen um eine reine Aufschiebung. Die Horizontalprojektion der P-Achse verläuft NW—SE.

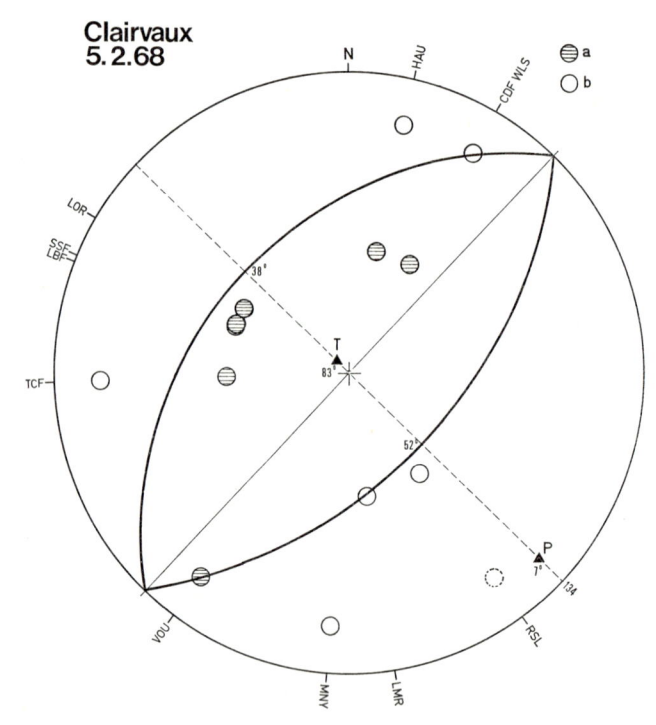

Abb. 3. Herddiagramm des Bebens von Clairvaux vom 5. Februar 1968. Legende wie Abb. 2.

III. Weitere Beben in der Gegend von Jeurre

Weitere Beben fanden in der sonst recht bebenarmen größeren Umgebung von Jeurre statt (Abb. 1).

Am 12. Juli 1960 wurde die Gegend von Nantua mit einer Intensität 5 erschüttert. Das Beben war fühlbar in einem Umkreis von über 30 km, so daß eine oberflächennahe Herdlage im Sediment auszuschließen ist. Mikroseismisch wurde in der selben Gegend ein Erdstoß am 9. Mai 1970 bestimmt. Mangels jeglicher makroseismischer Meldung kann es sich auch um eine größere Sprengung gehandelt haben.

Ein sehr schwaches Beben wurde in der Gegend von Clairvaux am 31. März 1972 mit maximaler Intensität 4 notiert. Der Herd, verschieden von demjenigen von 1968, ist in geringer Tiefe anzunehmen. Das mikroseismische Material dieses Bebens reicht für eine herdmechanische Untersuchung nicht aus.

IV. Herdmechanismen und lokale Tektonik

Die Epizentren der Beben liegen im Gebiet zwischen dem Faisceau lédonien, dem äußeren Faltenjura, und der Haute Chaîne, der inneren Hauptkette des Faltenjura. Das Epizentralgebiet des Bebens von Jeurre ist auf dem sehr schönen, neuen Blatt „Moirans-en-Montagne" der geologischen Karte 1:50 000 enthalten. Abb. 4 zeigt einen Ausschnitt aus dem tektonischen Kärtchen der Erläuterungen zu diesem Blatt (J. TRICART et al. 1970).

Der Verlauf der Antiklinal- und Synklinalachsen im Gebiet von Jeurre (Synklinale von Montcusel, Antiklinale von Martigna, Synklinale von Jeurre) und die Überschiebung am Ruisseau d'Heria lassen einen stärksten Zusammenschub des

Abb. 4. Lokale Tektonik des Faltenjura im Gebiet von Jeurre und Umgebung. Ausschnitt aus der tektonischen Karte von J. TRICART, M. CLIN & J. PERRIAUX (1970). Stern: Epizentrum des Bebens von Jeurre. Pfeile: Richtung der größten horizontalen Druckspannung. 10: Antiklinale von Martigna, 12: Antiklinale von Vaux-lés-Saint-Claude, i: Synklinale von Montcusel, j: Synklinale von Jeurre, H: Überschiebung am Ruisseau d'Héria. A: Ain, B: Bienne, S: Stausee, V: Staudamm von Vouglans.

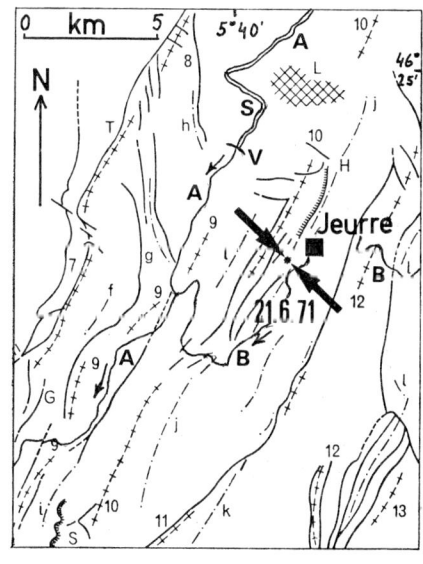

Faltenjura in diesem Gebiet in etwa Richtung N 60° W — S 60° E erkennen. Die Richtung der Horizontalprojektion der P-Achse, wie sie sowohl für das Beben von Jeurre wie auch für das Beben von Clairvaux bestimmt wurde, steht somit in guter Übereinstimmung mit der Richtung des stärksten Zusammenschubes im betreffenden Gebiet. Ein möglicher Zusammenhang zwischen neogener und rezenter Tektonik ist damit angezeigt.

V. Erdbeben und Tektonik des Faltenjura

Eine Untersuchung der Erdbeben im Jura als Ausdruck heute sich vollziehender tektonischer Bewegungen ist in mehrfacher Beziehung von Interesse für die genauere Kenntnis des Juragebirges und seines Untergrundes und ist durchaus geeignet, Hinweise für die Entstehung des Gebirges zu liefern. In bezug auf die Entstehung des Faltenjura bestehen heute bekanntlich zwei im Prinzip grundlegend verschiedene Hypothesen. Unbestritten ist, daß über einem Sockel eine vollständig abgescherte und in sich verfaltete mesozoisch-tertiäre Sedimentdecke liegt. Die entscheidende Frage lautet: Wurde der Sockel während der Faltung mitbewegt oder nicht? Steht die Faltung der Sedimentdecke mit Bewegungen im Sockel in ursächlichem Zusammenhang (Faltung durch Sockelzerscherung) oder verhielt sich der Sockel völlig passiv, unbewegt (Fernschub und Schweregleitung)?

In den folgenden Gesichtspunkten ist eine Untersuchung der Erdbeben von großer Bedeutung:

1. Die Verteilung der Epizentren: Sie liefert Hinweise über die geographische Verteilung der tektonischen Aktivität. Seismizität ist im Gebiet des Faltenjura eindeutig vorhanden. Die Errichtung zusätzlicher Stationen wird die Feststellung und genaue Lokalisierung auch schwacher Beben ermöglichen. Die Errichtung der Station Vouglans im Jahr 1967 bietet dafür ein gutes Beispiel.

2. Die Tiefenlage der Erdbebenherde: Eine wesentliche Frage in bezug auf die Juratektonik lautet: Gibt es Herde, die im Sockel liegen, oder liegen alle Herde in der Sedimentdecke oder im Bereich der Abscherungszone? Gibt es Herde, die tiefer als 2 km liegen? Aufgrund der Intensitätsverteilung kann geschlossen werden, daß zahlreiche Jurabeben einen sehr wenig tiefen Herd besitzen und somit innerhalb der Sedimentdecke oder im Bereich der Abscherungszone liegen. Es gibt anderseits eine ganze Reihe von Beben, deren Herde eindeutig im Sockel liegen (PAVONI 1966).

3. Die Untersuchung der Herdmechanismen: Die beiden Beben sind die ersten Beben im Faltenjura, die in bezug auf ihren Herdmechanismus hin untersucht wurden. Das Beben von Jeurre mit Horizontalverschiebungscharakter liegt vermutlich noch in der Sedimentdecke oder in der Abscherungszone. Das Beben von Clairvaux mit Aufschiebungscharakter liegt, aufgrund der Intensitätsverteilung zu schließen, im Sockel. Ist das unterschiedliche Bewegungsverhalten auf die unterschiedliche Tiefenlage zurückzuführen? Bemerkenswert ist jedenfalls die übereinstimmende Richtung der größten horizontalen Druckspannung bei beiden Beben mit der Richtung der stärksten Einengung des jungen Faltengebirges. Es wird von größtem

Interesse sein, für weitere Beben die Frage zu untersuchen, inwieweit im Faltenjura Zusammenhänge zwischen der Herdkinematik und der Kinematik abgeleitet aus dem tektonischen Deformationsbild bestehen.

4. Die zeitliche Abfolge des Bebens: Sie vermag Hinweise auf regionale tektonische Zusammenhänge zu liefern, Zusammenhänge des Faltenjuragebietes mit seiner weiteren Umgebung.

5. Eine eingehende Untersuchung der Laufzeiten und Phasen von Beben im Faltenjura und seiner Umgebung vermag Auskünfte über den Aufbau der Kruste unter dem Gebirge zu geben.

Es scheint, daß die Untersuchung der Erdbeben einige Argumente zugunsten einer tektonischen Mitbeteiligung des Sockels bei der Faltung des Faltenjura liefert. Im Falle einer Faltung durch Fernschub und Schweregleitung ist jedenfalls anzunehmen, daß der Faltungsvorgang längst abgeschlossen ist, da der Fernschub durch Schweregleitung infolge der starken Erosion und veränderten Geometrie heute tot ist. Seismische und damit tektonische Aktivität ist heute jedoch sowohl im Sockel wie in der Sedimentdecke vorhanden. Wie die Herdmechanismen der beiden Beben zeigen, entspricht der heutige Spannungszustand im Prinzip dem für die Faltung und Krustenverkürzung im Jura geforderten.

Die Untersuchungsergebnisse für die Beben von Jeurre und Clairvaux stehen in guter Übereinstimmung mit den Resultaten von Untersuchungen über Herdmechanismen von Erdbeben in Mitteleuropa, wie sie durch die Arbeiten von HILLER (1936), SCHNEIDER et al. (1966), SCHNEIDER (1967) und AHORNER (1967, 1970) bekannt geworden sind, die auf eine NNW—SSE gerichtete maximale horizontale Druckspannung in der Kruste Mitteleuropas schließen lassen. Im Vergleich zu dieser Richtung erscheint die Horizontalprojektion der P-Achsen der beiden Jurabeben im Gegenuhrzeigersinn in NW—SE-Richtung abgedreht. Hängt diese Abweichung mit der bogenförmigen Erstreckung des Gebirgszuges zusammen oder ist sie lokal bedingt? Für das Beben von Corrençon vom 25. April 1962 findet SCHEIDEGGER (1967) eine W—E, d. h. quer zum Westalpenbogen gerichtete größte horizontale Druckspannung. Die herdmechanische Untersuchung weiterer Beben im Jura und im Westalpenbogen wird zeigen, ein wie enger Zusammenhang zwischen rezenter und neogener Tektonik sich erhalten hat.

Neben der Verbindung zum Westalpenbogen ist die Beziehung des Faltenjura zum südlichen Rheingraben und zum Bressegraben von besonderer Bedeutung. Zwischen Bresse und Rheingraben wird heute von LAUBSCHER (1970) und ILLIES (1971) eine Transform Fault Zone postuliert. LAUBSCHER verlegt diese Blattverschiebungszone mit sinistralem Verschiebungssinn in das Gebiet zwischen Ognonlinie und Mont Terri-Landsberg-Lägern-Linie, somit in eine ziemlich genau W—E verlaufende Zone. ILLIES läßt, in Übereinstimmung mit AHORNER (1970), die Kruste beidseits der Gräben schief zum Verlauf der Grabenachsen in ENE—WSW-Richtung auseinanderweichen.

Eine W—E verlaufende, sinistrale Blattverschiebung am Nordrand des Faltenjuragebietes steht, falls heute aktiv, im Widerspruch zum Bewegungsbild abgeleitet aus den Herdmechanismen der Beben von Jeurre und Clairvaux und ebenso mit den Ergebnissen herdmechanischer Untersuchungen weiterer mitteleuropäischer Beben. Eine Streckung der Kruste in ENE—WSW- bis NE—SW-Richtung erscheint damit nicht im Widerspruch. Die Untersuchung der Herdmechanismen von Beben im Bereich der Burgundischen Pforte wird deshalb von ganz besonderem Interesse sein.

VI. Möglicher Zusammenhang des Bebens von Jeurre mit dem Stausee von Vouglans

Das Epizentrum des Bebens von Jeurre liegt nur etwa 5 km süd-östlich des Staudammes von Vouglans (Abb. 4). Diese geringe Entfernung zwingt zur Fragestellung, ob nicht ein Zusammenhang zwischen dem Beben und dem 1968 im Aintal geschaffenen Stausee besteht.

Die Wahrscheinlichkeit eines solchen Zusammenhanges wird aus folgenden Gründen herabgesetzt:

1. der Herd liegt außerhalb der Stauung, und zwar talwärts;

2. zwischen der ersten Füllung — 70 m, Frühjahr 1968 —, der maximalen erreichten Stauhöhe — 100 m, Januar, Mai und Oktober 1970 — und der Bebenauslösung — 26. VI. 1971 — liegt eine weit größere Zeitspanne als die bei solchen Zusammenhängen beobachteten;

3. es fanden keine Vorbeben statt;

4. die Wassereinpressung als mögliche Ursache läßt sich schwer mit der lokalen Geologie vereinbaren — Vorhandensein der Martigna-Antiklinale zwischen dem Stausee und dem Herdgebiet, Abb. 4;

5. in der größeren Umgebung von Jeurre sind innere Spannungen vorhanden, die durch die Beben von Clairvaux und Nantua, unabhängig von der Stauung, bezeugt sind (PETERSCHMITT 1971).

Es wurde in dieser Arbeit gezeigt, daß der Herdmechanismus des Jeurre-Bebens mit der geologischen Kinematik des Faltenjura gut übereinstimmt. Diese Übereinstimmung deutet darauf hin, daß die Stauung, wenn sie mit dem Beben im Zusammenhang steht, mit der größten Wahrscheinlichkeit nur zur Auslösung vorgegebener Spannungen beigetragen hat. Zu genau derselben Schlußfolgerung kommen HAGIWARA & OHTAKE (1972) bei ihrer herdmechanischen Bearbeitung der seismischen Aktivität im eindeutigen Zusammenhang mit dem Kurabe-Stausee, Japan.

Literatur

AHORNER, L., 1967: Herdmechanismen rheinischer Erdbeben und der seismotektonische Beanspruchungsplan im nordwestlichen Mittel-Europa. — Sonderveröff. Geol. Inst. Köln, **13**, 109—130.

— 1970: Seismo-tectonic Relations between the Graben Zones of the Upper and Lower Rhine Valley. — In: ILLIES, H. & MUELLER, ST. (eds.), Graben Problems, 155—166, Stuttgart.

BERSIER, A., 1934: Carte tectonique du Jura. — Geol. Führer d. Schweiz, Fasc. I, pl. III, Wepf, Basel.

HAGIWARA, T. & OHTAKE, M., 1972: Seismic activity associated with the filling of the reservoir behind the Kurabe Dam, Japan, 1963—1970. — Tectonophysics, **15**, 241—254.
HILLER, W., 1936: Das Oberschwäbische Erdbeben am 27. Juni 1935. — Württemb. Jb. f. Statistik u. Landeskde., 1934/35, 209—206.
ILLIES, H., 1971: Der Rheingraben. — Fridericiana, Z. Univ. Karlsruhe, H. 9, 17—32.
LAUBSCHER, H. P., 1970: Grundsätzliches zur Tektonik des Rheingrabens. — In: ILLIES, H. & MUELLER, ST. (eds.), Graben Problems, 79—87, Stuttgart.
PAVONI, N., 1966: Kriterien zur Beurteilung der Rolle des Sockels bei der Faltung des Faltenjura. — In: SCHAER, J.-P. (ed.), Etages Tectoniques, 307—314, A la Baconnière, Neuchâtel.
PETERSCHMITT, E., 1971: Informations sur le séisme de Jeurre, Jura, 21 juin 1971. Relations avec la séismicité locale du cours supérieur et moyen de l'Ain et la retenue de Vouglans. — Manuskript, S. 19.
SCHEIDEGGER, A. E., 1967: The tectonic stress in the vicinity of the Alps. — Z. Geophys., **33**, 167—181.
SCHNEIDER, G., 1967: Erdbeben und Tektonik in Südwest-Deutschland. — Tectonophysics, **5**, 459—511.
SCHNEIDER, G., SCHICK, R. & BERCKHEMER, H., 1966: Fault-plane solutions of earthquakes in Baden-Württemberg. — Z. Geophys., **32**, 383—393.
TRICART, J., CLIN, M. & PERRIAUX, ,., 1970: Notice explicative. Feuille "Moirans-en-Montagne", Carte Géologique au 1:50 000, XXXII-28. — B. R. G. M., Orléans.

Deep Seismic Sounding in the Limagne Graben[1]

by

A. Hirn and G. Perrier

With 7 figures in the text

Abstract: Some geological and geophysical facts about formation and present state of the Limagne region are given. Structure of the crust is derived from a 180 km long, partly reversed, seismic profile in the graben axis and compared with the structure found in the Rhinegraben region. The anomalous lower crust and upper mantle is characteristic of an extension.

The French Massif Central (Fig. 1), part of the hercynian belt of late Paleozoïc folding, is an antetriassic peneplain reactivated since Jurassic times. After a large Jurassic transgression covering most parts of the peneplain, Oligocene ages saw the appearance of subsidence and the formation of large depressions (Limagne d'Allier, basins of Montbrison and Roanne, . . .). Volcanic activity, beginning in the Limagne at this time migrated to the west later on with a Villafranchian paroxysm. A simultaneous rising of the whole Massif Central with tilting from east to west took place. In Quaternary times, volcanic activity migrated back to the east, on the western marginal fault of the Limagne with most recent volcanoes only 2500 years old.

---
[1] Contribution IPG NS No 50, Institut de Physique du Globe, Université de Paris VI.

The Limagne region

The Limagne d'Allier (to be noted simply Limagne hereafter) west of and the basins of Montbrison and Roanne east of the Forez horst form the Limagne region. The western and eastern depressions join in the region of Moulins, Oligocene sediments getting progressively thinner to the north and vanishing at the latitude of Nevers.

The mean thickness of sediments is greater in the western than in the eastern part of the Limagne, the greatest depth of the basement occurring just north of Clermont-Ferrand (Fig. 2). To the south (Brioude) the basement rises; a complex faulted area shows alternating outcrops of crystalline rocks and Oligocene sediments unaffected by erosion. The original extension of sediments in this region is not known.

The Bouguer anomaly map shows a mean value of —35 to —40 mgal. In the Rhinegraben, this anomaly is smaller (—25 mgal). However, the whole Massif Central is undercompensated, mean isostatic anomaly being 25 mgal for a depth of compensation of 30 km (Airy). Anomalies of total magnetic field computed for a height of 13 km show that, on the north of Vichy and Moulins anomalies, the Massif Central extends to the north-east as far as the south of Alsace (AUBERT & PERRIER 1972). Present earthquake activity is much smaller than in the Rhinegraben region, and seems to be confined to the graben itself (Vichy and Brioude).

Seismic profiles in the area of the Limagne

The results of a north—south seismic profile of 180 km length in the axis of the Limagne graben will be discussed here. It was recorded during the 1970 experiment of the "Opération Grands Profils Sismiques" (Institut National d'Astronomie et de Géophysique) which covered the western half of the Massif Central. Shotpoints were located near Moulins (northern end) and Langeac (southern end). Two shots at each end allowed to obtain about 50 records between 0 and 180 km from each shotpoint giving a partly reversed profile. Recording was achieved with 3-components 1 Hz geophones and frequency multiplex tape recording systems.

Some results from a similar profile on the western flank of the graben will be given in the discussion.

Profile 1 — Langeac (South) — Moulins (North)

0.4 and 0.8 tons borehole shots were fired in gneiss SW of Langeac (Fig. 3). The surface velocity of 4.88 km/s was derived by reducing the arrival times to the mean ground slope, difference in heights being of the order of 300 m. Between 6 and 19 km first arrivals may be approximated by the straight line $t(s) = 0.11 s \pm 0.01 + \triangle (km) / (5.47 \pm 0.03 [km/s])$ which corresponds to a mean thickness of the superficial layer of 0.6 km.

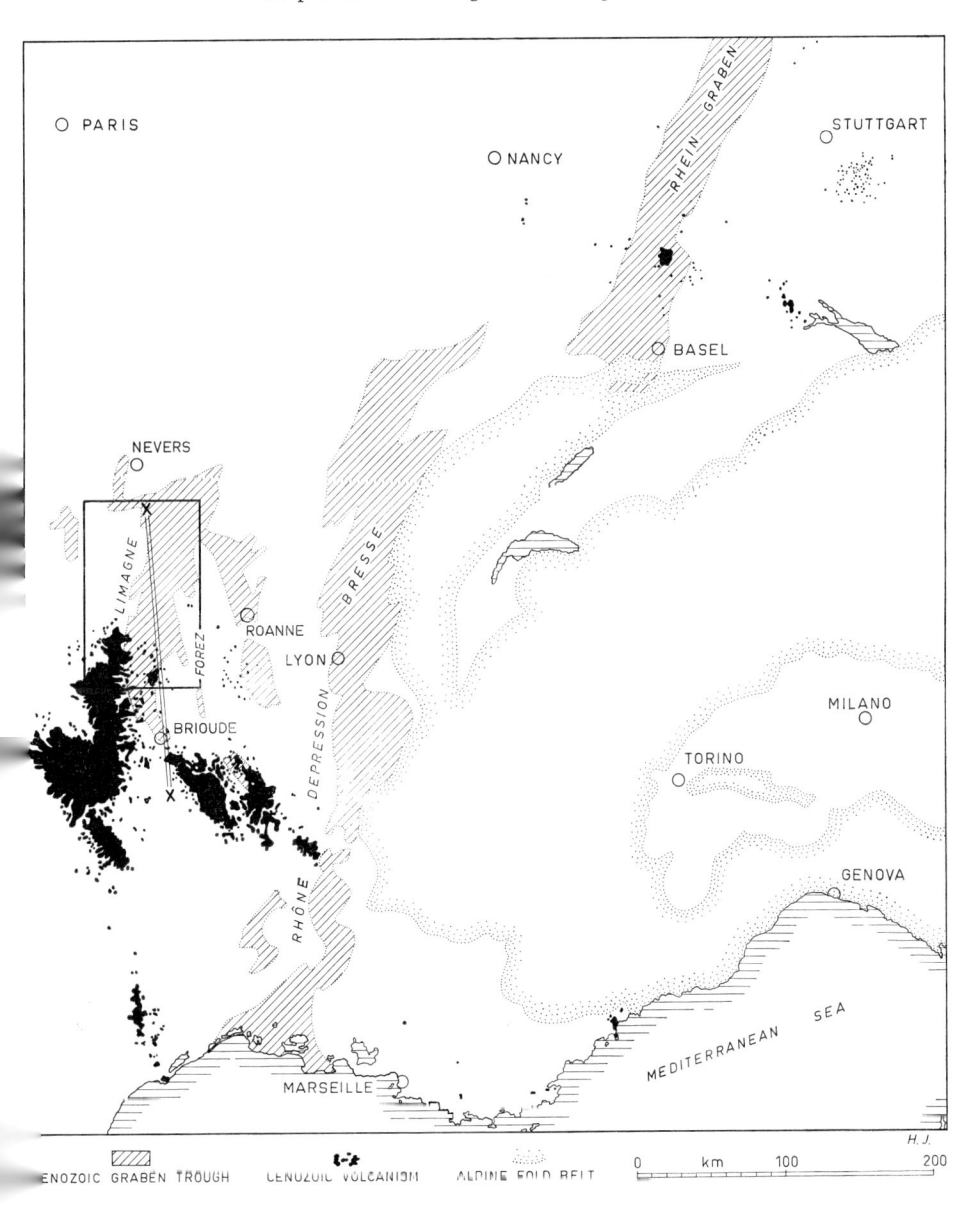

Fig. 1. Mid european grabens (after ILLIES) with recording line in the Limagne.

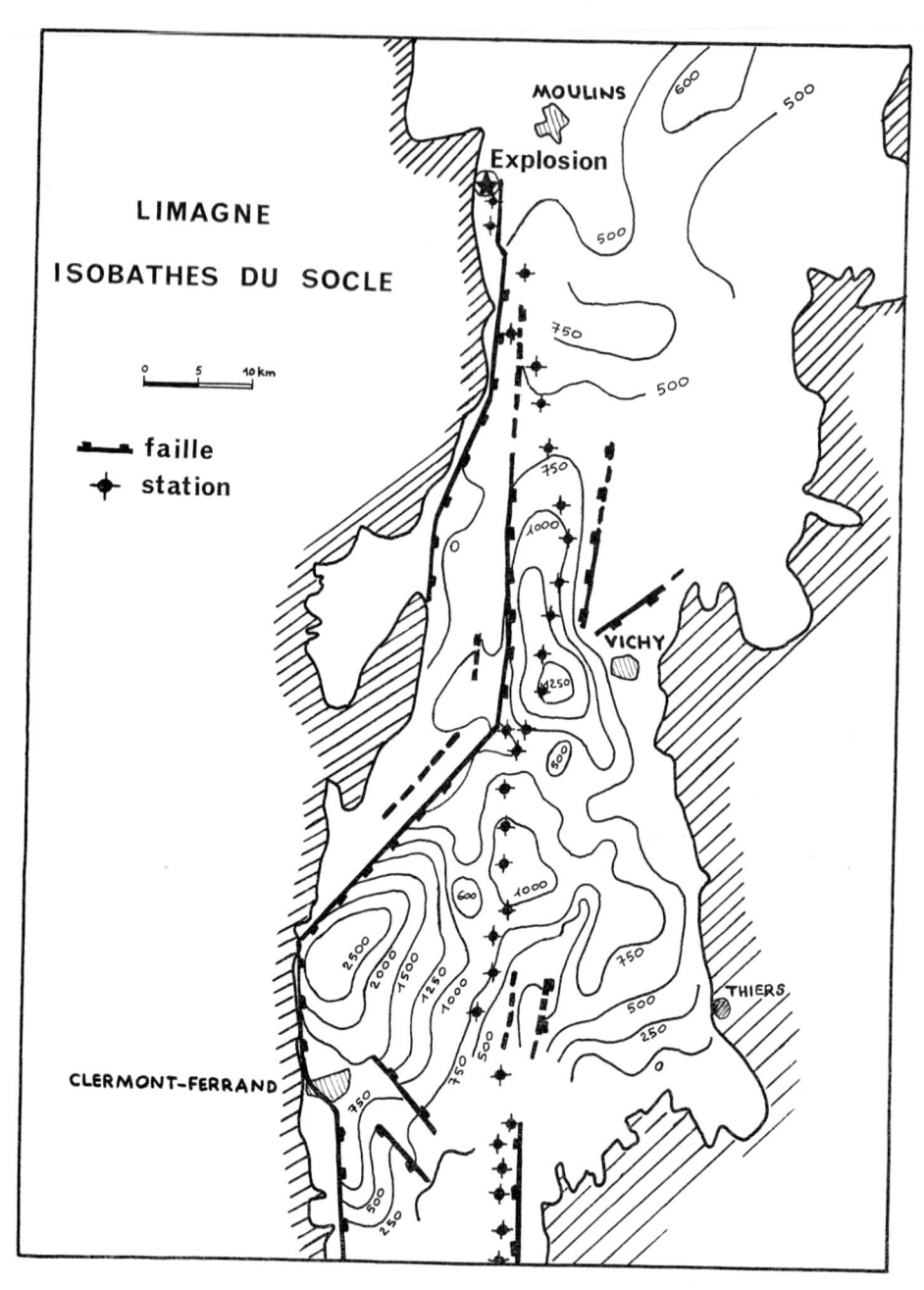

Fig. 2. Schematic isobaths of sediments-basement (after Morange et al.); northern shot-point and some recording points.

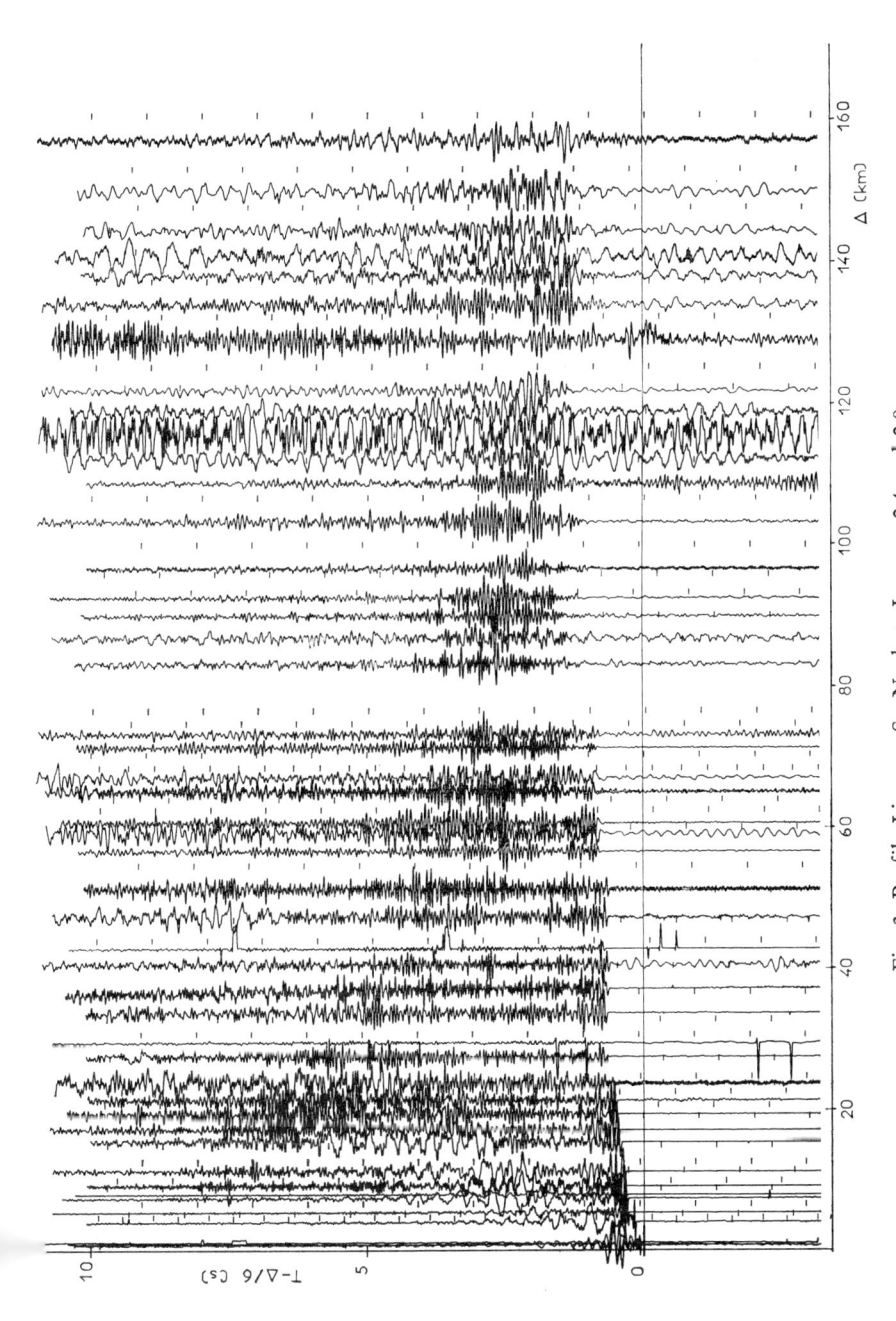

Fig. 3. Profile Limagne S—N, shots Langeac 0.4 and 0.8 tons.

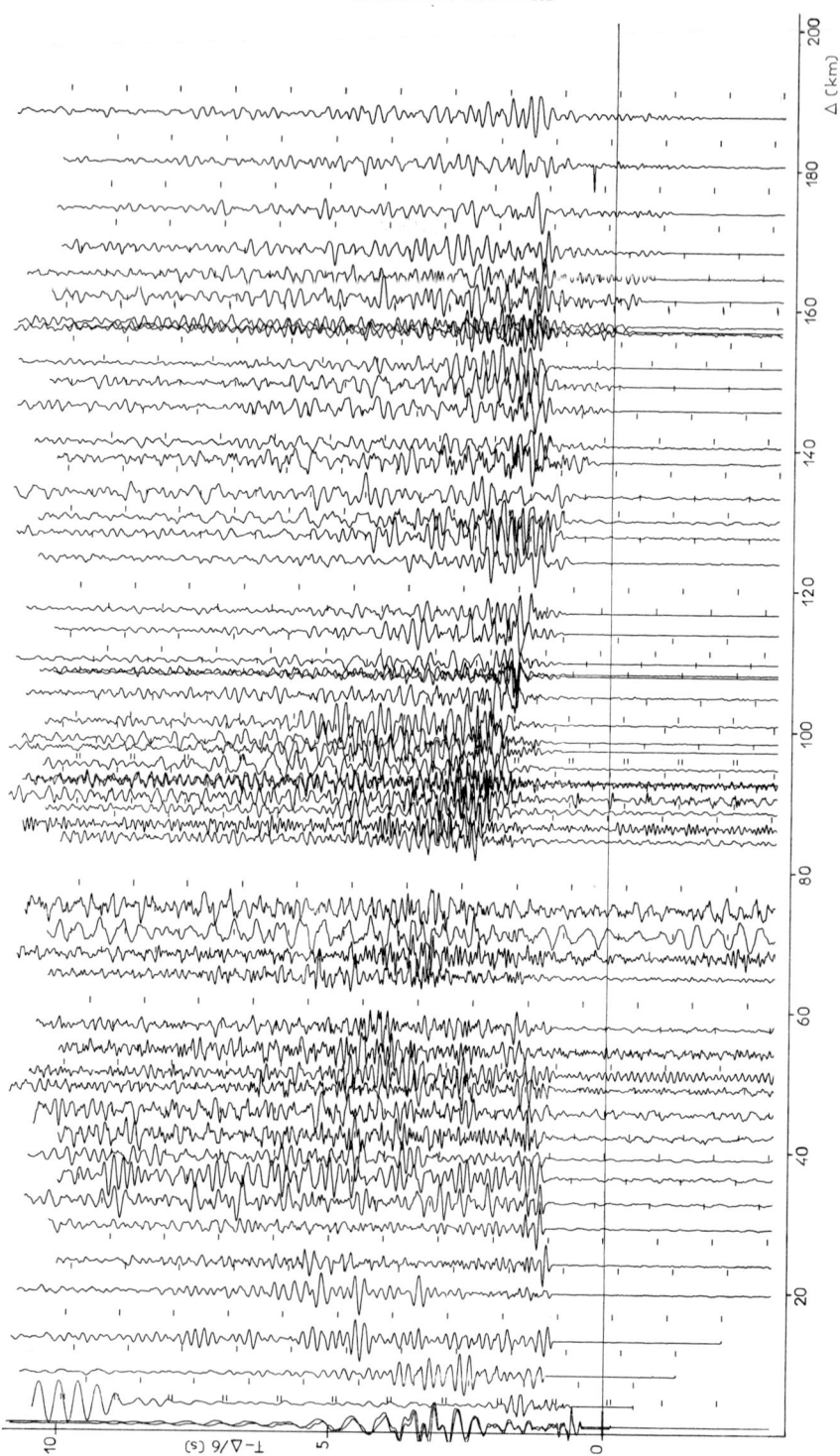

Fig. 4. Profile Limagne N—S, shots Moulins 0.7 and 1.9 tons.

— The travel time curve of P_g waves, recorded as first arrivals from 20 to 115 km shows strong variations caused by the alternating basement and Oligocene outcrops. After correcting for superficial layers, observed curves are:

30 to 51 km	$t = 0.52 \pm 0.02 + \Delta / (6.08 \pm 0.22)$
56 to 73 km	$t = 0.13 \pm 0.04 + \Delta / (5.77 \pm 0.02)$
92 to 115 km	$t = 0.59 \pm 0.02 + \Delta / (6.02 \pm 0.02)$

These waves travel with a true velocity of 6.04 km/s beneath a refractor at a mean depth of 3 km.

— At greater distances (133 to 157 km) the travel time curve of first arrivals assumed to be refracted waves with equation $t = 4.37 \pm 0.02 + \Delta / 7.39 \pm 0.02$ can be correlated with the aid of special filtering. The velocity being assumed as the true one (confirmed by reverse shooting), the depth to the refractor is 22.5 km.

— Strong later arrivals are to be seen from 20 km to the end of the profile. They are characterized by high frequency and long duration of oscillations, the onset of which is not very clear. $T^2 - \Delta^2$ method applied in the distance range 47 to 156 km gives strong support to the hypothesis that they are waves reflected beneath a layer of mean velocity 5.98 km/s at a depth of 21.8 km (with a strong scatter of about 2.5 km). The mean velocity and the depth are in good agreement with those found by the refraction method, the rather low value of mean velocity in the upper crust is caused by the sediments and superficial layers.

The profile is too short and the signal to noise ratio too low to give consistent data for deeper layers.

Profile 2 — Moulins — Langeac

Northern borehole shots of 0.7 and 1.9 tons were detonated in sediments, this accounting for the lower frequency contents of recorded signals (Fig. 4).

At small distances a very strong phase of apparent velocity 5.5 km/s is recorded, but in this case as a second arrival. This is an argument for considering the 5.5—6.04 boundary at a depth of 3 km as a sharp discontinuity.

— The P_g wave recorded as a first arrival from 8 to 115 km, again follows different crustal blocks:

8 to 24 km	$t = 0.85 \pm 0.03 + \Delta / (6.07 \pm 0.06)$
29 to 67 km	$t = 0.55 \pm 0.07 + \Delta / (6.08 \pm 0.03)$
104 to 117 km	$t = 0.81 \pm 0.03 + \Delta / (6.08 \pm 0.02)$

The upper part of Fig. 6 shows the restitution of the base of sediments (V — 2 km/s from soundings) (MORANGE et al. 1972) and superficial layer (V = 5.47 km/s).

— The apparent velocity of first arrivals between 90 and 190 km (Fig. 5) jumps from 6.04 to 7.5—7.8 km/s and then to more than 8.4 km/s, strong second arrivals being observed before 120 km with velocities of 6.7 and 7.5 km/s and between 130 and 190 km with an apparent velocity of 8.2 km/s.

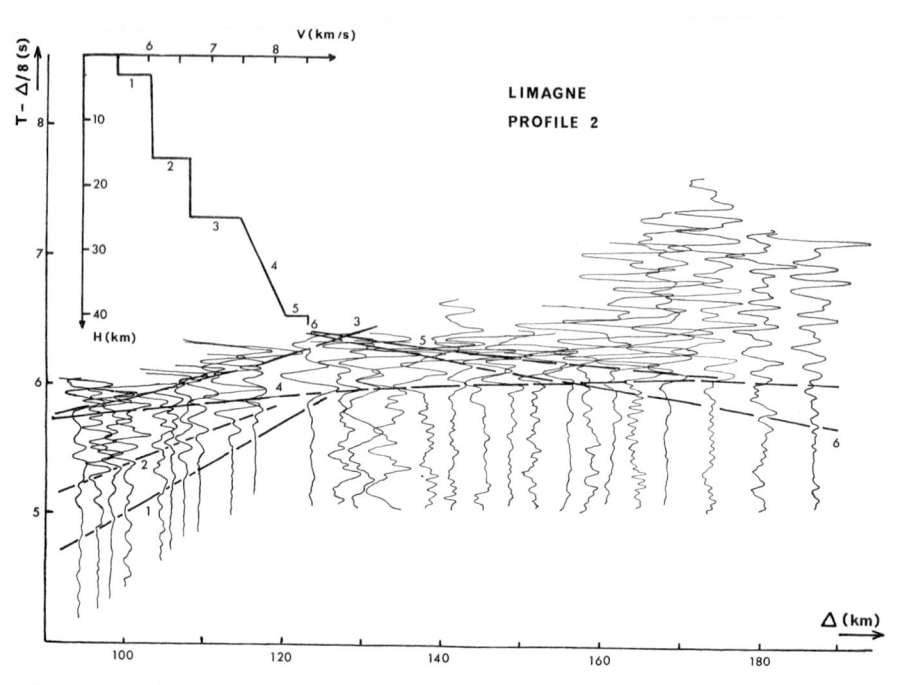

Fig. 5. Record section of profile 02 (1.9 t shot Moulins). Times reduced with velocity 8 km/s. Sensitivity constant from 95 to 120 km, from 120 to 140 km and from 140 to the end.

Assuming plane layers along the whole profile for depths greater than 10 km allows this complicated pattern of waves to be explained by a model in which the velocity jumps from 6.04 to 6.7 km/s at 16 km depth and from 6.7 to 7.5 km/s at 24 km. If the 7.5 branch as a second arrival before 125 km and the 7.6—7.9 as a first arrival afterwards are to be correlated, they may be interpreted as corresponding to a diving wave, beneath the 24 km deep 6.7 to 7.5 km/s discontinuity in a medium with strong gradient (about 0.05 s^{-1} to account for travel time curvature).

Another argument for the presence of such a gradient is the fact that, if the wave with apparent velocity above 8.4 km/s as a second arrival from 120 to 150 km and perhaps first arrival afterwards (although signal to noise ratio gets poor) is to be interpreted as a wave reflected and refracted at the upper limit of a medium of 8.4 km/s, it has a critical distance of 120 km which would be 90 km only, if the overlaying layer was of constant 7.4 km/s velocity. An increase of velocity from 7.5 km/s at 24 km depth to 8.2 km/s at about 40 km depth and then a jump to 8.6 km/s is a possible explanation for this travel time pattern. The 8.6 km/s velocity is only an estimate, taken from the reflection branch (second arrivals), of the apparent velocity under the deepest refractor, the dip of which is not known.

At last it must be said that the value of 6.7 km/s at 16 km depth is possibly an appearance of a layer the depth of which seems to vary along the profile (Fig. 6).

— From 65 km onward reflected waves may well be correlated, as in profile 1. They are of long duration but the lower frequency character does not allow them to be seen before 50 km. Mean depth of the reflector is again 21.0 ± 2.5 km.

Discussion

The results have to be compared with those obtained in the Rhinegraben.

1. If we consider superficial layers we find in the Limagne a strong reflector at 3 km depth (5.5—6.0 km/s). This feature of the uppermost crust is very different from the continuous increase of velocity in the first kilometers of the crust with strong gradients as is often found or assumed.

2. Considering the question of the presence of a velocity reversal in the upper 15 kilometers of the crust, as it is proposed in the Rhinegraben (ANSORGE et al. 1970), the following statements may be made for the Limagne:

— at great distances (150—180 km) the delay of reflected waves with respect to the prolongation of direct waves travel time curve (P_g) is very small, the delay of about 1 s with respect to the line $t - \triangle / 6 = 0$ is introduced by sedimentary and superficial layers;

— at distances smaller than 110 km, although some energy is visible in front of the main reflected phase, the correlation from one record to the next is difficult, indicating a rough or not well defined interface. These arrivals cannot be followed close to the shotpoints, their amplitudes being very small. However,

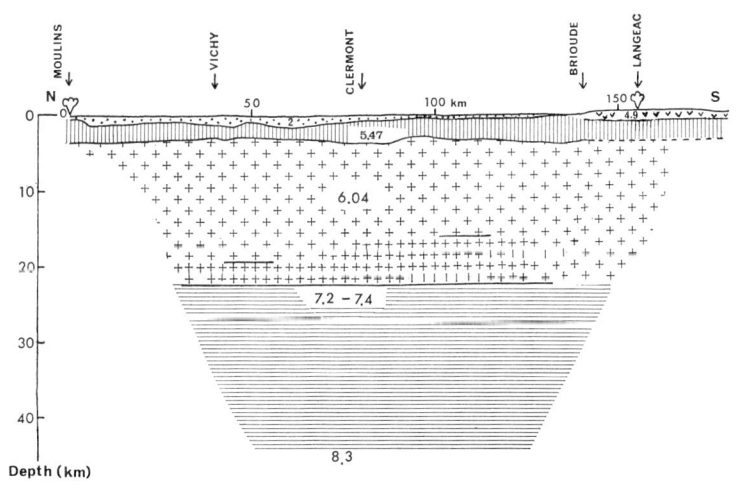

Fig. 6. North-South crustal section in the axis of the Limagne graben and mean velocity-depth distribution from deep seismic sounding.

from the lack of evidence for the presence of a thick low velocity layer with sharp boundaries in the sialic crust, it should not be concluded that velocity reversals are absent in the Limagne region. Such low velocity layers appearing together with high velocity ones as laminated medium might be responsible for the highly ringing characteristic of waves (FUCHS 1969) reflected at the top of the 7.4 km/s boundary and at different interfaces of the laminated medium lying underneath.

An important feature in a tentative comparison of the Limagne and the Rhinegraben structures is the existence in the Limagne of a boundary at 24 km depth clearly marked by reflected as well as by refracted waves. Depths in the Rhinegraben are somewhat higher, one reason for this could be the fact that profiles there are not in the axis of the graben. Still more important is the fair evidence in the Limagne of this reflector at 24 km depth being underlain by a layer with velocity increasing from 7.5 to about 8.2 km/s at 40 km depth and the presence of fairly well developped wave trains coming up from a medium of velocity over 8.4 km/s below this depth. Some of these last features were found in earlier work in the Rhinegraben region also (ANSORGE et al. 1970), a layer of 7.6—7.7 km/s between 25 and 40 km depth being called the "rift cushion" and proposed to be characteristic of regions having undergone a phase of extension.

A very recent experiment in the Rhinegraben region (see this book) where, as in the Limagne graben, profiles were shot in the graben axis, has somewhat changed the picture of the structure of the crust, derived from profiles transverse to this axis or on the flanks, by replacing the 7.6 km/s velocity at 24 km depth by a velocity higher than 8 km/s, the old upper limit of the cushion being now defined as the Mohorovičić discontinuity.

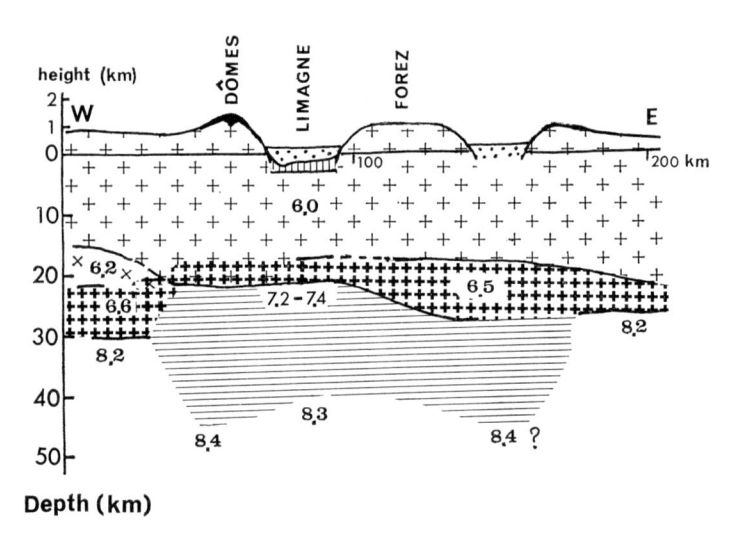

Fig. 7. West—East simplified structure of the crust of the graben region at the latitude of Clermont-Ferrand.

General conclusions

Results derived from a similar profile in the linear volcanic area, 15 km west of the Limagne graben and on an older profile in the Forez horst have been used for the schematic W—E crustal section at the latitude of Clermont-Ferrand (Fig. 7).

Under the volcanic region a reflector at 23 km depth, on top of a layer with velocity increasing from about 7.2 to 8.4 km/s at 45 km depth (PERRIER & RUEGG 1972), from which diving waves are recorded, shows a similar picture of crustal structure as that under the Limagne depression itself. Under the Forez horst, this main reflector is found at a somewhat greater depth (28 km) (PERRIER 1963); it is overlain by a 6.5 layer, the western part of which may be considered responsible for giving scattered reflections at 16 and 20 km depth beneath the graben. To the west, under the volcanic area waves possibly reflected at the top of 6.5 layer are even more scattered and the presence of an interface 6.1 to 6.5 is doubtful. No deeply refracted waves were recorded on the older Forez profile. But at the junction of the western and eastern depressions north of the Forez, waves reflected at a depth of 47 km (V > 8.3 km/s) (AUBERT & PERRIER 1972) give indications for a somewhat different nature of the 7.4 to 8.3 transition.

As a result of deep seismic sounding in the Limagne graben discussed here and of investigations in the neighbouring regions, the presence of a prominent reflector at depths of 20 to 24 km must be pointed out. In other parts of the continental crust the strongest reflector is most often situated at depths around 30 km with the exception of regions having undergone strong, tectonically activated, tension or deformation, for instance extension as in the Rhinegraben region. It seems now that the reflector situated at about 25 km depth beneath the Rhinegraben separates rocks with velocities under 7 km/s corresponding to a normal crustal velocity from rocks of velocity above 8 km/s corresponding to the upper mantle and may be considered as a Mohorovičić discontinuity in anomalous shallow position.

This anomalously high position of the major reflecting horizon found in both regions of extension may be associated with the graben structure. A major difference between the two graben structure is the presence of a layer with velocities ranging from 7.5 to more than 8 km/s between 24 and 40 km depth between the crust (V < 7 km/s) and the mantle (V > 8.2 km/s) beneath the Limagne. Possible causes for this difference may be related to the somewhat different geological history posterior to the formation of the graben, perhaps to the uplift of the Massif Central as a whole and more general occurrence of volcanic activity than in the Vosges — Rhinegraben — Black forest region.

References

ANSORGE, J., EMTER, D., FUCHS, K., LAUER, J. P., MÜLLER, S. & PETERSCHMITT, E., 1970: Structure of the crust and upper mantle in the Rift System around the Rhinegraben. — In: Graben Problems, IUMP, Sci. Rep. 27, Stuttgart.

AUBERT, M. & PERRIER, G., 1971: Structure profonde du Massif Central. — In: Symposium sur la Géologie, la Morphologie et la Structure profonde du Massif Central français, Editions Scientifiques, Clermont-Ferrand.

FUCHS, K., 1969: On the properties of deep crustal reflectors. — Z. Geophysik, **35**, 133—149.

MORANGE, A., HERITIER, F. & VILLEMIN, J., 1971: Recherches d'hydrocarbures en Limagne. — In: Symposium sur la Géologie, la Morphologie, et la Structure profonde du Massif Central français, Editions Scientifiques, Clermont-Ferrand.

PERRIER, G., 1963: Ondes séismiques enregistrées dans les monts du Forez. — C. R. Acad. Sci. Paris, **257**, 1321—1322.

PERRIER, G. & RUEGG, J. C., 1973: Structure profonde du Massif Central français. — Ann. Geoph., **4**.

Towards a model of crust and mantle structure of the Rhinegraben rift system

Petrologische Gedanken zum Untergrund des Rheingrabens

von

E. Althaus

Mit 1 Abbildung im Text

Abstract: There are two interesting features in the seismic velocity profiles observed beneath the Rhine Graben region: Inversions in the velocity gradient at depths of about 10—15 km and again at 20—22 km and a shallow crust- mantle boundary at 25 km. Following the model of RINGWOOD the material below the crust-mantle boundary can be interpreted as plagioclase pyrolite. For the occurrence of the velocity inversion a hypothesis is developed based on considerations about the water budget of rocks.

Das Problem

Die Resultate, welche aus der Zusammenarbeit zwischen Geologie und Geophysik im Bereich des Oberrheingrabens hervorgegangen sind, regen natürlich auch den Mineralogen an, sich an der Interpretation der Meßresultate zu versuchen. Zweierlei ist ihm hierbei jedoch hinderlich: Einmal die Tatsache, daß wir nicht allzuviel über den Stoffbestand in Tiefen unterhalb der Aufschlußbereiche wissen, zum anderen aber, daß experimentelle Untersuchungen, die zur Interpretation der Beobachtungen herangezogen werden könnten, bislang für die zur Diskussion stehende Region noch nicht durchgeführt worden sind. Es sind also Extrapolationen nach zwei Seiten hin nötig, um zu Aussagen über die chemische Zusammensetzung und die korrespondierende Mineralvergesellschaftung eines herausgegriffenen Gesteinsbereiches zu kommen. Zwar hat die experimentelle Petrologie uns in der letzten Zeit mit einer Fülle von Daten auch zum Problemkreis der tieferen Kruste und des oberen Mantels versorgt, doch es ist nach wie vor schwierig, Resultate, die für ein bestimmtes Gestein ermittelt wurden, nun auf andere Regionen mit möglicherweise nicht unwesentlich abweichender chemischer Zusammensetzung zu übertragen. Viele der im folgenden geäußerten Gedanken sind daher mit einem hohen Anteil an Spekulation behaftet. Es wird aber bereits mit Experimenten begonnen, die helfen sollen, unsere Wissenslücke über die Mineral- und Gesteinstypen im Untergrund des Rheingrabens zu schließen.

Bei der Betrachtung der seismischen Profile erregen insbesondere zwei Charakteristika das Interesse des Petrologen: Einmal die Inversionen der P-Wellengeschwindigkeit im Tiefenbereich von ca. 10—15 sowie 20—22 km, zum andere die mit 25 km vergleichsweise seichte Lage des Einsatzes von „Mantel-Geschwin digkeiten". Eine Interpretation dieser Beobachtungen soll versucht werden.

Daten über die Gesteinszusammensetzung

Die Gesteine unterhalb der Sedimentfüllung des Rheingrabens sind nicht ver schieden von denen der Umgebung, also von Schwarzwald und Vogesen. Dor treffen wir — insbesondere im höher herausgehobenen Südteil — neben Granite und Granodioriten verschiedener Zusammensetzung durchweg mittel- bis hoch metamorphe Gesteine an. Die Temperaturbedingungen der Anatexis sind häufi erreicht worden; viele Merkmale der granitischen Gesteine sprechen trotz des ein deutig intrusiven Charakters für eine anatektische Entstehung (EMMERMANN & REIN 1972).

Bei der Bildung granitoider Schmelzen aus metamorphen Gesteinen wird i der Regel nur ein Bruchteil des Muttergesteins aufgeschmolzen. Wenn dieses, wa nicht unwahrscheinlich ist, von grauwacke- bis tonschieferartiger Zusammensetzun war (MEHNERT 1953—64), müssen als Aufschmelzungsreste Gesteine übrig bleiber in denen diejenigen Komponenten angereichert sind, die von den Schmelzen nu in untergeordneten Mengen aufgenommen werden: Mg-Fe-Minerale wie Bioti und Cordierit, Ca-reiche Minerale wie anorthitreiche Plagioklase oder Horn blenden. Wir sollten also damit rechnen, daß in den tieferen Erdschichten, au denen die aufgestiegenen Granite abgeleitet werden müssen, diese Mineral ange reichert sind, daß also etwa biotitreiche Gneise bis Biotitite, Cordieritgneise un Amphibolite anzutreffen sind. Über noch tiefer liegende Krustenzonen könner wir nur Vermutungen anstellen; es ist aber nicht unwahrscheinlich, daß auch dor noch wasserhaltige Minerale anzutreffen sind.

Daten über die physikalischen Bedingungen

Die Frage nach dem Mineralbestand eines Gesteinskomplexes gegebener chemi scher Zusammensetzung ist gleichzeitig eine Frage nach den physikalischen Be dingungen, insbesondere den Faktoren Druck und Temperatur. Relativ einfad ist der Belastungsdruck aus der Mächtigkeit der Überdeckung ableitbar. De Belastungsdruck aber ist bei vielen Reaktionen nicht der thermodynamisch wirk same Parameter. Bei allen Reaktionen, an denen Wasser oder sonstige fluid Komponenten beteiligt sind, spielt die Fugazität der fluiden Komponenten unc deren Verhältnis zum Belastungsdruck die entscheidende Rolle (ALTHAUS 1968) Leider haben wir nur selten eindeutige Indikatoren für die Höhe etwa der H_2O Fugazität in Gesteinen. Wir sind hierbei meistens auf Schätzungen und Mut maßungen angewiesen.

Eine etwas bessere Vorstellung haben wir von den Temperaturen im tieferer Untergrund. Eine Zusammenstellung der in unserem Gebiet gefundenen Wärme flußdaten findet sich bei HAENEL (1971). Wenn man seinen Vorstellungen übel die Wärmeproduktion in der Kruste folgt, kann man mit Temperaturen von mindestens 550—580° C in ca. 25 km Tiefe, also an der Unterkante der Kruste, rechnen. In der Tiefenlage der ersten Inversion sollten mindestens ca. 300° C (bei ca. 3 kb Belastungsdruck), in der der zweiten ca. 500° C herrschen. In Gebieten mit relativ hohem Wärmefluß (z. B. bei Landau und Zabern, HAENEL 1970) können diese Temperaturen auch schon in seichteren Lagen erreicht sein. Für die Diskussion der Reaktionsschemata im folgenden Abschnitt spielt jedoch die absolute Höhe der Temperatur keine Rolle.

Mögliche Mineralreaktionen in der Kruste

Die Drucke und Temperaturen, die im vorigen Abschnitt umrissen wurden, entsprechen Bedingungen, bei denen eine Gesteinsmetamorphose stattfinden kann (WINKLER 1967). Die meisten Mineralreaktionen, die in diesem Bereich ablaufen (oder abgelaufen sein!) können, entsprechen einem sehr häufigen Typus: Bei Zufuhr von Wärme wird Wasser freigesetzt, etwa nach folgendem Schema:
Mineral A + Mineral B = Mineral C + Mineral D + Wasser.
Solche Reaktionen haben im P-T-Diagramm stets eine positive Steigung, d. h. bei höheren Drucken müssen höhere Temperaturen erreicht werden, um die Reaktion nach rechts ablaufen zu lassen.

Sehr häufig — wenn nicht bei den meisten gesteinsbildenden Mineralen — entspricht den Mineralkombinationen auf der linken Seite der Gleichung eine niedrigere P-Wellengeschwindigkeit als denen der rechten Seite (allerdings muß das Wasser das System verlassen haben). Dehydratisierungsreaktionen bewirken also in der Regel eine Zunahme der seismischen Geschwindigkeiten, haben also den gleichen Effekt wie eine Zunahme des „basischen" Charakters des Gesteins.

Diese Eigenschaft gibt uns also eine Alternativ-Möglichkeit, die Zunahme der seismischen Geschwindigkeit mit der Erdtiefe zu interpretieren. Darüber hinaus können wir aber mit ihrer Hilfe auch das Auftreten von „low-velocity"-Schichten innerhalb der Erdkruste auf eine andere Weise als durch — was natürlich unbestreitbar eine Erklärungsmöglichkeit darstellt — Verschiedenheit der chemischen Bruttozusammensetzung verstehen.

In Abb. 1a ist eine Reaktionskurve des oben angegebenen Typus schematisch im P-T-Diagramm dargestellt. Unter der Bedingung, daß der H_2O-Druck (P_f) gleich dem Belastungsdruck (P_s) ist, hat die Kurve die bereits diskutierte positive Steigung. Wenn nun aber P_f kleiner ist als P_s, dann gilt eine andere Reaktionskurve, z. B. die angegebene. Generell ist in einem solchen Fall die Reaktionstemperatur niedriger als bei $P_f = P_s$ oder, mit anderen Worten: Es ist die wasserfreie Mineralparagenese in solchen Tiefen stabil, in denen bei $P_f = P_s$ noch die wasserhaltige vorliegt. Das bedeutet, daß dann auch die seismische Geschwindigkeit höher sein kann.

Es gibt mehrere Möglichkeiten, durch welche bewirkt werden kann, daß P_f kleiner ist als P_s. Hauptursache dürfte die „Permeabilität" des Gesteins für Wasser sein. In Gesteinen mit einer hohen Permeabilität kann das Wasser leichter

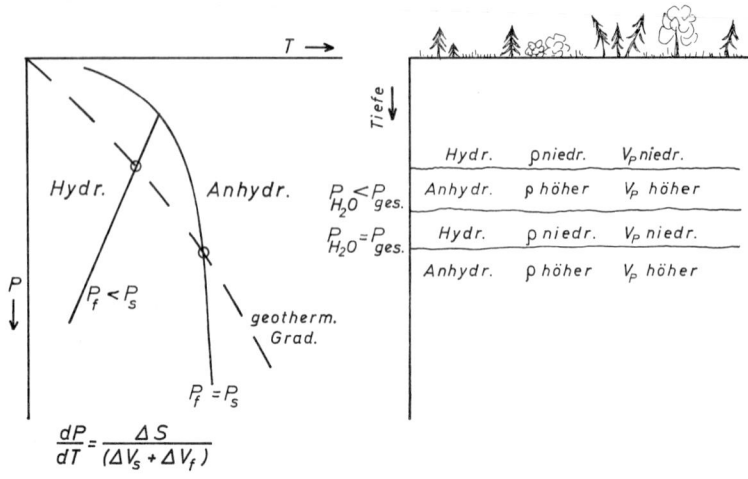

Abb. 1. Entstehung von low-velocity-Schichten durch Mineralreaktionen. Näheres im Text.

aus dem System entweichen als in solchen mit einer niedrigen. Entsprechend können die Mineralreaktionen in den Gesteinen vom ersteren Typus in seichteren Erdtiefen ablaufen als in denen vom letzteren Typus. Große Unterschiede in der chemischen Bruttozusammensetzung brauchen deswegen aber nicht aufzutreten. Wir können also die gleiche Reaktion in unterschiedlichen Erdtiefen antreffen, beeinflußt lediglich durch den Haushalt an einer mobilen fluiden Komponente.

Betrachten wir nun noch einmal Abb. 1a. Der Schnittpunkt des (ebenfalls schematischen) geothermischen Gradienten mit der Reaktionskurve ergibt die Bedingungen von Belastungsdruck und Temperatur, also die Erdtiefe, bei der die Reaktion von links nach rechts abläuft. Sie ist geringer, wenn die H_2O-Fugazität niedriger ist als der Belastungsdruck. Wenn wir nun unterhalb einer Zone relativ hoher Permeabilität eine Zone relativ geringer Permeabilität haben, kann in dieser die Mineralvergesellschaftung der linken Seite der Gleichung erhalten bleiben, obwohl in der darüberliegenden Zone bereits die Paragenese der rechten auftrat. Erst in noch größerer Tiefe, wenn der Gradient nun auch die Reaktionskurve von $P_f = P_s$ schneidet, wird auch in dieser Zone die Paragenese der rechten Seite der Gleichung stabil.

Das Resultat ist in Abb. 1b angedeutet: Eine Zone relativ niedriger Geschwindigkeit (mit Phasen der linken Seite der Reaktionsgleichung) ist eingelagert zwischen zwei Zonen höherer Geschwindigkeit, bewirkt einzig durch unterschiedliche Verhältnisse von P_{H_2O} zu P_s. In Gesteinsnamen übersetzt bedeutet das, daß wir Gesteine in Granulitfazies wechsellagernd mit solchen in z. B. Amphibolitfazies haben können, ohne daß die chemische Pauschalzusammensetzung sich wesentlich unterscheiden muß. Lediglich im Wassergehalt der Reaktionsendprodukte ist ein Unterschied zu sehen; dieser muß aber nicht bereits bei den Ausgangsprodukten angelegt gewesen sein, denn für den Ablauf der Reaktion ist nicht der Wasser g e h a l t , sondern die Wasser f u g a z i t ä t maßgeblich. Leider fehlen bislang noch die Meßdaten für Gesteine, die man möglicherweise im Untergrund unserer Region antreffen kann, sowie deren Reaktionsprodukte. Eine quantitative Überprüfung dieser Vorstellung ist also derzeit noch nicht möglich, jedoch ist die T e n d e n z recht plausibel.

Die Natur des oberen Mantels unter dem Rheingraben

Auf den meisten bisher publizierten Geschwindigkeits-Tiefen-Diagrammen, welche auf quer zum Rheingraben gelegenen seismischen Profilen basieren, erscheint zwischen Gesteinen mit normalen Krusten-Geschwindigkeiten und solchen mit normalen Mantelgeschwindigkeiten bei Tiefen ab 25 km eine Zone, für welche P-Geschwindigkeiten um 7,7 km/sec errechnet wurden (z. B. ANSORGE et al. 1970). Die Erklärung solcher Geschwindigkeiten mit chemisch und mineralogisch plausiblen Gesteinszusammensetzungen trifft auf einige Schwierigkeiten; man könnte allenfalls an granulitische bis charnocktitische Gesteine denken. Glücklicherweise wurde diese Schwierigkeit für die mineralogische Seite des Problems behoben durch neue Messungen innerhalb des Grabens (FUCHS, mündl. Mitt. 1973), welche zwar die relativ seichte Lage des Geschwindigkeitssprungs bestätigen, aber nunmehr „richtige" Mantelgeschwindigkeiten um 8,1 km/sec für die Zone unterhalb des Geschwindigkeitssprungs ergaben. Zu größeren Tiefen hin nehmen die Geschwindigkeiten, wie nicht anders erwartet, zu, doch scheinen die Übergänge allmählich, nicht diskontinuierlich zu sein. Die früher angenommene „kissenförmige" Struktur mit intermediären Geschwindigkeiten wäre demnach also zu interpretieren als eine Aufbuckelung des Mantels, und der Geschwindigkeits-Sprung entspricht der Kruste-Mantel-Grenze.

Die für den kontinentalen Bereich seichte Lage der Manteloberbergrenze hat natürlich Konsequenzen für die Mineralogie der Mantelmaterie. Wenn wir die RINGWOODsche (z. B. 1962, 1966) Modellvorstellung von der im wesentlichen ultramafischen Zusammensetzung des oberen Mantels zugrunde legen, können wir damit rechnen, daß bei normaler Tiefenlage der Moho-Diskontinuität von 35—40 km im obersten Bereich des Mantels ein Gemenge aus Olivin, Aluminiumhaltigem Orthopyroxen, Al-haltigem Klinopyroxen, Spinell und/oder Granat vorliegt; GREEN & RINGWOOD (1967) gab diesem Gestein den Namen Pyroxen- bzw. Granat-Pyrolit. Die seismischen Geschwindigkeiten, die diesem Gestein zugeordnet werden müssen, liegen jedoch (RINGWOOD 1969) in der Gegend von 8,2—8,4 km/sec, also höher als die im obersten Mantelbereich unterhalb des Rheingrabens gemessenen.

Die Al-haltigen Pyroxene sind jedoch nur innerhalb eines bestimmten Druck- und Temperatur-Intervalls stabil. Bei höheren Drucken werden sie zu normalen Pyroxenen + Granat abgebaut (GREEN & RINGWOOD 1967). Bei niedrigeren Drucken (oder höheren Temperaturen) jedoch wird aus einem Teil des Ca-, Na- und Al-Gehaltes der Pyroxene Plagioklas gebildet. Die Drucke müssen hierzu umso höher sein, je Natrium-reicher der neugebildete Plagioklas ist.

Diese Reaktion Al-haltiger Pyroxen = Al-armer Pyroxen + Plagioklas ist schon von mehreren Autoren untersucht worden (zusammengefaßt bei GREEN & RINGWOOD 1970 und GREEN 1970), aber leider nur bei wesentlich höherer Temperatur, als für unser Problem angenommen werden kann. Wenn wir zur Abschätzung der Temperaturen im oberen Bereich des Mantels wieder das HAENELsche Modell annehmen, erhalten wir Temperaturen um ca. 600° C. Höhere Temperaturen an der Kruste-Mantel-Grenze sind für diese Modellvorstellung noch günstiger. Trotz der Unsicherheit, die mit der notwendigen Extrapolation der experimentellen Daten auf diese Temperaturen verknüpft ist, kann dann angenommen werden, daß im oberen Mantelbereich unserer Region Plagioklas stabil ist und in nicht unbedeutenden Mengen vorliegt, das Gestein also Plagioklas-Pyrolit (nach GREEN & RINGWOOD 1967) ist. Für den Plagioklas-Pyrolit berechnet RINGWOOD (1969) eine V_p-Geschwindigkeit von 8,01 km/sec, also nahezu den gleichen Wert, der unter dem Rheingraben beobachtet wurde. Daß die Geschwindigkeit etwas höher liegt, als für reinen Plagioklas-Pyrolit postuliert, mag folgenden Grund haben: Die Umwandlung von Plagioklas in Pyroxen ist nicht eine einfache Phasenumwandlung, die bei einem bestimmten Druck bei einer bestimmten Temperatur stattfindet: Alle beteiligten Phasen sind Mischkristalle, deren Umwandlung sich anders als bei unären Phasen in einem relativ großen Druck- bzw. Temperaturintervall vollzieht. Dieses Intervall ist für die Endglieder der Plagioklasreihe etwa 10 kb weit. Bei der Umwandlung intermediärer Mischkristalle muß eine ganze Reihe von Zwischenstadien durchlaufen werden, ein Vorgang, der völlig analog der fraktionierten Kristallisation der Plagioklase erfolgt.

Diese Bivarianz der Reaktionen bewirkt, daß mit zunehmender Tiefe nicht abrupt Plagioklas durch Pyroxen ersetzt wird, sondern daß vielmehr diese Umwandlung über einen großen Tiefenbereich „verschmiert" ist. Es wird also der anorthitreiche Anteil des Plagioklas zuerst umgewandelt, anorthitärmerer bleibt zunächst noch neben neugebildeten Al-reichem Pyroxen stabil. Mit zunehmender Tiefe ändern sich nun die Mengenverhältnisse dieser beiden Phasen derart, daß die Pyroxenmenge auf Kosten der Plagioklasmenge ständig zunimmt. Das hat zur Folge, daß auch die Zunahme der seismischen Geschwindigkeit mit der Tiefe, die an die Mengenverhältnisse von Plagioklas zu Pyroxen gebunden ist, nicht diskontinuierlich, sondern kontinuierlich erfolgt. Im obersten Bereich des Mantels dürfte ein Teil des Plagioklases bereits umgewandelt und als „Ca-Tschermak's Molekül" in den Pyroxen aufgenommen worden sein. Es ist wegen der Bivarianz der Reaktion nicht weiter verwunderlich, daß eine scharfe Untergrenze des ehemaligen „Kissens" nicht beobachtet werden kann. Daß die Oberkante so scharf auftritt, rührt lediglich daher, daß hier eine s t o f f l i c h e Grenze vorliegt, an der die chemisch mehr oder weniger dioritischen Gesteinen der unteren Kruste an den ultramafischen Mantel stoßen. Dieser oberste Mantel besteht wahrscheinlich chemisch aus durchaus „gewöhnlicher" Mantelmaterie; ungewöhnlich, weil durch die ungewöhnlich seichte Lage bedingt, ist lediglich die Mineralvergesellschaftung.

Schlußbetrachtung

Wie schon eingangs gesagt, fehlen noch viele experimentelle Daten, mit denen die hier angesprochenen Aussagen und Vermutungen untermauert werden müßten. Die experimentell-petrologische Bearbeitung der in dieser Region vorkommenden und zu erwartenden Gesteine steht erst im Beginn. Aber einiges kann man wohl auch jetzt schon sagen: Für viele Effekte, wie z. B. das Auftreten von low-velocity-Schichten, kann man sehr einfache Lösungen finden, wenn man von herkömmlichen Hypothesen abgeht. Und obwohl unsere Vorstellungen von der Natur der Materie unterhalb der Moho nur hypothetisch sind, auch in Zukunft wohl kaum anders als hypothetisch sein werden, können sie mit Feld- und Laborbeobachtungen durchaus in Einklang gebracht werden.

Literatur

ALTHAUS, E., 1968: Der Einfluß des Wassers auf metamorphe Mineralreaktionen. — N. Jb. Miner. Mh., 1968, 289—306.
ANSORGE, J.; EMTER, D.; FUCHS, K.; LAUER, J. P.; MÜLLER, ST. & PETERSCHMITT, E., 1970: Structure of the crust and upper mantle in the rift system around the Rhinegraben. — In: ILLIES, J. H. & MÜLLER, ST. (Eds.), Graben Problems, 190—197, Schweizerbart, Stuttgart.
EMMERMANN, R. & REIN, G., 1972: Genesis of Granites by anatexis and differentiation. — In: Granite 71, Trans. Geol. S. A.
GREEN, D. H. & RINGWOOD, A. E., 1967: The stability fields of aluminous pyroxene peridotite and garnet peridotite and their relevance in upper mantle structure. — Earth Planetary Sci. Letters 3, 151—160.
— 1970: Mineralogy of peridotitic compositions under upper mantle conditions. — In: RINGWOOD, A. E. & GREEN, D. H. (Eds.), Phase Transformations and the Earth's Interior, 359—370, North Holland Publishers, Amsterdam.
GREEN, T. H., 1970: High pressure experimental studies on the mineralogical constitution of the lower crust. — Ibid., 441—450.
HAENEL, R., 1970: Interpretation of the terrestrial heat flow in the Rhinegraben. — In: ILLIES, J. H. & MÜLLER, ST. (Eds.), loc. cit. 116—120.
— 1971: Heat flow measurements and a first heat flow map of Germany. — Z. Geophys. 37, 975—992.
MEHNERT, K. R., 1953—1963: Petrographie und Abfolge der Granitisation im Schwarzwald 1—4. — N. Jb. Miner. Abh., 85, 59—140 (1953); 90, 39—90 (1957); 98, 208 bis 249 (1962); 99, 161—199 (1963).
RINGWOOD, A. E., 1966: A model of the upper mantle. — J. Geophys. Res. 67, 857.
— 1966: Composition and evolution of the upper mantle. — In: HART, P. J. (Ed.), The Earth's Crust and Upper Mantle, A. G. U. Geophys. Monogr. 13, 1—17, A. G. U. Washington D. C.
WINKLER, H. G. F., 1967: Petrogenesis of Metamorphic Rocks. — 2nd ed., Springer, Berlin — Heidelberg — New York.

Ein Krusten-Mantel-Modell für das Riftsystem um den Rheingraben, abgeleitet aus der Dispersion von Rayleigh-Wellen[1]

von

H. Reichenbach und St. Mueller

Mit 5 Abbildungen im Text

Abstract: Over the past few years a station network of nine long-period vertical seismographs was in operation covering the region of the rift system around the Rhinegraben. Special efforts were made to calibrate these stations as accurately as possible. The observed phase velocity dispersion for the fundamental mode of Rayleigh waves shows regional differences depending on the orientation of the observation line as compared to the strike of the Rhinegraben. For a center profile along the strike of the rift system a crust-mantle model (KA 141) was deduced which exhibits rather pronounced shear velocity inversions in the lower crust and uppermost mantle. The crustal model (KA 20) appropriate for a transverse profile across the central part of the Rhinegraben requires an increase in shear velocity within the lower crust as the margin of the rift system is approached.

Bei der Ausbreitung in einem geschichteten Medium sind seismische Oberflächenwellen einer geometrischen Dispersion unterworfen. Aus den Dispersionskurven solcher Wellen (Phasengeschwindigkeit c in Abhängigkeit von der Periode T) lassen sich Rückschlüsse auf die Verteilung der P- und S-Wellen-Geschwindigkeiten sowie auf die Dichteverteilung des durchlaufenen Mediums in Abhängigkeit von der Tiefe ziehen, die bis in den oberen Erdmantel hineinreichen.

Die vorliegende Arbeit bezieht sich auf das Riftsystem um den Rheingraben. Zur Registrierung von Rayleigh-Wellen wurde ein Stationsnetz mit langperiodisch abgestimmten Seismographen betrieben (SEIDL, REICHENBACH & MUELLER 1970). Die einzelnen Stationen waren wie folgt verteilt (Abb. 1):

TNS auf dem Kleinen Feldberg im Taunus, HEI Heidelberg, PIR Pirmasens, KRW Karlsruhe, FWS Furtwangen, HAU Haudompre südwestlich Epinal, und NEU Neuchâtel. Seismogramme der Stationen STR Strasbourg und STU Stuttgart konnten mitbenutzt werden. Damit war es möglich, im Streichen des Grabens liegende Wellenwege wie TNS—NEU, TNS—HAU, TNS—FWS und PIR—NEU, wie auch solche, die quer zum Graben verlaufen, wie STU—HAU, zu verwenden.

Als Eigenperiode der langperiodischen Seismometer (SPRENGNETHER) wurde $T_S = 30$ sec gewählt, um auch große Perioden mit genügender Vergrößerung zu registrieren. Die Eigenperiode des Spiegelgalvanometers (UED) betrug $T_G = 100$ sec. Die Berechnung der Dispersionskurven erfolgte auf Großrechenanlagen

---

[1] Beitrag Nr. 94 Geophysikalisches Institut, Universität Fridericiana Karlsruhe
Mitteilung Nr. 65 Institut für Geophysik, Eidgenössische Technische Hochschule Zürich

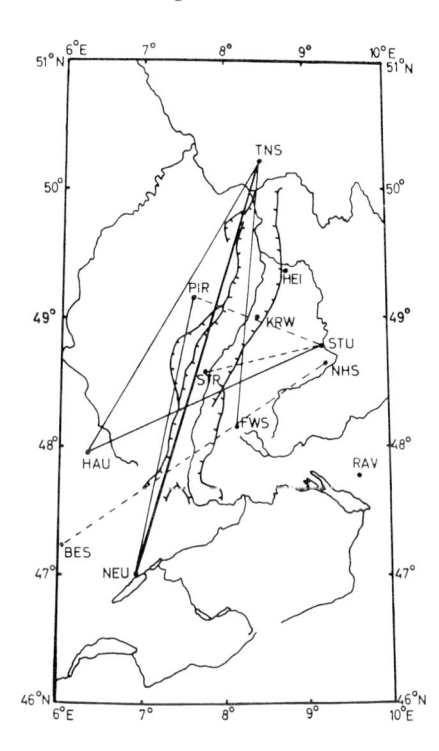

Abb. 1. Schematische Karte des Stationsnetzes.
Fig. 1. Schematic Map of the Station Network.

mit einem modifizierten und erweiterten Programmpaket von KNOPOFF, MUELLER & PILANT (1966).

Die vier im Streichen des Grabens verlaufenden Linien zeigen sehr ähnliche Dispersionskurven (Abb. 2). Die Streuung untereinander ist geringer als die Meßungenauigkeit. Deshalb wurden in Abb. 3 Mittelwerte der Phasengeschwindigkeiten dieser vier Linien eingetragen und für sie das einheitliche Krusten-Mantel-Modell KA 141 abgeleitet. Als Startmodell diente das von ANSORGE, EMTER, FUCHS, LAUER, MUELLER & PETERSCHMITT (1970) aus refraktions- und reflexionsseismischen Messungen gewonnene Modell KA 30 (siehe Abb. 4). Da die Dispersion von Rayleigh-Wellen auf Änderungen der P-Wellen-Geschwindigkeit wenig empfindlich reagiert, wurden diese Werte in den einzelnen Schichten nicht verändert und nur die S-Wellen-Geschwindigkeiten und die Schichtmächtigkeiten korrigiert. Die gemessenen Dispersionskurven sind mit den in der Kruste aus reflexions- und refraktionsseismischen Messungen abgeleiteten Inversionen der P- und S-Geschwindigkeit („Kanal") verträglich. Wegen des steilen Abfalls der Dispersionskurve und der damit niedrigen Phasengeschwindigkeitswerte im Periodenbereich von 20—30 sec mußten die S-Wellen-Geschwindigkeiten in Tiefen von 18—25 km jedoch wesentlich reduziert und die Mächtigkeit des Kanals in der unteren Kruste erweitert werden.

Abb. 2. Experimentelle Phasengeschwindigkeitswerte c(T) mit mittleren quadratischen Fehlern η(T) und theoretischen Dispersionskurven.

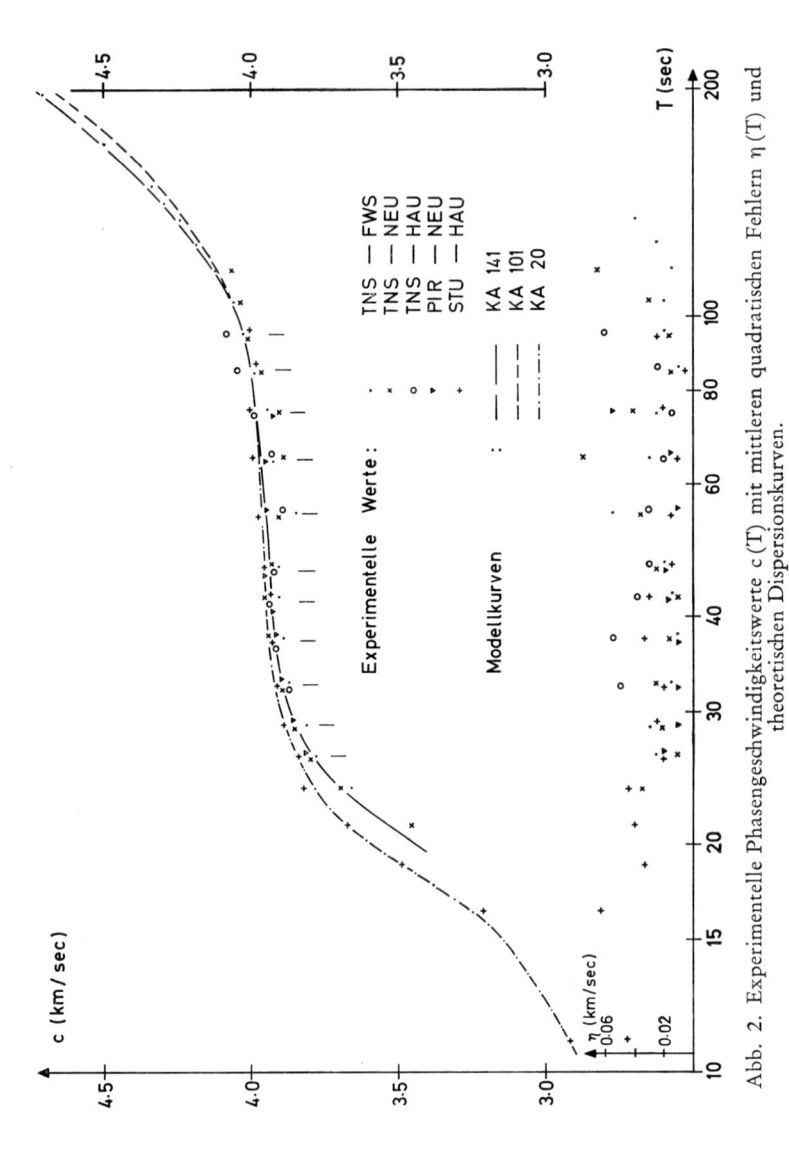

Fig. 2. Experimental Phase Velocity Values c(T) with Mean Square Errors η(T) and Theoretical Dispersion Curves.

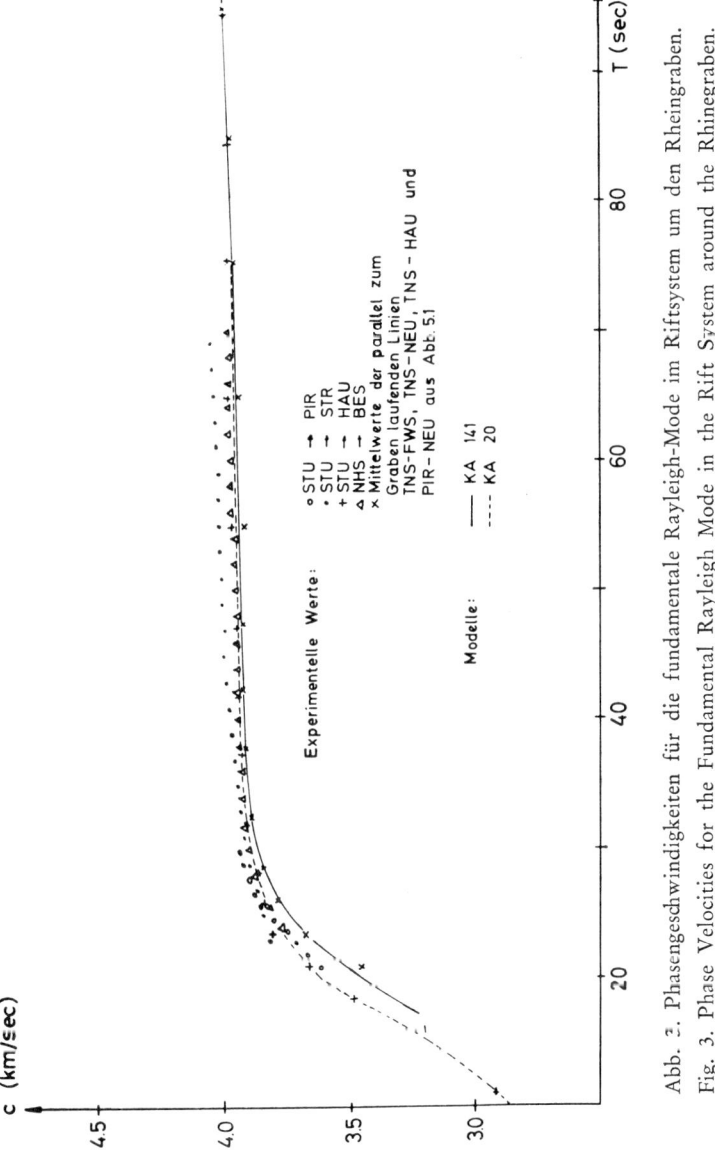

Abb. 3. Phasengeschwindigkeiten für die fundamentale Rayleigh-Mode im Riftsystem um den Rheingraben.
Fig. 3. Phase Velocities for the Fundamental Rayleigh Mode in the Rift System around the Rhinegraben.

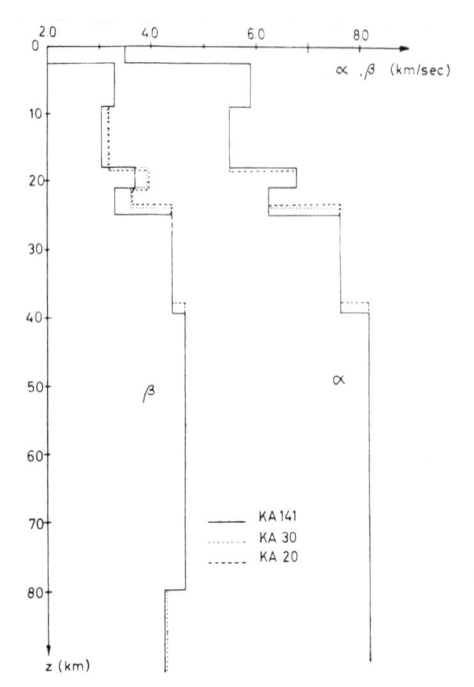

Abb. 4. Theoretische Geschwindigkeits-Tiefen-Funktionen (Erdkruste).

Fig. 4. Theoretical Velocity-Depth Functions (Crust).

Für die quer zum Graben verlaufende Linie STU—HAU wurde das Krustenmodell KA 20 abgeleitet. Da die gemessenen Phasengeschwindigkeitswerte im Periodenbereich von 15—40 sec über denen des Modells KA 141 liegen, wurden die Mächtigkeiten der Schichten mit niedriger S-Geschwindigkeit in der unteren Kruste reduziert. Der durchlaufene Wellenweg schließt aber sowohl Teile der Grabenmitte als auch Teile des äußeren Randes des Riftsystems ein. Das Modell KA 20 stellt also ein Mittelmodell verschiedener Strukturen dar. Da für die Grabenmitte bereits das Modell KA 141 gilt, bedeutet dies, daß die S-Geschwindigkeiten in der unteren Kruste gegen den Rand des Riftsystems hin zunehmen müssen. Eine Bestätigung dafür ergibt sich aus der Zusammenfassung der bisher gemessenen Dispersionskurven in Abb. 3. Nähert man die Stationslinien der Streichrichtung des Grabens an, so vermindern sich die gemessenen Phasengeschwindigkeiten kontinuierlich. Die niedrigsten Werte ergeben sich für die in Streichrichtung in der Grabenmitte gelegenen Linien.

Im Periodenbereich oberhalb 40 sec überwiegt der Einfluß des oberen Mantels. Hier beginnen sich die auf verschiedenen Linien ermittelten Dispersionskurven einander anzugleichen und sind oberhalb 70 sec praktisch identisch. Dies deutet auf einen homogenen oberen Mantel hin. Im Krusten-Mantel-Modell KA 141 wird die Unterkante der Asthenosphäre in eine Tiefe von 280 km verlegt (siehe Abb. 5), gegenüber 220 km in den bisherigen Modellen STU 1 (SCHNEIDER, MUELLER & KNOPOFF 1966) und STU 3 (SEIDL, MUELLER & KNOPOFF 1966). Zum

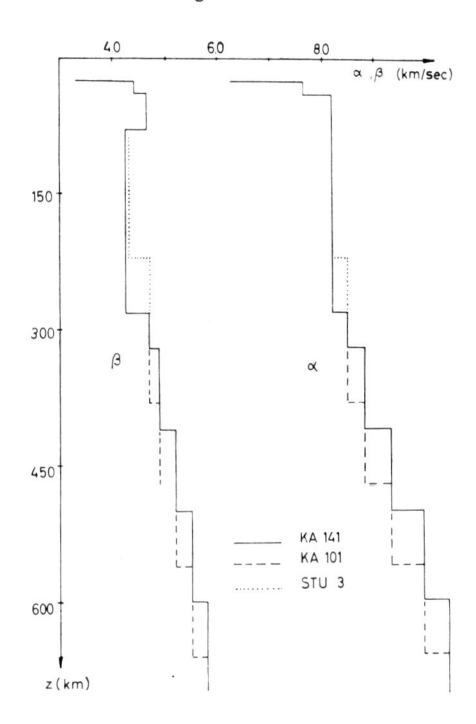

Ein Krusten-Mantel-Modell für den Rheingraben 353

Abb. 5. Theoretische Geschwindigkeits-Tiefen-Funktionen (Erdmantel).
Fig. 5. Theoretical Velocity-Depth Functions (Mantle).

gleichen Ergebnis gelangte SEIDL (1971) bei Messungen auf einem transeuropäischen Profil von Malaga bis Kopenhagen.

In der folgenden Zusammenstellung der abgeleiteten Modelle KA 141 und KA 20 sind die Schichtindizes n, die Schichtmächtigkeiten d_n [km], die Tiefen der Schichtuntergrenzen z_n [km], die Kompressions- und Scher-Wellen-Geschwindigkeiten α_n und β_n [km/sec] sowie die Dichten ϱ_n [g/cm³] angegeben:

Modell KA 141	n	d_n	z_n	α_n	β_n	ϱ_n
	1	2,5	2,5	3,50	2,00	2,50
	2	6,5	9,0	5,90	3,30	2,82
	3	9,0	18,0	5,50	3,05	2,70
	4	3,0	21,0	6,80	3,70	3,00
	5	4,0	25,0	6,25	3,30	2,90
	6	14,5	39,5	7,65	4,41	3,20
	7	40,5	80,0	8,20	4,65	3,40
	8	200,0	280,0	8,20	4,25	3,40
	9	40,0	320,0	8,49	4,77	3,53
	10	90,0	410,0	8,81	4,89	3,60
	11	90,0	500,0	9,32	5,19	3,76
	12	100,0	600,0	9,97	5,49	4,01
	13	Halbraum		10,48	5,79	4,32

23 Approaches to Taphrogenesis

Modell KA 20

1	2,5	2,5	3,50	2,00	2,50
2	6,5	9,0	5,90	3,30	2,82
3	9,5	18,5	5,50	3,15	2,70
4	2,5	21,0	6,80	3,93	3,00
5	2,5	23,5	6,25	3,61	2,90
6	14,5	38,0	7,65	4,41	3,20
7	42,0	80,0	8,20	4,65	3,40
8	200,0	280,0	8,20	4,25	3,40

Dieses Modell ist nur bis in 280,0 km Tiefe festgelegt.
Als Fortsetzung können die Schichten Nr. 9—13 des Modells KA 141 eingesetzt werden.

Literatur

ANSORGE, J., EMTER, D., FUCHS, K., LAUER, J. P., MUELLER, ST. & PETERSCHMITT, E., 1970: Structure of the Crust and Upper Mantle in the Rift System around the Rhinegraben. — In: ILLIES, H. & MUELLER, ST.: Graben Problems, Schweizerbart, Stuttgart, 190—197.

KNOPOFF, L., MUELLER, ST. & PILANT, W. L., 1966: Structure of the Crust and Upper Mantle in the Alps from the Phase Velocity of Rayleigh Waves. — Bull. Seism. Soc. Amer., **56**, 1009—1044.

SCHNEIDER, G., MUELLER, ST. & KNOPOFF, L., 1966: Gruppengeschwindigkeitsmessungen an kurzperiodischen Oberflächenwellen in Mitteleuropa. — Z. Geophysik, **32**, 33—57.

SEIDL, D., 1971: Spezielle Probleme der Ausbreitung seismischer Oberflächenwellen mit Beobachtungsbeispielen aus Europa. — Dissertation, Univ. Karlsruhe, 105 S.

SEIDL, D., MUELLER, ST. & KNOPOFF, L., 1966: Dispersion von Rayleigh-Wellen in Südwestdeutschland und in den Alpen. — Z. Geophysik, **32**, 472—481.

SEIDL, D., REICHENBACH, H. & MUELLER, ST., 1970: Dispersion Investigations of Rayleigh-Waves in the Rhinegraben Rift System. — In: ILLIES, H. & MUELLER, ST.: Graben Problems, Schweizerbart, Stuttgart, 203—206.

Hypothese zur Erklärung des Fehlens von P_n-Einsätzen im Bereich von Grabenzonen, abgeleitet aus zweidimensionalen modellseismischen Untersuchungen

von

J. Strowald

Mit 7 Abbildungen im Text

Hypothesis to Explanate the Missing P_n-arrivals in Rift-valleys Derived from Two-dimensional Seismic Model Studies

Abstract: With the help of two-dimensional seismic model study amplitude measurements were carried out at a model of cushion shaped structure (aluminium) embedded in a half-plane (plexiglas). Assuming that the waves spreading out have no plane wavefronts the cushion-shaped body acts as a screen. The upper and lower boundaries are lying in a shadow-zone. They will be reached only by diffracted waves with strongly decreasing amplitudes. It is, therefore, supposed that no refracted waves will be generated inside the cushion while a wave-system generating at the point of the wedge-shaped part of the cushion will send measurable refracted waves into the half-plane.

Einführung

Nach den bisher durchgeführten refraktionsseismischen Messungen im Oberrheintalgraben und anderen Grabenbruchsystemen, vor allem in Ostafrika (Kenia), muß als erwiesen angesehen werden, daß unter den Gräben in ca. 20—25 km Tiefe im Grenzbereich zwischen Kruste und Mantel eine kissenförmige Struktur mit einer P-Wellengeschwindigkeit von 7,5—7,7 km/s liegt [St. Müller et al. (1969), Meissner & Berckhemer (1967), Griffith et al. (1971) und Khan & Mansfield (1971)].

Die seitliche Ausdehnung der Struktur reicht weit über die sichtbaren Grabengrenzen hinaus und konnte mit Hilfe gravimetrischer und seismischer Methoden sicher erfaßt werden, nicht jedoch die untere Grenzfläche.

Hier soll mit den Mitteln der zweidimensionalen Modellseismik versucht werden, eine mögliche Erklärung für das Fehlen von P_n-Einsätzen aus dem Bereich unterhalb der Kissenstruktur zu geben.

Versuch einer Deutung

Abb. 1 zeigt den Teil eines gedachten Krustenschnitts durch den Oberrheintalgraben, vereinfacht zu einem 3-Schichtenmodell. Unter der Voraussetzung, daß die sich ausbreitenden Wellen keine ebenen Wellenfronten besitzen, ergibt sich, bei den auch im Rheintalgraben vorhandenen Geschwindigkeitsverhältnissen, VpII/VpI = 0.915, VpIII/VpII = 0.862 und VpIII/VpI = 0.789, der in Abb. 1 dargestellte zeitliche Ablauf der P-Wellenausbreitung.

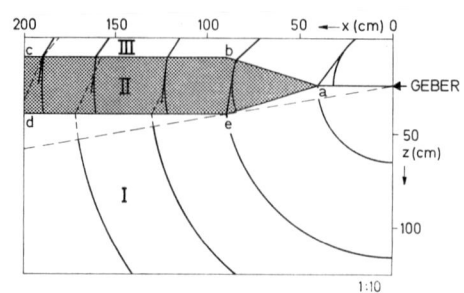

Abb. 1. P-Wellenausbreitung in einem modellhaften Krustenschnitt durch eine Grabenzone.

Fig. 1. P-waves spreading out in a simplified model of a rift-valley.

Die direkte, im oberen Mantel laufende P-Welle wird von der unteren Seitenfläche der Kissenstruktur so abgeschirmt, daß die Kissenunterkante in einer Schattenzone liegt. An dem Übergang Seitenfläche—Unterkante (e) entsteht eine gebeugte Welle, die sich in der Schattenzone ausbreitet und deren Amplitude rascher als die der direkten Welle abnimmt. Sie erzeugt im Kissen eine Kopfwelle, deren Intensität in gleichem Maße schwächer wird und deshalb an der Erdoberfläche nicht mehr in Erscheinung treten kann, zumindest dann nicht mehr, wenn sie als Ersteinsatz im Seismogramm auftauchen sollte.

Die Kissenspitze ist gleichfalls ein Beugungszentrum. Das hier erzeugte P-Wellensystem breitet sich im Kissen mit entsprechender Materialgeschwindigkeit aus und überlagert sich dabei mit der an der unteren Kissenseitenfläche refraktierten P-Welle. Die Überlagerung führt aufgrund des Geschwindigkeitsverhältnisses V_{pI}/V_{pII} zu einer einzigen Wellenfront im Kissen. Ihren Einsätzen an der Erdoberfläche wird die Geschwindigkeit 7,5—7,7 km/s zugeordnet.

Die Modelle

Die modellseismischen Untersuchungen wurden an zwei, durch unterschiedliche Materialkomposition gekennzeichneten Modellen, der in Abb. 2 gezeigten geometrischen Konfiguration, durchgeführt. Als Vorlage diente das von ST. MÜLLER et al. (1967) aus feldseismischen Daten konstruierte Modell der Kissenstruktur im Oberrheintalgraben.

Das Modell 101, Plexiglashalbraum (V_p = 2380 m/s) — Aluminiumkissen (V_p = 5400 m/s) weist einen starken Geschwindigkeitskontrast auf und erlaubt die ungestörte Untersuchung der Wellenausbreitung im Kissen. Der Nachteil dabei ist, daß die Wellen beim Übertritt von einem Material in das andere erheblich an Energie verlieren.

Zur Untersuchung der an den Kissengrenzen geführten Wellen wurde deshalb das Modell 201, PVC-Halbraum (V_p = 1870 m/s) mit eingeschlossenem Plexiglaskissen gewählt.

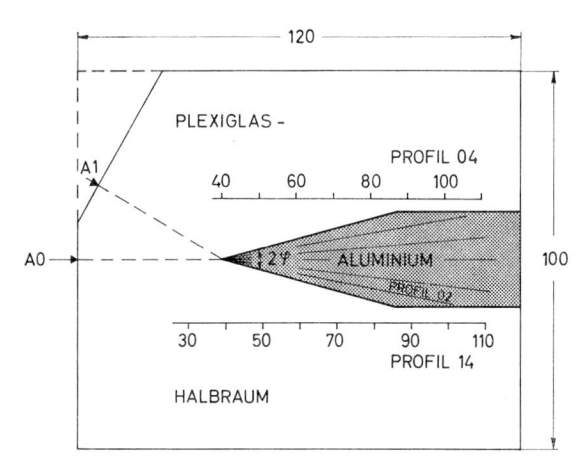

Abb. 2. Kissenmodell für die zweidimensionalen modellseismischen Untersuchungen. Plattendicke 2H = 3 mm, Keilwinkel 2φ = 30°, A0, A1 — Orte der Anregung.

Fig. 2. Cushion-model for two-dimensional seismic model studies. Thickness of plate 2H = 3 mm, wedge — angle 2φ = 30°, A0, A1 — sources of seismic waves.

Als Quelle diente ein Piezokristall PZT 5 der Dicke 6,4 mm mit einem Durchmesser von 12,5 mm. Der elektrische Anregungsimpuls ist ein Rechteck. Seine Länge wurde so gewählt, daß im Modell ein Signal der in Abb. 7 angegebenen Form entstand. Die Empfangsseite bestand aus einem richtungsempfindlichen Biegeschwinger.

Die Wellenausbreitung im Kissen

Die Wellenausbreitung im Kissen, d. h. im keilförmigen Teil des Kissens hängt von der Strahlrichtung der einfallenden Wellen ab. Solange der kritische Einfallswinkel noch unterschritten wird, werden die einfallenden Wellen an den Keilseiten reflektiert, und es entsteht nur an der Keilspitze, nach dem Huygenschen Prinzip, ein sekundäres Wellensystem, welches sich im Kissen ausbreitet. Dieses Wellensystem besteht aus einer P-Welle mit kreisförmiger Wellenfront, an den Keilseiten geschleppten PS-Wellen und der S-Welle. In Abb. 3 sind die Wellenfronten mit ihren Hauptschwingungsrichtungen eingetragen.

Die Keilspitze wirkt bei symmetrischer Anregung (Geberort A0) für Meßpunkte in Entfernungen größer als 20 cm angenähert wie eine Punktquelle.

Die Amplitudenabnahme der P-Welle mit der Entfernung gehorcht für die Profile φ = −10°, 0°, +10° dem Exponentialgesetz $A = A_0 R^n$, $n = -0.96 \pm 0.07$. Da die Absorption im Aluminium vernachlässigbar gering ist, kann die starke Amplitudenabnahme nur dadurch erklärt werden, daß das an der Keilspitze entstandene Wellensystem gebeugt ist. Ein weiteres Indiz für die Beugung liefert das Frequenzverhalten des P-Signals. Alle anfangs vorhandenen Anteile hoher Frequenzen verschwinden mit zunehmender Entfernung von der Keilspitze.

KEILPROFIL 02

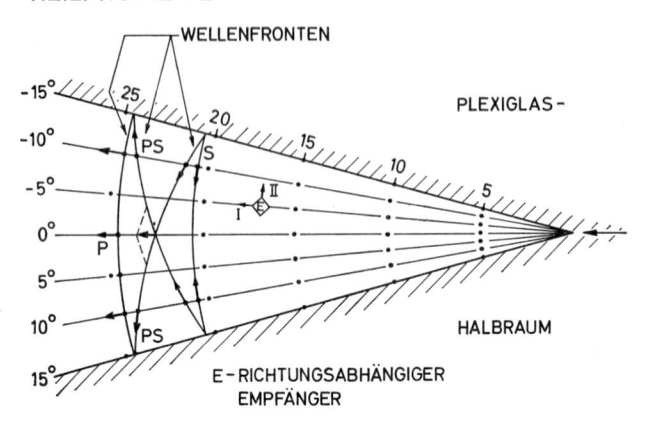

Abb. 3. An der Keilspitze entstandenes Wellensystem — Wellenfronten und ihre Hauptschwingungsrichtungen.

Fig. 3. Wave-system generated at the point of the wedge. Wave-fronts and their main directions of motion.

Wellenausbreitung im Halbraum

Auf den Profilen 04 (Anregung A 0) und 14 (Anregung A 1) parallel zur Kissenober- bzw. -unterkante wird
1. das Verhalten der direkten, im Plexiglashalbraum laufenden Plattenwelle und
2. das vom Keil nach außen abgestrahlte Wellensystem, die geführte PP_bP-Welle beobachtet.

Abb. 6 (oben) zeigt die bezüglich der Abstrahlcharakteristik des Gebers korrigierten und mit R (R = wahre Entfernug Quelle—Meßpunkt) multiplizierten Amplituden der direkten Plattenwelle des Profils 04. Die Amplituden der beiden Signalphasen P 01 und P 12 (s. Abb. 7) weisen im Bereich zwischen 30 und 70 cm, abgesehen von einigen Schwankungen, eine durch materialbedingte Absorption hervorgerufene Amplitudenabnahme auf. In allen Meßpunkten für $X \geq 75$ cm wird eine stärkere Amplitudenabnahme registriert. Das ist noch deutlicher bei den Amplituden des Profils 14 zu sehen.

Beide Profile führen in eine vom Kissen gebildete Schattenzone hinein, Abb. 4. Es tritt Beugung auf. In den Schattenraum dringen nur an den Kissenkanten von der geradlinigen Ausbreitung abgelenkte Anteile der direkten Plattenwelle ein. Dabei vergrößert sich die mittlere Periode des in Abb. 7 dargestellten Signals. Die Beugung ähnelt dem Typ der Fresnelbeugung.

Die an der Kissengrenze im Halbraum geführte Welle PP_bP bildet auf dem Profil 04 ab X = 62 cm den Ersteinsatz (Abb. 5). Die dazugehörige Laufzeitkurve liefert zwischen X = 62 cm und 90 cm, wegen der zum Profil hin ansteigenden Keilseiten des Kissens, eine hohe Scheingeschwindigkeit. Im weiteren Verlauf gibt

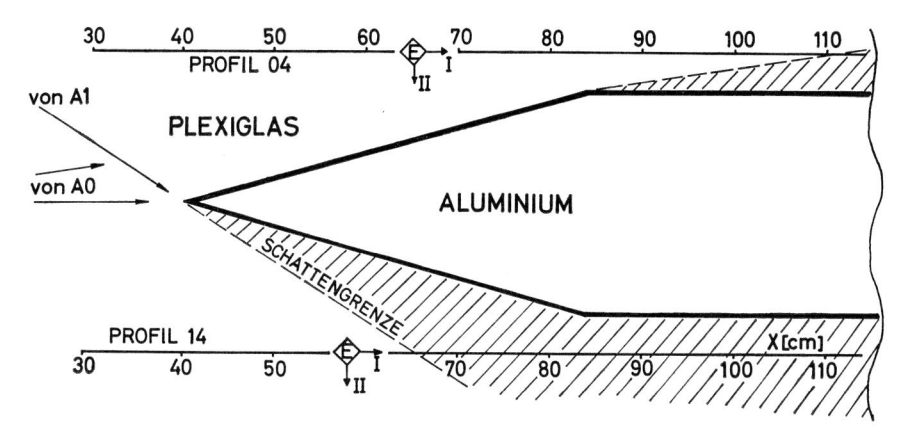

Abb. 4. Schattenzonen im Kissenbereich bei verschiedenen Einfallsrichtungen der direkten Plattenwelle.

Fig. 4. Shadow-zones generated at the cushion by plate-waves with different ray paths.

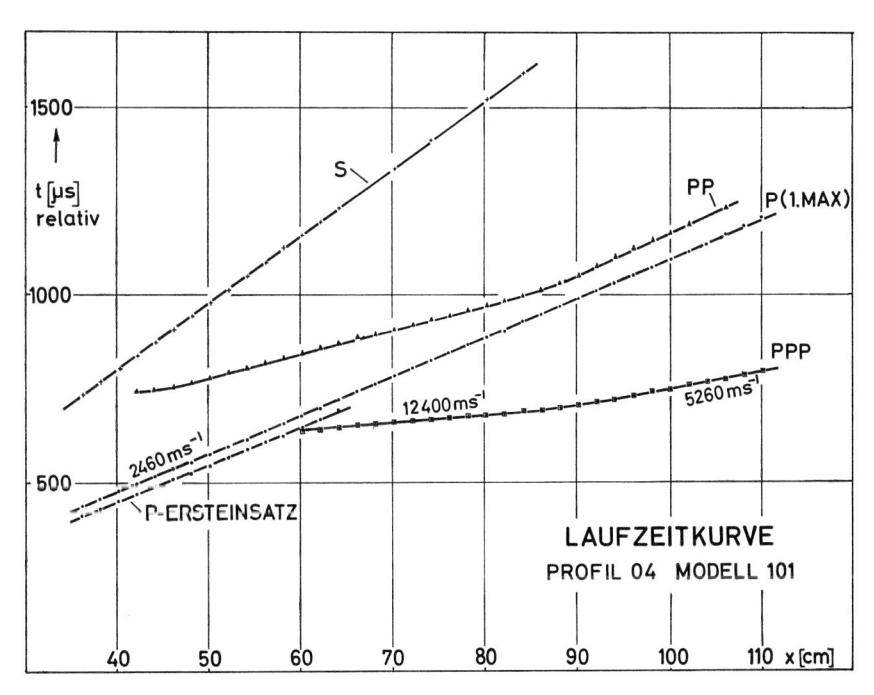

Abb. 5. Laufzeitkurven für die registrierten Welleneinsätze auf Profil 04, Anregung AO, Modell 101.

Fig. 5. Travel-time-curves of wave arrivals registered on profile 04, source AO, model 101.

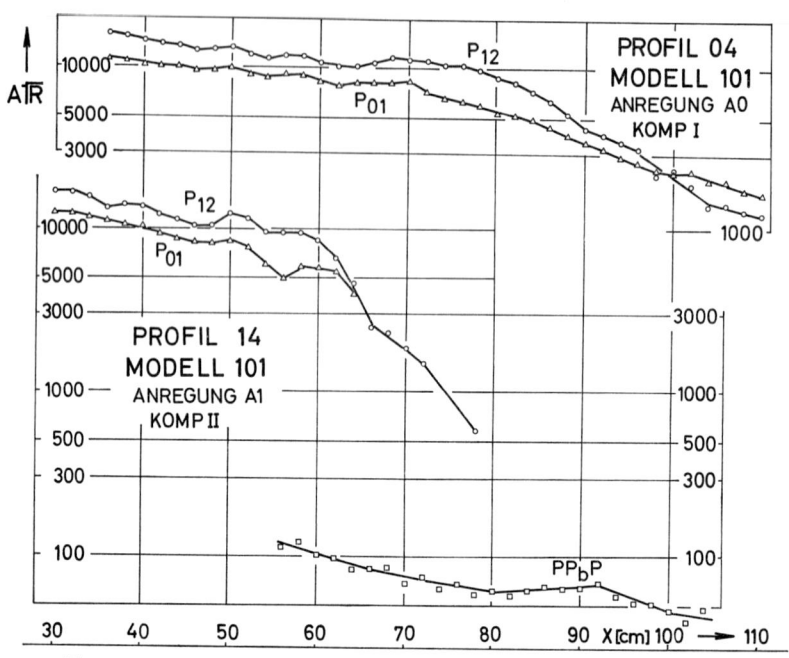

Abb. 6. Amplitudenkurven für das Modell Plexiglashalbraum-Aluminiumkissen.

Fig. 6. Amplitdue curves-model: plexiglas half-plane-aluminium-cushion.

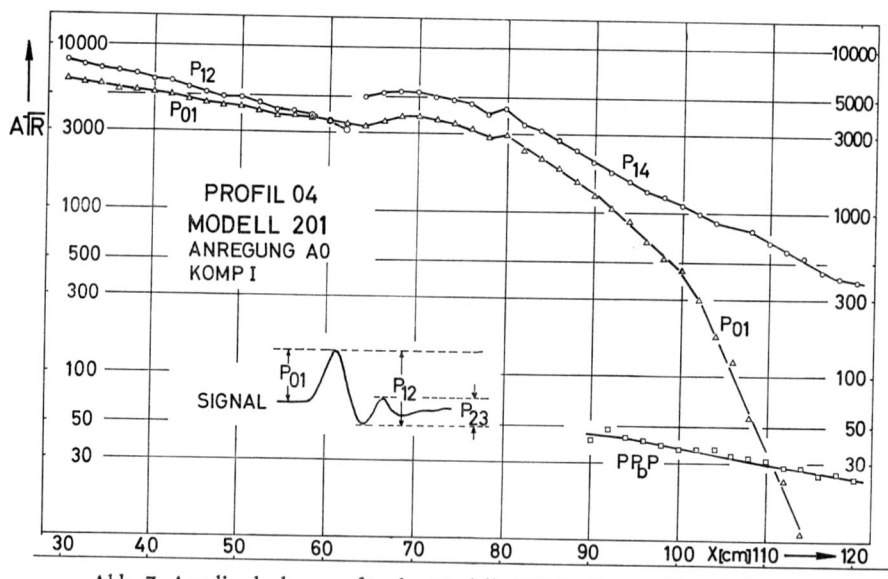

Abb. 7. Amplitudenkurven für das Modell PVC-Halbraum-Plexiglaskissen.

Fig. 7. Amplitude curves-model: PVC-half-plane-plexiglas-cushion.

die Neigung der Laufzeitkurve etwa die wahre P-Wellengeschwindigkeit des Kissenmaterials wieder, d. h. die PP_bP-Welle wird an der zum Profil parallelen Grenze geführt.

Für das Profil 14, Abb. 6 (unten) wurden die Amplituden der geführten PP_bP-Welle denen der direkten Plattenwelle gegenübergestellt. Die Amplitudenabnahme ist deutlich geringer.

Auf dem etwas längeren Profil 04 des Modells 201, PVC-Halbraum-Plexiglaskissen (Abb. 7), nimmt die Phase P 01 der direkten Plattenwelle nahezu vollständig ab. Die Phase P 12 wechselt durch das Verschwinden des ersten Minimums im Signal über in die langperiodische Phase P 14. Da für Wellen mit größer werdender Wellenlänge sich die Übergangszone vom „Licht" in den „Schatten" verbreitert, dringt die P 14-Phase tiefer in die Schattenzone ein.

Die PP_bP-Welle verhält sich ähnlich wie auf Profil 14. Sie hat jedoch relativ zur direkten P-Welle größere Amplitudenwerte.

Zusammenfassung

Mit den hier durchgeführten Amplitudenuntersuchungen kann nachgewiesen werden, daß die „horizontalen" Grenzflächen des Kissens, für Wellen aus den im Experiment benutzten Richtungen, im Schatten liegen. Die Amplitudenabnahme der gebeugten Plattenwellenphasen zeigen weiter, daß keine bzw. nur geringe Wellenenergie zur Erzeugung einer geführten Welle im Kissen zur Verfügung steht und folglich von der unteren Grenzfläche zum Beispiel keine Information in Form von Refraktionseinsätzen zu erwarten ist. Das in der Kissenspitze entstandene gebeugte Wellensystem besitzt jedoch, wie Abb. 7 zeigt, genügend Energie, um oberhalb des Kissens eine Refraktionswelle anzuregen.

Diese Arbeit wird von der Deutschen Forschungsgemeinschaft unterstützt.

Literaturverzeichnis

GRIFFITH, D. H., KING, R. F., KHAN, M. A. & BLUNDELL, D. J., 1971: Seismic Refraction Line in the Gregory Rift. — Nature, Physical Science, 229, 69—71.

KHAN, M. A. & MANSFIELD, J., 1971: Gravity Measurements in the Gregory Rift. — Nature, Physical Science, 229, 72—75.

MEISSNER, R. & BERCKHEMER, H., 1967: Seismic Refraction Measurements in the Northern Rhinegraben. — Rhinegraben Progress Rep. 1967, 105.

MUELLER, ST., PETERSCHMITT, E., FUCHS, K. & ANSORGE, J., 1967: The Rift Structure of the Crust and Upper Mantle beneath the Rhinegraben. — Rheingraben Progress Rep 1967, 108.

— 1969: Crustal Structure beneath the Rhinegraben from Seismic Refraction and Reflection Measurements. — Tectonophysics, 8, 529—542.

Models for the Resistivity Distribution from Magneto-Telluric Soundings

by

I. Scheelke

With 4 figures in the text

Abstract: From magneto-telluric measurements in the area of the Rhinegraben a resistivity distribution is obtained by applying twodimensional model calculations for the sediments. This distribution agrees essentially with known observational results. A complete interpretation of the measurements is, however, not possible before a well-conducting intermediate layer is assumed in the deeper structures.

In the years 1968 to 1970 measurements were made on two profiles perpendicular to the Rhinegraben at altogether 20 stations (Fig. 1). These measurements consisted in determining the variations in time of the electromagnetic horizontal

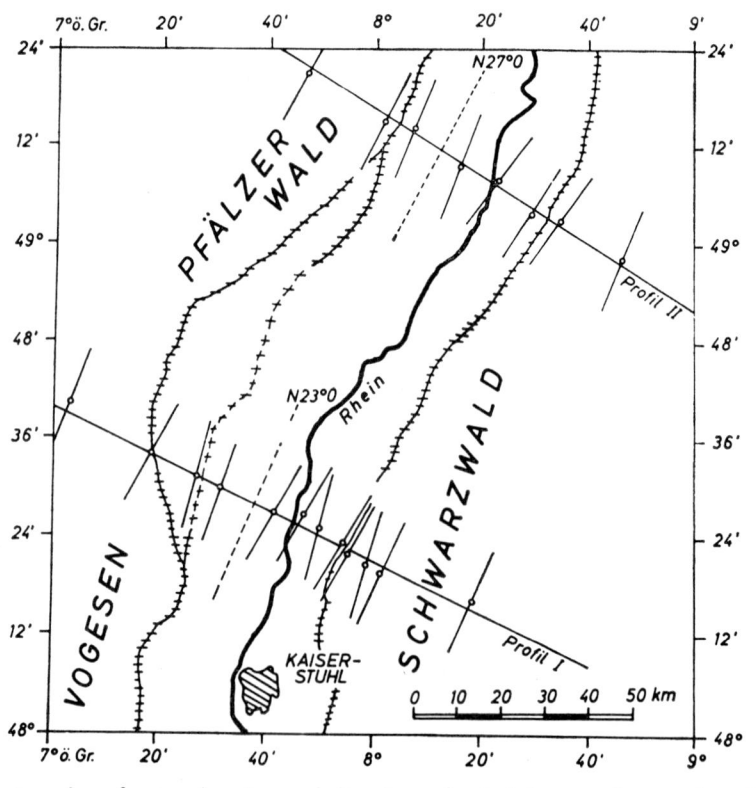

Fig. 1. Location of measuring sites and directions of principle axes of tensor impedances.

components $H_x(t)$, $H_y(t)$, $E_x(t)$, $E_y(t)$ for periods of a few seconds to about 1000 seconds. By certain procedures (SCHEELKE 1972) involving the so-called impedance tensor, both the apparent resistivities $\varrho(T)$ and the phase functions $\Phi(T)$ can be determined from these variations (T is the period). Altogether the following functions are obtained: $\varrho_\parallel$, $\Phi_\parallel$ in the case of "E-polarization", which means the E-component in the direction of the structure, with a perpendicular H-component, and $\varrho_\perp$, $\Phi_\perp$ in the case of "H-polarization", which means the H-component in the direction of the structure, with a perpendicular E-component.

The evaluation aimed at interpreting the behaviour of these functions along the profiles by an approximation with two-dimensional numerical models. As can also be seen from Fig. 1, it is legitimate to assume a distribution of conductivity which is in the main two-dimensional in the area of the Rhinegraben. For each station the mean direction of the principal axis of the impedance tensor (shown in Fig. 1 by small lines) was determined for the range of periods which is considered here. The mean value for the profiles agrees quite well with the direction of structure of the Graben.

Two-dimensional models generally have quite a large number of independent parameters, for example the number of layers and their thicknesses, the appropriate resistivities, as well as the number of horizontal discontinuities in conductivity and the distances between them. Therefore it is advisable to diminish the ambiguity of a model by laying down as many parameters as possible with the observational results already known.

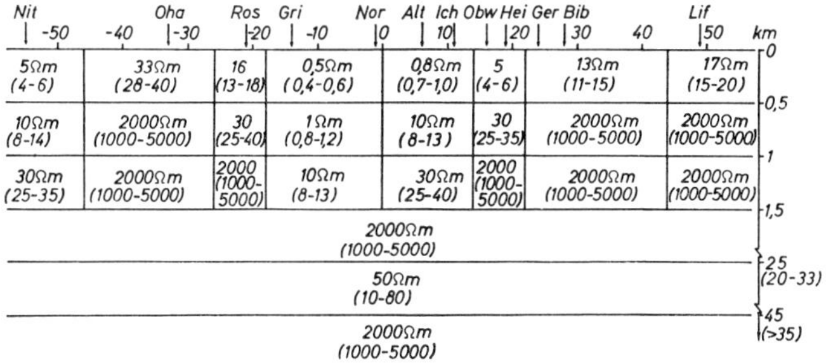

Fig. 2. Model of resistivity distribution for the surface layers along Profile I.

As was to be expected, the behaviour of the short-period variations is influenced nearly exclusively by the resistivity distribution in the layers near the surface. Therefore it was first of all tried to adapt the measured functions for a period of 10 sec as well as possible. Before calculations were begun, the horizontal and vertical boundaries of the layers were laid down in such a way as to agree with the geological conceptions in the main points. Subsequently, only the resistivities were varied in the particular ranges. The result of the model calculations for Profile I is shown in Fig. 2. This model which confirms in the main the observational evidence concerning the resistivity distribution of the sediments and the depth of the basement, is to be considered as optimum because it interprets all the four measured functions ($\varrho_{\shortparallel}$, $\varrho_{\perp}$, $\Phi_{\shortparallel}$, $\Phi_{\perp}$) simultaneously in the best way. The well-conducting intermediate layer which is shown between depths of 25 and 45 km is not necessary for the interpretation of the short-period variations; no contradiction arises, however, if this layer is assumed. The errors in the experimental data permit some variation of the resistivities without destroying the adaptation.

Only after the structures near the surface have been laid down in the way described, it can be hoped to arrive at reliable statements about the behaviour of the long-period variations. For this purpose the first three layers of the surface-model (Fig. 2) were combined to form a single layer with the same integrated conductivity; the intervals between the horizontal discontinuities remained unchanged. In the following calculations only the resistivity distribution in the deeper structures was varied. The final model is shown in Fig. 3. To interpret the observed data for the period of $T = 100$ sec, it is unavoidable to assume a well-conducting layer beginning at a depth of about 25 km and having a resistivity of about 25 Ωm. This is, however, to be regarded only as an intermediate layer of about 20 km in thickness, which is followed again by material of high resistivity. Otherwise an adaptation of the measured functions for the period of $T = 1000$ sec would have been impossible. The conductosphere cannot be expected to begin at a depth of less than 200 km.

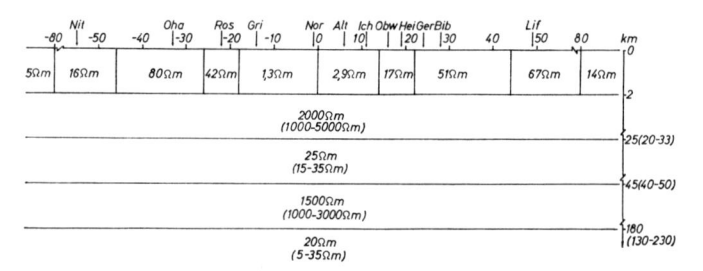

Fig. 3. Model of resistivity distribution for the deep structure along Profile I.

Because of the finite length of the profile, the horizontal extension of the well conducting intermediate layer can be reliably determined only to a distance of about 50 km to each side of the middle of the graben.

The model of resistivity of the deeper structures on Profile II is shown in Fig. 4. The measurements were adapted in the best way with this model. The surface-layer was determined in the same way as in the case of Profile I; because of the different thickness of the sediments, a different distribution is obtained. This model too shows a well-conducting intermediate layer. This layer begins in the same depth as in the case of Profile I, it has about the same resistance, but seems to be slightly thicker. Again at a depth of about 200 km the conductosphere appears.

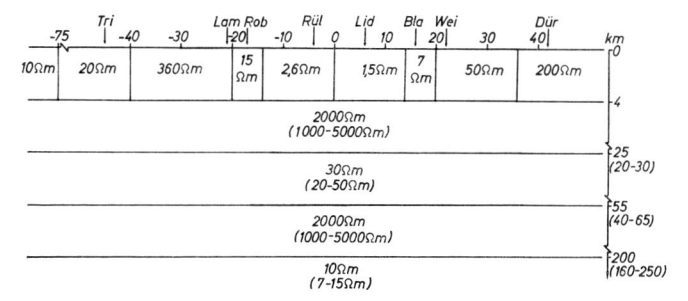

	Tri		Lam	Rob	Rül	Lid	Bla	Wei		Dür	
-75	-40	-30	-20	-10	0	10	20		30	40	km
10Ωm	20Ωm	360Ωm	15 Ωm	2,6Ωm	1,5Ωm	7 Ωm	50Ωm			200Ωm	0

2000Ωm (1000-5000Ωm)	4
30Ωm (20-50Ωm)	25 (20-30)
2000Ωm (1000-5000Ωm)	55 (40-65)
10Ωm (7-15Ωm)	200 (160-250)

Fig. 4. Model of resistivity distribution for the deep structure along Profile II.

To summarize, the following can be said: The magneto-telluric measurements can be interpreted in a satisfactory way only if a well-conducting intermediate layer is assumed under both profiles. This layer begins at a depth of about 25 km, it has a resistivity of about 25 Ωm, and a thickness between 20 and 30 km. — Geomagnetic measurements (WINTER 1972) and studies of telluric variations (HAAK 1970) also suggest the existence of this layer.

References

HAAK, V., 1970: Das zeitlich sich ändernde erdelektrische Feld, beobachtet auf einem Profil über dem Rheingraben; eine hiervon abgeleitete Methode der Auswertung mit dem Ziel, die elektrische Leitfähigkeit im Untergrund zu bestimmen. — Diss., Univ. München.

SCHEELKE, I., 1972: Magnetotellurische Messungen im Rheingraben und ihre Deutung mit zweidimensionalen Modellen. — Diss., Techn. Univ. Braunschweig.

WINTER, R., 1972: A Model of the Resistivity Distribution from Geomagnetic Depth Soundings. — In: ILLIES, H. & FUCHS, K.: Approaches to Taphrogenesis. Schweizerbart, Stuttgart.

The Distribution of Electrical Resistivity in the Rhinegraben Area as Determined by Telluric and Magnetotelluric Methods

by

V. Haak and G. Reitmayr

With 2 figures in the text

Abstract: The local variation of the thickness of the sediments along a profile across the Rhinegraben running from the West to the East has been determined from the time-varying electric field. The computation of two-dimensional models of resistivity is still in progress, but at two stations situated at some distance from the graben an interpretation by a one-dimensional resistivity model could readily be carried out. It yielded a low resistivity layer at a depth of about 80 km. This is in agreement with the results by many other authors in other areas, but it is much deeper than the shallow layer directly beneath the Rhinegraben which SCHEELKE and WINTER (in this volume) deduced from their measurements.

During the last three years the time-varying electric and magnetic fields have been measured by the Institut für Angewandte Geophysik der Universität München along a profile crossing the Rhinegraben south of Strasbourg. It coincides in part with a profile measured by SCHEELKE (c. f. his article in this volume), but it has been extended eastwards as far as Lake Constance and to the West up to the Moselle valley. This was in order to get the depth distribution of the electrical resistivity not only in the Rhinegraben but also in the neighbouring region which possibly may not have been affected by the graben tectonics.

1. Lateral distribution of the integrated subsurface conductivity

There is a simple relationship between the amplitudes of the electric field and the thickness of the sediments: The amplitudes of the electric field along a profile are inversely proportional to the integrated conductivity (i. e. the sum of the products of conductivity and thickness of each layer). If one assumes a constant resistivity of the sediments along the profile as well as a resistivity of the underlying crystalline basement which is an order of magnitude larger than the resistivity of the sediments, it follows that the amplitudes of the electric field are small at those locations where the sediments are thick, and that the amplitudes are large where there are more or less no sediments at all. Furthermore, if one considers the topography along the profile to be effectively flat, then the amplitudes of the electric field are proportional to the morphology of the crystalline basement. In Fig. 1 the local variation of the amplitudes of the electric field (component perpendicular to the graben axis) along the profile is shown. It displays the rather strong variation of the integrated electric conductivity at the borders of the Rhinegraben. This makes it difficult to deduce the depth distribu

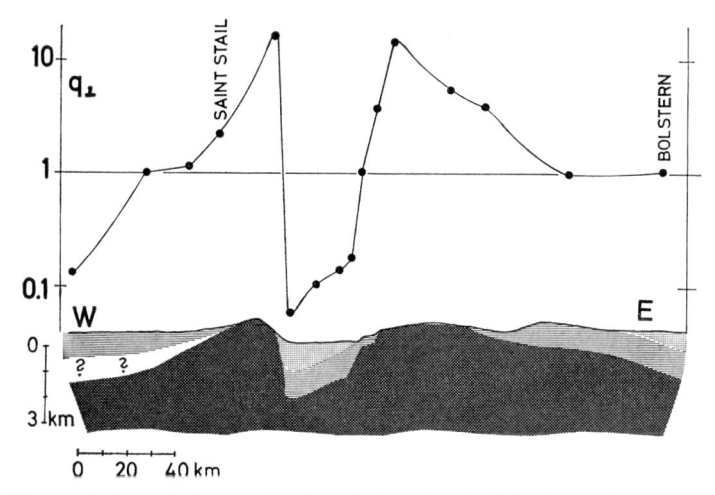

Fig. 1. The variation of the amplitudes of the electric field from site to site (d o t s) along the profile (component of the electric field perpendicular to the graben). B e l o w : A simplified, vertically exaggerated cross section through the subsurface (after BUNDES-ANSTALT FÜR BODENFORSCHUNG). G r e y : sediments of low resistivity; d a r k : crystalline basement of large resistivity. From the figure the high correlation of the local variation of the electric field and of the morphology of the crystalline basement below the sediments is obvious. This is caused by the varying thickness of the sediments which means a varying integrated conductivity. In the figure both of the stations **Bolstern** and **Saint Stail** are shown where the resistivity as a function of depth has been determined (see Fig. 2).

tion of the resistivity from the electric and magnetic fields in this area. Two-dimensional models of resistivity should match the data in a first order approximation, but the computation of such models is still in progress.

2. Depth distribution of resistivity at two selected stations

However, towards the eastern and western ends of the profile the lateral variation of the electric field is greatly reduced as it readily can be seen from Fig. 1. This points to a more constant integrated conductivity along those parts of the profile. Therefore one has to deal there with a resistivity which is to a first order approximation only a function of depth. At these sites a much more simple method of interpretation is applicable. Two stations have been selected where the electric and magnetic fields have been recorded in a rather broad spectrum of periods in order to determine the resistivity at great depths: Bolstern, 120 km east, and Saint Stail 45 km west of the Rhinegraben. Two methods have been applied to calculate the resistivity as a function of depth: The classical method of CAGNIARD (1953) and that proposed by SCHMUCKER (1969). The latter method is an inversion technique applied to the observed data which yields directly a smoothed variation of the resistivity with depth. In Fig. 2 the results at these two stations (calculated with both methods of analysis) are presented. At both stations

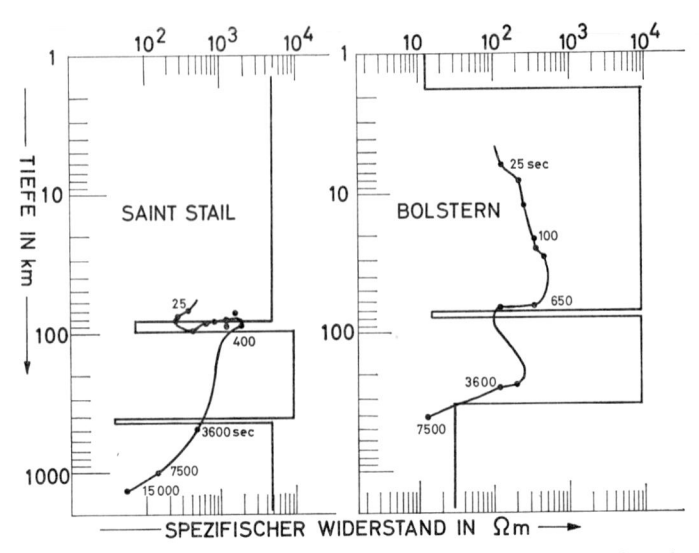

Fig. 2. The resistivity as function of depth at the site **Saint Stail** at the left and at **Bolstern** at the right. V e r t i c a l   a x i s : Depth in kilometers, h o r i z o n t a l   s c a l e : Resistivity in Ohm·m (both scales, the horizontal and the vertical one are logarithmic). S t r a i g h t   r e c t a n g u l a r   l i n e s : Models of resistivity as determined with the method of CAGNIARD. They coincide more or less with the depth distribution of resistivity from the direct method which is shown by the c o n n e c t e d   d o t s : Result of the inversion method of SCHMUCKER. The dots are thus directly calculated estimates of the resistivity. With increasing periods (small numbers at some dots) this method yields values of the resistivity with increasing depth due to the skin effect of transient electric and magnetic fields. The accumulation of dots at the depth of 80 km means a low resistivity layer at that depth which the inducing magnetic field is unable to penetrate, unless the period becomes longer than 2000 sec. At the right hand figure (**Bolstern**) a sedimentary cover appears as a low resistivity layer at the upper surface till the depth of 1.8 km. The shielding effect of this layer is the reason that the resistivity estimates for the 25 sec and longer periods occur at shallower depths than at **Saint Stail** where there exists no sedimentary layer. Therefore also the effect of the layer at a depth of 80 km is not as significant as in the left hand figure.

one finds a low resistivity layer, whose upper surface is between 70 and 90 km in depth and which is about 20 km thick. Such a layer is considered by many authors to exist more or less deep and thick on a global scale. WINTER and SCHEELKE (c. f. their contributions in this volume) deduced from their measurements the existence of a low resistivity layer which lies at a shallower depth of only about 25 km. However, their result applies only to the area of the Rhinegraben itself, whereas the layer in a depth of 80 km inferred from the present measurements lies some distance from the Rhinegraben. Hence the different depths of the low resistivity layers dont necessarily contradict each other but could mean an upwelling, resp. swelling of the low resistivity layer beneath the Rhinegraben.

Acknowledgement: We would like to thank Prof. Dr. G. ANGENHEISTER for his support during this project, and Dr. A. BERKTOLD and Dr. S. GREINWALD for their help in the fieldwork. This project was sponsored by the Deutsche Forschungsgemeinschaft.

References

CAGNIARD, L., 1953: Basic theory of the magnetotelluric method of geophysical prospecting. — Geophysics 18, 605—635.

SCHMUCKER, U., 1969: Zwei neue Verfahren zur Bestimmung des elektrischen Widerstandes als Funktion der Tiefe aus erdmagnetischen und erdelektrischen Beobachtungen. — Protok. Koll. Erdmagn. Tiefensondierung Reinhausen, 134—147.

A Model for the Resistivity Distribution from Geomagnetic Depth Soundings

by

R. Winter

With 5 figures in the text

Abstract: The analysis of geomagnetic data from a profile across the Rhinegraben and the interpretation with two-dimensional models consisting of a non-uniform thin surface sheet and a sequence of uniform plane layers below leads to the following best fitting model for the resistivity distribution of the crust and the uppermost mantle: Under the well-conducting sediments (2—3 Ωm for the centre of the graben) extends the high resistivity zone of the crystalline basement ($\simeq 1000\ \Omega$m) to a depth of about 25 km. The data fitting requires an intermediate conducting layer of approximately 30 Ωm at subcrustal depth. A second zone of high resistivity ($\simeq 1000\ \Omega$m) follows at a depth of about 75 km. An ultimate conducting layer of approximately 10 Ωm begins between 250 and 330 km in the upper mantle.

The anomalous behaviour of the geomagnetic field variations at the Rhinegraben has been studied at 17 temporary geomagnetic observation points on a 400 km profile, extending from the Mosel valley to Munich (WINTER 1967, 1970).

In order to compare the observations with model calculations complex-valued transfer functions between the Fourier spectra of the vertical and horizontal variations have been derived for 0.5 to 6.0 cycles per hour (cph) with statistical methods.

Let H_{az} be the anomalous internal part of the vertical component correlated with the observed horizontal component H_p, tangential to the earth's surface and projected into the direction perpendicular to the strike of the graben structure. Then a transfer function $W_p{}'$ is calculated for a single station in the frequency domain:

$$H_{az} = W_p{}' H_p. \tag{1}$$

In this relation it is assumed that the normal part of H_z is negligible. The in-phase and out-of-phase part of these transfer functions for all stations on the profile and for five groups of frequencies are shown in Fig. 1. The strike of the graben anomaly has been assumed to be N 5° E and the transfer functions have been transformed by rotating the system of coordinates into this new direction. Inside the Rhinegraben area there are fairly enough stations to interpret the systematical behaviour of the data. The edges of the graben become clearly apparent by maxima and minima, which are at a small distance outside of the well-conducting graben filling. A distinct frequency dependence can be seen in particular in the in-phase part. For the comparison of the transfer functions with numerical model values it is necessary to eliminate the anomalous part of the horizontal component in (1). Let H_{np} be the hypothetical normal value of $H_p = H_{ap} + H_{np}$ at a large distance from the Rhinegraben. Then

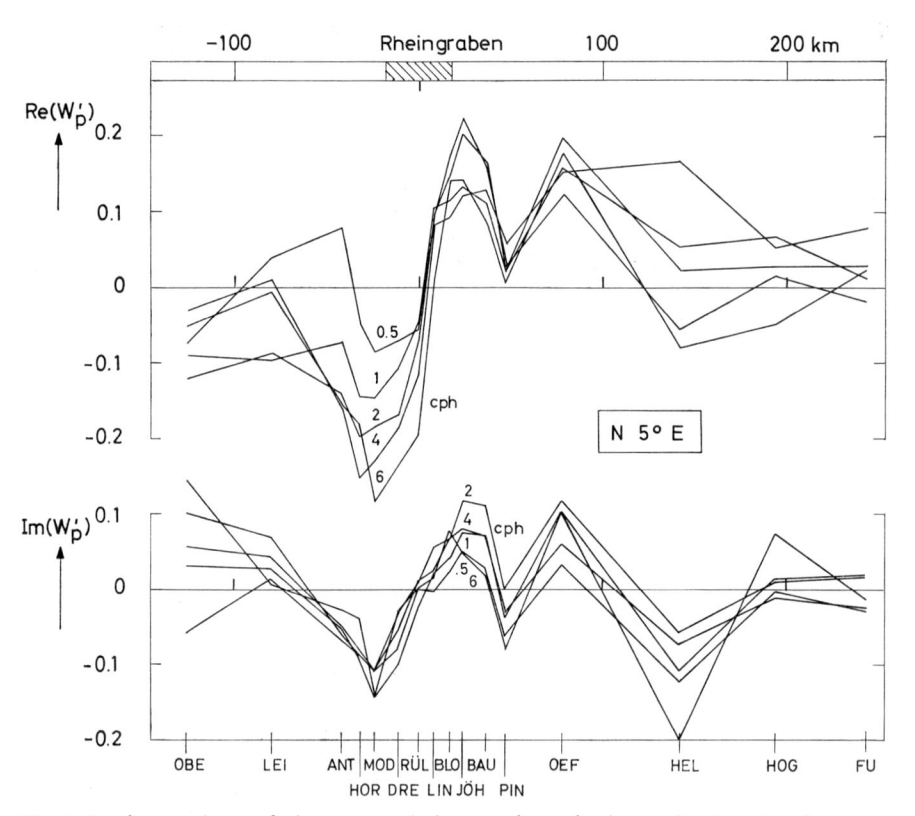

$$W_p = H_{az}/H_{np} \qquad (2)$$

Fig. 1. In-phase and out-of-phase part of the complex-valued transfer functions between the internal anomalous part of the vertical component and the observed horizontal component of the magnetic field variations for different frequencies along a profile across the Rhinegraben.

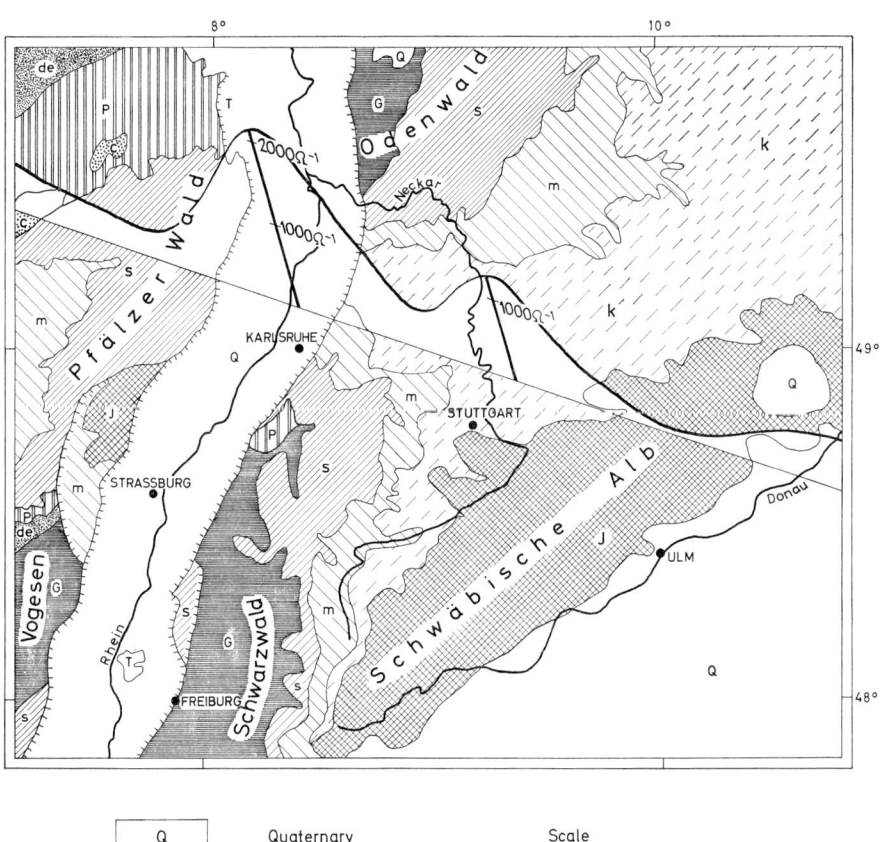

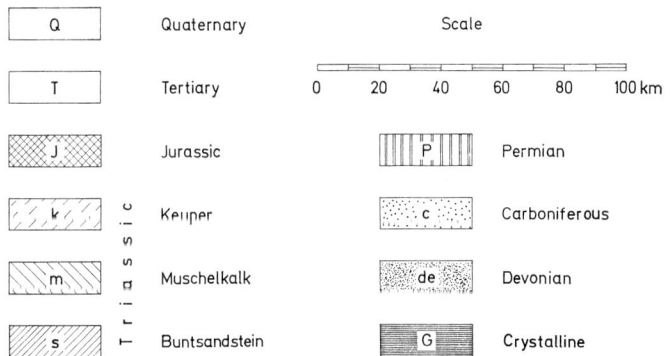

Fig. 2. A perspective display of the conductance τ of the surface layer above the crystalline basement as derived by direct inversion of the empirical transfer functions. Notice the prominent amplitudes in accordance with well-conducting geological structures.

has been calculated from (1) and the fact that H_{az} and H_{ap} are connected by a Hilbert transform and by solving a system of linear equations.

The transfer functions W_p are interpreted by two-dimensional numerical models using formulae given by SCHMUCKER (1970, 1971). These models consist of a thin surface layer with lateral inhomogeneities of the conductivity and a sequence of uniform plane layers below.

The integrated conductivity or conductance τ of the surface layer above the crystalline basement is derived by direct inversion of the empirical transfer functions W_p and a given model for the substructure. All available frequencies have been used. The inversion has been performed in such a manner that a central Rhinegraben conductance of 2100 Ω^{-1} was obtained according to resistivity measurements in boreholes. The smoothed curve of the conductance in Fig. 2 shows — in a perspective display — its change with different material of geological formations. Besides the two maxima at well-conducting sediments in the centre of the graben and in the north of Stuttgart, there are minima at the flancs of the graben and at the Schwäbische Alb, where the influence of high resistivity of

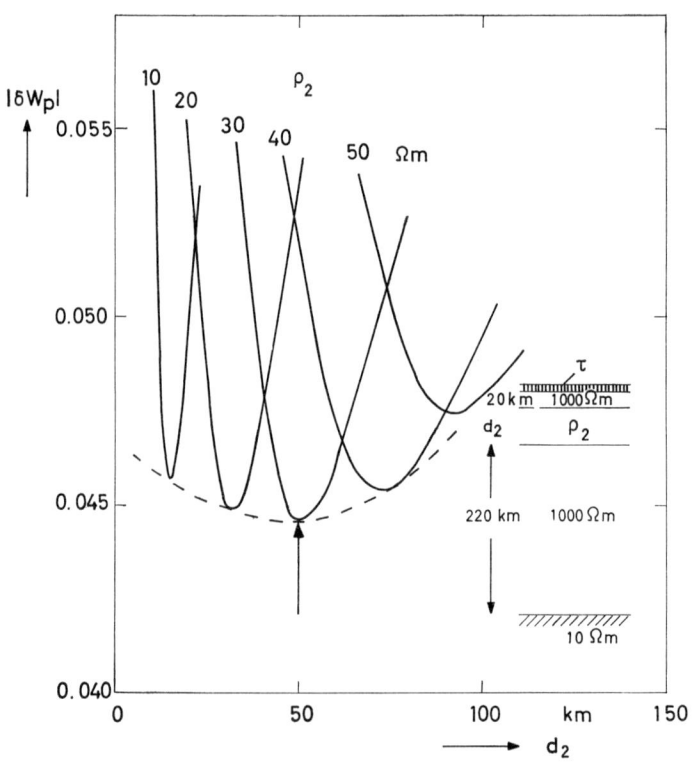

Fig. 3. The mean absolute difference $< |\delta W_p| >$ as a function of different thickness and resistivity of the second layer in a four-layered model. The best fitting parameters ϱ_2, d_2 correspond to the minimum of the envelope.

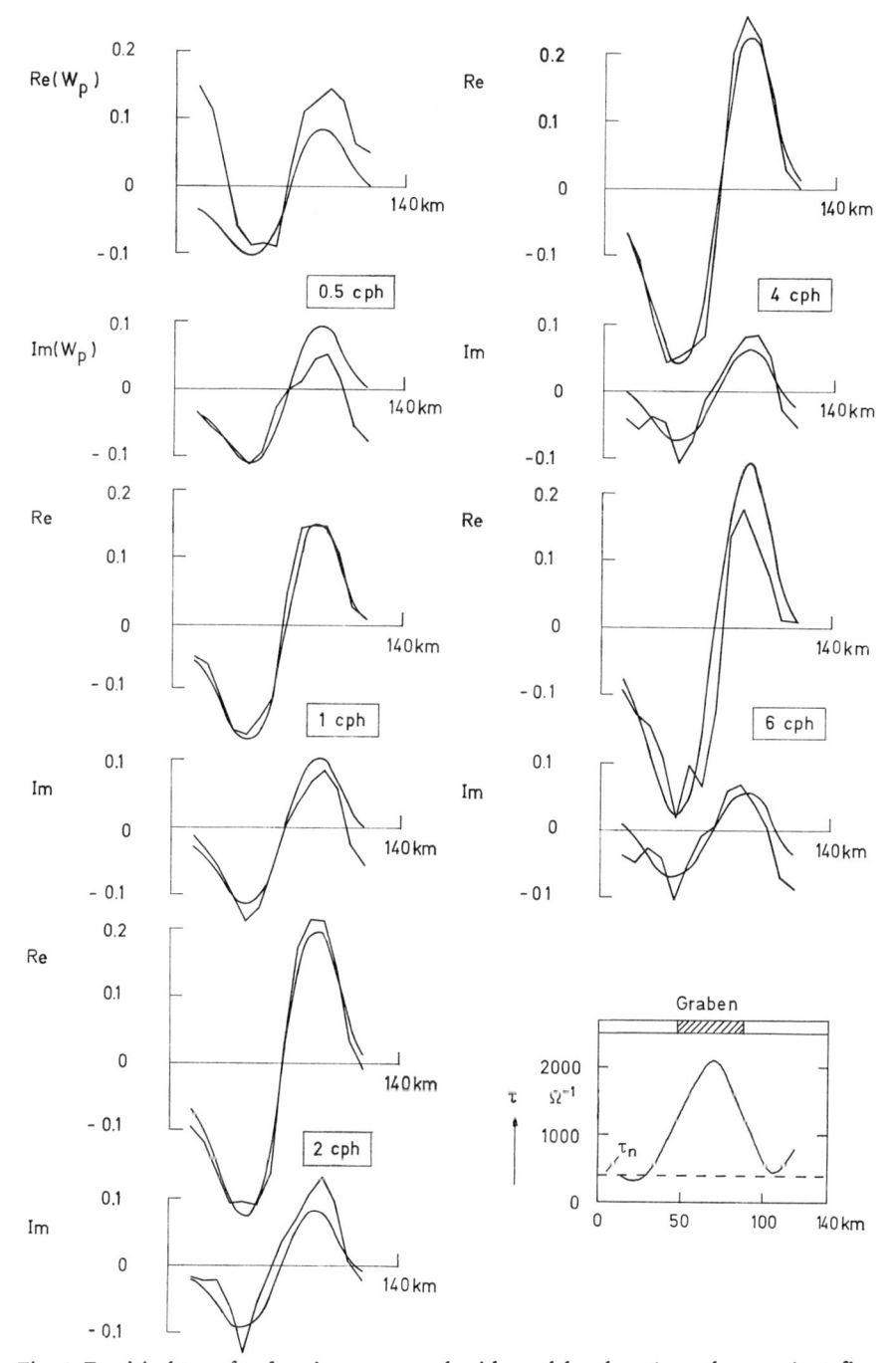

Fig. 4. Empirical transfer functions compared with model values (smooth curves) at five frequencies. Best agreement between observed and calculated W_p-profiles is evolved for 1 cph.

surface material is visible. For the normal conductance at large distance from the anomaly 400 Ω^{-1} was automatically obtained by this method.

The two-dimensional models of the substructure have a large number of independent parameters: the number of layers, their resistivities and thicknesses. Varying these parameters, but keeping the distribution of the surface conductivity (Fig. 2) fixed, a best fitting model is obtained with four layers shown at the right-hand side of Fig. 3. In order to characterize the deviation of the empirical values W_p from model values the mean absolute difference $<|\delta W_p|>$ averaged over all frequencies and grid points near the Rhinegraben has been calculated. As an example for this method $<|\delta W_p|>$ is shown in Fig. 3 by varying the thickness d_2 and the resistivity ϱ_2 of the second layer in a four-layered model for the deep structure. The minimum of the envelope provides the interesting parameters: $d_2 = 50$ km, $\varrho_2 = 30$ Ωm.

Fig. 4 gives an idea of the agreement between observed and calculated W_p-profiles, underlying the best fitting model from Fig. 3 and displaying the curves for different frequencies. The behaviour of the surface conductance τ derived by

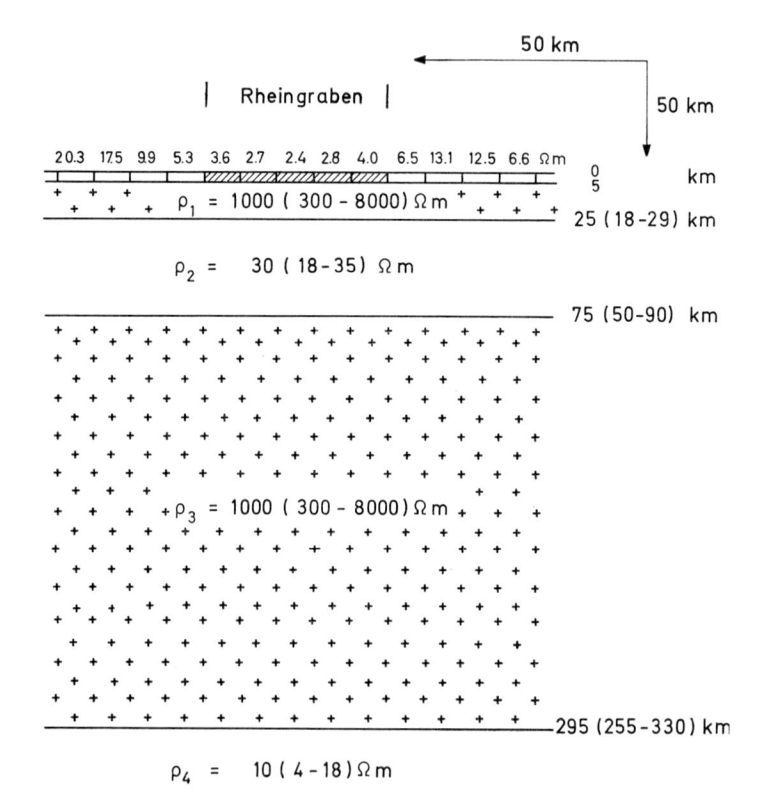

Fig. 5. Display of the best fitting model with lateral change of the resistivities in the surface layer. The parameters of the uniform plane layers below are shown with their estimated ranges of variability.

inversion is drawn at the bottom right of Fig. 4. Best agreement between observed and calculated W_p-profiles evolves for 1 cph.

The final model of the resistivity distribution is shown in Fig. 5. The lateral change of the resistivity of the surface layer is calculated for each segment with $\varrho = d/\tau$ from the conductance τ (Fig. 2) and fixing the thickness d by 5 km — corresponding approximately to the deepest situation of sediments in the graben filling. The values in parentheses give the estimated range of variability of the substructure parameters. Under the highly resistent crystalline basement ($\simeq$ 1000 Ωm) it is necessary to assume a channel of low resistivity beginning at a depth of about 25 km and having a resistivity of approximately 30 Ωm. This intermediate layer of about 50 km in thickness is followed again by a zone of high resistivity ($\simeq$ 1000 Ωm). Between 250 km and 330 km the conductosphere appears with a resistivity of approximately 10 Ωm.

The existence of a well-conducting layer in the uppermost mantle under the Rhinegraben has been confirmed also by magnetic-telluric investigations (SCHEELKE 1972) and telluric studies (HAAK 1970).

References

HAAK, V., 1970: Das zeitlich sich ändernde erdelektrische Feld, beobachtet auf einem Profil über dem Rheingraben; eine hiervon abgeleitete Methode der Auswertung mit dem Ziel, die elektrische Leitfähigkeit im Untergrund zu bestimmen. — Diss., Univ. München.

SCHEELKE, I., 1972: Magnetotellurische Messungen im Rheingraben und ihre Deutung mit zwei-dimensionalen Modellen. — Diss., Techn. Univ. Braunschweig.

SCHMUCKER, U., 1970: Anomalies of geomagnetic variations in the southwestern United States. — Bull. Scripps Inst. Oceanogr., **13**.

— 1971: Interpretation of induction anomalies above nonuniform surface layers. — Geophysics, **36**, 156—165.

WINTER, R., 1967: Geomagnetic deep-sounding at the Rhinegraben: 1. Measurements of the magnetic variations. — Abh. Geol. L.-Amt Baden-Württ., **6**, 127—130.

— 1970: Erdmagnetische Tiefensondierung im Gebiet des Oberrheingrabens. — Graben Problems, Schweizerbart, Stuttgart.

The Electrical Conductivity of Minerals and the Temperature Distribution in the Upper Mantle

by

A. Schult

With 1 figure in the text

Abstract: Combining laboratory measurements on olivines with resistivity profiles of the Earth's interior one can get some information upon temperature distribution in the Earth because temperature is one of the main factors controlling resistivity. Recent results indicate that the resistivity of olivine is also strongly influenced by the Fe^{3+} content. Therefore the method is subjected to some uncertainty at the present stage. — For the top of the low resistivity layer (in general at a depth of 80 km and of 25 km in the area of the Rhinegraben) a temperature of about 800° C can be computed.

There exist several experimental data on the electrical conductivity of minerals and rocks under pressure and temperature conditions of the upper mantle for plausible materials. The temperature dependence of the electrical conductivity of minerals and rocks can be described in the relevant range by

$$\sigma = \sum_i \sigma_i \exp(-E_i/kT)$$

where E_i are activation energies, T is the absolute temperature, k BOLTZMANN's constant and σ_i constants. The subscript i takes up to three values depending on the temperature range and indicates that several conduction mechanisms may be present. The parameters σ_i and E_i are in most cases only sligthly pressure dependent. Thus the conductivity is mainly controlled by temperature at least in the upper few hundred kilometers.

In Fig. 1 resistivity data in the olivine system are summarized with pressure as parameter. — The electrical resistivity of other mineral systems including basic rocks which may form a constituent of the upper mantle is similar to that of the olivine system (SCHULT 1974). — In nearly all cases pressure has the tendency to decrease the resistivity in the temperature ranges shown. When pressure is raised from near atmospheric to about 50 kbar the resistivity decrease is in most cases less than an order of magnitude. The effect of pressures above 50 kbar on the resistivity is not known.

From Fig. 1 it follows that the fayalite (Fe_2SiO_4) content of the olivine system for typical upper mantle composition (5—20%) is not the only important factor controlling the resistivity. Olivine samples with approximately the same fayalite content have sometimes electrical resistivities that differ by 2 to 3 orders of magnitude (see Fig. 1, e. g. curve 1 and 5). The most resistive sample is a single crystal (9,4% fayalite content) with practical no Fe^{3+} content, all other samples in Fig. 1 have detectable Fe^{3+} contents. Therefore it was proposed by DUBA (1972) that the resistivity in the olivine system is strongly controlled by the Fe^{3+} content.

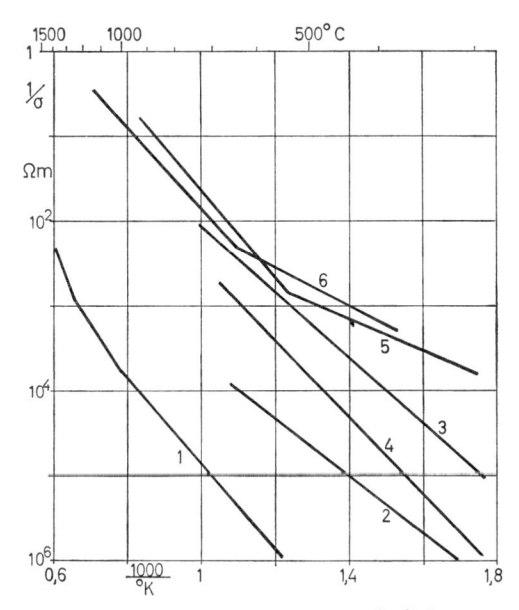

Fig. 1. The electrical resistivity of olivines.
Curve 1: single crystal, 9,4 mol % fayalite, no detectable Fe_2O_3, 8 kbar (DUBA 1972).
Curve 2: single crystal, 8,2 % fayalite, 0,16 wt % Fe_2O_3, 7,5 kbar (DUBA 1972).
Curve 3: single crystal 26,4 % fayalite, 7,5 kbar (DUBA 1972).
Curve 4: polycrystal, 10 % fayalite, 0,6 % Fe_2O_3, 23 kbar (BRADLEY et al. 1964).
Curve 5: single crystal, 11 % fayalite, 0,5 % Fe_2O_3, 30 kbar (SCHOBER 1971).
Curve 6: polycrystal, 8,6 % fayalite, 0,4 % Fe_2O_3, 35 kbar (SCHOBER 1971).

— No systematic dependence of the resistivity on the Fe^{3+} content can be derived from Fig. 1 but one should have in mind that the measurements were carried out under various experimental conditions.

The scatter of the results in Fig. 1 is large and therefore the calculation of the temperature distribution in the earth from electrical resistivity profiles is subject to large uncertainties. However, some useful information can be drawn from the data.

From geomagnetic depth sounding many authors derived a layer of low resistivity (in the order of 30 Ωm) at a depth of about 80 km. This was also found at some distance East and West of the Rhinegraben (HAAK & REITMAYR, c. f. their article in this volume). The low resistivity layer in the area of the Rhinegraben lies at a shallower depth of only about 25 km (WINTER, SCHEELKE, c. f. their contributions in this volume). Basing a temperature calculation on results for olivines with approximately 10 % fayalite content and a high content of ferric iron one obtains a temperature of 800° C for the above mentioned depths of the low resistivity layer. The zone of higher resistivity which is found below the low resistivity layer (HAAK et al., SCHEELKE, WINTER, c. f. their contributions in this volume) may consist mainly of olivines with a minor or zero ferric iron content which have a high resistivity of the order of 1000 Ωm (see Fig. 1) at temperatures above 800° C.

References

Bradley, R. S., Jamil, A. K. & Munro, D. C., 1964: The electrical conductivity of olivine at high temperatures and pressures. — Geochim. Cosmochim. Acta **28**, 1969—1678.

Duba, A., 1972: Electrical conductivity of olivine. — J. Geophys. Res. **77**, 2483—2495.

Schober, M., 1971: The electrical conductivity of some samples of natural olivine at high temperatures and pressures. — Z. Geophys, **37**, 283—292.

Schult, A., 1974: Some petrophysical aspects of deep structure. — (in preparation).

Dynamic models for taphrogenesis

Crustal Dynamics in the Central Part of the Rhinegraben[1]

by

St. Mueller and L. Rybach

With 6 textfigures

Abstract: An attempt is made to summarize all geophysical results that have been obtained for the central part of the Rhinegraben. In a synoptic interpretation of seismic refraction and reflection data, focal depths of earthquakes, gravity and geothermal observations as well as recent crustal movements determined by precise levelling it is shown that the driving mechanisms of the observed dynamic processes must be sought in a velocity and density inversion within the upper crust. Based on an experimental relationship between heat production and compressional velocity in crustal rocks the temperature field of layered models can be calculated. For the central part of the Rhinegraben the temperature structure is governed by pronounced vertical and lateral temperature gradients.

The primary aim of the International Geodynamics Project is to determine the nature and origin of the dynamic processes in selected regions of the earth. Among the areas chosen for a more detailed study is the Rhinegraben of which the exact relation to the world rift system has not yet been elucidated. Thorough geological and geophysical studies have therefore been carried out concurrently over the past seven years in order to obtain information on the relationship between surface features, recent crustal movements and the underlying structure of the crust and upper mantle. The amount of data now available is considerable and it seems appropriate to appraise and synthesize the existing information.

To this end a crustal model for the central part of the Rhinegraben rift zone has been developed which incorporates all known results of seismic refraction measurements and deep reflection observations (MUELLER et al. 1967, 1969, 1973; ANSORGE et al. 1970, 1973). In Fig. 1 a detailed WNW—ESE cross-section through this model is presented. Its geographical location is indicated by the solid line A—A in Fig. 2. The length of the section — the surface geology of which is, of course, oversimplified — does not reach quite 200 km.

Information about seismic velocities in the Rhinegraben proper is available from a large number of well logs obtained during oil and gas exploration work. In the upper part of the crystalline basement with prevailing P-velocities of 5.9 to 6.0 km/sec a positive velocity gradient exists. At a depth of about 10 km under the

---

[1] Mitteilung Nr. 80, Institut für Geophysik, Eidgenössische Technische Hochschule, Zürich (Schweiz).

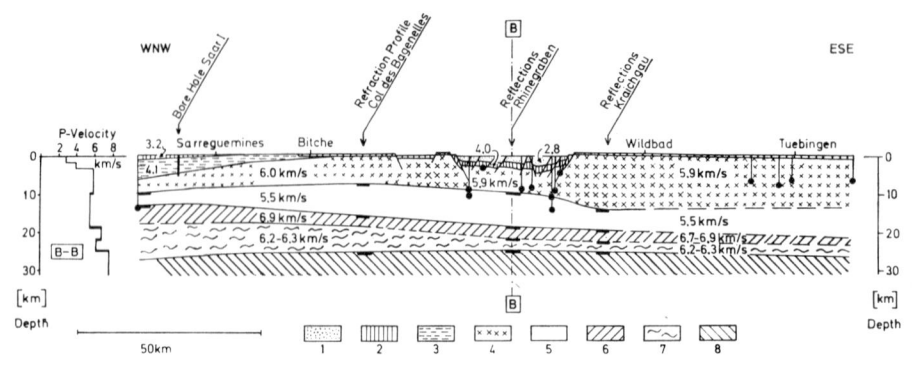

Fig. 1. Crustal cross-section through the central part (profile A—A in Zone I, see Fig. 2) of the Rhinegraben rift system based on seismic data. The P-velocity-depth function in the center of the rift (B—B) is shown on the left. Reflecting horizons are marked by heavy bars. Hypocenters of earthquakes are indicated as solid dots at the appropriate depth. Legend: 1 = Quaternary and Tertiary, 2 = Jurassic, Triassic and Permian, 3 = Carboniferous and Devonian, 4 = Crystalline basement, 5 = Sialic low-velocity channel, 6 = Intermediate crustal layer, 7 = Intermediate low-velocity zone, 8 = Crust-mantle transition zone.

rift a velocity reversal must occur (ref. MUELLER & LANDISMAN 1966). A constant velocity of 5.5 km/sec was taken to be the most plausible approximation to the velocity distribution likely to be found in the sialic low-velocity channel.

Beneath the low-velocity layer in the upper crust an intermediate layer with P-velocities between 6.7 and 6.9 km/sec has been found consistently in the rift system. A second velocity reversal underneath this layer must be assumed in order to explain the observed offset of the corresponding travel-time branches. The most likely minimum P-velocity in this intermediate low-velocity zone is 6.2 to 6.3 km/sec. Below these two layers a sharp discontinuity marks the top of the crust-mantle transition zone for which in recent experiments a true velocity of about 8 km/sec has been determined.

The predominant echoes found in deep crustal reflection work are indicated in the cross-section of Fig. 1 by heavy bars. They check the seismic refraction results quite well. A P-velocity-depth profile (B—B) through the crust in the center of the rift system is shown in Fig. 1 on the left. Heavy lines again designate velocity (and density) changes which give rise to normal-incidence reflections. In the depth range between 19 and 25 km a laminated transition structure must be assumed judging from the striate character of the observed deep reflections. This fine structure is represented only schematically in the section of Fig. 1.

The most conspicuous feature of the crustal section through the central part of the Rhinegraben is its asymmetry which reaches to depths of the order of 20 km (Fig. 1). For the sedimentary filling of the graben the asymmetry has been convincingly demonstrated by DOEBL (1967, 1970). It is interesting to compare the crustal structure in Fig. 1 with the distribution of focal depths determined from shallow earthquakes in the same region by macroseismic methods (SPONHEUER 1969). The depths of foci in the graben proper with one exception do not exceed

12 km, while under the eastern flank of the rift focal depths of 6 to 8 km seem to prevail. SCHNEIDER (1968) has pointed out that the earthquakes in the Black Forest always lie close to the contacts between gneisses and granites. These observations would place most of the hypocenters close to the top or roof of the sialic low-velocity channel (MUELLER 1970) in accordance with similar findings in Central Japan. Some foci seem to fall into the depth range where the main faults meet the sialic low-velocity zone. The discrepancy observed for earthquakes on the eastern side of the rift must be ascribed to the fact that the structure of the upper crust in that region is not yet known in sufficient detail. It is, therefore, marked by dashed lines in the right part of Fig. 1.

At the beginning of the detailed crustal investigations in and around the Rhinegraben MUELLER & PETERSCHMITT (1966) pointed out that a distinct asymmetry was also evident in the gravity anomalies observed within the graben. In Fig. 2 the map of Bouguer anomalies of the Rhinegraben area is shown. This map has

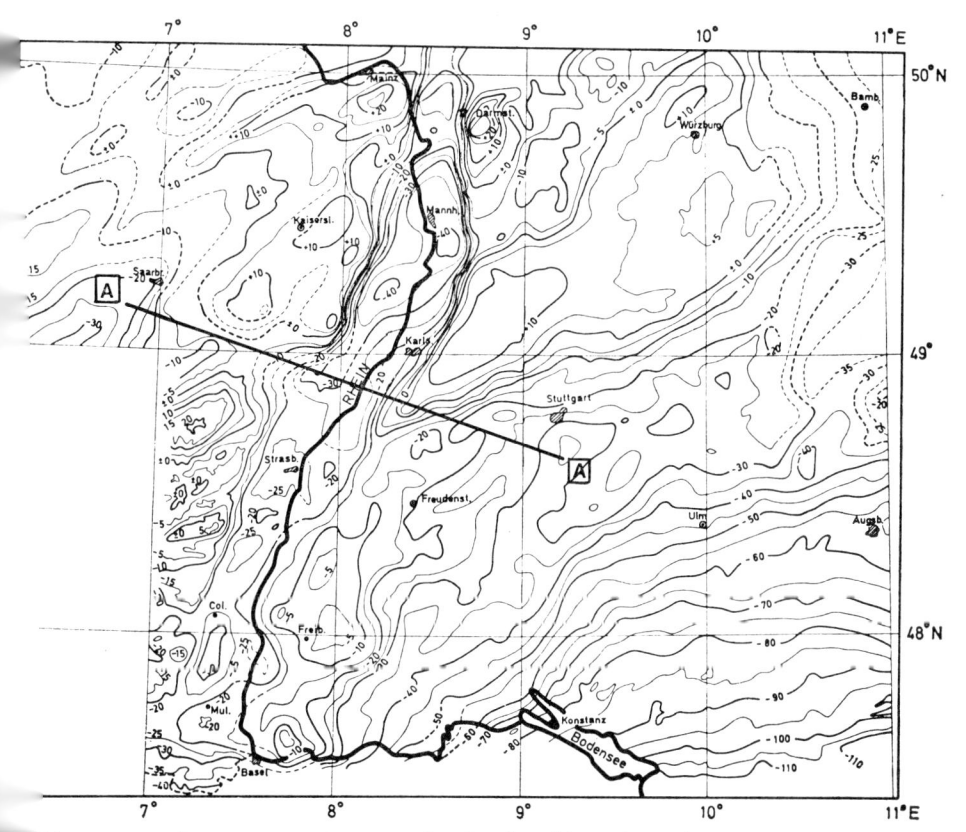

Fig. 2. Map of Bouguer gravity anomalies for the Rhinegraben rift system (compiled by W. LOPAU). Contour interval is 5 mgal. The profile A—A lies close to the southern boundary of Zone I. It marks the position of the crustal cross-sections in Figs. 1 and 3.

been compiled from the data published by GERKE (1957) and LECOLAZET (1970). In the northern part the outline of the graben can be clearly seen. To the south of profile A—A in Zone I it becomes more difficult to recognize the eastern margin of the rift. The chain of gravity minima lies close to the eastern flank of the rift in the northern portion of the Rhinegraben. About midway between Mannheim and Karlsruhe the axis of negative Bouguer gravity anomalies shifts from the eastern to the western half of the graben and remains there all the way down to Mulhouse close to the southern end of the Rhinegraben. It is most interesting to note that this chain of negative anomalies roughly follows the axis of local horsts and regions of slightly elevated temperatures. HAEGELE & WOHLENBERG (1970) have mentioned the fact that the regions of highest seismic energy flux in the graben coincide with the axis of the negative Bouguer anomalies — an observation which has also been confirmed in other parts of the world, as e. g. in the Alps and in the East African rift system.

In Fig. 3 a profile of Bouguer anomalies is shown through Zone I along the profile A—A (see Fig. 2). The graben is characterized by an asymmetric negative gravity anomaly of about 30 mgal. It had been pointed out by MUELLER et al. (1967) that most of the anomaly can be accounted for by the sedimentary filling of the Rhinegraben. A closer inspection, however, showed that the gravity minimum observed in the western portion of the graben did not coincide with the

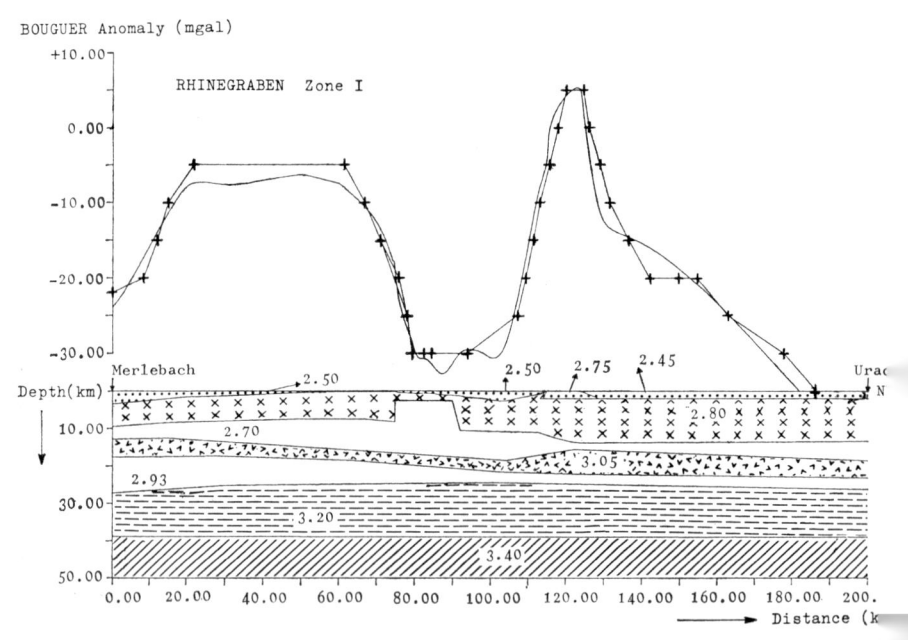

Fig. 3. Gravimetric section along the profile A—A in Fig. 2. Density values of the crustal model are given in g/cm³. Plus signs mark the observed Bouguer anomaly, while the solid line represents the calculated two-dimensional gravity anomaly for the model in the lower half of the figure, which should be compared to Fig. 1.

minimum of the gravity effect caused by the sedimentary cover which in Zone I is located close to the eastern margin of the Rhinegraben.

After ascribing reasonable density values to the layers of the seismic model in Fig. 1 following the NAFE & DRAKE relation (1959) depicted in Fig. 5 a number of gravity models were calculated. The results demonstrated clearly that the negative graben anomaly cannot be explained simply by the mass deficiency of the sediments. Its cause must rather be sought at some depth within the crystalline basement. Led by the observed P-velocity decrease and the associated density inversion in the upper crust MUELLER (1970) proposed that at the corresponding depth a change in composition of the basement material must occur. It is suggested that a possible explanation for the inferred velocity and density reversal is the intrusion of acidic rock into the adjacent crystalline country rocks. Granitic material with densities between 2.65 and 2.70 g/cm^3 has a density contrast of about 0.1 to 0.3 g/cm^3 as compared to the surrounding and overlying rocks. It would tend to rise wherever it is given a chance.

A satisfactory fit of the model calculations with the observed values of the Bouguer anomaly in Fig. 3 can be achieved if an upward indentation of the sialic low-velocity channel into the overlying basement is assumed. The thinning of the upper crust due to the density instability might be a first step towards a more advanced phase in the rifting process. It should be noted that this result is not influenced by the subdivision of the uppermost mantle into two layers with densities of 3.2 and 3.4 g/cm^3. This gradation will only cause a shift in the average gravity level and is, therefore, irrelevant in this discussion.

Several mechanisms have been proposed for the formation of graben structures in continents. The suggestion of KNOPOFF (1970) that a rift may be produced by strike-slip motion along an irregular contact between two crustal plates offers an explanation for the vertical crustal movements in Zone I of the Rhinegraben. In the generalized seismicity pattern an "en échelon" offset is observed north of Karlsruhe (HAEGELE & WOHLENBERG 1970) corresponding to the switch-over of the chain of gravity minima in Fig. 2. If it is assumed — as sketched in Fig. 4 a — that the active fault zone changes over from the eastern master fault in the northern graben to the western boundary in the southern graben (MUELLER et al. 1969) and if a left-lateral relative displacement is inferred for which there is considerable evidence (see e. g. AHORNER 1970), then earthquakes of strike-slip character should take place along the straight north—south trending portions of the plate contact. In the offset region local earthquakes will have a more or less pure compressional (dip-slip) character, and the entire offset region (Fig. 4 a + b) will either move up or down.

KUNTZ, MAELZER & SCHICK (1970) have found a relative uplift of about 0.5 mm/year for the region Kandel—Karlsruhe—Bruchsal (just north of profile A—A in Fig. 2) from precise levelling. It is thus possible to fix the sign of the relative compressional motion (Fig. 4 b). The earthquake mechanisms described by SCHNEIDER (1968) for the neighbourhood of Karlsruhe indeed show the predominant dip-slip character which is to be expected if the hypothesis presented above

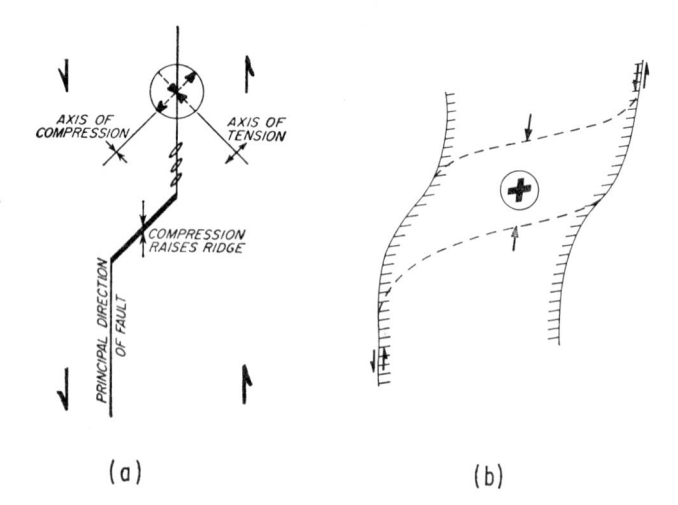

Fig. 4. (a) Schematic diagram illustrating the regional and local stress pattern in the Rhinegraben. (b) Enlargement of the compressional zone just north of the profile A—A in Fig. 2. The uplift (+) is in agreement with focal mechanisms of earthquakes and precise levelling measurements.

is correct. It is probably no accident that the region of maximum uplift in the graben coincides with the zone where the sialic low-velocity channel reaches its maximum thickness (Fig. 3).

For some time it has been suggested that there may be an strong correlation between heat flow data and tectonic features. The remarkable linear relation between terrestrial heat flow and the radioactive heat production in rocks (see e. g. ROY et al. 1968) has helped to produce quantitative data on the thermal regime in the upper few tens of kilometers of the continental crust. It has also led to important restraints on the geochemistry and the formation of plutons.

Since there exists now a large number of continental heat flow measurements but no corresponding determinations of heat production, an attempt has been made to derive an experimental relationship between density (ϱ) and heat production (A) of crustal rocks from data given in the compilation of CLARK (1966). By utilizing the density-velocity relationship of NAFE & DRAKE (1959) it is then possible to give approximate values of heat production (A) as a function of compressional velocity (V_P) and density (ϱ). These relations are depicted in Fig. 5. They have been used by RYBACH (1973) to calculate for average values of heat flow the stationary temperature—depth distributions beneath the Alpine Foreland and under the central massifs of the Swiss Alps where the crustal structure is fairly well known. For these model calculations reasonable assumptions must, of course, be made about the surface temperature and the heat conductivity of the various crustal layers.

By the same method (Fig. 5) the temperature—depth distribution has been determined for selected crustal models at three locations along profile A—A

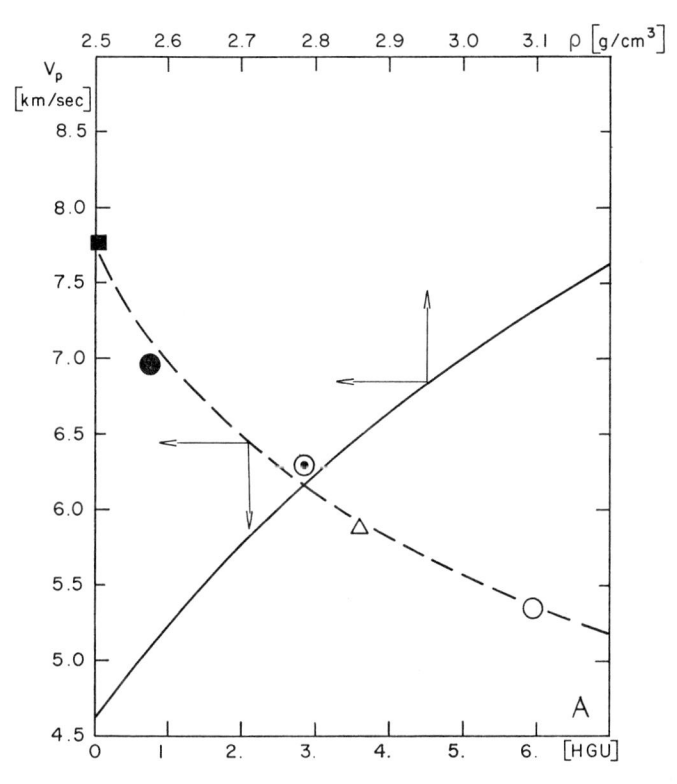

Fig. 5. Experimental relationships for crustal rocks. Dashed line: Relation between radioactive heat production A (1 HGU = 10^{-13} cal/cm³ · sec) and compressional wave velocity V_P. ○ = granite, △ = syenite, ⊙ = diorite, ● = gabbro, ■ = peridodite. Data from CLARK (1966). Solid line: Density-velocity relation according to NAFE & DRAKE (1959).

through the central part of the Rhinegraben rift system (see Fig. 2). Based on the crustal section in Fig. 1, the surface heat flow values (q, measured in HFU, where 1 HFU = 10^{-6} cal/cm² · sec) published by HAENEL (1970, 1971) for Zone I of the Rhinegraben and the data given by WERNER (1968, 1970) for the upper crust of that region three temperature—depth profiles were constructed which are shown in Fig. 6. On the eastern flank of the rift an average surface heat flow of q = 1.7 HFU (curve A in Fig. 6) was considered to be representative for the Black Forest area. This value is close to the mean heat flow found by HAENEL (1971) for Central Europe. The new data on temperature gradients compiled by WERNER & DOEBL (1973) for the Rhinegraben permit to deduce an average surface heat flow of q = 2.0 HFU (curve B in Fig. 6) in rather good agreement whit the measured value of q = 2.2 HFU near Legelshurst east of Strasbourg (HAENEL 1970). In an attempt of interpreting the 100°C geoisotherm published by ILLIES (1965) a reasonable explanation was given by WERNER (1968, 1970) for the spectacular geothermal anomaly of q = 4.0 HFU found near Soultz/Alsace (curve C in

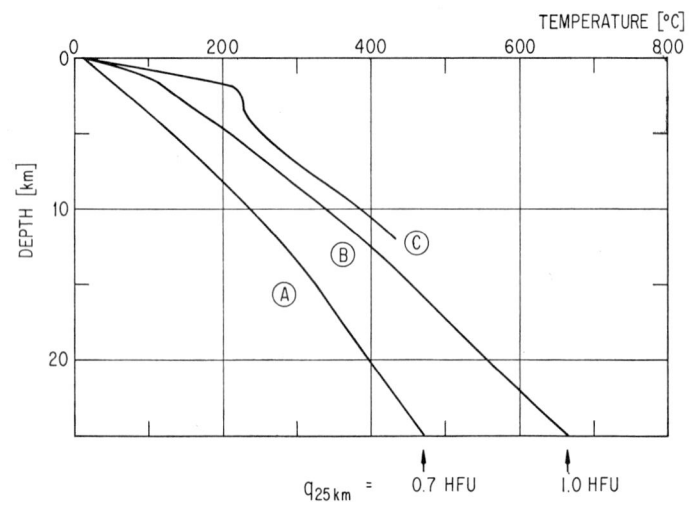

Fig. 6. Temperature-depth profiles for three locations along profile A—A through the central part of the Rhinegraben (see Fig. 2) based on model calculations assuming steady-state conditions. A = Black Forest, q = 1.7 HFU (after RYBACH 1973). B = Rhine-graben (central part), q = 2.0 HFU (based on data by WERNER & DOEBL 1973). C = Geothermal anomaly near Soultz/Alsace, q = 4.0 HFU (after WERNER 1968).

Fig. 6). Thermal waters circulating within the fracture zone of the master fault close to the western graben margin may be a likely cause of this singular anomaly.

The highest temperatures in the upper crust beneath the central graben segment are associated with the anomaly near Soultz. Under most of the graben and particularly in the center (see section B—B in Fig. 1) the temperatures at the same depth are on an average 25 to 30 per cent higher than in the adjacent Black Forest area. It should be realized that except for shallow anomalies like the one near Soultz the dominating factor in determining the temperature—depth distribution is the surface heat flow and not the inherent heat production or the contrast in heat conductivity. The heat flow ($q_{25\,km}$) coming from depths greater than 25 km (compare Fig. 6 to Figs. 1 and 3) is 0.7 HFU under the Black Forest mountains, and 1.0 HFU under the Rhinegraben. Both values do not differ too much from the average mantle value of about 0.9 HFU for Central Europe as calculated by HAENEL (1971).

In an interpretation of the elevated terrestrial heat flow in the Rhinegraben HAENEL (1970) has postulated that the material in the sialic low-velocity channel must be in a state of fusion or at least partial fusion. The deduced decrease in the P-velocity of about 7 per cent (Fig. 1) is in fairly good agreement with the argument that the decrease in velocity and density is primarily caused by relatively high temperatures in the upper crust. The geothermal maps of WERNER & DOEBL (1974) show that there exist considerable local variations in temperature which are superimposed on the regional field of increased temperatures (see Fig. 6). If these local "hot spots" are explained by cooling bodies of acidic magma which have intruded into the crystalline basement from the sialic low-velocity zone below (see Fig. 3), their age can be estimated to be about 1 to 2 million years (HAENEL 1970).

The systematic compilation and analysis of existing geophysical data for the central part of the Rhinegraben have given some clue as to the most recent dynamics and relative movements observed in that segment of the continental rift which traverses Central Europe. It is suggested that the driving mechanisms of these motions must be sought in a velocity and density inversion within the upper crust. Pronounced vertical and lateral temperature gradients must also play an important role in the rifting process.

The authors would like to thank W. LOPAU (Karlsruhe) for his valuable help in compiling the Bouguer anomaly map and for calculating numerous gravity models. They also wish to acknowledge fruitful discussions with D. WERNER (Zürich).

References

AHORNER, L., 1970: Seismo-tectonic Relations between the Graben Zones of the Upper and Lower Rhine Valley. — In: ILLIES, J. H. & MUELLER, ST. (eds.), Graben Problems, 155—166, Schweizerbart, Stuttgart.

ANSORGE, J. et al., 1970: Structure of the Crust and Upper Mantle in the Rift System around the Rhinegraben. — In: ILLIES, J. H. & MUELLER, ST. (eds.), Graben Problems, 190—197, Schweizerbart, Stuttgart.

— 1973: Seismic refraction investigations in the central and southern part of the Rhinegraben. — In: GIESE, P. & STEIN, A. (eds.), Results of Deep Seismic Sounding in the Federal Republic of Germany and Some Other Areas, in press, Physica, Würzburg.

CLARK, S. P. jr., 1966: Handbook of Physical Constants (revised edition). — Geol. Soc. Amer. Memoir **97**, 587 p.

DOEBL, F., 1967: The Tertiary and Pleistocene sediments of the northern and central part of the upper Rhinegraben. — Abh. geol. Landesamt Baden-Württ., **6**, 48—54.

— 1970: Die tertiären und quartären Sedimente des südlichen Rheingrabens. — In: ILLIES, J. H. & MUELLER, ST. (eds.), Graben Problems, 56—66, Schweizerbart, Stuttgart.

GERKE, K., 1957: Die Karte der Bouguer-Isanomalen 1 : 1 000 000 von Westdeutschland. — Dt. Geodät. Komm., Reihe B, **46**, Teil I, 13 S.

HAEGELE, U. & WOHLENBERG, J., 1970: Recent Investigations on the Seismicity of the Rhinegraben Rift System. — In: ILLIES, J. H. & MUELLER, ST. (eds.), Graben Problems, 167—170, Schweizerbart, Stuttgart.

HAENEL, R., 1970: Interpretation of the Terrestrial Heat Flow in the Rhinegraben. — In: ILLIES, J. H. & MUELLER, ST. (eds.), Graben Problems, 116—120, Schweizerbart, Stuttgart.

— 1971: Heat flow measurements and a first heat flow map of Germany. — Z. Geophys., **37**, 975—992.

ILLIES, J. H., 1965: Bauplan und Baugeschichte des Oberrheingrabens. — Oberrhein. geol. Abh., **14**, 1—54.

KNOPOFF, L., 1970: Problems of Continental Rift Structures. — In: ILLIES, J. H. & MUELLER, ST. (eds.), Graben Problems, 1—4, Schweizerbart, Stuttgart.

KUNTZ, E., MAELZER, H. & SCHICK, R., 1970: Relative Krustenbewegungen im Bereich des Oberrheingrabens. — In: ILLIES, J. H. & MUELLER, ST. (eds.), Graben Problems, 170—177, Schweizerbart, Stuttgart.

LECOLAZET, R., 1970: La Carte gravimétrique de l'Alsace. — In: ILLIES, J. H. & MUELLER, ST. (eds.), Graben Problems, 233—234, Schweizerbart, Stuttgart.

MUELLER, ST. & LANDISMAN, M., 1966: Seismic studies of the earth's crust in continents; Part I: Evidence for a low-velocity zone in the upper part of the lithosphere. — Geophys. J. R. A. S., **10**, 525—538.

MUELLER, ST. & PETERSCHMITT, E., 1966: Die Geschwindigkeitsverteilung seismischer Wellen im tieferen Untergrund um den Oberrheingraben. — DFG-Kolloquium Oberrheingraben, Wiesloch, 3 S.

MUELLER, ST. et al., 1967: The rift structure of the crust and upper mantle beneath the Rhinegraben. — Abh. geol. Landesamt Baden-Württ., **6**, 108—113.

— 1969: Crustal structure beneath the Rhinegraben from seismic refraction and reflection measurements. — Tectonophysics, **8**, 529—542.

MUELLER, ST., 1970: Geophysical Aspects of Graben Formation in Continental Rift Systems. — In: ILLIES, J. H. & MUELLER, ST. (eds.), Graben Problems, 27—37, Schweizerbart, Stuttgart.

MUELLER, ST. et al., 1973: Crustal structure of the Rhinegraben Area. — In: MUELLER, ST. (ed.), Crustal Structure Based on Seismic Data. Tectonophysics, in press.

NAFE, J. E. & DRAKE, C. L., 1959: Experimental relationship between compressional wave velocity and density. — In: TALWANI, M. et al., A crustal section across the Puerto Rico trench, J. Geophys. Res., **64**, 1545—1555.

ROY, R. F. et al., 1968: Heat generation of plutonic rocks and continental heat flow provinces. — Earth Planet Sci. Lett., **5**, 1—12.

RYBACH, L., 1973: Wärmeproduktionsbestimmungen an Gesteinen der Schweizer Alpen. — Beitr. Geol. Schweiz, Geotechn. Serie, Nr. 51.

SCHNEIDER, G., 1968: Erdbeben und Tektonik in Südwestdeutschland. — Tectonophysics, **5**, 459—511.

SPONHEUER, W., 1969: Die Verteilung der Herdtiefen in Mitteleuropa und ihre Beziehung zur Tektonik. — In: STILLER, H. (ed.), Seismologische Arbeiten aus dem Institut für Geodynamik. Veröff. Inst. f. Geodynamik Jena, Reihe A, Heft 13, 82—103, Akademie-Verlag, Berlin.

WERNER, D., 1968: Anomalien des Temperaturfeldes im Untergrund des Oberrheingrabens. — Diplomarb., Univ. Karlsruhe, 132 S.

— 1970: Geothermal Anomalies of the Rhinegraben. — In: ILLIES, J. H. & MUELLER, ST. (eds.), Graben Problems, 121—124, Schweizerbart, Stuttgart.

WERNER, D. & DOEBL, F., 1974: Eine geothermische Karte des Rheingrabenuntergrundes. — In: ILLIES, J. H. & FUCHS, K. (eds.), Approaches to Taphrogenesis, 182—191, Schweizerbart, Stuttgart.

Model of Crustal Stresses in the Region of the Rhinegraben and Southwestern Germany

by

K. Strobach

With 3 figures in the text

The motivation of the following consideration and the proposed (certainly speculative) model of taphrogenesis of the Rhinegraben arised from the following, at first sight c o n t r a d i c t i n g facts:
i) Based on earthquake mechanisms, especially for the earthquakes in Southwestern Germany and for the Lower Rhine Basin, SCHNEIDER (1971) and AHORNER (1970) concluded that the results are consistent with an u n i f o r m s t r e s s f i e l d with the principal compressive stress directed approximately from SE to NW.
ii) This in mind we must ask the following question: How could the Rhinegraben structure, which is caused by t e n s i o n a l f o r c e s, develop in the midst of an uniform compressive stress field which is nearly p e r p e n d i c u l a r to this structure and should rather c l o s e the lateral gap (of 4.8 km after ILLIES 1970)?

Gravity gliding?

A possible cause of the t e n s i o n a l s t r e s s e s in the graben axis, and of the c o m p r e s s i o n a l s t r e s s e s in the region of the Hohenzollerngraben (earthquake region) could be the effect of g r a v i t y g l i d i n g due to the uplift of the crustal blocks along both sides of the graben structure. To be sure the force which is responsible for gravity gliding cannot be calculated from a model where a solid block is simply gliding down on a sloping plane; only the much s m a l l e r f o r c e c a u s e d b y d e v i a t i o n f r o m i s o s t a t i c e q u i l i b r i u m could be transformed into a horizontal driving force. But even though the slope angle is in the order of $5/80 = 3°,6$, the d e n u d a t i o n of the uplifting crustal block acts in the s e n s e o f r e s t o r i n g the isostatic equilibrium. The resulting forces can be calculated as follows:

From the isostatic equilibrium condition $h_o = h_w \frac{\varrho_2 - \varrho_1}{\varrho_1}$, and the weight of the crustal load $\zeta \varrho_1 g = \varrho_1 g \left(\Delta h - h_w \frac{\varrho_2 - \varrho_1}{\varrho_1} \right)$, follows the total v e r t i c a l force K_v acting by gravity on the gravity centre of the block

$$K_v = g \varrho_1 \int_{x=0}^{1} \Delta h \, dx - g (\varrho_2 - \varrho_1) \int_{x=0}^{1} h_w \, dx. \tag{1}$$

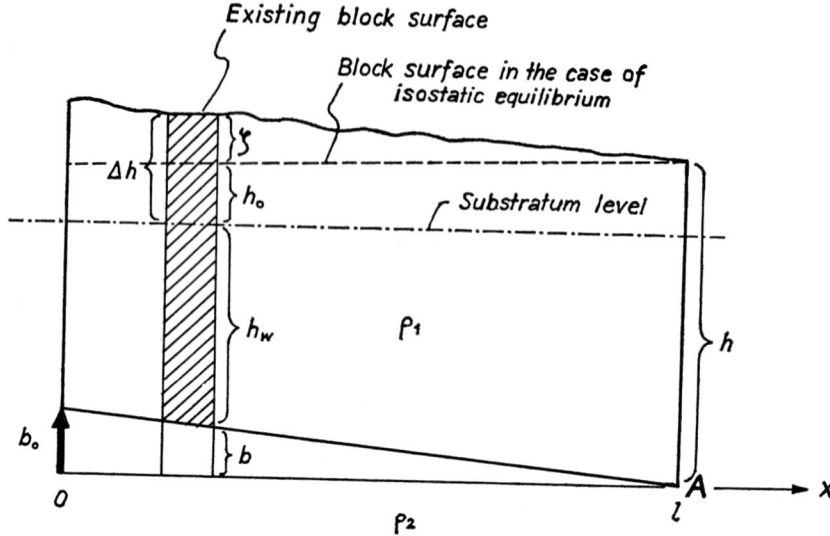

Fig. 1.
- ζ = height of crustal load caused by block elevation minus denudation.
- h_o = distance of block surface in the case of isostatic equilibrium from substratum level.
- $\triangle h$ = $\zeta + h_o$ = distance of existing block surface from substratum level.
- h_w = distance of lower block boundary from substratum level.
- b = elevation of lower block boundary (assumed to be a linear function of X).
- l = block length.
- A = axis of block rotation caused by uplift.
- ϱ_1, ϱ_2 = densities of crustal block and substratum respectively.
- g = gravity.

Using the densities $\varrho_1 = 2{,}7$ and $\varrho_2 = 3{,}0$, a crustal thickness $h = 30$ km at the right-hand side of the block, $b_o = 5$ km, $l = 80$ km, and $\triangle h$-values from the cross section of a geological map, we obtain $K_v = 0{,}24 \cdot 10^{15}$ dyn. Hence it follows for the maximum possible h o r i z o n t a l driving force due to gravity gliding $K_v \cdot 5/80 = 1{,}5 \cdot 10^{13}$ dyn, or a corresponding horizontal stress in the crustal block, 80 km east of the graben axis, in the order of o n l y 5 b a r. This small stress can neither act as a gap-forming force nor it can be responsible for the earthquakes.

Subcrustal currents

Instead of p a s s i v e f o r c e s such as gravity forces, we have to look for a c t i v e f o r c e s. Forces of a sufficient order of magnitude can be transmitted to a rigid crust by subcrustal flow of mantle material:

$$\tau = \eta \cdot \frac{\delta v}{\delta z}. \qquad (2)$$

- τ = horizontal shear stress acting on the crust from below.
- η = viscosity of a Newtonian viscous substance.
- $\dfrac{\delta v}{\delta z}$ = vertical velocity gradient of flow.

It is important to note that the velocity gradient $\frac{\delta v}{\delta z}$ depends not only on the velocity of flow at some depth, but also on the accompanied velocity gradients beneath the adjoining crustal blocks. An active driving velocity gradient causing shear stresses is only possible if the driven block is pushed against a retarded, resting, or in an opposite direction driven block.

Therefore it is necessary that the subcrustal flow and/or the involved viscosities are **inhomogenious**. Otherwise the crust would swim like a boat on a streaming river and the velocity gradient becomes **zero**.

If the spatial distribution of the velocity gradient is given, we obtain **the mean horizontal stress p** in the crustal layer by the formula

$$p = \frac{1}{h} \int_{x=0}^{l} \eta \cdot \frac{\delta v}{\delta z} \, dx, \qquad (3)$$

h = thickness of the crustal layer, l = distance from the point x = 0 where p is assumed to be zero. The mean stress p is the **integral** of the shear stresses caused by subcrustal flow, and hence considerable compressive stresses may develop inside the crustal layer as indicated by the stress trajectories of Fig. 2. Fig. 2 shows a schematic section through a crustal block of 30 km thickness and of a horizontal extension of 150 km.

This may tentatively represent the crust in Southern Germany east of the earthquake region or nearly the eastern boundary of the Rhinegraben-cushion respectively. We further assume the adjoining block on the left-hand side to act as a strong pier, that is without yielding. If we introduce a **velocity gradient** $\frac{\delta v}{\delta z} = 1.8 \frac{cm/year}{100 \text{ km}}$ and a viscosity of $\eta = 5 \cdot 10^{22}$ poise, we obtain by formula (3) a horizontal compressive stress p in the crustal layer with its maximum at the left-hand boundary of the block of $p_{max} = 1.4$ kbar. This is the compressive strength for granite in the upper continental crust.

Considering also Fig. 3, which is based partly on seismic refraction work (ANSORGE et al. 1970; EMTER 1971), two questions are to be answered in connection with this model:

i) We have to seek for the forces which prevent yielding of the adjoining block on the left-hand side (i. e. the eastern boundary block of the graben structure), and, closely related herewith:

ii) What prevents this block from being affected by the subcrustal flow in the same way as the eastern block under consideration?

Let us answer the second question first: It is possible, that the cushion material has a **lower viscosity** as the surrounding mantle material. In this case the cushion would act as a **decoupling mechanism** with respect to the transfer of shear stresses into the upper crustal layer. This implies that the subcrustal flow may continue beneath the Rhinegraben structure without affecting the upper crustal blocks noteworthy.

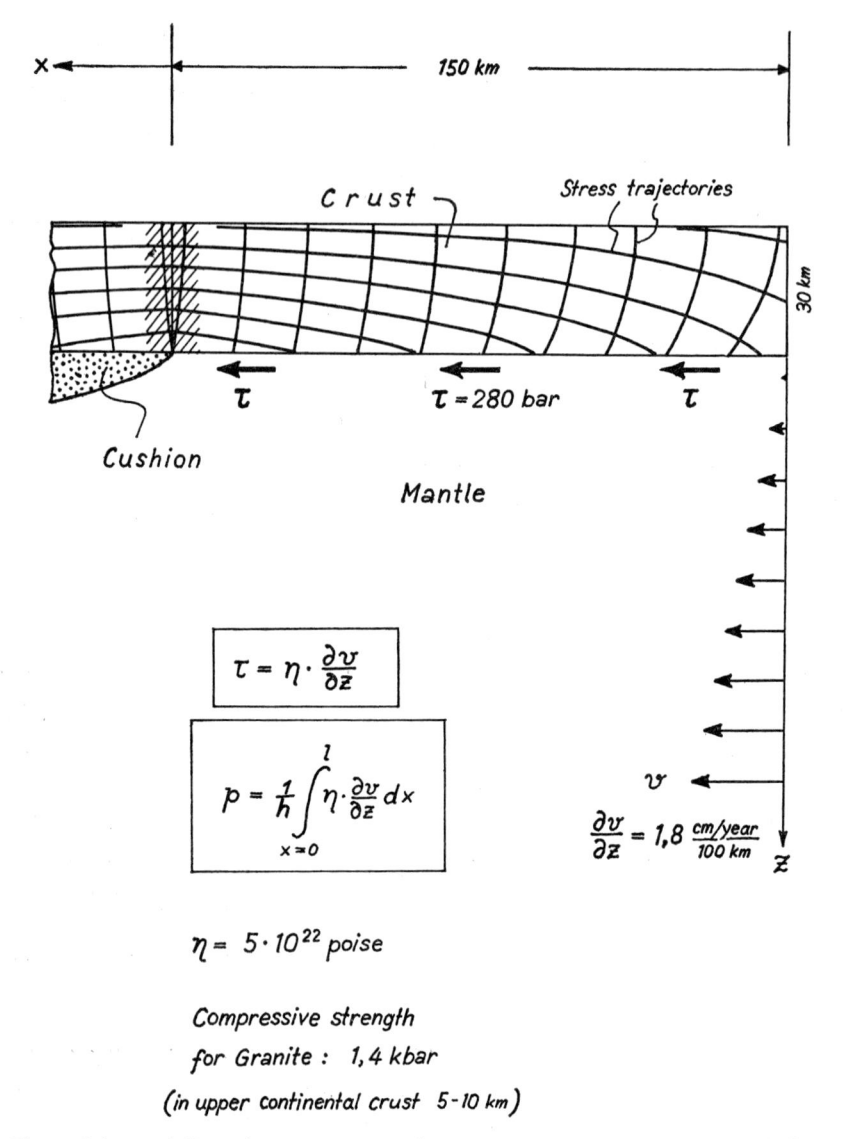

Fig. 2. Subcrustal flow, shear stresses τ and stress trajectories inside a crustal block.

As to the first question: "What prevents yielding of the left-hand block", the answer is given when we consider the model of the whole Rhinegraben mechanism presented in Fig. 3.

The tensional stresses in the middle axis of the graben are supposed to be due to a flow mechanism of upper mantle material which forms the cushion. The divergent flow produces tensional forces about the graben axis, and compressive

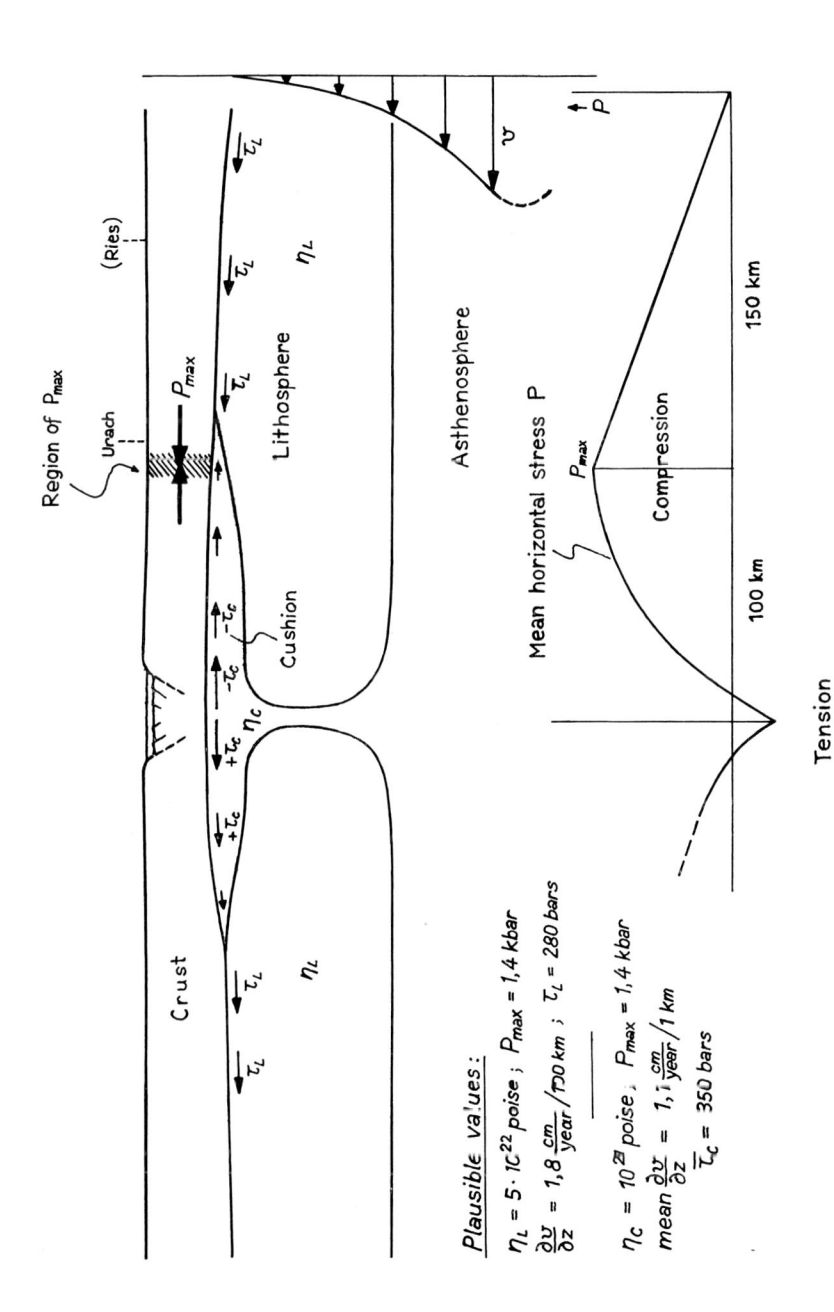

Fig. 3. Above: Supposed model of the Rhinegraben structure; crust and cushion after seismic refraction measurements. Below: Tentative crustal stress distribution.

forces which are directed away from the axis. In the region which was termed as a "strong pier" (Fig. 2 and Fig. 3, hatched area), maximum compressive stresses are built up. Plausible values for the mean velocity gradient and the viscosity of the cushion material are noted in Fig. 3. The lower part of this figure represents a more or less qualitative curve of the spatial distribution of the horizontal compressive stress p inside the crust and which is a consequence of formula (3). The maximum compressive stress, also figured by the hatched area in Fig. 2 and 3, coincides with the earthquake region of the Swabian Jura. Negative anomalies on gravity profiles in the same area show further indications of a w e a k c r u s t a l z o n e probably caused by strong breaking and desintegrating forces (JENSCH 1972).

Considering the presented model which involves the concept of continued genesis of the cushion by ascending and spreading mantle material, attention should be drawn to the tectonic model experiments of RAMBERG (1972), especially to fig. 7 of his paper, which shows the generation of a cushion by ascending silicone putty.

Conclusions

Gravity gliding cannot be considered as the main cause of forces which are strong enough to open the crust for taphrogenesis, and to built up earthquake generating compressional stresses. Estimations using plausible values for velocity gradients and viscosities of upper mantle material show that considerable tensional and compressional stresses may develop inside the crust. The model removes the contradiction which concerned the tensional graben structure situated just in the interior of an extended field of compressional stresses nearly perpendicular to the graben axis. Further, the region of considerable earthquake activity in the Swabian Jura coincides with the theoretically expected maximum stresses in the crust.

References

AHORNER, L., 1970: Seismo-tectonic Relations between the Graben Zones of the Upper and Lower Rhine Valley. — Graben Problems, 155—166, Stuttgart.
ANSORGE, J., EMTER, D., FUCHS, K., LAUER, J. P., MÜLLER, ST. & PETERSCHMITT, E., 1970: Structure of the Crust and Upper Mantle in the Rift System around the Rhinegraben. — Graben Problems, 190—197, Stuttgart.
EMTER, D., 1971: Ergebnisse seismischer Untersuchungen der Erdkruste und des obersten Erdmantels in Südwestdeutschland. — Dissertation Stuttgart.
ILLIES, J. H., 1970: Graben Tectonics as Related to Crust-Mantle Interaction. — Graben Problems, 4—27, Stuttgart.
JENSCH, A., 1972: Die Entwicklung eines Algol-Programms für zweidimensionale Schweremodelle. — Mit Berechnungsbeispielen zu gravimetrischen Langprofilen aus dem Hohenzollerngebiet. — Dissertation Stuttgart.
RAMBERG, H., Theoretical Models of Density Stratification and Diapirism in the Earth. — J. Geophys. Res., **77**, Nr. 5, 877—889.
SCHNEIDER, G., 1971: Seismizität und Seismotektonik der Schwäbischen Alb. — Stuttgart.

Thin-skinned Graben, Plastic Wedges, and Deformable-plate Tectonics

by

Barry Voight

With 16 figures and 1 table in the text

Introduction

Small-scale graben associated with the Alaskan earthquake are compared with the continental-scale Rhine graben on geometric and mechanistic grounds. Detailed consideration is given to the geometry and mechanical significance of a structural discontinuity at the base of the graben zone, leading to a concept of "thin-skinned" taphrogenesis. The state of stress within graben zones is examined, found to be typically compressive, and locally of structural significance in the evolution of graben.

A consideration of a gravitational gliding hypothesis shows it to be reasonable in association with the Anchorage graben but inapplicable to the Rhine structure. Rather, an argument is presented in which the Rhine graben is related to Greenland-Eurasia plate separation; a deformable-plate tectonic model is proposed, involving ductile deformation of mantle and consequent localized crustal stretching; this "working hypothesis" seems to adequately account for graben development and related phenomena, e. g., the origin of microcontinents.

The writer is grateful to Professor H. ILLIES for his kind invitation to prepare a brief article for the Graben Symposium on the development of graben associated with landslides induced by the Alaskan earthquake. A few comments on the Rhine graben were to be included as an appropriate analogy, in appreciation both of the hospitality shown to me at Karlsruhe and a long-standing communication with the vines of Riesling. This, however, proved difficult, inasmuch as my understanding of the origin of the Rhine graben immediately led to several inconsistencies. This paper is the consequence of an attempt to unravel the problems as I envisaged them and as they arose; I have been carried far afield from my initial purpose, as the reader will soon discover. The consequence thus may seem a rambling, heterogeneous discourse, but it is one which accurately reflects the nature of the search. Because I have been instructed that "Les poètes et quelques naturalistes vivent dans une patrie sans frontières, un univers an delà du rationalisme", no apology seems necessary. The writer is indebted to many discussions with his able colleagues, Messrs. R. SCHOLTEN, M KEITH, A. BOETTCHER, R. GREENFIELD, E. K. GRAHAM, C. W. BURNHAM, J. RYBÁŘ, and S. ALEXANDER; to Professor J. DOZY of Delft TH, who provided an introduction to the rifts in the province of Noord-Brabant; and to Mrs. TEMPLE, who provided much of the impetus for study of the greater Baltic region[1].

---

[1] Portions of this paper were presented in a lecture on "The Prince William Sound (Alaskan) Earthquake of 1964, with special emphasis on 'Landslides in Anchorage'", presented at the University of Karlsruhe on 19 May, 1972. At that time the author was associated with the Geological Institute, TH Delft, Netherlands; present address: Department of Geosciences, The Pennsylvania State University, University Park, Penna. 16802, U.S.A.

Origin of graben associated with the Anchorage, Alaska, landslides

As has been pointed out in detail by HANSEN (1965, p. Q—38) each of the highly destructive landslides in the Anchorage area associated with the Alaskan Earthquake of 27 March, 1964, was of a single structural-dynamical family, despite rather wide variations in size, complexity, and overall appearance. All moved chiefly by translation rather than rotation, gliding laterally on one or several virtually-horizontal slip zones following earthquake-induced strength loss.

In a few slides[2], the slide mass was preserved as a single intact block throughout slide development (Figs. 1, 2; cf. HANSEN 1965, p. 41—53; WILSON 1967, p. 268—282, 291—293); in contrast, the Turnagain Heights slide complex, the largest and most devastating at Anchorage, involved enormous internal disruption in the form of a repeated sequence of block glide and lateral spreading events (VOIGHT 1973; HANSEN 1965, p. 59—66; WILSON 1967, p. 282—291; SEED & WILSON 1968).

Each of the slide areas was characterized by the formation of graben. In the simpler slides, a single graben developed behind intact, displaced glide blocks. Two graben developed at the eastern portion of the Fourth Avenue slide (Fig. 2), and three occurred at Government Hill; successive development of numerous graben characterized Turnagain Heights deformation.

Fig. 1. U.S. Army photograph of Native Hospital slide area, Anchorage, Alaska, showing graben and pressure ridge. The scar of an older graben of an earlier slide is transected by the 1964 slide.

---

[2] Notably the Fourth Avenue, L. Street, and Native Hospital slides.

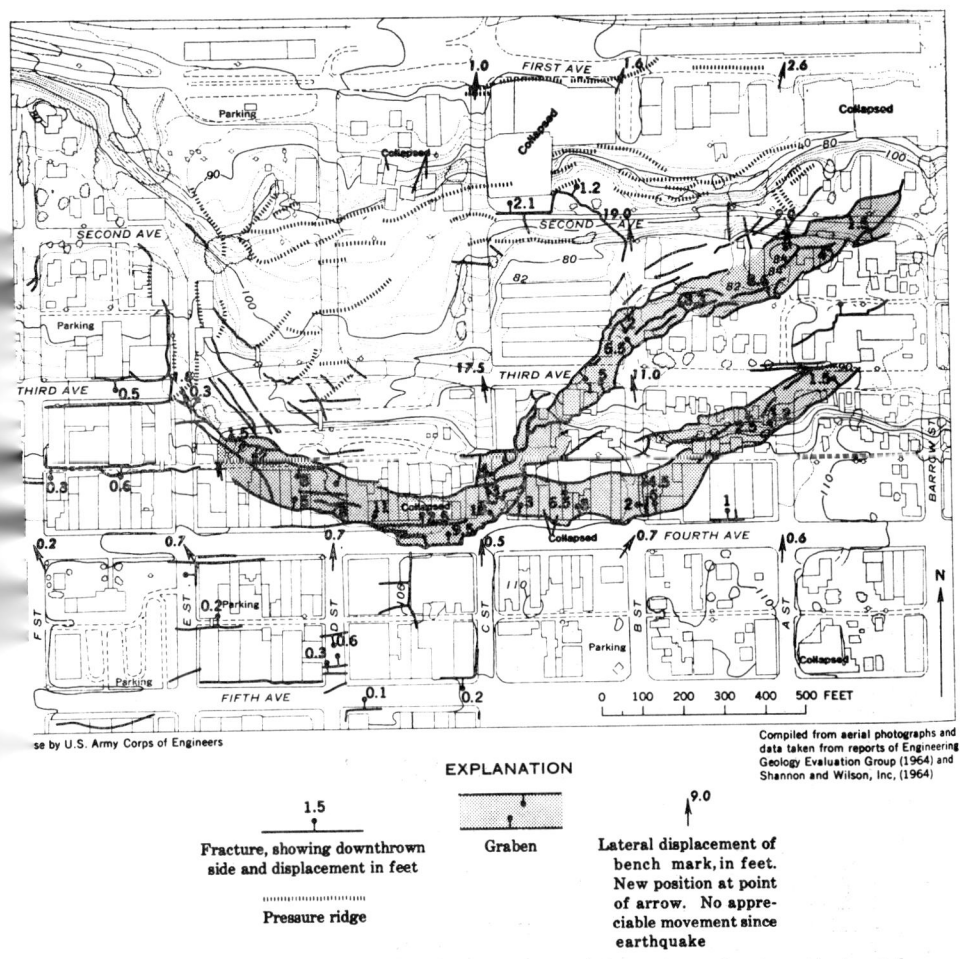

Fig. 2. Graben associated with Fourth Avenue landslide area, Anchorage, Alaska (after HANSEN 1965, Fig. 25).

The nature of the displacements occurring in the Anchorage slides is given in Fig. 3. These curves, and others given elsewhere (VOIGHT 1973, Fig. 7), are based on reconstructed sections of the slide areas.

A summary of reconstruction methods is provided in Fig. 4. The lower diagram illustrates a section across the L-street slide area (cf. HANSEN 1965, Fig. 3); a reconstruction of "pre-slide" conditions is given in the upper diagram. The reconstruction problem can generally be resolved by detailed field observations, even though certain kinds of data may not be obtainable, even by thorough exploratory investigations. For example, positive identification of the slip surface is commonly difficult to achieve by coring methods. It is, of course, necessary to establish this element for the reconstructed diagram, and indirect means must

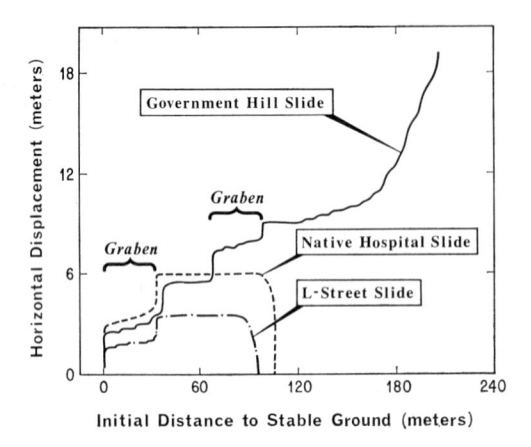

Fig. 3. Displacement diagrams for some Anchorage, Alaska, landslides.

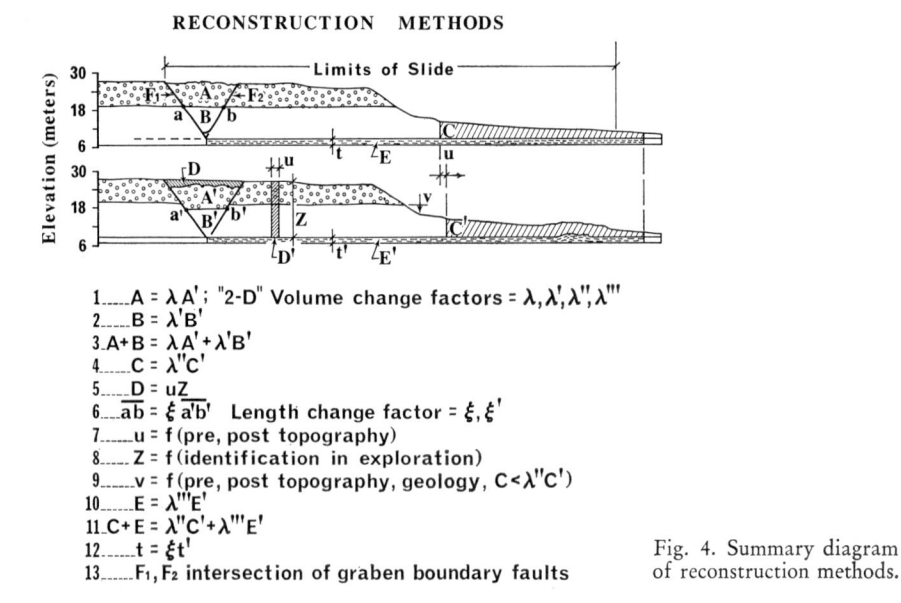

RECONSTRUCTION METHODS

1. $A = \lambda A'$; "2-D" Volume change factors $= \lambda, \lambda', \lambda'', \lambda'''$
2. $B = \lambda' B'$
3. $A+B = \lambda A' + \lambda' B'$
4. $C = \lambda'' C'$
5. $D = uZ$
6. $\overline{ab} = \xi \overline{a'b'}$ Length change factor $= \xi, \xi'$
7. $u = f(\text{pre, post topography})$
8. $Z = f(\text{identification in exploration})$
9. $v = f(\text{pre, post topography, geology, } C < \lambda'' C')$
10. $E = \lambda''' E'$
11. $C+E = \lambda'' C' + \lambda''' E'$
12. $t = \xi t'$
13. F_1, F_2 intersection of graben boundary faults

Fig. 4. Summary diagram of reconstruction methods.

therefore commonly be employed; some of these procedures are obvious, e. g., palinspastic restoration of faulted blocks to their original positions; others may be less well known, e. g., volumetric restoration (typically in two dimensions) of stratigraphic or structural units (e. g. Fig. 4, A—E). Volume changes may occur in deformation and this effect may have to be considered in analysis. Consideration of these procedures leads to several independent measures of the depth to the sliding surface, and indeed, is relevant to the important question of whether a "slip surface" in fact exists, as compared to the alternative possibility of a zone of extrusion or shear flow beneath the slideblock.

Fig. 5, modified from HANSEN (1965, Fig. 24) is a schematic block diagram in which the essential structural elements, characteristic of all major Anchorage slides, are shown; these include the graben zone at the head of the slide, the horizontal slip "zone", and the pressure ridge at the toe.

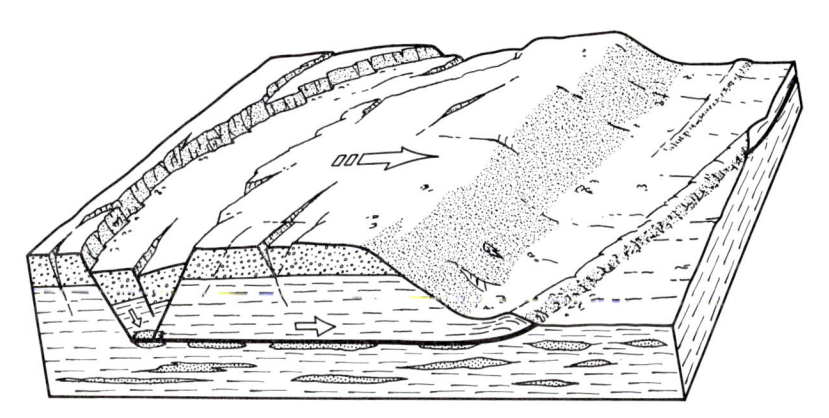

Fig. 5. Schematic diagram of characteristics for the 1964 Anchorage, Alaska, landslides.

Evolution of the slides was as follows:
(1) Earthquake vibrations caused widespread strength loss within a nearly horizontal zone of sediments. The specific mechanism of strength loss was pore-pressure enhancement and local fluidization (liquefaction) of saturated clay, silt, and sand, a consequence of vibrational grain structure modification; this mechanism resulted in a nearly frictionless surface upon which the glide blocks could move.
(2) Widespread quasi-vertical, systematic tension fractures developed; these features were also induced by elastic wave vibrations, probably assisted by hydraulic fracture mechanisms associated with subsurface liquefaction.
(3) Instability, hence initiation of sliding, occurred when the propulsive (driving) force components for the slide block exceeded frictional and cohesive resistive force components of the slip zones and the toe.
(4) A graben developed at the head of the slide in association with outward movement of the glide block[3].
(5) Forward motion was retarded when resistive forces exceeded propulsive forces (both forces were by no means constant during the deformational process).

The mechanisms associated with the propulsive force are of particular interest to this discussion. Propulsive forces were due to a net downslope component of

---
[3] The geometry of the slides suggests, in general, that the graben developed initially in response to downward motion on the rear scarp (Fig. 2; HANSEN 1965, Figs. 30, 33, 37, 44). Net rotation of the ground surface within graben resulted in a rearward inclination of the graben surface with the result that the rear graben scarp was generally the greater. The nature of multiple pressure ridges at some slides suggests multiple slip horizons.

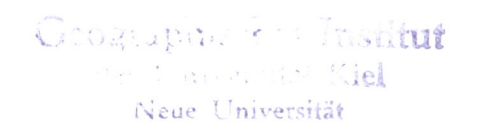

gravitational acceleration in the direction of motion, quake-induced inertial forces, and/or lateral pressure mechanisms. Inertial forces could have been significant locally, but were not essential because, at some localities, block-glide regression and slide development occurred long after earthquake motions had ceased (HANSEN 1965, p. 64; VOIGHT 1973, Fig. 8). Apparently the essential role of the earthquake in relation to landsliding at Anchorage was in the development of low-resistance glide horizons. HANSEN (1965, p. 41) had believed that sliding was precipitated by the downslope component of gravity on the slip surface; however, it has not been demonstrated that the average slope of all slip surfaces at Anchorage were, in point of fact, inclined in the direction of glide motion; a horizontal slip surface, obviously, would produce a zero tangential gravity component, and a negative slope would produce a gravity component acting in a direction to oppose motion.

Yet the flatness of the slip surface is supported by sound field evidence; a few degrees of slope in the direction of motion appears to be insufficient to overcome the resistance due to surface roughness and sediment strength on the slip surface (these are small but non-negligible forces, VOIGHT 1973) and the toe resistance.

By process of elimination, therefore, some lateral pressure mechanisms seem required. At least two can be envisaged, and they are related; (1) open fractures, filled wholly or in part with water or water-sediment mixtures, exert significant lateral hydrostatic or hydrodynamic pressures against the fracture walls; in the hydrostatic case, the force arising from this "fluid", or "viscous-wedge" mechanism is proportional to the square of the height of the fluid prism; (2) the weight of material within the graben exerts lateral pressure on the boundary walls. Thus as a graben subsides, a continuous (but decreasing) lateral pressure is applied to the moving glide block; the subsiding ground under the graben behaves as a "plastic wedge" which actively contributes to block propulsion.

Horizontal force components resulting from both mechanisms (1) and (2) are significant. The principal distinction between them appears to be in their respective abilities to sustain lateral pressures once slide motion commences; in general, fluid-filled fractures can be extremely transient sources of propulsion inasmuch as the fluid typically escapes with relatively small lateral block motion. Plastic wedges exhibit a more delayed form of transient behavior, principally because the "volume" is larger. A case has been presented concerning the significance of the plastic wedge mechanism in the development of the Anchorage landslides (VOIGHT 1973).

Attention will now be given to the possibility of application of these "wedge" mechanisms to geometrically analogous structures on continental and global scales. Are the classic continental-scale graben and rift valleys necessarily passive features produced by crustal extension which permit the blocks to fall into place? Or could they be ductile wedges of stone, hammered into the earth's crust by the hand of gravity, helping to pry apart the crustal blocks with an unrelenting horizontal pressure?

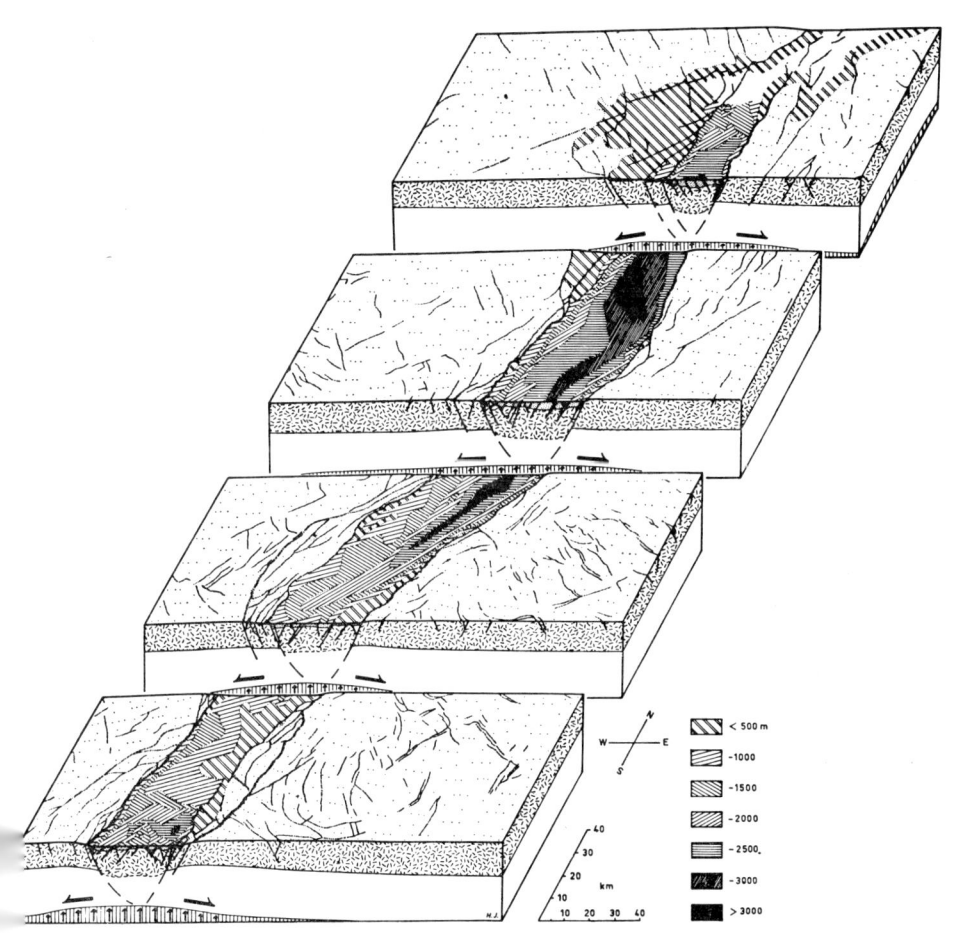

Fig. 6. Block diagrams of the Upper Rhine graben (after ILLIES 1971). Sediment isopachs in graben area given by symbols shown in lower right. Brittle sialic crystalline basement shown by cross-hatch pattern, lower crust unpatterned; mantle-derived "rift cushion" underlies crust.

Origin of the Rhine graben

The Rhine graben, for example, has always been attributed to crustal tension, either as a gradually increasing horizontal tension, or to tension produced along the crest of a rising dome (Fig. 6). The question of "tension", however, needs to be critically examined, and in so doing the distinction between extensional strains and tensional stresses must be kept in mind. It has been claimed that recent tectonics and seismic activity indicate that tensional stress is still existent (e. g., ANSORGE et al. 1970), but in general such references pertain to deformations, i. e. displacements and strains, and not to the "sign" of the associated stress field.

Barring the imposition of special boundary conditions at depth, tensile stresses can exist only above some limiting depth Z^* given by

$$Z^* = C_0/\varrho g, \tag{1}$$

"total" stresses considered, where C_0 is the unconfined crushing strength of the rock mass, ϱ is density, and g is gravitational acceleration. Assuming $C_0 \cong 10^3$ kg/cm², a reasonable value for a crystalline rock mass[4], and $\varrho = 2.7$ g/cm³, $Z^* \cong 4$ km; for a "submerged" water saturated condition, $Z^* \cong 6$ km.

The base of the faulted zone which bounds and underlies the graben may be about an order of magnitude deeper (e. g., a figure of 30 km is suggested by ILLIES 1970, Figs. 4, 5), and hence it seems certain that the lateral stress at graben fault boundaries at depth will be compressive.

The lateral stress σ_x (compression reckoned positive) produced by crustal extension at a point at depth Z, is given by

$$\sigma_x = k\varrho g\, Z - 2\, k^{1/2} S_0, \tag{2}$$
$$k = \tan^2(45° - \Phi/2),$$

when S_0 and Φ are Coulomb strength parameters for the rock mass, the other terms having been previously defined. Coulomb parameters are rather imprecisely known for the in-situ PTX conditions inferred from geophysical data; values given in Tab. 1 seem reasonable to a first approximation, and are appropriate for the temperature range 500—800° C; these temperatures seem appropriate, with qualification, for the Rhinegraben area in depth range 10—40 km (HÄNEL 1970, Fig. 2). The resulting lateral pressure can be estimated from rock strength data and is given in Fig. 7. Note that the strength data have been obtained under conditions

Tab. 1. "Total stress" Coulomb parameters for anhydrous crustal rocks.

Temperature (°C)	So (kb)	Φ	Material
25°	0.1	55°	"typical" crystalline rock
500°	0.3	19°	anhydrous mica schist[*]
800°	1.2	13°	anhydrous granite[*]
800°	—0.5	17°	anhydrous basalt[*]
700°	0.6	0°	serpentinite[**]
800°	—2.1	40°	anhydrous peridotite[*]

[*] Coulomb parameters based on coupled extension and compression triaxial experiments at 5 kb confining pressure employing strain rates of 2—4 percent per minute by GRIGGS, TURNER & HEARD (1960), using data published by HANDIN (1966, Tab. 11-3). Employment of these parameters will give reasonable values for ultimate strength for depths equivalent to 5 kb.

[**] Coulomb parameters for serpentinite based on triaxial compression experiments by RALEIGH & PATERSON (1965); confining pressures in the range 35 — 5 kb, strain rate 7×10^{-4} sec^{-1}. Dehydration reactions released water during these experiments leading to embrittlement and weakening. The corresponding effective stress Coulomb parameters S_0', Φ', assuming pore pressure equal to confining pressure, are $S_0' = 0.15$ kb, $\Phi' = 33°$.

[4] Strength properties of intact rocks may be regarded as upper-bound values with respect to the strength exhibited by the "rock mass"; the distinction between the two becomes less important at great depth.

of rapid loading with dry materials; all temperatures were thus in the sub-solidus range during experimental deformation, and ultimate strengths can be regarded as upper bound values[5]. The lateral stress curve thus differs from the vertical stress by (at most) a few kilobars, with this difference diminishing greatly in zones of partial melting or enhanced creep strains (Fig. 7). The resulting lateral force to a depth $\bar{Z}$ is given simply by the integration of the pressure curve[6].

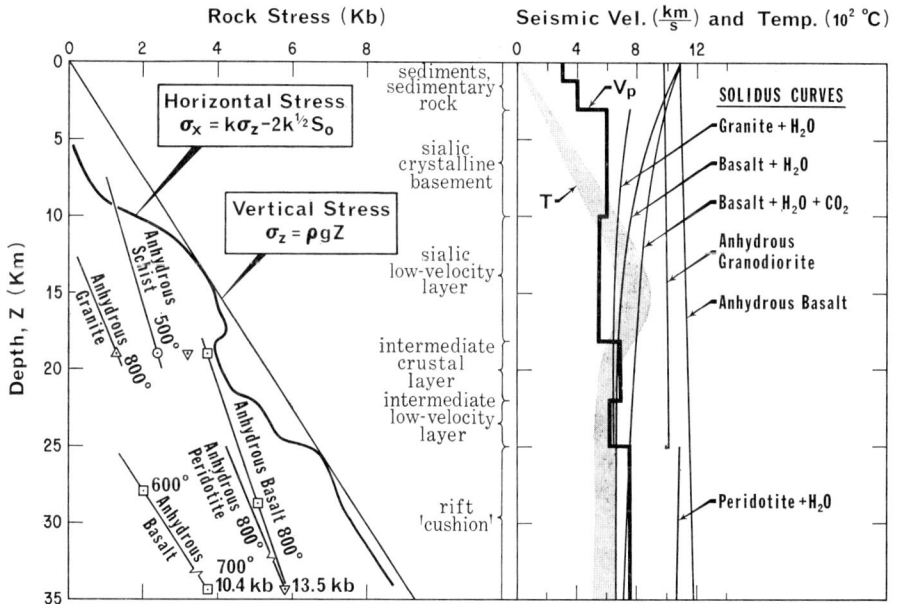

Fig. 7. Rock stress, seismic (P) velocity, and temperature versus depth for Rhine graben. Vertical stress reckoned as pressure due to overburden weight. Horizontal stresses crudely estimated within limits suggested by equation (2), rock strength estimates, and seismic profiles. Seismic profile after ANSORGE et al. 1970; temperature profile after HÄNEL 1970. Solidus curves refer to compositions, not necessarily rock types; data summarized from MERRILL, ROBERTSON & WYLLIE 1971; HILL & BOETTCHER 1970.

[5] The effect of pore fluids such as H_2O and CO_2 are twofold. On one hand, pore fluids cause a reduction of the intergranular confining pressure according to the "effective stress" principle introduced by KARL TERZAGHI. The mechanical effect of this process is to decrease frictional resistance between grains by an amount equal to the fluid pressure; thus the strength can be profoundly decreased and "embrittlement" enhanced if fluid pressure is large. The second group of effects are physico-chemical and concern such phenomena as profound reduction of the solidus temperature by fluids, the reduction of cohesive resistance, and dramatic hydrolytic weakening of silicate minerals (e. g., GRIGGS 1967; BLACIC 1970). The combined effect of available fluids, at PT conditions of interest here, would be profound strength reduction. These conditions may be applicable in the sialic low-velocity zone where partial melting and/or hydrolytic weakening seems quite likely (Fig. 7); here the long-term creep strength (under continued loading) may virtually vanish, and the horizontal stress would thus approach the value of the vertical stress. Phenomenologically such materials would be regarded as viscoplastic.

[6] For depth-independent strength parameters, $F_x = \frac{k}{2} \varrho g \bar{Z}^2 - 2k \frac{1}{2} S_0 \bar{Z}$.

This equation dates back to COULOMB's (1773) classic essay.

$$F_x = \int_O^{\bar{Z}} \sigma_x \, dZ \tag{3}$$

A similar horizontal force can be envisaged from a "viscous wedge" mechanism; such a mechanism can be related to widespread dike intrusions often associated with large-scale graben evolution. In the limiting case of a spatially continuous intrusion,

$$F_x = 1/2 \, \varrho_m \, g \, \bar{Z}^2 \tag{4}$$

where ϱ_m is the density of the magma.

A net compressive force may therefore be presumed to be active along the graben boundary faults. Having thus considered the "sign" of the force field, the next question which arises concerns the **importance** of this force field in the evolution of the structure.

Such a question can perhaps be resolved if the geometry and boundary conditions of the real earth can be deduced with sufficient accuracy. For example, with respect to the Rhine graben, broad regional uplift in association with the graben trend is apparent, which "anyone would recognize who had ever travelled from the Black Forest toward Swabia, and from the Vosges toward the Paris Basin". The association of uplift with graben formation is, of course, a different circumstance than found at the landslide sites previously considered. In Fig. 6, arrows are given which schematically denote the interpreted movement of crustal blocks away from the crest of the uplift "cushion", and relative movement of crustal blocks over subcrustal material[7]. Are these movements restricted to the zone of uplift? If so, could the structure be a hybrid feature, with induced displacements analogous to slip in flexural folds with clamped boundaries? The graben in this case could be interpreted as a passive extensional feature, developing as the direct consequence of boundary conditions imposed by uplift. Nonetheless, of course, the net pressure exerted by the wedge would be compressive, even though the strains are extensional.

The amount of relative displacement can be used to establish whether this movement can be totally accounted for by the geometry of the uplift: if more displacement has occurred than can be accounted for by this mechanism, then consideration must be given to the possibility that uplift may not be the direct cause of the deformation, although it might indirectly be involved, perhaps e. g., by providing gravitational potential for gravity gliding. The amount of relative movement can in fact be deduced by dividing the volume of graben "space" beneath some appropriate horizon, in a section of unit width, by the thickness of the crustal block over the slide horizon, if the position of the slide horizon is definitely known (relationship 5, Fig. 5), or by palinspastic restoration of fault blocks over the Hercynian crystalline basement. Following the latter procedure, for example, ILLIES (1967) has determined the amount of post middle Eocene exten-

---

[7] Movement directions along the Rhine graben can be inferred from asymmetry of graben form, as was the case at Anchorage.

sion. This amount of extension seems too large to be accounted for by the "doming" mechanism (ILLIES 1970, p. 9).

Among the possibilities that can be envisaged in order to "explain" this crustal extension are the following: (1) a substantial volume of subcrustal material has extended, causing the crust, which remains volumetrically intact and relatively brittle, to spread apart in localized zones of concentrated extension; (2) a substantial volume of crustal material has compressed elsewhere to precisely "balance" graben extension; (3) crustal material has more or less rigidly slipped over subcrustal material with extension at one end balanced by "overlap" at the other[8]. Each of these alternatives requires a basal "structural boundary" (possibly, but not necessarily, a décollement); the boundary could perhaps occur within the crust, at the base of the crust, and/or possibly at the base of the lithosphere[9].

For example, the fact that several low seismic-velocity zones, and probably low-strength zones (Fig. 7), seem to be present within the crust is suggestive of the possibility that an intracrustal décollement could be involved. It seems possible, however, to resolve this difficulty; by dividing the 4.8 km extension deduced from palinspastic fault restoration into the cross-sectional area of the graben trough, I have obtained an estimate of the depth Z to the structural boundary[10] (relationship 5, Fig. 5; cf. HANSEN 1965, p. A 41). The depth thus obtained is 24.5 km, as measured beneath the top of the Hercynian basement at the graben boundary (Fig. 8); this figure almost exactly coincides with the 25 km depth (ANSORGE et al. 1970, p. 194) to the depth to the top of 7.7 km/sec "rift cushion" layer, presumably a mantle derivative, separating "ordinary" upper mantle material (8.2 km/sec) from "ordinary" crust (6.3—6.9 km/sec). The interpretation of ILLIES (1970, p. 8), which suggested that graben fault structures extended to the crust-mantle boundary, thus seems confirmed[11].

Earthquake focal depths under the graben rarely exceed 8 km, whereas under the eastern escarpment somewhat deeper focal depths, rarely to 20 km, are common (MÜLLER 1970, Fig. 4; p. 33). The hypocenters occur close to the roof of the sialic low velocity channel, which deepens eastward. The graben fault wedge, which continues to 25 km, therefore presumably consists of two zones characterized by a significant change in the mode of deformation. In the depth range 0—10 km,

---

[8] Gravitational gliding or subcrustal tractions associated with the spreading of the "blister" have been suggested as mechanisms (ILLIES 1970, p. 12).

[9] With reference to the currently-favored views concerning plate tectonics, it is mentioned that multiple-horizon décollement horizons are possible; each level could possess a degree of kinematic independence separate from subjacent plates.

[10] The quantitative measurement of extension across the graben zones, based on geometric reconstructions, provides fundamental evidence for the existence and geometry of this structural boundary; thus this kind of data should be evaluated with special care on a regional basis.

[11] Straight line extrapolation of graben boundary faults suggest a depth of 30.6 km (ILLIES 1970, Fig. 4); this predictive procedure is, however, uncertain because of possible fault curvature and because of the inherent unproved assumption that the faults converge to a point at a décollement. At Anchorage, similar constructions provided upper-bound values, which however were, on the whole, reasonable approximations. The same apparently can be said for the Rhine graben.

deformation is possible in the fracture mode; in the range 10—25 km, flow is predominant (cf. GOETZE 1971).

In any case, we could be dealing with "thin-skinned"[12] graben. If mantle extension [hypothesis (1)] seems at first glance unlikely or insufficient, the hypotheses which remain suggest a crustal plate which behaves, in some respects, like the Anchorage landslides; i. e., a graben zone exists at the head, a décollement exists at the base, and somewhere a foreward edge or "toe" must exist, where the décollement either dies out in a zone of local compression [a variant of hypothesis (2)] or breaks through to the surface [hypothesis (3)]. The crustal blocks therefore must have the capability for lateral motion. This can be termed the "toe problem"

---

[12] Further consideration of this question may lead to what may be termed "thick-skinned" and "thin-skinned" taphrogenetic hypotheses, wholly analogous to "décollement" versus "basement involvement" hypotheses for Appalachian and Jura thrust and fold belts. The graben which developed in association with the Anchorage slides, for example, are clearly of the "thin-skinned" variety, requiring a basal structural discontinuity (e. g., décollement) for their formation. Similar features have been reported by RYBÁŘ (1971) from the Tertiary brown coal basins of Bohemia, Bulgaria and Germany. Analogous extension zones, on a larger scale, occur in association with the Heart Mountain "fault" of western Wyoming, U.S.A. (PIERCE 1973; VOIGHT 1973). That structure, which seems to be a great rockslide (10^3 km³), is bounded in the area of slab release by multiple-level décollement horizons which follow stratigraphic boundaries. Graben zones extend to décollement level and no further (Fig. 9), and were instrumental in the disruption of the glide mass. To extend the structural association to yet another order of magnitude, reference is made to Fig. 10, which illustrates the RUBEY & HUBBERT (1959) concept of a major bedding-plane glide surface on the flank of a geosyncline. Section A shows the central and deepest part of the geosyncline in which thick, rapid sedimentation has caused abnormal fluid pressures to develop at depth; at some depth at which the critical relationship between fluid pressure and the lateral component of stresses acting on the glide surface is exceeded, a thick section of sedimentary rock would part from its foundation and begin to move (RUBEY & HUBBERT 1959, p. 194). Rocks would buckle at the forward edge of the moving plate (Section B), causing increased resistance which could be overcome by subsequent erosion, hence raising a possibility of renewed movement. The status of the hypothesis can be tested for a given area by consideration of body and surface forces acting on a free-body diagram, in which a force balance is taken which includes resistive components of toe and glide horizon, and propulsive components, e. g., the net downslope weight component of the glide mass. The same question arises that has been previously posed: is the graben zone an area of tensile resistance, or one of compressive propulsion in association with the plastic wedge mechanism? For the geosynclinal area of western Wyoming, Z may be taken as approximately 6.0 km, $\varrho = 2.3$ g/cm² (RUBEY & HUBBERT 1959, p. 196—200). The crushing strength of the rock mass, particularly as influenced by shale sections, cannot have been large, e. g., $C_0 \cong 10^2$ bars. Under the circumstances, even with hydrostatic pore pressure, tension could not be sustained by the rock mass; a pull-apart zone such as illustrated in Section B can be regarded as a physical impossibility, and would have to be replaced by a zone of plastic deformation; this zone could be a regionally-extensive domain of plastic deformation, or a localized zone in association with well-defined grabens. In either case, the plastic wedge force would develop and would be additive to other "propulsive" forces considered in the force balance. As movement progressed, graben areas would tend to fill with younger sediments, with the consequence that the active pressure value of plastic wedges could be maintained. In any event, consideration of the plastic wedge mechanism decreases the fluid pressure-overburden ratio theoretically required for a gravitational sliding hypothesis[13].

[13] It should be noted that the existence of great rift areas had (and have) not been positively identified in association with the western Wyoming Thrust belt, which led RUBEY & HUBBERT to regard a hybrid mechanism of regional compression and gravity sliding as a more likely explanation in the light of existing data.

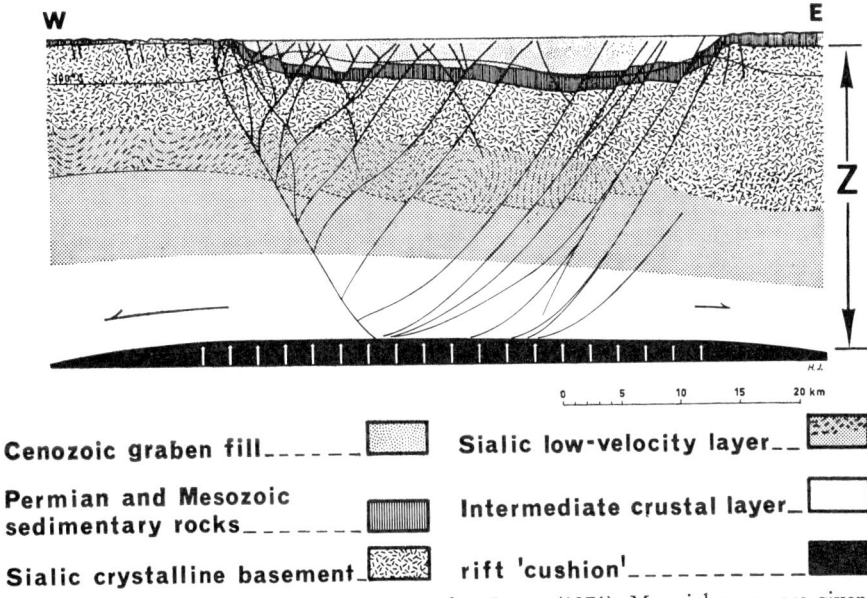

Fig. 8. Cross section through Rhine graben, after ILLIES (1971). Material types are given in legend. The position of the 100° C isotherm is shown. Depth Z refers to a calculation based on the graben trough area and the measured extension.

for gravitational gliding hypotheses; it has not been resolved. Where, indeed, would the "toe" be located? In the Pyrenees? Pyrenean tectogenesis represents an overall shortening related to collision between Iberian and European crustal blocks (CHOUKROURE & SÉGURET 1973). The timing of major Pyrenean events seems to coincide with Rhine graben development, i. e. middle Eocene, but the Pyrenean displacements required are an order of magnitude greater and displacement directions are not in adequate accord. Slip of the entire western European block (i. e., west of the Rhine graben) cannot therefore account for this movement if the Iberian block is considered to have been "fixed"; the more likely possibility is that

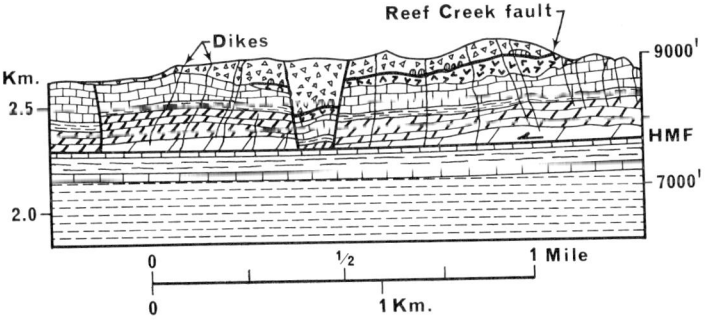

Fig. 9. Cross-section of the Cathedral Cliffs fault block complex on the Heart Mountain décollement, western Wyoming, U.S.A. (modified after PIERCE 1963). The graben zone terminates at the décollement boundary. HMF refers to "Heart Mountain Fault".

the Iberian block was mobile (see, e. g., CAREY 1958; PITMAN & TALWANI 1972), and it is thus to the interaction of two mobile blocks that the Iberian problem must be resolved.

The required movements for the Rhine graben seem, in any case, predominantly westward, so that, if sliding is presumed to have had occurred, the solution to a "toe problem" must be sought in the area now occupied by the North Atlantic Ocean. The obvious question which next arises concerns the configuration of the Atlantic Ocean at the time middle Eocene rifting commenced in the Rhine graben area. From magnetic anomaly interpretations, considered in the light of sea-floor spreading and plate tectonics hypotheses, it has been concluded that the final separation of Eurasia (Spitsbergen), Greenland, and North America also occurred during the middle Eocene (PITMAN & TALWANI 1972)[14]. It will be of interest to ascertain whether these arguments withstand the test of more closely-spaced magnetic profiles; nonetheless, this apparent coincidence of deformation dates, and the comparative geometry of the Rhine graben and related extensional structures with particular reference to the Greenland—Eurasia plate boundary, raises the strong possibility of a genetic relationship. The relative middle Eocene positions of Europe, Africa, Greenland, and North America for the 53 my position is shown in Fig. 11, together with their current positions[15].

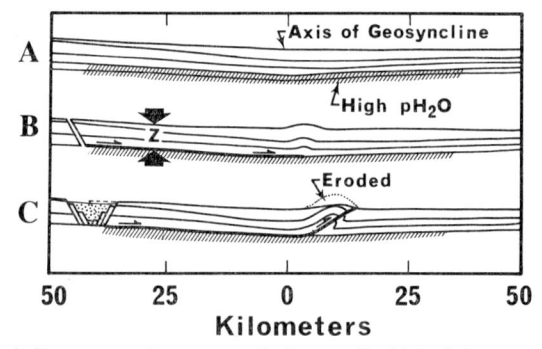

Fig. 10. Idealized diagram to show zone of abnormally-high fluid pressure, bedding plane glide surface on flank of a geosyncline, buckling at forward edge and graben development at the rear edge of a glide block, and erosion and "break-through" of the frontal thrust (after RUBEY & HUBBERT 1959).

---

[14] The initiation of rifting between Africa and North America occurred earlier, e. g. 180—200 my.

[15] Certain geologic features are superimposed in this illustration; these include the Rhine—Bresse—Limagne rift system; the English channel graben (NAFE & DRAKE 1967, Fig. 4; CAREY 1958, Fig. 35), the Norwegian Channel; deep sedimentary basins below the Norwegian shelf (GRØNLIE & RAMBERG 1972; TALWANI & ELDHOLM 1972); the Oslo graben; east Greenland block faults, and the Scoresby Sound flexure (HALLER 1971); Spitsbergen block faults; North and St. Georges channels in the Irish Sea (Tectonic Map of Great Britain and Northern Island 1966); the Rockall-Lousy "graben" (?) (NAFE & DRAKE 1969, Fig. 4); the Baltic depression (SEDERHOLM 1913). Some of these structures also have a pre-Tertiary history, but all for which data are available seem characterized by significant early Tertiary events; e. g., the Irish sea channels are parallel to dike swarms of Eocene age, suggesting that they are Eocene extensional features; the Oslo graben is a Permian structure reactivated in the early Tertiary (BEDERKE 1966, p. 217); East Green-

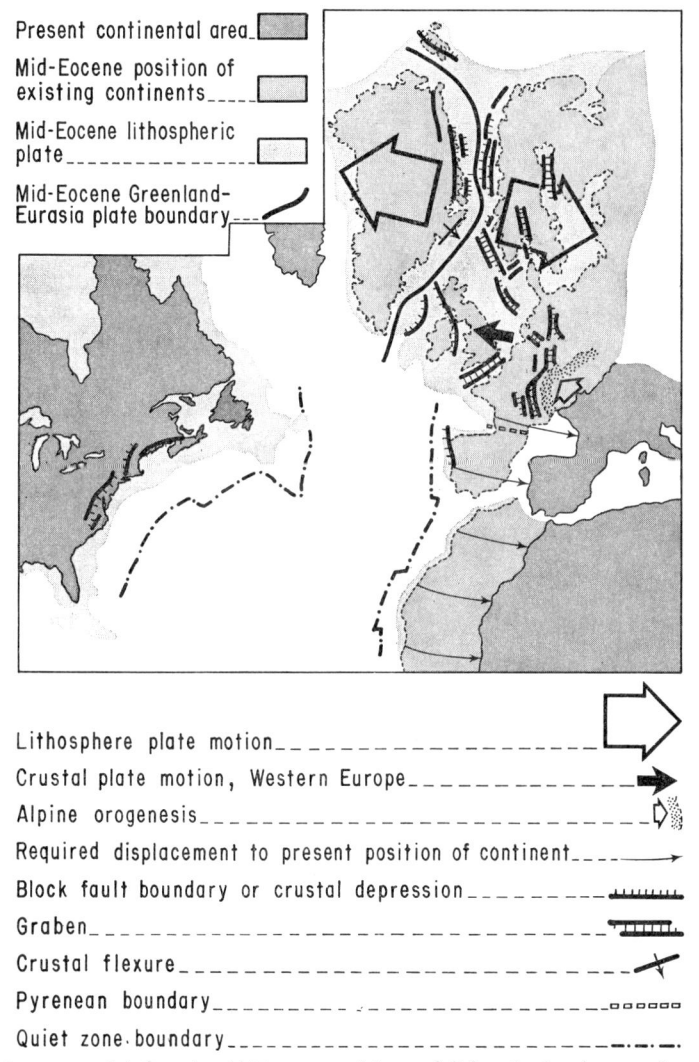

Fig. 11. Present and inferred mid-Eocene positions of lithospheric plates and continental blocks in the North Atlantic region. Mid-Eocene positions from PITMAN & TALWANI 1972. Individual structural elements are discussed in footnote 12 and elsewhere in text. Westward crustal plate motion of Western Europe is relative motion superimposed upon presumed eastward motion for the Eurasian lithospheric plate.

land was subjected to early Tertiary Hebung - Spaltung - Vulkanismus (HALLER 1971, Fig. 145, p. 342); the important crustal flexure (suggesting intra- or sub-crustal extension) south of Scoresby Sound occurred during the middle Eocene (Haller 1969, Fig. 3; 1971, p. 343); volcanic ash of Eocene age in Denmark may be derived from volcanism in the Norwegian channel (WILSON 1966, p. 9; cf. BEDERKE 1966, p. 217), although it must be mentioned that some workers consider this structure not to be of tectonic origin (TALWANI & ELDHOLM 1972); granitic intrusions of 50—60 my occurred in Rockall (MILLER 1965, p. 187).

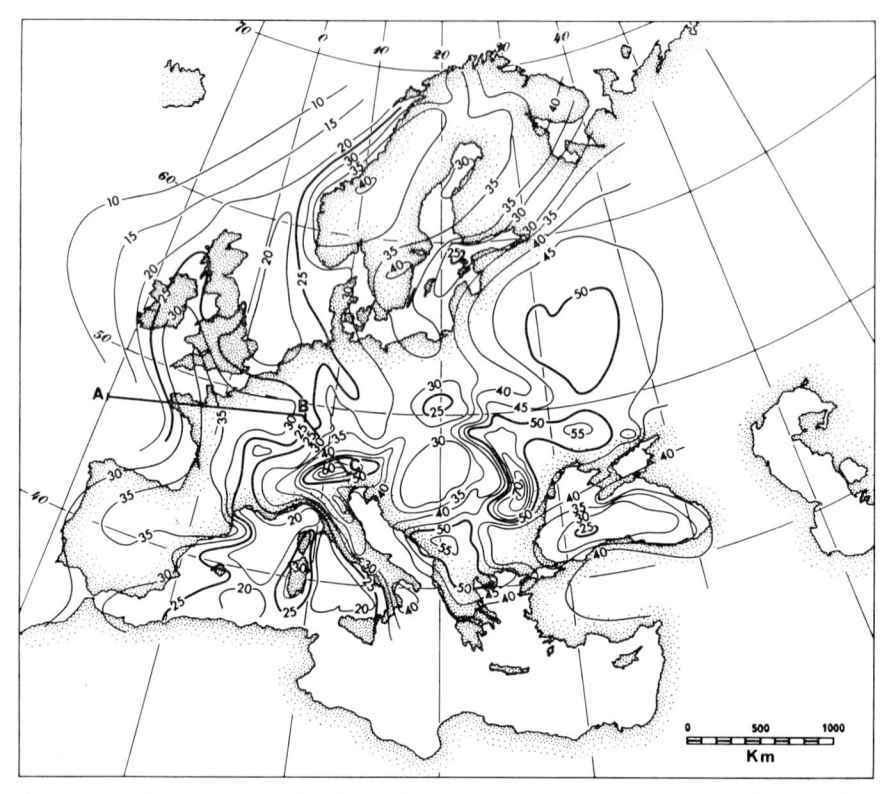

Fig. 12. Depth contour map for the M-discontinuity in Europe, based on preliminary data of MORELLI et al. 1967, with minor revisions.

To pursue the gravity sliding hypothesis, it can be concluded that westward mid-Eocene sliding would be rendered "possible" as a consequence of plate separation during that time in the North Atlantic. However, the resistance to gravity sliding, even if a frictionless crust-subcrust boundary is assumed, seems too large to be accounted for by a plastic wedge mechanism; toe resistance would be significant, and the gravitational component of the crustal block does not seem capable of rendering propulsive assistance. "Uphill" block gliding would have to be assumed if the present configuration of the M-discontinuity is assumed (Figs. 12, 13). Furthermore, although it seems likely, to me at least, that the mid-Eocene configuration of the M-discontinuity differed from the present one (e. g., as one possibility, a relatively constant crustal thickness on the order of 25—30 km may have been present prior to continental plate separation), it seems reasonable to assume that tilting of this boundary away from the incipiently-developed Eocene mid-oceanic ridge axis would have occurred. Hence the net gravitational component on any potential slip surface would act in a direction so as to oppose motion. A gravitational sliding hypotheses therefore is rejected.

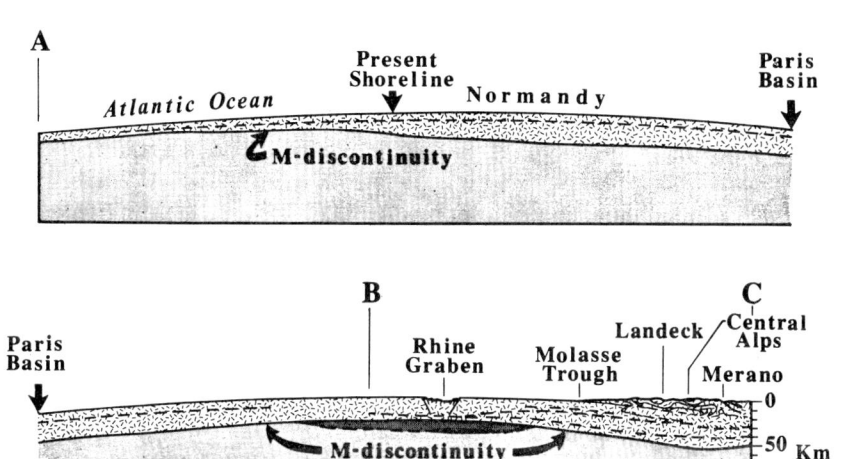

Fig. 13. Cross-section from the Atlantic Ocean to the Alps (cf. Fig. 12) showing position of M-discontinuity and crustal structure (cf. ILLIES 1972, Fig. 20).

Deformable-plate tectonics and sea-floor spreading and stretching

We return therefore to hypothesis (1), namely, that subcrustal material has been subjected to lateral extension, and has stretched the less-ductile crustal "skin". This phenomenon can be accounted for by largescale global expansion, or within the context of the sea-floor spreading scheme, by merely assuming that the lithospheric[16] plates are not, in point of fact, "rigid plates", as has been tacitly assumed in the sea-floor spreading hypothesis (e. g., MORGAN 1968, PITMAN & TALWANI 1972, p. 62), but are ductile and thus deformable, particularly in association with sub-crustal PTX conditions associated with the general vicinity of axial ridges. The viscoplastic[17] properties of sub-crustal mantle material at the time of plate disruption, as inferred from the ductile behavior of peridotite and other olivine-rich rocks subjected to experimental deformation at high temperature, and consideration of the hydrolytic weakening mechanism and high temperature creep

---

[16] ALEXANDER & SHERBURNE (1972) have recently suggested that the moving plates may extend to 400 km rather than the approximately 100 km depth commonly assumed. This point is, however, not crucial to our argument, although it may be presumed that plate dynamics would be profoundly influenced.

[17] Olivine and orthorhombic pyroxene deformed in the upper mantle flow by the same dislocation slip and syntectonic recrystallization mechanisms as those induced by high temperature experimental deformation (GREEN & RADCLIFFE 1972, p. 155). Flow in the upper mantle is thus probably not Newtonian as inferred by CAREY 1953; the power-law relation experimentally established between stress and strain is more likely a closer approximation (compare GREEN & RADCLIFFE 1972; CARTER et al. 1972; WEERTMAN 1970; RALEIGH & KIRBY 1970).

effects (e. g., GRIGGS et al. 1960, p. 50—63; GRIGGS 1967; BLACIC 1971, 1972; PHAKEY et al. 1972; GREEN & RADCLIFFE 1972; CARTER et al. 1972; RALEIGH & KIRBY 1970; CARTER & AVÉ LALLEMANT 1972; POST 1970; HILL & BOETTCHER 1970), render nearly "rigid" behavior of the lithosphere as highly unlikely. Instead of "rigid" behavior, it seems far more reasonable to suppose that the lithosphere has been and continues to be subjected to ductile extension in association with plate separation mechanisms (Fig. 14). Tractions resulting from mantle extension would be in part transferred to the crust, which would, in deforming as a whole in a less ductile manner[18], produce such observed features as grabens[19] and crustal flexures. The M-discontinuity under continental areas therefore seems to be of structural significance in that it represents an interface (one of several) separating zones of contrasting rheology under existing PT conditions, i. e., it is a rheologic discontinuity. Furthermore, there seems no reason to suppose that stretching should have ceased with the commencement of the process of sea floor spreading. Hence the "stretching phase" is presumably followed by a deformational phase characterized by seafloor spreading and stretching of both old and new crust and mantle material.

The above explanation, a hybrid hypothesis between "continental stretching", the "sea floor stretching" of HOLMES and the "sea floor spreading" of HESS, seems capable of resolving the genetic link implied by the presumably synchronous development of graben and other extensional structures, and oceanic ridges, and seems to provide as well an explanation for the origin of the microcontinents such as Rockall, Lomonosov Ridge and the Mascarene Plateau[20].

According to this hypothesis both mantle and crust may be presumed to have thinned during and subsequent to plate separation; thus protocontinental reconstructions based upon existing "plate" boundaries in map view, generally taken at some arbitrary but presumably appropriate isobath (CAREY 1958; BULLARD et al. 1965), are held to be consistently in error because the vertical (thickness) dimension is characteristically ignored. It is proposed that attempted reconstructions should account for distortions which may have occurred. Insufficient information is presently available concerning mantle distortion to permit this, so that plate reconstructions would presumably be based upon crustal reconstruction for a given time period, in which subsequent additions (e. g. volcanics) are subtracted and crustal

---

[18] The lower portion of the sialic crust would presumable chiefly deform by ductile faulting and flow, the upper portion by brittle fracture. Strain discontinuities would not by unexpected within the crust, and it seems likely that the greatest amount of stretching would occur in the lower half of the crust.

[19] The question of why any particular graben develops at a particular location remains to be answered; several explanations are possible, and it is likely that no single explanation will serve adequately for all continental grabens. These include such factors at imposed strain and stress gradients, strength variability as a function of local PTX conditions, and geometric variations. The Rhine graben, for example, may be related to localized zones of thin and/or weak crust (e. g., Fig. 13), either of which would tend to effect localization of crustal strain (cf. ILLIES 1972, p. 33).

[20] A similar interpretation can be offered for graben and related features (FAILL 1973) in eastern North America and western Africa, which according to this hypothesis presumably evolved in association with plate separation in Mesozoic time.

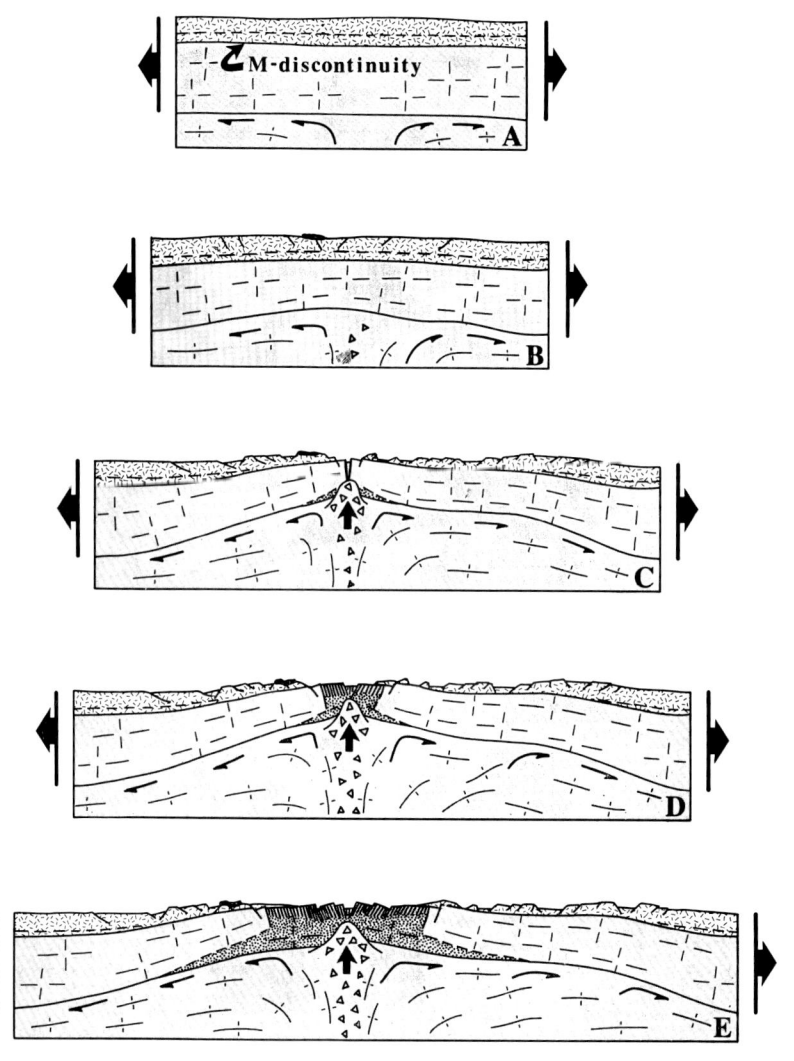

Fig. 14. Schematic cross sections of the evolution of an oceanic rift. The initial condition involves a sialic crust overlying layered mantle, each zone having relatively constant thickness (A). The crustal brittle-ductile transition is given by a dashed line. Aesthenospheric convective motion has just begun, imposing tractions on overlying ductile mantle; mantle stretching commences, with some transference of extension to the crust, resulting in grabens and crustal flexures (B, C). Finite strain trajectories are suggested by "cross" patterns. At some limiting value of strain, rupture of mantle occurs with consequent intrusion of mantle derivatives and generation of new oceanic crust and mantle (C). Separation of lithospheric and continental plates occurs due to both sea floor spreading and to continuation of stretching processes (D). Stretching continues in mantle under continental areas, with the consequence that continental graben evolution continues; the rate of extension decreases, however, as separation proceeds inasmuch as mantle strain gradients decrease with distance from the axial ridge (E). Stretching also occurs in newly generated crust and mantle, producing extensional phenomena on the ocean floor.

distortions approximately accounted for. Appropriate reconstruction methods have been used in structural geology (e. g., VOIGHT 1964).

The general result of these corrections would be to make the continental plates somewhat smaller (although approximately of the same configuration) due to reconstructed "compression" of existing (i. e., previously stretched) continental boundaries. The precision of "fit" of pre-existing continental positions, based in part upon calculated pre-existing isopachs of crustal thickness, is thus changed in detail. Detailed consideration of the new boundary relationships may offer new solutions to the "room problem" involved in such areas as the Greenland-Norwegian Sea, where Jan Mayen, the Faeroe Islands, and Iceland can be claimed to have certain continental affinities (e. g., JOHNSON & HEEZEN 1967, p. 769, Fig. 5), and yet do not fit the BULLARD or analogous jig-saw puzzle patterns of reconstructed continental plates. A reasonable case can be presented for the origin of a part of the crust in these areas as a consequence of crustal stretching. The accuracy of reconstruction methods depends, of course, on accurate geophysical identification of crustal boundaries, particularly in the vicinity of continental margins and microcontinental areas. This information is known only in a general way at the present time (for an exception, refer to TALWANI & ELDHOLM 1972).

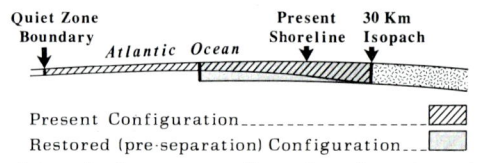

Fig. 15. Restoration of (stretched) present configuration of continental crust to pre-separation configuration, assuming that the stretched boundary extends to the magnetic quiet zone.

Fig. 15 illustrates an attempt at reconstruction in a section parallel to the presumed direction of extension. The 30 km isopach is presumed to be a reasonable approximation of resulting crustal thickness, and for purposes of example the boundary of the magnetic quiet zone is taken as a possible[21] original boundary of the continental plate; isopach data, taken in conjunction with these assumptions suggest stretching from 30 to about 10 km thickness occurred prior to crustal separation[22]. This restoration suggests that the "pre-stretching" plate boundary lay just landward of the present position of the 20 km isopach, and it is suggested that it is this boundary that should be employed in continental drift reconstructions. Neither the "arbitrary isobath" method employed by CAREY, BULLARD and others, nor the magnetic discontinuity or quiet zone boundary methods can directly provide the appropriate plate boundary configurations prior to the initiation of the separation mechanism.

---

[21] The quiet-zone boundary can, in general, be taken as a limiting boundary, seaward of which the existence of continental material is unlikely.

[22] Stretching would presumably not have been uniformly distributed throughout the crust; the lower crust was presumably more ductile and the majority of crustal strain would have occurred there.

Examination of crustal isopachs, particularly those based on seismic profiling (LYONS 1970, Fig. 13; KING 1970, Fig. 8) suggest that crustal plate stretching at the boundary from 30—40 km to about 10 km could be typical. In this case, the original plate boundary may be presumed to have been located approximately at the present position of the 20 or 25 km isopach. This isopach typically occurs at approximately the continental rise, a fortuitous circumstance which renders the results of isopach-based reconstructions nearly the same as those based on the 0.5 or 1.0 km isobath method at most locations. Thus, on the basis of existing isopach data, the BULLARD reconstructions seem approximately correct, for fortuitous reasons, in providing "pre-stretching" reconstructions; the DRAKE & NAFE (1968) reconstruction, based upon quiet zone magnetic boundaries (Fig. 16), could be interpreted to be approximately accurate (if their basic assumption is correct) with respect to the time of incipient plate separation (as distinct from "pre-stretching"), although details north of the 50° latitude remain to be adequately treated. In any case, there is a specific time interval which must be allotted to the stretching process; e. g., assuming that (1) active initiation of the rift process between Africa and North America may have commenced 180—200 m. y. ago, (2) that the quiet-zone boundary represents the 155 m. y. isochron, and that (3) the initial sialic crust thickness was about 30 km, an estimate of 25—45 m. y. is provided for the duration of the "stretching phase" of separation which could have occurred prior to the initiation of sea-floor spreading. The third of these assumption seems questionable for North American—African rifting, but may not be greatly in error with respect to separation of Greenland and Europe. Under the circumstances, the separation stage in the Norwegian-Greenland Sea may be approximated by Fig. 14 D, with the consequent expectation that much of the sea floor in the region would be underlain by inhomogeneously-thinned and isostatically-submerged sialic crust[23].

Summary and conclusions

Examination of the features of graben which developed in association with the 1964 Alaskan earthquake landslides suggest many elements in common with continental-scale graben. Both seem typically to be "thin-skinned" structures in the sense that a basal structural discontinuity seems to be involved in their evolution. Examination of the state of stress within graben zones suggests further similarities; lateral stresses seem typically compressive, although of a magnitude at any point

---

[23] Recent Russian oceanographic seismic investigations apparently have confirmed[24] this expectation (A. A. MEYERHOFF, oral communication, 1973).

[24] Note added in press:
The results of these investigations may in point of fact not be "confirmation" as inferred in footnote 23. Submerged sialic crust appears to be present; indeed, except for two small (10,000 km²) oceanic areas, A. A. MEYERHOFF (private communication, July, 1973) reports that the North Atlantic from Rockall and the Faeroe-Iceland-Greenland Ridge to Lena Strait (west of Spitsbergen) overlies "continental" crust. However, crustal thickness is reportedly 35—40 km (to the 8.0—8.2 km/sec layer), with a 16—20 km granitic layer. Thus "thinning" of the crust toward the ocean center has apparently not been observed by these workers.

Fig. 16. Atlantic Ocean Floor, showing position of quiet zone boundary (white dashed lines); this is inferred to be the seaward limit of possible continental-derived crustal material (base map from The National Geographic Society 1971).

less than that of the pressure due to overburden weight. These lateral stresses can be of significance with respect to the mechanics of structural evolution, as can be demonstrated at Anchorage, or may be incidental, as is considered more likely for the upper Rhine graben. The gravitational slide hypothesis for the origin of graben seems applicable in the Anchorage examples but not for the Rhine graben.

The crustal extension associated with the Rhine graben and related features can, however, be temporally and mechanistically related to a plate tectonic model involving separation of Greenland and Eurasian plates. The concept of "rigid plate tectonics", however, seems inadequate to this task; a "deformable plate" hypothesis is thus suggested here, which involves stretching of both mantle and crustal materials, the generation of new oceanic mantle and crust, and continued stretching of both old and new lithospheric materials. Ductile behavior for lower crust and mantle seems realistic in view of measured strength and flow law properties of peridotites and other crystalline rocks subjected experimentally to high-temperature, high-pressure deformation.

The plate tectonics model suggested here seems to be capable of accounting for the origin of the microcontinents and continental margin features, in addition to continental grabens. Distortions in the vertical as well as lateral dimensions occur in the separation process; these should be accounted for in global plate reconstructions. Resolution of this distortion requires better detailed subsurface information, especially near continental boundaries, than is generally available at present, but there is some prospect in this procedure of accounting for such anomalous areas as Iceland and the Greenland-Norwegian Sea, which display continental affinities heretofore regarded as anomalous.

References cited

AHORNER, L., 1970: Seismo-tectonic Relations between the Graben Zones of the Upper and Lower Rhine Valley. — In: Graben Problems, Inter. Upper Mantle Project, Sci. Rept. 27, 155—166, Schweizerbart, Stuttgart.

ALEXANDER, S. S. & SHERBURNE, R. W., 1972: Crust mantle structure of shields and their role in global tectonics. — Trans. Amer. Geophys. Union, 53, no. 11, p. 1043.

ANSORGE, J. et al., 1970: Structure of the Crust and Upper Mantle in the Rift System around the Rhinegraben. — In: Graben Problems, Inter. Upper Mantle Project, Sci. Rept. 27, 190—197, Schweizerbart, Stuttgart.

BEDERKE, E., 1966: The development of the European rifts. — Geol. Soc. Canada, 66—14, 213—219.

BLACIC, J. D., 1971: Hydrolytic weakening of quartz and olivine. — Ph. D. Thesis, Univ. Calif. Los Angeles, 205 p.

— 1972: Effect of water on the experimental deformation of olivine. — In: HEARD, H. C., et al. (eds.), Amer. Geophys. Union, Geophys. Mon. 16, 109—116.

BULLARD, E. C., EVERETT, J. E. & SMITH, A. G., 1965: The fit of the continents around the Atlantic. — Roy. Soc. London Phil. Trans., Ser. A., 258, 41—51.

CAREY, S. W., 1953: The rheid concept in geotectonics. — J. Geol. Soc. Australia, 1, 67—118.

— 1958: A tectonic approach to continental drift. — In: CAREY, S. W. (ed.), Continental Drift: A symposium. 177—355, Univ. Tasmania, Hobart.

CARTER, N. L. et al., 1972: Seismic anisotropy, flow, and constitution of the upper mantle. — In: HEARD, H. C. et al. (eds.), Amer. Geophys. Union, Geophys. Mon. 16, 167—190.

CARTER, N. L. & AVÉ LALLEMANT, H. G., 1970: High-temperature flow of dunite and peridotite. — Geol. Soc. Amer. Bull., 81, 2181—2202.

CHOUKROUNE, P. & SÉGURET, M., 1973: Tectonics of the Pyrenees: Role of Compression and Gravity. — In: DE JONG, K. & SCHOLTEN, R., Gravity and Tectonics, J. Wiley & Co., New York.

COULOMB, C. A., 1773: Essai sur une application des règles de maximis et minimis à quelques problèmes de statique relatifs à l'architecture. — Mém. de la mathematique et de physique, Acad. Roy. Sci., Paris, 7, 343—382.

DRAKE, C. L. & NAFE, J. E., 1968: The transition from ocean to continent from seismic refraction data, in Continental margins and island arcs. — Amer. Geophys. Union Geophys. Mon., 12, 174—186.

FAILL, R. T., 1973: Tectonic development of the Triassic Newark-Gettysburg Basin in Pennsylvania. — Geol. Soc. Amer., Bull., **84**, 725—740.

GOETZE, C., 1971: High-temperature rheology of Westerly granite. — J. Geophys. Res., **76**, 1223—1230.

GREEN, H. W. & RADCLIFFE, S. V., 1972: Deformation processes in olivine. — In: HEARD, H. C. et al. (eds.), Amer. Geophys. Union, Geophys. Mon. **16**, 117—138.

GRIGGS, D. T., 1967: Hydrolytic weakening of quartz and other silicates. — Geophys. J. Roy. Astron. Soc., **14**, 19—31.

GRIGGS, D. T., TURNER, F. J. & HEARD, H. C., 1960: Deformation of rocks at 500—800 C. — In: GRIGGS, D. T. & HANDIN, J., Rock Deformation, Geol. Soc. Amer. Mem., **79**, 39—104.

GRØNLIE, G. & RAMBERG, I. B., 1972: Gravity indications of deep sedimentary basins below the Norwegian continental shelf and the Vøring Plateau. — Norsk Geol. Tidsskr., **50**, 375—391.

HALLER, J., 1969: Tectonics and neotectonics of East Greenland — review bearing on the drift concept. — In: KAY, M. (ed.), Amer. Geophys. Union Geophys. Mon. **12.**, 852—858.

— 1971: Geology of the East Greenland Caledonides. — Interscience, N. Y., 413 p.

HANDIN, J., 1966: Strength and ductility. — Section 11 in CLARK, S. P., Handbook of Physical Constants, Geol. Soc. Amer. Mem. **97**, 587 p.

HÄNEL, R., 1970: Interpretation of the Terrestrial Heat Flow in the Rhinegraben. — In: ILLIES, J. H. & MÜLLER, S. (eds.), Graben Problems, Inter. Upper Mantle Project. Sci. Rept. 27, 116—120, Schweizerbart, Stuttgart.

HANSEN, W. R., 1965: Effects of the earthquake of March 27, 1964, at Anchorage, Alaska. — U. S. Geol. Surv. Prof. Paper **542 A**, 68 p.

HILL, R. E. T. & BOETTCHER, A. L., 1970: Water in the earth's mantle; melting curves of basalt-water and basalt-water-carbon dioxide. — Science, **167**, 980—982.

ILLIES, H., 1967: An attempt to model Rhinegraben tectonics. — Abh. geol. Landesamt Baden-Württemberg, **6**, 10—12.

— 1970: Graben Tectonics as Related to Crust-Mantle Interaction. — In: ILLIES, J. H. & MÜLLER, S. (eds.), Graben Problems, Inter. Upper Mantle Project, Sci. Rept. 27, 4—26, Schweizerbart, Stuttgart.

— 1971: Der Oberrheingraben. — Fridericiana, Z. d. Univ. Karlsruhe, **9**, 17—32.

JOHNSON, G. L. & HEEZEN, B. C., 1967: Morphology and evolution of the Norwegian-Greenland Sea. — Deep-Sea Research, **14**, 755—771.

KEITH, M. L., 1972: Ocean-floor convergence: a contrary view of global tectonics. — J. Geol., **80**, 249—276.

KING, P. B., 1970: Tectonics and geophysics of eastern North America. — In: JOHNSON, H. & SMITH, B. L. (eds.), The Megatectonics of continents and oceans, 74—112, Rutgers U. Press.

LYONS, P. R., 1970: Continental and Ocean geophysics. — In: JOHNSON, H. & SMITH, B. L. (eds.), The Megatectonics of continents and oceans, 147—166, Rutgers U. Press.

MERRILL, R. B., ROBERTSON, J. K. & WYLLIE, P. J., 1971: Melting reactions in the system $NaAlSi_3O_8$—$KAlSi_3O_8$—SiO_2—H_2O to 20 kb compared with results for other feldspar—quartz—H_2O and rock—H_2O systems. — J. Geol. (in press).

MILLER, J. A., 1965: Geochronology and continental drift. — Proc. Roy. Soc. London Phil. Trans., Ser. A, **258**, 41—51.

MORELLI, C., BELLEMO, S., FINETTI, I. & DE VISINTINI, G., 1967: Preliminary depth contour maps for the Conrad and Moho discontinuities in Europe. — Bull. Geofis. Teor. Appl., **IX**, 34, 142—157.

MORGAN, W. J., 1968: Rises, trenches, great faults, and crustal blocks. — J. Geophys. Res., **73**, 1957.

MUELLER, S., 1970: Geophysical Aspects of Graben Formation in Continental Rift Systems. — In: ILLIES, J. H. & MÜLLER, S. (eds.), Graben Problems, Inter. Upper Mantle Project, Sci. Rept. 27, 27—36, Schweizerbart, Stuttgart.

NAFE, J. E. & DRAKE, C. L., 1969: Floor of the North Atlantic — summary of geophysical data. — In: KAY, M. (ed.), North Atlantic Geology and Continental Drift, 59—87, Amer. Assoc. Petrol. Geol. Mem., **12**.

PHAKEY, P. et al., 1972: Transmission electron microscopy of experimentally deformed olivine crystals. — In: HEARD, H. C. et al. (eds.), 117—138, Amer. Geophys. Union, Geophys. Mon. **16**.

PIERCE, W. G., 1963: Reef Creek detachment fault, northwestern Wyoming. — Geol. Soc. Amer. Bull., **74**, 1225—1236.

— 1973: Principal features of the Heart Mountain Fault and the Mechanism problem. — In: DE JONG, K. & SCHOLTEN, R. (eds.), Gravity and Tectonics, J. Wiley and Sons, New York.

PITMAN, W. C. & TALWANI, M., 1972: Sea Floor spreading in the North Atlantic. — Geol. Soc. Amer. Bull., **83**, 3, 619—646.

POST, R. L., 1970: The flow laws of Mt. Burnette dunite at 750—1150° C. — Eos. Trans. AGU, **51**, 424.

RALEIGH, C. B. & PATERSON, M. S., 1965: Experimental deformation of serpentinite and its tectonic implications. — J. Geophys. Res., **70**, 3965—3985.

RALEIGH, C. B. & KIRBY, S. H., 1970: Creep in the upper mantle. — Miner. Soc. Amer. Spec. Pap. **3**, 113—121.

RUBEY, W. W. & HUBBERT, M. K., 1959: Role of fluid pressure in mechanics of overthrust faulting, part II. — Geol. Soc. Amer. Bull., **70**, 167—206.

RYBÁŘ, J., 1971: Tektonisch beeinflußte Hangdeformationen in Braunkohlenbecken. — Rock Mechanics, **3**, 139—158.

SEDERHOLM, J., 1913: Weitere Mitteilungen über Bruchspalten mit besonderer Beziehung zur Geomorphologie von Fennoskandia. — Bull. Commiss. Géol. Finlande, **37**.

SEED, H. B. & WILSON, S. D., 1967: The Turnagain Heights landslide, Anchorage, Alaska. — Amer. Soc. Civ. Eng. Proc., **93**, n. SM-4, 325—353.

TALWANI, M. & ELDHOLM, O., 1972: Continental margin off Norway. — Geol. Soc. Amer. Bull., **83**, 3575—3606.

VOGT, P. R., SCHNEIDER, E. D. & JOHNSON, G. L., 1969: The crust and upper mantle beneath the sea. — Amer. Geophys. Union, Geophys. Mon. **13**, 735 p.

VOIGHT, B., 1964: Boudinage: a natural strain and ductility gauge in deformed rocks. — Geol. Soc. Amer. Ann. Meet. Program Abs., 213—214.

— 1973: The mechanics of retrogressive block-gliding, with emphasis on the evolution of the Turnagain Heights landslide, Anchorage, Alaska. — In: DE JONG, K. & SCHOLTEN, R. (eds.), Gravity and Tectonics, J. Wiley and Sons, New York.

WEERTMAN, J., 1970: The creep strength of the earth's mantle. — Rev. Geophys. Space Phys., **8**, 145—168.

WILSON, J. T., 1966: Some rules for Continental drift. — In: GARLAND, G. D. (ed.), Continental Drift, 3—17, Roy. Soc. Canada Spec. Pub. 9, U. Toronto Press.

WILSON, S. D., 1967: Landslides in the city of Anchorage. — In: WOOD, F. J. (ed.), The Prince William Sound earthquake, Alaska, 1964, and aftershocks: U.S. Dept. Commerce, Coast and Geodetic Surv. 10—3, **2**, 253—297.

Geophysical Contributions to Taphrogenesis[1]

by

K. Fuchs

With 2 textfigures

Graben formation, taphrogenesis, is closely linked to the dynamic processes within the earth's interior which drive the tectonic activities at the earth's surface. Therefore, the study of graben formation is one of the windows through which we may attempt to gain a more direct insight into these deep seated processes which are not accessible to direct observations. Earthquakes and observed vertical motion provide evidence that the graben is still alive. The history of its past formation, however, must be obtained indirectly from those observations of all disciplines of geosciences which reflect the processes involved. Continental grabens have the advantage of direct access in comparison with rifts in oceanic areas. This is especially true for the Rhinegraben which has been studied both interdisciplinarily and internationally during recent years.

In this paper, presently available geophysical observations which may provide information related to the formation of the graben are examined. Which forces initiated taphrogenesis? Which forces are driving the formation of the graben during later stages? In the area of the Rhinegraben, such an analysis is complicated by the close vicinity of the Alpine orogenic belt. The latter clearly impresses its influence on the Rhinegraben, especially on its southern part.

The process of graben formation is reflected in geophysical observations related both to the structure and to the physical properties of the crust and upper mantle in the region of the Rhinegraben. Studies of seismic activity and recent crustal movements provide the most relevant information on the present development of the graben. In the present discussion, after a review of geophysical observations from explosion seismology, and from gravity, geothermal and magnetotelluric surveys, three hypothetical models of the uplifted crust-mantle boundary will be discussed and finally a model of taphrogenesis is proposed in which the sialic low velocity zone introduces a gravitational instability into the crust of the graben.

Geophysical observations relevant to the formation of the Rhinegraben

Explosion Seismology. — One of the most important findings of explosion seismology in the area of the Rhinegraben was the detection of an elevated crust-mantle boundary at a depth of only 25 km in the southern part of the graben (ANSORGE et al. 1970; ANSORGE et al. 1973). This uplifted boundary is not

---

[1] Contribution No. 165 within a joint research program of the Geophysical Institutes in Germany sponsored by the Deutsche Forschungsgemeinschaft (German Research Association). — Contribution No. 103 Geophysical Institute, University Karlsruhe.

confined to the graben proper but reaches beneath the mountains of the Black Forest and of the Vosges (BONJER & FUCHS 1970; EMTER 1971). The elevation of the crust-mantle boundary is somewhat smaller in the northern part of the graben where it is restricted essentially to the graben proper (MEISSNER et al. 1970; MEISSNER & VETTER, this vol.).

Recent seismic refraction investigations (RHINEGRABEN RESEARCH GROUP FOR EXPLOSION SEISMOLOGY, this vol.) in the southern part of the Rhinegraben and adjacent areas confirmed the uplifting of the crust-mantle boundary and contributed to the definition of its relief. An unexpected result of this experiment was the revision of the velocity for P-waves below the elevated boundary. The first reversed profile within the graben, and along its strike, led to a velocity of 8.1 km/sec in contrast to the previously estimated velocities of 7.6—7.7 km/sec. This high velocity at a depth of only 25 km is even more surprising since in the neighbouring Limagne graben (HIRN & PERRIER, this vol.) a velocity of 7.6 km/sec was derived at the same depth on a reversed profile. — The velocity of 7.2 km/sec reported for the northern part of the graben (MEISSNER et al. 1970) cannot be revised by the new experiment since no reversed observations are available in this part of the graben.

With the revised P-wave velocity of 8.1 km/sec in the uppermost part of the uplifted mantle the question of cushion structure at the base of the crust and especially its lower boundary is open to discussion again. Such a rift cushion was proposed previously by a number of authors (ANSORGE et al. 1970; MEISSNER et al. 1970). However, no reversed observations were available before the 1972 seismic experiment. The earlier decision to consider the observed apparent velocity of 7.6—7.7 km/sec as the true velocity within the cushion was influenced also by considerations of gravity anomalies. Since the uplifting of the crust-mantle-boundary could not be detected in the Bouguer anomaly, the elevation of high density material must have a compensating mass deficiency at depth. This could be achieved by assuming a cushion of intermediate velocity and corresponding intermediate density floating in the normal upper mantle: the masses above about 30 km — the normal depth of the crust-mantle boundary — produced an excessive gravity which was compensated by the mass deficiency of that part of the cushion immerged into the upper mantle.

Seismic arrivals with an apparent velocity of 8.2 km/sec on the profile Taben Rodt—SE traversing the Rhinegraben were observed to the east of the graben (ANSORGE et al. 1970). These arrivals could be interpreted as headwaves from the bottom of the cushion at a depth of about 40 km. No arrivals from the bottom part of a cushionshaped body have been observed during the 1972 seismic experiment. Implications of these new findings will be discussed below.

Another important finding by various groups (ANSORGE et al. 1970; MEISSNER et al. 1970) of explosion seismology is the presence of the sialic low velocity channel (MUELLER & LANDISMAN 1966; LANDISMAN & MUELLER 1966) in the area of the Rhinegraben rift system. The 1972 experiment also confirmed its presence in this area (EDEL, GELBKE, personal communication). The importance of this channel for the process of taphrogenesis will be reviewed below.

Gravity. — The Bouguer gravity in the area of the Rhinegraben (GERKE 1957) is closely related to its crustal and upper mantle structure. Its analysis in combination with seismic models provides evidence for a mass deficiency in the upper mantle compensating the uplift of the crust-mantle boundary and for the presence of a low velocity channel in the upper crust.

A number of authors (MUELLER et al. 1967; CLOSS & PLAUMANN 1967, 1968) have pointed out that the sedimentary filling of the graben accounts for the observed minimum in the Bouguer anomaly within the Rhinegraben proper. If the Bouguer gravity is corrected for the presence of sediments, the crustal uplift determined from seismic observations is not apparent in the corrected Bouguer gravity.

This observation is supported by a regional study of Bouguer anomalies $\Delta g''$ in which MAKRIS (1971) extrapolated the depth T_M to the crust-mantle boundary from $T_M(\Delta g'')$-relations derived in areas where both gravity and seismic observations were available. Applying in the Rhinegraben the same $T_M(\Delta g'')$-relation as in the Molasse basin of the Bavarian foreland of the Alps, he found a depth of 32 km. The uplifted crust-mantle boundary below the Rhinegraben does not appear in the Bouguer-anomaly. The 1972 seismic experiment in the Rhinegraben (RHINEGRABEN RESEARCH GROUP FOR EXPLOSION SEISMOLOGY, this vol.) confirmed the uplift of the crust-mantle boundary. However, the velocity of P-waves in the uppermost mantle had to be revised to be 8.1 km/sec. This larger velocity must be related to an increased density requiring therefore an even larger mass deficiency at depth for compensation.

In a closer analysis of the gravity minimum in the Rhinegraben a typical asymmetry of the axis of the gravity low and the axis of thickest sediments was pointed out by MUELLER & PETERSCHMITT (1966) and CLOSS & PLAUMANN (1967, 1968). MUELLER (1968, 1970) suggested that this asymmetry is caused by the presence of a zone of lower density in the upper part of the crust corresponding to the zone of low P-wave velocities. A model including low density intrusions from this zone into the basement layers allows a close fit of observed and computed gravity (MUELLER & RYBACH, this vol.).

Geothermal observations. — The highest heat flow values in Germany are encountered in the Rhinegraben. A 5th order trend surface reaches a broad maximum northwest of Karlsruhe where the peak-value of the trend surface of 2.6 µcal/cm²/sec (= HFU) exceeds the mean value for this area by 0.8 HFU (HAENEL 1971). Local anomalies even attain values of 4.0 HFU. The studies of WERNER (1970) and WERNER & DOEBL (this vol.) suggest that these excessive local values can partly be attributed to heat transport by circulating water. HAENEL (1971) relates the regional high to large heat production within a layer of 16 km thickness centred in the depth range of the low velocity zone in the upper crust. Relating seismic velocities with heat production and density MUELLER & RYBACH (this vol.) link both the asymmetry in the Bouguer anomaly and sedimentary thickness and the high heatflow values in the Rhinegraben west of Karlsruhe with the intrusion, from the low velocity channel, of material with low density and high heat production. With this model, they find, in the Rhinegraben proper, a heatflow of 1.0 HFU from below a depth of 25 km which exceeds only slightly the average value of 0.9 HFU deduced by HAENEL (1971) for the contribution to the heatflow from below the MOHOROVIČIĆ discontinuity in Germany. Although this is nearly a normal mantle heatflow, it should be noted that the temperatures at the crust-mantle boundary below the graben proper exceed by nearly 200° C the temperature at the same depth below the Black Forest. The temperature at the crust-mantle boundary below the Rhinegraben would be even higher if less weight were given to the heat production within the low velocity zone. —

Furthermore, it is quite likely that the maximum of the regional heat flow would be shifted to the South if the contribution from near surface heat sources were subtracted from the excessive values West of Karlsruhe. In this case the trend surface of the regional surface would tend to correlate with the surface of the uplifted crust-mantle boundary in the southern part of the graben. This would indicate that a substantial part of the regional high heatflow in the area of the Rhinegraben could originate from below the crust-mantle boundary and that this elevated uppermost mantle is excessively heated.

Electrical resistivity. — Geomagnetic (WINTER, this vol.) and magnetotelluric observations (SCHEELKE, this vol.) have been inverted into depth-resistivity distributions using advanced inversion techniques. The geomagnetic observations have been fitted to a model over a frequency range from 0.5—6 cph; the magnetotelluric observations cover periods from 10—1000 sec. Both methods indicate a layer with a low resistivity of about 25—35 Ωm at a depth of about 25 km below the Rhinegraben proper. The thickness of this low resistivity layer is estimated to be between 50—65 km from geomagnetic, and between 20—30 km from magnetotelluric data. Outside the graben proper, at distances of 45 km to the West and 120 km to the East of the graben axis. HAAK & REITMAYR (this vol.) do not find indications of this shallow low resistivity layer on their magnetotelluric data. Instead they propose for a flat layered earth a rather thin conducting layer at a depth of about 80 km.

The most important findings of these resistivity soundings seem to be the presence of a layer of high conductivity at a depth of 25 km below the Rhinegraben proper and its absence in neighbouring areas. From laboratory measurements on olivines, SCHULT (this vol.) estimates the temperature at the top of this low resistivity zone to be 800° C. This value exceeds the temperature of 660° C deduced by MUELLER & RYBACH (this vol.) by 140° C. Even if the temperature estimate of SCHULT is on the high side one should regard the results of resistivity soundings as strong indication of excessive temperatures in the upper mantle below the Rhinegraben proper.

Implications of the elevated mantle in the region of the Rhinegraben

The confirmation of the uplift of the crust-mantle boundary in the Southern part of the Rhinegraben together with the revision of the P-wave velocity Vp in the uppermost mantle raises some questions as to the nature of the elevated material.

The velocity of 8.1 km/sec below the Rhinegraben must be considered as large for a depth of only 25 km. Its value is very close to the normal Pn-velocity of the uppermost mantle. However, no abnormal origin of the material below the uplifted and heated crust-mantle boundary can be deduced from the revised velocity value alone.

If the normal material below the continental crust-mantle boundary with a Pn-velocity of about 8.2 km/sec is heated by about 200° C (MUELLER & RYBACH, this vol.; SCHULT, this vol.) and depressurized by about 1.25 kbar, corresponding to an uplift of 5 km, and if no phase changes are encountered during this process, the velocity decrease can be estimated to be less than 1.1% from

$$\frac{\partial v_p}{v_p} = \left(\frac{1}{v_p}\frac{d v_p}{\partial p}\right)_T dp + \left(\frac{1}{v_p}\frac{\partial v_p}{\partial T}\right)_p dT$$

with $\left(\frac{1}{v_p}\frac{\partial v_p}{\partial p}\right)_T = 0.001 \text{ kbar}^{-1}$ and $\left(\frac{1}{v_p}\frac{\partial v_p}{\partial T}\right)_p = -0.5 \cdot 10^{-4}/°C$

(FORSYTH & PRESS 1971). The heated and depressurized upper mantle material would therefore still have a P-velocity of 8.1 km/sec which is within the range of values observed in the Rhinegraben below the crust-mantle boundary.

More consideration must be given to the elevated crust-mantle boundary when the increased density is estimated from the revised velocity v_p. The velocity v_p in the upper mantle of 8.1 km/sec is to be contrasted with a velocity of about 6.8 km/sec in the overburden. According to the BIRCH (1961) density-velocity relation — $\Delta \varrho = 0.302 \Delta v_p$ — a velocity difference of 1.3 km/sec corresponds to a density-contrast of about 0.39 g/cm³. The uplifting of the crust-mantle boundary in the Southern Rhinegraben by about 5 km (RHINEGRABEN RESEARCH FOR EXPLOSION SEISMOLOGY, this vol.) contributes about a 40—60 mgal increase to the Bouguer gravity. Such an increase is not observed in the Rhinegraben region. Therefore, it must be expected that the masses of the elevated crust-mantle boundary are compensated by mass deficiency within the upper mantle. A compensation within the crust due to increased temperatures is not regarded as very likely since the average velocity of 6.25—6.35 km/sec within the crust of the Rhinegraben is a normal crustal average.

Since no further information is available on the properties of the compensating masses at the present stage of investigations, three hypothetical possibilities will be suggested in the following sections.

A reduction in the density of the lower lithosphere by heating alone could already provide an appreciable compensation of the gravity anomaly of the uplifted mantle. The density reduction by heating may be estimated from the coefficient of thermal expansion

$$\alpha = \frac{1}{V}\left(\frac{\partial V}{\partial T}\right)_P = -\frac{1}{\varrho}\left(\frac{\partial \varrho}{\partial T}\right)_P = 3 \cdot 10^{-5}/°C \quad \text{(FORSYTH \& PRESS 1971)}.$$

Assuming an average density of about 3.4 g/cm³ for the lower lithosphere the density would be reduced by about 10^{-4}/°C. If the temperature throughout a column of the lower lithosphere below the Rhinegraben is raised by 200°—300° C above normal geothermal temperatures the density reduction would amount to 0.02—0.03 g/cm³. Depending on its diameter such a heated column lying between a depth of 30—130 km would reduce the gravity by about 30—40 mgal and thus nearly compensate the gravity increase due to the uplift. Such a column would be roughly in a state of isostatic equilibrium (Fig. 1a). Such a compensation would correspond to a lower lithosphere heated by conduction. The large time constant (several 10^8 years) makes this mechanism seem rather unlikely. Heating connected

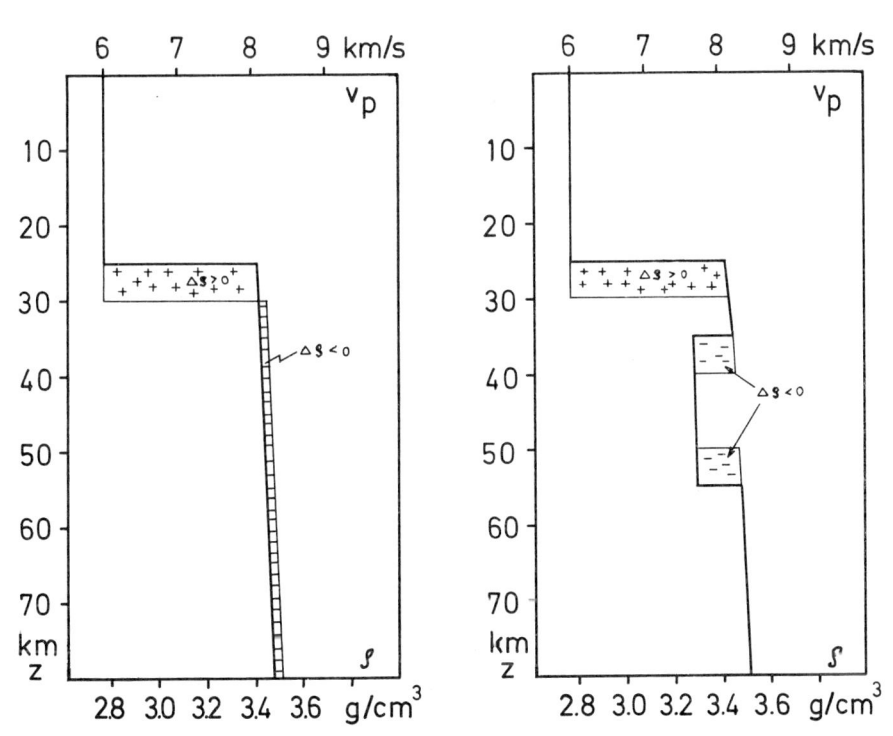

Fig. 1. Compensation of gravity excess by mass deficiency in the lower lithosphere (see text).
   a) left side: compensation by a column of low density throughout the lower lithosphere.
   b) right side: compensation by a thickening of the low velocity channel in the lower lithosphere.

with mass transport seems to be more likely, since heat would rise faster, and in a more localized manner, from the asthenosphere to the base of the crust. Two different processes may be imagined: a diapiric rise of material from the base of the lithosphere or the formation of a low density zone below the uplifted crust-mantle boundary.

If material from the lower lithosphere is raised adiabatically by about 100 km to a depth of 25 km, and if we assume 0.5° C/km for the adiabatic and 3° C/km for the geothermal gradient, the temperature of the material will be lowered by only 50° C and will stay 250° C above normal temperatures which is only a little above the temperatures estimated by MUELLER & RYBACH (this vol.) for this depth. Taking the pressure gradient in the lower lithosphere as 0.33 kbar/km, the total velocity decrease of the adiabatically cooled and depressurized material would be about 0.25 km/sec. This would bring material with an original velocity of 8.35 km/sec within the lower lithosphere into the range of the observed velocity in the uplifted region. The velocity-density-depth distribution would be similar to Fig. 1 a. Such an estimate is only valid if neither phase change nor melting are involved in the diapiric process.

Instead of compensation by a mass deficiency distributed uniformly over the whole column within the lower lithosphere compensation could also be achieved by a pocket of low density material immediately below the crust-mantle boundary. This could be connected with a zone of low velocity material (Fig. 1 b) similar to that which has been found at a depth range between 40—50 km with a velocity of about 7.7 km/sec (HIRN et al. 1973) during a recent seismic refraction experiment in France. A thickening of this zone of low velocity and, correspondingly, of low density by about 20 km would compensate the effect on gravity of the uplifted crust-mantle boundary. It must be admitted, however, that, apart from some late reflections (DOHR 1970), no seismic evidence has been found for such a zone of low velocity in the lower lithosphere below the Rhinegraben. If it exists it could serve as the intermediate magma chamber required from the analysis of volcanic rocks (ILLIES, this vol.). Partial melting would be the most plausible cause for the velocity reduction at this depth.

One further comment must be added concerning the significance of the P_n-velocity observed during the 1972 seismic experiment in the Rhinegraben. In a recent time-term analysis of all available P_n observations in the Federal Republic of Germany BAMFORD (1973) found evidence for a pronounced anisotropy of the P-wave velocity in the uppermost mantle with a maximum velocity in a direction 20° East of North. The anisotropy is probably greater than 0.5 km/sec, the mean P_n-velocity amounts to 8.1 km/sec. It must be noted that the direction of the reversed profile within the Rhinegraben nearly coincides with the azimuth of the maximum velocity. Should this anisotropy also occur in the Rhinegraben with the same azimuth dependence, then the observed P_n-velocity of 8.1 km/sec could be regarded as maximum velocity belonging to a mean velocity of 7.8—7.9 km/sec. This would reduce the effect of the uplifted crust-mantle boundary on the gravity, and, therefore, a smaller mass deficiency would be required for compensation.

Taphrogenesis and the role of the sialic low velocity channel

The classical model of graben formation as proposed by VENING-MEINESZ (1950) is incompatible with a number of observations in the Rhinegraben. His model requires a minimum of the Bouguer anomaly caused by the sinking of low density material of the lower crust into the high density upper mantle. However, the observed minimum in the Rhinegraben may be attributed solely to the presence of sediments. The presently observed width of the Rhinegraben of 36 km is less than twice the crustal thickness, even if the reduced crustal thickness is taken into account. The most prominent departure from the classical model is the development of an uplifted crust-mantle boundary.

Crustal thinning seems to be a general feature of graben formation. This can be achieved either by the development of a rift cushion or by updoming of the crust-mantle boundary. Therefore, many authors have regarded the active growth of the updoming as the primary driving force of taphrogenesis. However, the widening of the Rhinegraben by 4.8 km from its original width of 31.2 km (ILLIES 1972) is larger than would be expected from crustal arching (ARTEMJEV & ARTYUSHKOV 1971). ILLIES (1972) proposes gravity sliding of the whole crust down the flanks of the updomed crust-mantle boundary as mechanism for the additional opening of the Rhinegraben. STROBACH (this vol.) challenges the concept of gravity sliding by estimating the effective horizontal pressure to be about 5 bar only. As an alternative, he introduces horizontal tensions induced by opposite viscous flow within the upper mantle and the rift cusion. The observed high velocity of P-waves below the uplifted crust-mantle boundary throws some doubts on the feasibility of the required horizontal flow velocities within the rift cusion.

In both models taphrogenesis is initiated by the active growth of a rift cushion or updoming of the crust-mantle boundary which in turn must receive its energy from the asthenosphere by heating and magmatic mass transport. It is an amazing observation that the formation of the Rhinegraben occurred along an old European line of weakness, the Mediterranean—Mjoesa zone. How did the uprising material from the asthenosphere "know" that this old line of weakness is present in the crust on top? The most likely link of communication between the upper crust and the upper mantle seems to be the hydrostatic pressure. If part of this pressure is reduced in the old zone of weakness, this could trigger the rise of magma from the upper mantle.

In the following discussion a different mechanism of graben formation is proposed in which the low velocity channel in the sialic part of the crust plays a prominent role, especially during the early stages of formation. Graben formation by crustal necking in a horizontal stress field has been proposed for the first time by BUCHER (1933). This mechanism has been discussed and applied to the Baikal rift by ARTEMJEV & ARTYUSHKOV (1971). Their model will be modified by the inclusion of the sialic low velocity channel. With this decrease of velocity there is an associated lowering of density (MUELLER 1968, 1970; MAKRIS 1971). This introduces a gravitational instability into the upper crust which in zones of weakness could become a driving force of tectonic movements. The petrological composition of the lower velocity layer is not so important. A granitoid material can be assumed which possesses a reduced viscosity compared to the metamorphic basement on top and to the lower crust beneath the channel (RAMBERG 1972).

The model of graben formation is sketched in Fig. 2. If a horizontal stress is applied to a stratified viscous medium the various layers will react with different creep velocities. If the stress is less than the tensile strength the layer with the lowest viscosity tends to produce a neck (Fig. 2a). The development of a neck in the low velocity layer is enhanced by the gravitational instability introduced by the low density of this layer. From a model of the upper crust very similar to that derived for the Rhinegraben (RAMBERG (1972) calculates the dominant wavelength which develops during the uprise of the low density material to be 4.5 times the thickness of the low density layer. Thus a wavelength of about 36 km corresponds to a thickness of 8 km of the low velocity channel. This is in accordance with the width of the Rhinegraben and with the thickness of the low velocity layer. Noticeable movements of the upper surface of the low velocity layer occur already after 10 Ma.

The yielding of the layer of low viscosity concentrates the stress in the more viscous brittle layer in the upper crust. When the tensile strength is reached a fault develops in the brittle part of the crust — most likely in an old zone of weakness — whereupon the creep rate and in turn the necking increases in the low viscosity layer (Fig. 2b). The fractured upper crust is bent down into the neck until finally a second fault develops and the fractured graben wedge is sliding (Fig. 2c) into the opening between the two uplifted flanks of the graben formed by the metamorphic basement layer and the sediments on top. Now, these flanks can yield to the horizontal tension rapidly. Compared to other models of graben formation there are a number of important differences. The sinking graben segment is only formed by the basement layer and not by the whole crust. This segment sinks into the less dense material of the low velocity layer. The buoyancy of the graben segment is less than its weight. Therefore, the segment moves downwards as long as the gap between the two flanks is not closed. It will start its downward motion again whenever new gaps develop by a sideward motion of the flanks. It should be noted that in this model, up to this stage, the driving force of taphrogenesis is the horizontal tension and the gravitational instability caused by the low velocity layer. An uplift of mantle material is not yet required.

As soon as the gap between the uplifted basement flanks has been closed by the sinking segment, an arch is formed by the two basement flanks and the segment. The span of this arch is about 150—200 km. Below this arch the lower crust is released from the pressure of the basement and sediment layer. If the arch were totally supported by the abutments the pressure release would amount to 3.1 kbar (2 km of sediments $\varrho = 2.5$ and 10 km of basement $\varrho = 2.75$ g/cm³), i. e. about 40 % of the whole crustal load. However, part of the arch weight is also supported by its buoyancy within the low density ($\varrho = 2.67$ g/cm³) channel. Therefore, the whole pressure of the overburden of the channel ist not released. But even part of

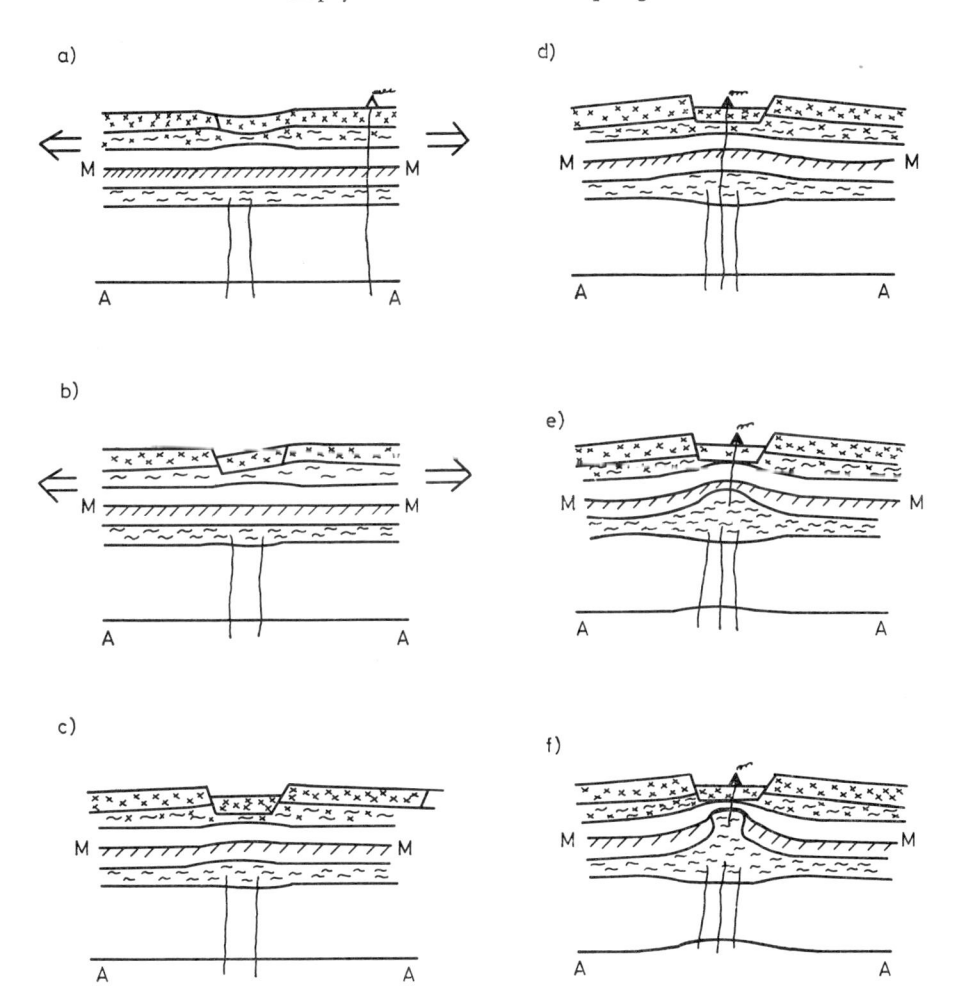

Fig. 2. Graben formation by crustal necking (see text). M - M: Crust-mantle boundary; A - A: Asthenosphere.

a) A neck is formed in the low velocity, low density, low viscosity layer of the sialic crust by horizontal stress.
b) The brittle upper crust fractures and is bent into the neck.
c) A second fault develops, the fractured graben wedge slides into the opening between the two flanks; pressure reduction by arching of the uppermost crust.
d) Pressure reduction invokes growing of the thickness of the low velocity layer in the lower lithosphere and thus generates elevation of the crust-mantle boundary.
e) This causes a regional uplift.
f) Material from the lower lithosphere has penetrated the lower crust and prevents the graben wedge from further sinking.

this pressure reduction could facilitate the uprising of hot, light magma from the upper mantle. The pressure release could lead to a lowering of melting temperature and thus invoke partial melting. It is suggested that the thickness of the low velocity channel in the lower lithosphere is increased and therefore the crust-mantle boundary elevated below the span of the arch (Fig. 2 d).

From now on the tectonic activity in the region of the graben is governed by the mass and energy transport from the upper mantle. The raising of the crust-mantle boundary causes a regional uplift of the graben area (Fig. 2 e). As soon as gaps develop between the flanks either by sideward movement of the abutments or by uplifting of the whole crust, the graben segment is free to continue its subsidence into the low velocity layer since no floating equilibrium can be reached. The gaps are closed when the arching is reestablished.

In Fig. 2 f a possible future stage of the graben is displayed. The low velocity material from the lower lithosphere has penetrated the original crust-mantle boundary. An intermediate velocity of 7.6—7.7 km/sec will now be observed within the graben. Crustal updoming continues. As soon as the low velocity channel is squeezed away the graben segment cannot sink below its floating position. Therefore, on further sideward movement of the flanks the gaps cannot be closed any more and volcanic activity will increase strongly in the region of the graben.

The model of graben formation by crustal necking attempts to answer the question as to why the updoming of the crust-mantle boundary develops below the old lineament of crustal weakness. The new model does not exclude magmatic activity from the asthenosphere prior to graben formation. This must be attributed to a general heating of the asthenosphere during the new Mesozoic cycle of plate tectonics which finally led to the collision of Africa and Europe. Precursors of this rising magma penetrate the lower lithosphere and the crust and start the first volcanic activity 95—65 Ma ago in the region of the Rhinegraben (LIPPOLT et al., this vol.). The main magma reaches the zone of low velocity in the lower lithosphere raising its temperature. The plate collision sets the European plate into a state of stress where tension normal to the graben axis dominates. This horizontal stress field led to the formation of a neck, updoming of the flanks and to the beginning of rifting around 45 Ma. The pressure release by arching of the uppermost crust facilitates the growing of the low velocity layer in the lower lithosphere either by additional melting, phase changes or increased supply of magma. This in turn causes an uplift of the crust-mantle boundary and the crust.

Conclusions

This paper has focussed on two important features of the Rhinegraben: the nature of the updoming of the crust-mantle boundary and the role of the sialic low velocity layer during the initial stage of graben formation. A number of models has been proposed for the uplift of the crust-mantle boundary which could explain the absence of a positive gravity anomaly. These models must be regarded as hypothetical and should be tested by further investigations. One of the most important experiments is a reversed N—S-refraction profile on the Eastern flank of the graben to determine the velocity of the uppermost mantle outside the graben proper and to test the possibility of the low velocity channel in the lower lithosphere.

The author appreciates many helpful discussions with Prof. ILLIES and with his colleagues at the Geophysical Institute. Drs. D. BAMFORD and C. PRODEHL kindly read the manuscript.

Literature

ANSORGE, J.; EMTER, D.; FUCHS, K.; LAUER, J. P.; MUELLER, ST. & PETERSCHMITT, E., 1970: Structure of the Crust and Upper Mantle in the Rift System around the Rhinegraben. — In: ILLIES, H. & MUELLER, ST.: Graben Problems, Schweizerbart, Stuttgart, 190—197.
ANSORGE, J.; EMTER, D.; FUCHS, K.; LAUER, J. P.; MUELLER, ST.; PETERSCHMITT, E. & PRODEHL, C., 1973: Seismic-refraction investigations in the central and southern part of the Rhinegraben. — In: GIESE, P. & STEIN, A.: Results of deep seismic sounding in the Federal Republic of Germany and some other areas (in press).
ARTEMJEV, M. E. & ARTYUSHKOV, E. V., 1971: Structure and isostasy of the Baikal rift and the mechanism of rifting. — J. Geophys. Res., 76, 5, 1197—1211.
BAMFORD, D., 1973: Refraction data in western Germany — a time-term interpretation. — Z. Geophys., 39 (in press).
BIRCH, F., 1961: The velocity of compressonal waves in rocks to 10 kilobars, 2. — J. Geophys. Res., 66, 219.
BONJER, K.-P. & FUCHS, K., 1970: Crustal Structure in Southwest Germany from Spectral Transfer Ratios of Longperiod Bodywaves. — In: ILLIES, H. & MUELLER, ST.: Graben Problems, Schweizerbart, Stuttgart, 198—206.
BUCHER, W., 1933: Deformation of the earth crust. — London, 518 p.
CLOSS, H. & PLAUMANN, S., 1967: On the gravity of the upper Rhinegraben. — Abh. geol. Landesamt Baden-Württ., 6, 92—93.
— 1968: Gedanken zur Tektonik der Kruste im Oberrheingraben aufgrund von Schweremessungen. — Geol. Jb., 85, 371—382.
DOHR, G., 1970: Reflexionsseismische Messungen im Oberrheingraben mit digitaler Aufzeichungstechnik und Bearbeitung. — In: ILLIES, H. & MUELLER, ST.: Graben Problems, Schweizerbart, Stuttgart, 207—218.
EMTER, D., 1971: Ergebnisse seismischer Untersuchungen der Erdkruste und des oberen Erdmantels in Südwestdeutschland. — Dissertation, Univ. Stuttgart, 108 p.
FORSYTH, D. W. & PRESS, F., 1971: Geophysical tests of petrological models of the spreading lithosphere. — J. Geophys. Res., 76, 7963—7979.
GERKE, K., 1957: Die Karte der Bouguer-Isanomalen 1 : 1 000 000 von Westdeutschland. — Dt. Geodät. Komm., Reihe B, 46, Teil I, 13 p.
HAAK, V. & REITMAYR, G. (this vol.): The Distribution of Electrical Resistivity in the Rhinegraben Area as Determined by Telluric and Magnetotelluric Methods.
HAENEL, R., 1971: Heat flow measurements and a first heat flow map of Germany. — Z. Geophys., 37, 975—992.
HIRN, A. & PERRIER, G. (this vol.): Deep Seismic Sounding in the Limagne Graben.
HIRN, A.; STEINMETZ, L.; KIND R. & FUCHS, K., 1973: Long range profiles in western Europe: II. Fine structure of the lower lithosphere in France (Southern Bretagne). — Z. Geophys., 39, 3, 363—384.

ILLIES, J. H., 1972: The Rhinegraben rift system — plate tectonics and transform faulting. — Geophys. Surv., **1**, 27—60.
— (this vol.): Taphrogenesis and Plate Tectonics.
LANDISMAN, M. & MUELLER, ST., 1966: Seismic studies of the earth's crust in continents; part II: analysis of wave propagation in continents and adjacent shelf areas. — Geophys. J. R. A. S., **10**, 539—554.
LIPPOLT, H. J.; TODT, W. & HORN, P. (this vol.): Apparent potassium argon ages of Lower Tertiary Rhinegraben volcanics.
MAKRIS, J., 1971: Aufbau der Kruste in den Ostalpen aus Schweremessungen und die Ergebnisse der Refraktionsseismik. — Dissertation, Univ. Hamburg, 65 p.
MEISSNER, R.; BERCKHEMER, H.; WILDE, R. & POURSADEG, M., 1970: Interpretation of Seismic Refraction Measurements in the Northern Part of the Rhinegraben. — In: ILLIES, H. & MUELLER, ST.: Graben Problems, Schweizerbart, Stuttgart, 184—190.
MEISSNER, R. & VETTER, U. (this vol.): The Northern End of the Rhinegraben due to some Geophysical Measurements.
MUELLER, ST. & LANDISMAN, M., 1966: Seismic studies of the earth's crust in continents; part I: evidence for a low velocity zone in the upper part of the lithosphere. — Geophys. J. R. A. S., **10**, 525—538.
MUELLER, ST. & PETERSCHMITT, E., 1966: Die Geschwindigkeitsverteilung seismischer Wellen im tieferen Untergrund um den Oberrheingraben. — DFG-Kolloqu. Oberrheingraben, Wiesloch.
MUELLER, ST.; PETERSCHMITT, E.; FUCHS, K. & ANSORGE, J., 1967: The rift structure of the crust and the upper mantle beneath the Rhinegraben. — Abh. geol. Landesamt Baden-Württ., **6**, 108—113.
MUELLER, ST., 1968: Low velocity layers within the earth's crust and mantle. — 10th General Assembly Europ. Seismol. Comm. (Leningrad).
— 1970: Geophysical Aspects of Graben Formation in Continental Rift Systems. — In: ILLIES, H. & MUELLER, ST.: Graben Problems, Schweizerbart, Stuttgart, 27—37.
MUELLER, ST. & RYBACH, L. (this vol.): Crustal Dynamics in the Central Part of the Rhinegraben.
RAMBERG, H., 1972: Theoretical models of density stratification and diapirism in the earth. — J. Geophys. Res., **77**, 877—889.
RHINEGRABEN RESEARCH GROUP FOR EXPLOSION SEISMOLOGY (this vol.): The 1972 Seismic Refraction Experiment in the Rhinegraben. — First Results.
SCHEELKE, I. (this vol.): Models of the Resistivity Distribution from Magneto-telluric Soundings.
SCHULT, A. (this vol.): The Electrical Conductivity of Minerals and the Temperature Distribution in the Upper Mantle.
STROBACH, K. (this vol.): Model of Crustal Stresses in the Region of the Rhinegraben and Southwestern Germany.
VENING-MEINESZ, F. A., 1950: Les «graben» africains, résultat de compression ou de tension dans la croûte terrestre? — Bull. Inst. Roy. Colonial Belge, **21**, 539—552.
WERNER, D., 1970: Geothermal Anomalies of the Rhinegraben. — In: ILLIES, H. & MUELLER, ST.: Graben Problems, Schweizerbart, Stuttgart, 121—124.
WERNER, D. & DOEBL, F. (this vol.): Eine geothermische Karte des Rheingrabenuntergrundes.
WINTER, R. (this vol.): A Model of the Resistivity Distribution from Geomagnetic Depth Soundings.

Taphrogenesis and Plate Tectonics

by

J. H. Illies

With 13 figures in the text and on 3 folders

The crust of Central and Western Europe is slashed by a system of rift valleys, fault troughs and fault scarps. Some of these rupture zones have been active during the Quaternary and moderate seismic activity indicates the continuance of crustal mobility up to Recent times. The Eurasian macro-plate reveals instability in the gusset between the power-fields of the Atlantic ocean and the Alpine orogenic belt. Active sub-plate boundaries traverse the foreland of the Alps and permit a restrained freedom to the individual micro-plate units for relative motions and deformations (VOIGT 1954).

The most spectacular scissure intersects the European crust from the Netherlands via Cologne, Mainz and Basel, from there, after a shift to the west, the rift system continues southward along the Bresse and Limagne depressions, and the graben zone of Alès in Southern France. The whole feature with its accompanying graben splays and appendant fault troughs and fracture zones is called the Rhinegraben rift system. The author (ILLIES 1972) made a first attempt to synthesize the structure and development of this rift system from the point of view of the plate tectonics hypothesis. Much new data and many ideas have been presented by several authors of this book and elsewhere. They justify the effort for a more wide-ranging synopsis from the aspect of plate tectonics. The concept will be presented in the way of elucidating captions for a sequence of figures drawn by the author.

Fig. 1 illustrates both the thickness distribution of the sedimentary fill of the Rhinegraben and the contours of simultaneous shoulder uplift. The isopach lines indicating the thickness of the Eocene to Recent graben fill demonstrate an asymmetrical pattern in transverse as well as in axial directions. As shown by DOEBL & OLBRECHT (p. 72) the trough axis in most parts of the graben comes closer to its eastern rim. Only near both graben ends the axis of maximum subsidence follows the western margin. This asymmetry has been explained by ILLIES (1972) as a preferred clockwise rotation of the graben segment in consequence of an enhanced yielding of the plate area west of the graben during the taphrogenic dilatation. The south—north asymmetry with a maximum thickness near the northernmost part is explained by gradual shifting of the center of enhanced subsidence from the southern part to the northern end during the graben development and a persistent subsidence in the northern part. Thus the total amount of subsidence dominates the northern half of the graben, whereas the frames of the southern graben area are subjected to a preferred shoulder uplift with a culmination at both flanks of the southernmost end of the graben. The shoulder elevation

Fig. 1

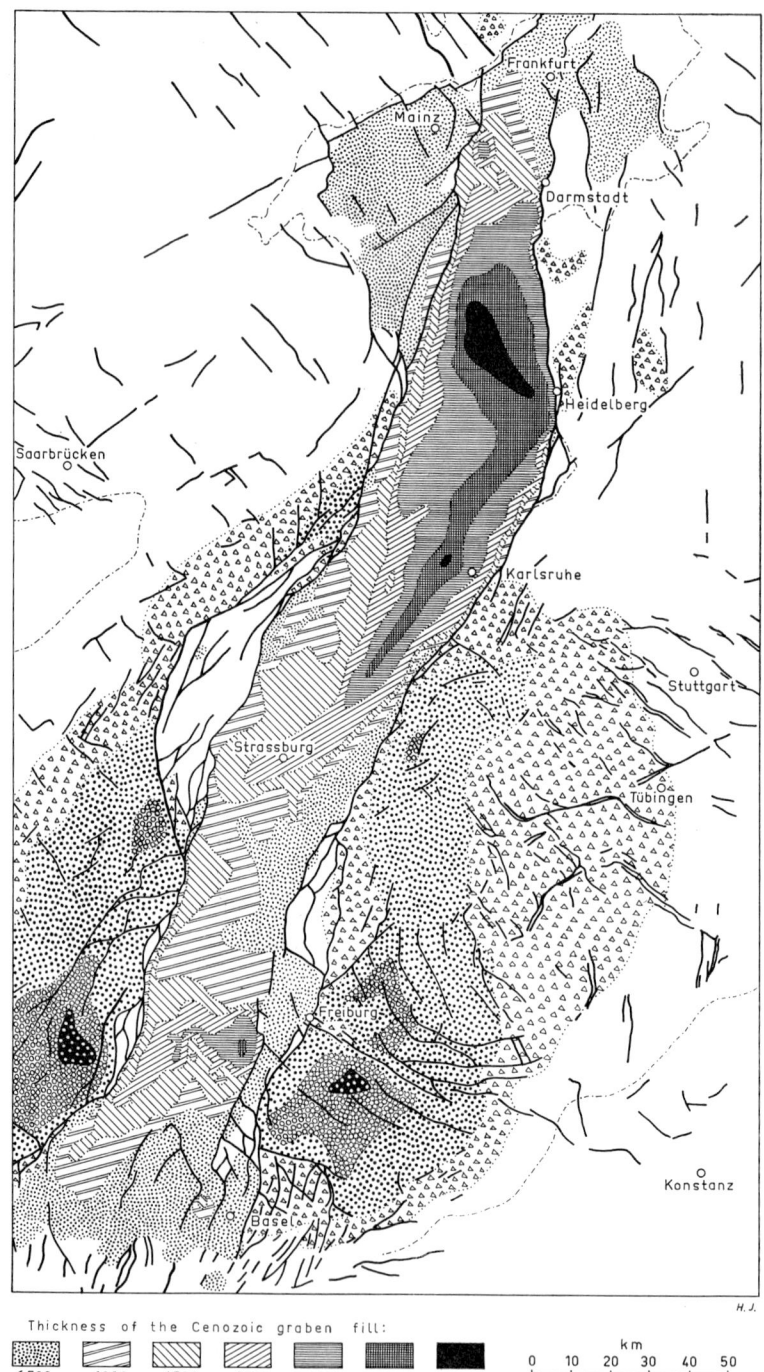

Thickness of the Cenozoic graben fill:
<500 −1000 −1500 −2000 −2500 −3000 >3000 m

0 10 20 30 40 50 km

Post Cretaceous elevation of the graben shoulders:
500 m − isobase <1000 −1500 −2000 −2500 >2500 m

is expressed by isobases (i. e. lines of equal uplift) as referred to the pre-Eocene peneplain. The Black Forest in the east and the Vosges mountains in the west form a bilateral-symmetric feature with corresponding degrees of shoulder elevation. The Rhinegraben is engraved into the vertex line of a largescale crustal vault. The author has mentioned in an earlier paper (ILLIES 1972) that the upwarping of the shoulders as well as the contemporaneous processes of rifting and graben subsidence were induced and controlled by the upsurging of a subcrustal swell of mantle material. As a consequence of the increase of crustal updoming towards the southern end of the graben, it seems likely that the mantle swell may be interpreted as in counterbalance to the Alpine subduction zone which follows immediately south of the graben (GOGUEL 1957).

Fig. 2. Graben formation in its initial Eocene stage did not occur uniformly over the entire area from Basel to Francfort. The crust was ripped up successively from the south to the north inducing a wandering of shoulder elevation and rift valley subsidence in the same direction. The thickness distribution of the Rhinegraben sedimentary fill is illustrated by a generalized longitudinal section. It shows a gradual wandering of the maximum subsidence center from the south in the Eocene to the northern end of the trough in the Plio-Pleistocene. In the same way, the recent crustal movements culminate near the northern end of the graben area (SCHWARZ, p. 262).

The Rhinegraben sedimentary cycle started 45 million years ago during the Middle Eocene with marly deposits of freshwater lakes. Contemporaneously with the sedimentation of the Upper Eocene *Lymnaea* marls, the graben trough developed now well-defined contours while the subsiding graben floor was covered by a brackish lagoon as a consequence of an episodic marine ingression from the Alpine geosyncline. During the Lower Oligocene, when the Pechelbronn beds were deposited, taphrogeosynclinal processes affected nearly the whole trough. The subsidence center with approximately 1600 m of sediments temporarily remained in the southern part. During the Middle and the Upper Oligocene the maximum subsidence center had shifted to the middle part of the graben where more than 1000 m of shaley to sandy marls were accumulated. This sequence is divided in the Grey Beds series and the Niederroedern beds, deposited partly under marine and partly under brackish to freshwater conditions. With the beginning of the Lower Miocene (Aquitanium) the subsidence in the southern graben area slackened, whereas the center of taphrogeosynclinal action was still moving farther northward, now with a focal point in the Mannheim—Heidelberg area where 1600 m of sediments have been deposited. That was the period when the last marine ingressions occurred, probably coming from the Lower Rhine embayment. During the Upper Miocene and Lower Pliocene the subsidence discontinued as is indicated in some localities by a fluvial erosion of the graben floor. With the beginning of the Upper Pliocene taphrogenesis became active again and the sedimentary accumulation, now under a fluvial regime, is concentrated in the northernmost segment of the graben where Upper Pliocene and Pleistocene strata attain a maximum thickness of 1000 m. The recent subsidential and seismo-tectonic activi-

436   J. H. Illies

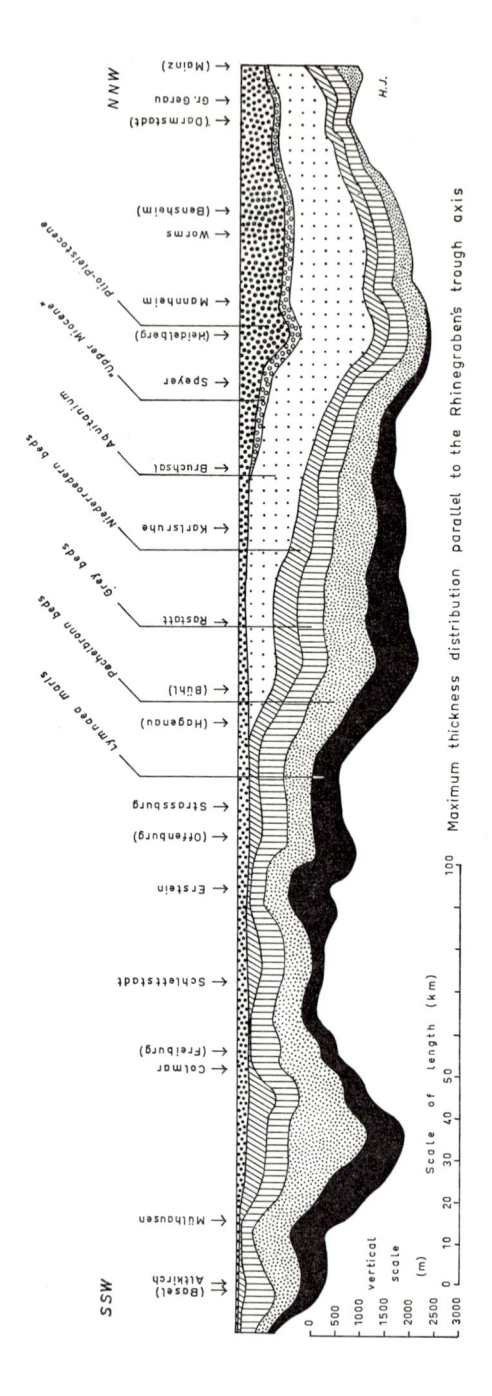

Fig. 2. Maximum thickness distribution parallel to the Rhinegraben's trough axis

ties are concentrated in the same area. However, additional subsidence started again during Middle Pleistocene time in local basins of the southern graben area, especially southwest of Freiburg and south of Strasburg (for details, see BARTZ, p. 80).

Fig. 3. North of Mainz the Rhinegraben is barred by the Hercynian mountain range of the Taunus hills. With a slight eastward offset, however, the fault trough

Fig. 3

continues north of Francfort into the small Nidda and Horloff grabens which link near Giessen to the Hessen depression. This graben segment combines the Rhinegraben with the Tertiary North Sea basin in the area of Hanover. The Hessen depression follows, like the Rhinegraben rift system, an old zone of weakness in the Hercynian basement which has been reactivated in the Upper Permian and Lower Triassic times, already forming then a narrow epeirogenic trough (BOIGK & SCHÖNEICH, p. 64). The same trough element, during its latest revival in Eocene/Oligocene times, became a northern extension of the Rhinegraben rift belt. The Rhinegraben lagoon continued into the Hessen depression, though with reduced sedimentary thickness and local facies changes. During the Middle Oligocene Rupel stage a narrow graben sea united the waters of the Alpine molasse foredeep in Switzerland and the enlarged North Sea basin in Northern Germany; a Red Sea in statu nascendi had been formed splitting Western Europe apart from the rest of the continent (Fig. 3). During the late Oligocene the subsidence of the Hessen depression decreased and the flat lagoon dried up. Only the Nidda graben, the Horloff graben and the Amoeneburg basin, situated in the southernmost part of this rift segment, reveal subsidence through Neogene and Pleistocene times. Thus, the marine ingression inundating the Rhinegraben trough during Neogene time, took another route and came probably from the Lower Rhine embayment.

Fig. 4. Near its northern end the Rhinegraben reveals a trumpet-like widening. The inner trough becomes flanked by wide areas with a flat Tertiary and Pleistocene cover: the Mainz basin in the west and the Hanau basin in the east. The inner trough continues north-northeastward to the Hessen depression, its subsidence had started nearly contemporaneously with the Rhinegraben in late Eocene times. However, since during end-Oligocene time the subsidence decreased and major parts of the Hessen depression became mainland, another branch of the rift system, the Lower Rhine embayment, enhanced its taphrogenic activity. The fault pattern of this rift segment (R. TEICHMÜLLER, p. 281) follows a bundle of cross fractures, normal to the fold structures of the Hercynian basement. This old zone of weakness caused epeirogenic subsidence during various stages of the Mesozoic and Paleogene. In early Miocene the activity of subsidence increased and the original normal faulting changed its character to dip-slip movements with rotational tendencies of the tilting single blocks. A strong neotectonic activity continued until Recent time forming a system of fault scarps which crosses and displaces the Pleistocene terrace plains of the Rhine river.

The Lower Rhine embayment extends northward in the Dutch Central Graben and further north in a fault trough system which traverses the central North Sea basin (R. TEICHMÜLLER, p. 270). Its southern extension, however, ends near Bonn. The rift belt connecting the Lower Rhine embayment with the Rhinegraben seems interrupted by the Rhenish Schiefergebirge mountains. Only the intervenient small basin of Neuwied appears as a missing link and its Upper Oligocene to Lower Miocene and Pleistocene sedimentary fill indicates paleogeographic interrelations between both major rift segments. The Hercynian basement, however, remained predominantly stable along the intervening segment. The only physiographical

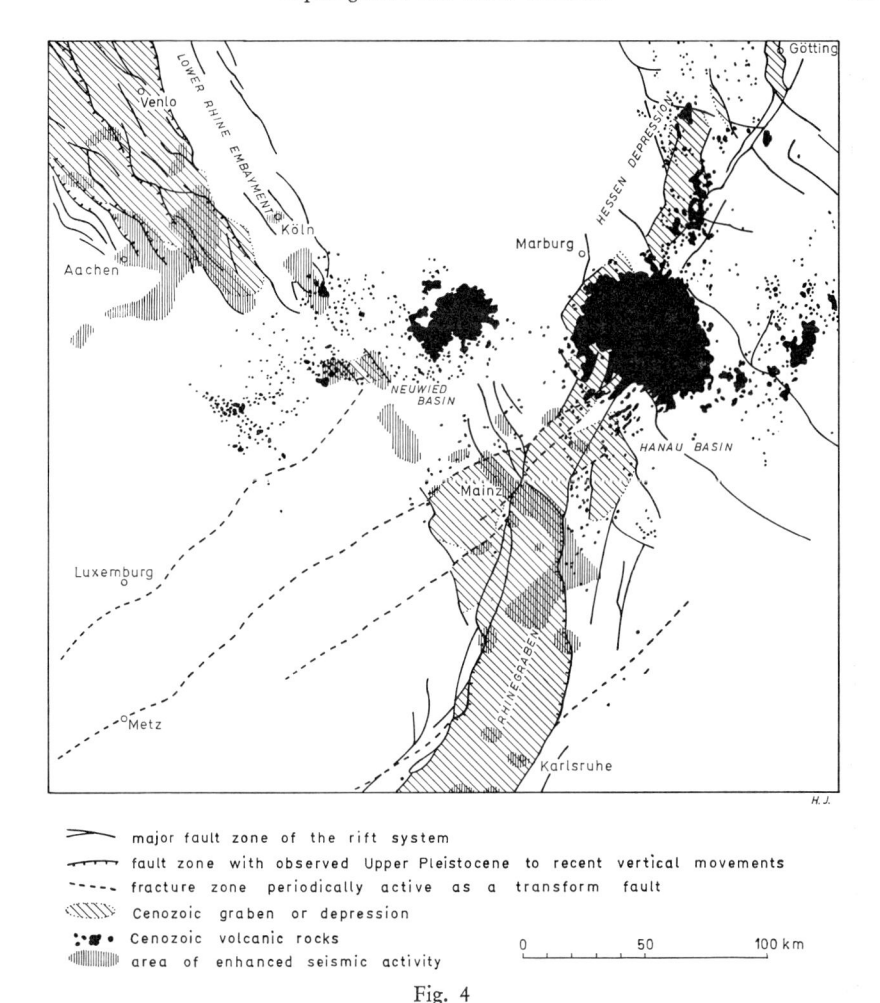

```
         major fault zone of the rift system
         fault zone with observed Upper Pleistocene to recent vertical movements
         fracture zone periodically active as a transform fault
         Cenozoic graben or depression
         Cenozoic volcanic rocks
         area of enhanced seismic activity
```

Fig. 4

reaction we can observe, is an axial post-Pliocene uplift of about 200 m encouraging the Rhine river to erode its deep epigenetic canyon between Bingen and Bonn.

The structural map offers the impression that the Rhinegraben, the Hessen depression, and the Lower Rhine embayment join at a triple rift junction with a triple point some twenty kilometers south of Mainz near the town of Oppenheim. Neotectonic activity, however, is confined to the western, still incomplete branch. A nearly uninterrupted belt of historical seismic epicenters forms a transition from the Rhinegraben to the Lower Rhine embayment. No seismic event has been recorded from the Hessen depression so far. A westward shift of rift activity is additionaly indicated by a corresponding displacement of volcanism. The rift belt is crossed by a transverse zone of enhanced Neogene volcanism. According to DUNCAN et al. (1972) the climax of volcanic activity is younger the more westerly the individual centers have been found. The Vogelsberg at the southern end of the Hessen depression is the largest neovolcano of Central Europe. Its activity culminated in the Upper Miocene. The east—west volcanic trace is then indicated by the Westerwald lava field which is situated between Hessen depression and Neuwied basin. A Pliocene age of this volcanism had been ascertained. West of a Pleistocene volcanic center in the Neuwied basin we find the Eifel maar crater province, the last eruptions of which took place about 10,000 years ago. An east wandering of the crust over a stable mantle plume seems possible.

Fig. 5. As mentioned before, the formation of the Mainz basin is the result of two phenomena, the trumpet-like widening of the northern end of the Rhinegraben and the (theoretical) triple junction of the Rhinegraben with its northern continuations, the Hessen depression and the Lower Rhine embayment. The present structural map is based on previous publications of ANDERLE (1970), KUBELLA (1951), SONNE (1970), STRAUB (1962) and many others. It illustrates a grating and interference of different fault systems. The master faults of the inner graben bifurcate north of Darmstadt and Oppenheim. The general trend, however, continues north of the Main river with minor faults framing the Nidda and the Horloff grabens north of Francfort. Another fault system with a northeasterly strike separates the Mainz basin from the Taunus mountains, traversing the towns of Bingen, Wiesbaden, Homburg and Nauheim. To the south the Alzey fracture zone runs almost parallel, and to the northeast it continues to the area of the Main trap lava fields and the Vogelsberg neovolcano. The Upper Miocene Vogelsberg extrusions are lineated along fissure systems of the same trend. There is a third fault system preferring a northwestern or north-northwestern direction. This is the prevalent trend of the internal block tectonics of the inner graben along which a major fracture zone crosses the graben rim near Ruesselsheim and forms the narrow graben splay of Idstein which divides the Taunus mountain range.

The most spectacular intersection of the different fracture zones can be observed in the area of Ruesselsheim where considerable recent crustal movements caused damages to buildings founded on the active fault zones (see SCHMITT, p. 257). South of Ruesselsheim, in the area of Gross Gerau, the neotectonic activity is supported by precise levellings indicating rapid vertical movements (see SCHWARZ, p. 264). The town of Gross Gerau and its vicinity is characterized by a cluster of epicenters of the most conspicuous earthquake swarm which occurred in the Rhinegraben area during historic times (LANDSBERG 1931). The seismic activity in this area started in January 1869 when several slight earthquakes were felt in the town of Darmstadt. The first event in Gross Gerau was recorded on October 24 of the same year. After this event a series of earthquakes was registered many times a week and often daily, sometimes with damages to houses. From February 11, 1871 on, the local events began to move southeastward to the area between Darmstadt and Reichenbach, whereas the activity in Gross Gerau decreased and finally came to an end on October 14, 1871. During the next years the activity relaxed but remained concentrated around the Odenwald fault scarp and has been especially felt in the village of Reichenbach. Only a few events had touched Gross Gerau during this interval. From mid-February 1874 on, the events became more seldom until January 10, 1877 when the last earthquake of the swarm was registered. Some events during the last months of the activity have been especially observed in the village of Lindenfels, the southernmost point along the trace, which has been touched by a swarm of very local and flat-seated earthquakes. The phenomenon may be explained by strain accumulation in the tectonic unit of the northern graben end and may also be a consequent progressive displacement combined with stress transmission along an active fault zone oblique to the graben.

The major part of the Cenozoic rift faults in this area followed the framework of joint systems and directional instability in the Hercynian basement. The northeasterly strike corresponds to the fold axes and the limitations of lithological units within the Hercynian fold belt. The northwestern system followed the crossjointing normal to the linear structures. The northnortheast orientated main faults of the graben resulted from shear or strike-slip elements of the same Hercynian orogeny. But rifting did more than to revive or rejuvenate a pre-existent pattern of crustal instability. Only such zones of weakness revived which were adequate to the taphrogenic kinematics. Others remained stable and most of them have been reactivated in another or opposite sense of their original movement. The crustal instability and anisotropy facilitated and influenced graben formation and development, but the tectonic regime and style of taphrogenesis has been grossly different.

Fig. 6. MEISSNER & VETTER (p. 242) have drawn the contours of the subcrustal mantle swell underneath the northern end of the Rhinegraben. According to their results a considerable flattening of the "cushion" surface towards the northern end of the graben is evident compared with the southern part. However, no buckling of the Moho has been observed being consistent with the surficial fault scarp which separates the northern end of the Rhinegraben from the Taunus mountain range with the Feldberg massif. The crest area of the "cushion" is characterized by two kinds of morphological surface appearances: by block units with a remarkable tendency of Plio-Pleistocene elevation (KUBELLA 1951), e. g. Feldberg, Koehler Berg and Nierstein horst, and by zones with a preferred down-faulting along the western master fault of the inner graben. An asymmetrical arrangement of the graben axis relative to the crest line of the "cushion" seems to be obvious. If we imply that the formation of the subcrustal mantle swell has induced both, crustal spreading in terms of graben formation and crustal warping in terms of shoulder uplift (ILLIES 1972), we may recognize a contrary behaviour north and south of the fracture zone which separates the Rhinegraben from the Taunus mountains. South of this scarp there is the graben field with its characteristical dip-slip and tilted block tectonics. Here an amount of lateral dilatation of about 5 km has been observed. This kind of crustal spreading is favoured by a sideward gravity slide of the flanking plate units which have been uplifted by the upsurging "cushion" (ILLIES 1972). North of the Taunus fault scarp the crust remained more or less stable, only narrow tensional features like the Idstein graben are visible. The crust, here consisting of Paleozoic quartzites and shales, had reacted coherently and

442   J. H. Illies

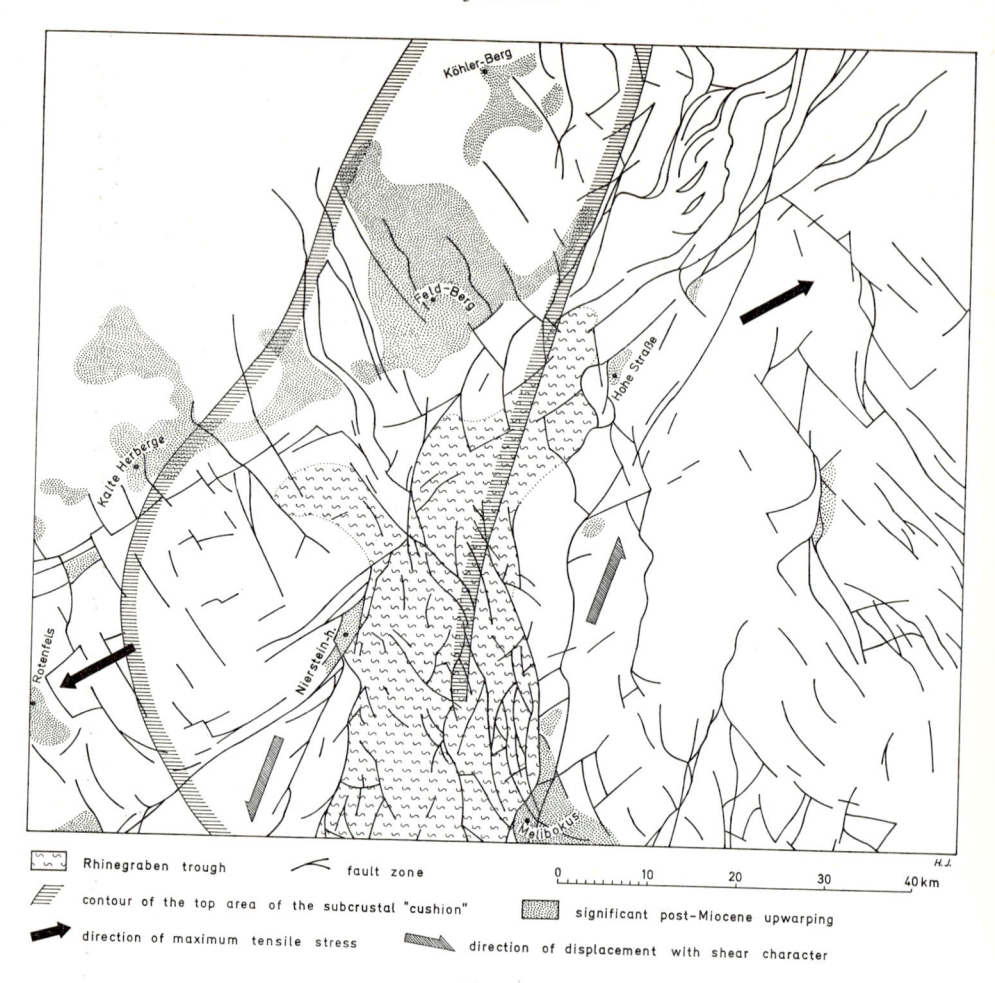

Fig. 6

followed passively the upsurging tendency of the mantle swell underneath. Crustal spreading contra crustal stability in neighbouring crustal units presumes mobile zones which transmit like slide bars the opposing tendencies of movement of the involved crustal units corresponding to transform faulting in terms of the related ocean-floor structures. The present author (ILLIES 1972) has just mentioned the transform fault nature of the transversal structures which intersect the northern end of the Rhinegraben. ANDERLE (p. 249) has observed horizontal slickensides along these fracture zones made by dextral strike-slip movements. The fault breccia of the same fracture movements has been already described by GOETHE in 1817, who found rocks of the quartzitic breccia at the fault scarp near Bingen (see W. WAGNER 1926).

For a better understanding of the observed tectonic movements we have inserted in the map the relevant components of the regional stress field calculated by AHORNER & SCHNEIDER (p. 107) by means of fault-plane solutions of local earthquakes. According to these investigations the axial direction of the Rhinegraben corresponds to the active s_2 shear component with a sinistral strike-slip tendency. The conclusion of AHORNER & SCHNEIDER agrees with geological investigations (ILLIES 1962). It also explains the NNW-striking internal ruptures of the inner graben which frequently reveal a sigmoidal trend and are formed in consequence of the observed shear movements comparable to feather joints in the realm of micro-tectonics. The direction of minimum stress or maximal dilatation is orientated obliquely to the graben. It corresponds to the northeast—southwest fracture zones of the Taunus scarp and of Alzey.

The axial directions of the regional stress field deduced from the symptoms of the neotectonic taphrogenesis correspond almost (not exactly) to the stress field under which the Hercynian fold belt has developed during the Carboniferous. The approximate congruence of the stress directions of the Cenozoic (Alpine) tectonics with the Paleozoic (Hercynian) movements seems significant for some parts of Central Europe. Under the contrary signs of taphrogenesis the pre-existent fault patterns of orogenesis have been used again, but the resulting deformations are grossly different.

Fig. 7. In the same way as the Rhinegraben extends northward, there is also a continuation of the rift system to the southern direction, e. g. the Bresse and the Limagne grabens. The Rhinegraben s. s. is 300 km long and ends in the south near Basel where the graben is crossed by the Swiss Jura mountain range. The Pliocene fold axes have plugged the southern opening of the graben which is called the Sundgau basin. The continuing rift segments seem to escape the front of the nappe units of the Western Alps: the next rift segment, the Bresse graben, appears with an westward offset of about 150 km at nearly the same latitude. But the subsidence and tectonic activity of the Bresse graben had slackened since the early Miocene. The eastern fault scarp, formerly a system of dip-slip faults, has been overthrusted by westward moving quasi-nappes or thrust sheets of the Jura fold belt (LEFAVRAIS-RAYMOND 1962). One hundred and fifty km more to the west an analogous fault trough sets in, the Limagne graben (VITARD 1969). There is, however, no recent tectonic activity within this fault trough. This may be observed along the western shoulder of the graben where the contiguous chain of Pleistocene to sub-Recent volcanoes of the Chaîne des Puys as well as a moderate seismic activity are obvious.

It seems evident that the lateral shift between the Rhinegraben and its southern extensions is marked by a transverse tectonic element or more precisely by two parallel traces of it. The transverse traces are distinguished by fault zones as well as by basin-like widenings and corresponding narrowings of the three grabens. The northern trace begins with the Swabian lineament east of the Rhinegraben, crosses the Rhinegraben at the height of the Kaiserstuhl volcano, cuts off the Bresse and Limagne grabens at their northern end and terminates at the Montluçon basin. The

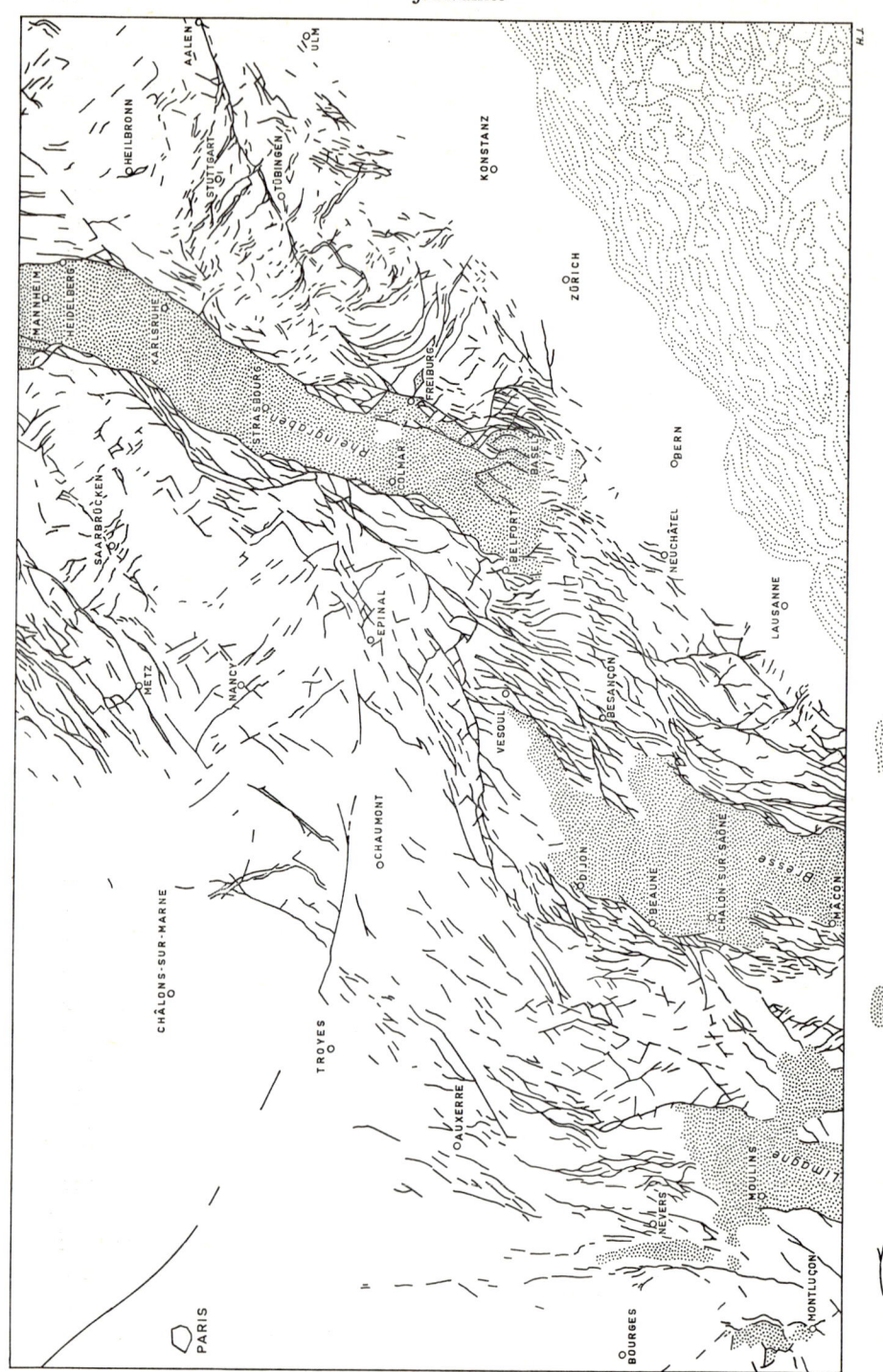

Fig. 7

southern trace is characterized by the Sundgau basin near Belfort, the corresponding widening of the Bresse between Besançon and Chalon and in the Limagne graben area by corresponding embayments near Charolles and Gannat. It seems reasonable to interpret the transverse structures as transform faults, a problem we will discuss further below. Disregarding the transform fault concept, it is evident that the fault patterns along the shoulder areas of the three grabens are strongly related to mutual plate movements which we must presume to explain crustal spreading along the axes of the concerned graben segments. Fault directions, relative displacements along the individual faults, and fault pattern densities make it possible to analyze the plate movements relative to the rift system and its offset along the transverse elements. Beyond that, it should be mentioned that the fault patterns controlled by graben tectonics can be followed for 150 km into the framing plate units. These units have not acted as totally rigid blocks, but differentiated displacements along a dense framework of ruptures have enabled the marginal plate areas for a more or less mobile reaction to the dilative movements caused by the rift system.

Fig. 8. This map shows the Cenozoic fault pattern and the distribution of Cenozoic volcanics in Southwestern Germany, drawn by the author in accordance with many published and unpublished geological maps. The area comprises the plate unit between the Rhinegraben and the sub-Alpine molasse trough. It is the region of the Black Forest which has been affected by strong shoulder elevation along the rim of the Rhinegraben. The plate upwarped along the fault scarp dips gently with 2—4° away from the graben. But the plate has been additionally affected when downdipping to the Molasse foredeep along the course of the Danube river. The complicated fault pattern also may be influenced by torsional effects caused by the simultaneous attacks along the western rim as the result of the upsurging rift "cushion" and at its non-parallel southern rim by means of downwarping to the Alpine subduction zone (ILLIES 1972).

The most conspicuous fracture zones must be regarded as transverse or transform elements related to the lateral offset of the rift system as already mentioned above. The dominating structure is the Swabian Lineament which combines dip-slip and sinistral strike-slip displacements (SEIBOLD 1951, WIEDEMANN 1968). According to SEIBOLD the fault zone extends to the Rhinegraben near Freiburg where an offset of the fault scarps corresponding to the strike-slip tendency is observed, i. e. the Freiburg embayment of the graben area. The Miocene Kaiserstuhl neovolcano is inserted on the intersection of marginal graben faults and transverse ruptures. Parallel to the Swabian lineament, but only slightly indicated, follows 55 km to the south the rupture zone of Waldshut which branches from the southern end of the Rhinegraben. The southern Black Forest area, bordered by both fracture zones, is traversed by a system of subcircular faults, first described by W. PAUL (1955). The swarm of arched faults which is grouped around a center near the town of Villingen indicates that the affected block unit underwent a state of rotational strain which caused an external solid-block rotation. Farther east the same block unit is traversed by the Hohenzollern graben, a narrow but very complex fault structure. The small graben structure is the site of a dense cluster of earthquake epicenters observed during the last 60 years. It is actually the area of maximum earthquake frequency in the whole Rhinegraben rift system. Some of these events resulted in considerable damage to houses and have been felt in the entire southern part of Germany. According to fault-plane solutions of some of these earthquakes the maximal compressive stress directions is NNW or NW orientated, and active left-lateral strike-slip movements along a NNE striking shear component have been ascertained (SCHNEIDER 1971). Although the depths of foci are shallow, ranging from 2 to 22 km, no active faulting in the calculated direction could be observed. As proposed by the author (ILLIES 1972), the tectonical inventory of the Hohenzollern graben may be interpreted as a feather fracture attributed to the supposed shear movements. This is also in congruence with the displacements calculated by earthquake activity. Subsequent rotational movements of smaller block elements apparently caused the special seismic competency of the fault system concerned.

Northeast of the Hohenzollern graben, still within the same tectonic unit which is bordered northward by the Swabian lineament, is the location of the so-called Swabian volcano. Within a nearly subcircular area of about 40 km in diameter there are 335 olivine-melilite necks and tuff funnels of Miocene age (MÄUSSNEST 1969). As CARLÉ (p. 208) has pointed out, the area is combined with a geothermal anomaly which is stretched along the whole southern rim of the Swabian lineament.

The area of subcircular faulting, the cluster of seismic epicenters around the Hohenzollern graben, the volcanic necks south of Kirchheim, and the geothermal anomaly intervene with both the traces of fault elements transverse to the Rhinegraben, i. e. the Swabian lineament and the Waldshut fracture zone. If we consider that the rate of displacement along both the fault zones is increasing eastward at the Swabian lineament but decreasing in the same direction at the Waldshut fracture zone, a rotational shear between both the fault systems seems evident facilitating a progressive taking-over of the compensatory movements from the southern to the northern trace of transform faulting.

Regardless of the described compensatory movements of the transform fault nature, special displacements and block rotations are superimposed by a regional stress field. Its components seem rather untouched by the majority of the local fault zones. This is shown by the fault-plane solutions of different earthquakes (AHORNER & SCHNEIDER, p. 107) as well as by micro-tectonic observations, especially by the evaluation of horizontal slickensides along major joint systems (ILLIES 1962). According to the analysis of earthquake data, the neotectonic direction of maximum compressive stress is approximately northwest to northnorthwest directed over the entire area delineated in Fig. 8. The corresponding strain ellipse makes it easy to understand most of the observed fault structures induced by shear displacements corresponding to the prevalent stress components.

Fig. 9. From the interpretation of the more detailed fault pattern of the Black Forest rift shoulder we return to large scale reflections on the rift system. For a better understanding of the map compare with Fig. 7 where the geographical names are indicated. If we consider the interrelation of the general trend of the Cenozoic fault zones to the components of the active regional stress field, a slight curving of the major fault zones is evident. It may especially be observed in the transition zone between Rhinegraben, Bresse graben, and Limagne graben where the fault zones accompanying the rift system seem to join the convex bend of the northwestern front of the Alpine orogene. A corresponding directional interdependence of fabric elements between the Alps and its immediate foreland is derived from evaluating the axial directions of horizontal stylolites in the Jura fold belt (PLESSMANN 1972). The younger generation of horizontal stylolites shows a radial orientation normal to the front of Alpine nappe units. The same radial divergency is ascertained if we compare the fault plane solutions of local earthquakes (a) for the plate area east of the Rhinegraben as presented by AHORNER & SCHNEIDER (p. 107) with (b) those of PAVONI & PETERSCHMITT (p. 323) concerning the southern Jura fold belt. A divergency of the direction of maximal stress normal to the curved rim of the Alps seems evident.

The nonhomogeneity of the regional stress field in the Alpine foreland is superimposed by differentiated displacements of plate units induced by crustal spreading along the individual rift segments. For the Rhinegraben an amount of lateral spreading of 4.8 km had been ascertained (ILLIES 1972). Provided that sideward gravity sliding of the framing plate units controlled the subsidence and dilatation of the Rhinegraben, an adequate mechanism must be ascribed to the formation of the corresponding grabens in eastern France. According to fault-plane solutions of earthquakes and fault-pattern analyses of the fault and joint systems, the creeping plate movements followed a northeasterly direction oblique to the axis of the Rhinegraben. If we attribute a comparable stress field to the Bresse and Limagne grabens, we also must expect sideward plate movements from which the lateral space required for rifting can be derived. Consequently, relative plate movements in an opposite sense would be involved inducing transverse features which are indicated by the lateral offset of the three graben segments concerned. The author (ILLIES 1972) proposed to make use of the term transform faults for this kind of displacements in accordance with the corresponding structures of oceanic rift systems. However, contrary to the striking features of oceanic transform faulting, its continental pendant seems to be veiled by the enhanced rigidity and anisotropy of the continental crust. CONTINI & THÉOBALD (p. 310) gave an excellent analysis of tectonic details in the area enclosed between the southern end of the Rhinegraben and the northern end of the Bresse. According to their investigations the observed strike-slip movements are in coincidence with the transform-fault concept. More spectacular than strike-slip faults is a bundle of innumerable tension faults and narrow local grabens with a general trend parallel to the main rift segments. The southern foothills of the Black Forest, the southern slope of the Vosges, the area between Belfort, Vesoul, and Besançon as well as the Mesozoic platform between Dijon and Nevers bordering the Morvan massif are traversed by many small grabens (MÉGNIEN et al. 1971). The individual grabens reveal frequently a sigmoidal configuration, and a staggered or en échelon arrangement is evident. The structures are comparable to the feather joints of micro-tectonics and are related to northeasterly striking traces of shear movements with a sinistral strike-slip tendency. The observed displacements neutralized the strain difference between the northern plate unit drifting away from the Rhinegraben and the southern unit influenced by the dilatation along the Bresse and Limagne grabens.

In addition, contours of the surface of the subcrustal "cushion" are indicated as referred to the seismic refraction data of MEISSNER & VETTER (p. 242) for the northern Rhinegraben area, while for the central and southern part we have used the data of ANSORGE et al. 1972 (the latest results of this team presented in p. 122, could not yet be considered). To outline the buckling of the Moho underneath the Limagne graben we referred to the data of HIRN & PERRIER (p. 338). If we suppose a causal connection between the mantle-derived swells and the formation of this kind of grabens, an asymmetric arrangement of the rift "cushion" relative to the physiographic features of the grabens is indicated. Already, when explaining Fig. 4 we noticed a westward shifting of taphrogenic action, e. g. the tectonic activity of the Hessen depression decreased at the end of the Paleogene while the Lower Rhine embayment received the corresponding function of rifting. Now we observe the same westward displacement concerning mantle features relative to the crust. The question that the ascertained displacement could be interpreted under the scope of a general westward shift of the European plate we have to discuss in the caption for Fig. 12.

Fig. 10. Several authors of this book have confirmed by new observations that the development of the Rhinegraben rift system was accompanied by left-lateral strike-slip movements. There are observations of horizontal slickensides and there is support by fault-plane solutions of earthquakes, though the total amounts of horizontal displacements in most cases are very small. The observed shear or strike-slip movements occurred mostly in a parallel direction to the rift segments, while other oblique faults operate as transform faults. Apart from this, much data on strike-slip faulting in other parts of Central Europe have been published, some of which pertain to the Hercynian basement, others relate to the Mesozoic cover as well as to neotectonic activity. In some cases repeated reactivations of old shear zones or ground-fissures during various tectonic cycles has been observed.

According to the data published by COGNÉ et al. (1966), GOSPODARIČ (1969), ILLIES (1962), LAUBSCHER (1971), PAVONI (1969), SCHMIDT-THOMÉ (1972), TOLLMANN (1969) and many others, we have outlined the distribution of strike-slip faults and those faults which indicate an additional strike-slip component. It is obvious that there is a coincidence of the main fault directions and the relative sense of displacement within the Alpine fold belt as well as outside in corresponding segments of the Alpine foreland. Regional stress interrelations between the orogenic belt and its northern foreland are evident. The pattern of strike-slip faults in the foreland is characterized to the north by a convergence of the major fracture zones: the Sillon Houiller and the Rhinegraben as left-lateral strike-slip elements and the Pfahl and Elbe lines with right-lateral displacements. The described main faults enclose a block area of triangular configuration north of the

Fig. 10

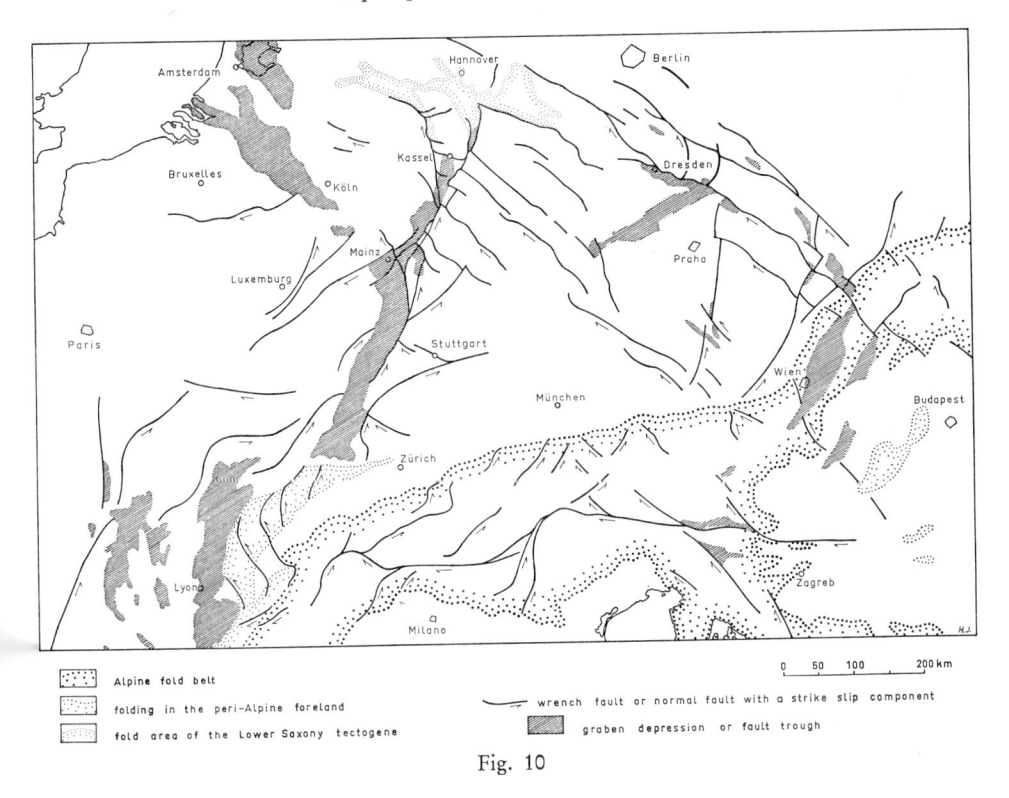

- Alpine fold belt
- folding in the peri-Alpine foreland
- fold area of the Lower Saxony tectogene
- wrench fault or normal fault with a strike slip component
- graben depression or fault trough

Alps. The northern apex of the inner triangular block with its center in the area of Munich coincides with the Vogelsberg neo-volcano, the most extensive eruption center of Central Europe. According to SCHENK (p. 295) the Miocene lava flows erupted along a fissure system induced by the intersection of active lineaments with strike-slip tendencies. The outer triangular block area has its northern corner near Hanover, where the end-Mesozoic folded Lower Saxony tectogene is located. In accordance with BOIGK (1968) the observed fold structures of the Mesozoic cover are the result of displacements along lineaments engraved in a deeper stockwork of the crust.

The formation and dilatation of grabens and fault troughs as well as their spatial arrangement in the Alpine foreland seem to be harmonically incorporated into the framework of sliding block movements. The nappe units of the northern Alps are overthrusted to the north, impelling the triangular crustal block northward which is framed by the Rhinegraben and the Bohemian and Sudetic massifs, by using pre-existent ground-fissures as slide-bars and pioneering the way for grabens and fault troughs. Old shear zones in the basement of the mobile belt controlled the gliding of Alpine nappe units, while their continuations into the foreland operated more in order to open the lateral space required for graben formation.

29 Approaches to Taphrogenesis

450   J. H. Illies

Fig. 11. For better interpretation of the rift-shear system in the Alpine foreland we attempted to find out the correspondant directions of maximal compressive stress of the neotectonic regional stress field. In the caption for Fig. 9 we have already referred to a divergence of the maximum stress directions in relation and normal to the curviform northern rim of the Alpine fold belt. To calculate the anisotropism of the regional stress field, the following indications are available: (1) fault-plane solutions of local earthquakes (AHORNER & SCHNEIDER, p. 104, PAVONI & PETERSCHMITT, p. 322, and AHORNER, MURAWSKI & SCHNEIDER 1972); (2) in-situ stress measurements carried out by GREINER (1973); (3) interpretation of neotectonic strike-slip movements in terms of shear components of the regional stress field (e. g. ILLIES 1962); (4) evaluation of horizontal stylolites in the Mesozoic cover (e. g. PLESSMANN 1972) and (5) horizontal displacements observed by geodetic investigations (BANKWITZ 1971, BOZORGZADEH & KUNTZ, p. 95). In order to obtain the specific maximum stress directions, all available data have been evaluated and the results or, in case of ambiguity of data, the most plausible results have been plotted in the figure.

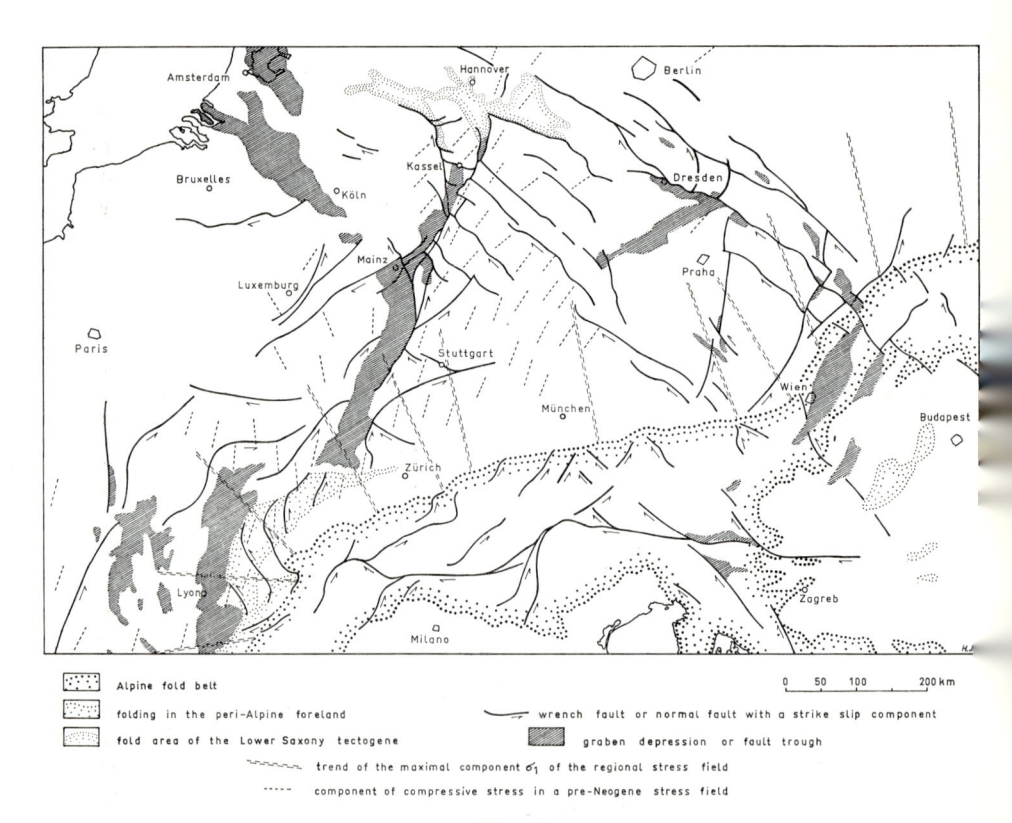

Fig. 11

An arrangement of the maximum compressive stress axes radial to the Alpine fold belt is evident. A similar orientation of the actual stress-field in the Alpine area has been calculated earlier by RITSEMA (1969) by means of seismo-tectonic implications. The question remains open, if the stress-field originated by active fold pressure due to Alpine orogeny (cf. PAVONI 1969), or vice versa, if sea-floor spreading in the northern part of the Mid-Atlantic ridge is responsible for the present stress-field as proposed by other authors. Provided that the behaviour of the enclosed lithospheric plate would be coherently rigid, the question would be unanswerable, especially because the sea-floor spreading at the Reykjanes Ridge and the fold deformation in the Central Alps are of a parallel orientation. Under this scope the tectonic development of the northern Alps could be seen as the result of the downgoing slab of a plate unit impelled southward by sea-floor spreading in the North Atlantic area. But the continental crust in the Alpine foreland behaved more differentiated. It is weakened by systems of ground fissures and shear zones which enable it to reduce its strength by internal displacements. A dense pattern of faults, fractures and joints disintegrated this crustal plate to many small sub-plate units causing the variety of structures and lithostratigraphic provinces in the geological maps of Central Europe. The question, if the crustal stress in the Alpine foreland is actively induced by Alpine orogeny or not, might be answered by means of the increase or decrease of deformations in approach to the Alps.

Till now the following symptoms are available. (1) The subsidence of the Rhinegraben has initiated at its southern end close to the Alps and has moved continuously northward (Fig. 2). (2) The shoulder elevation of the Vosges and Black Forest escarpments culminates at the frames of the southernmost part of the Rhinegraben (Fig. 1) and is explained to be a counterpart to the Alpine subduction zone (GOGUEL 1957). (3) The subcrustal rift "cushion" is considerably more ascended underneath the southern than the northern Rhinegraben area. (4) The geographical distribution of earthquake epicenters and the seismic energy flux indicate a stronger seismo-tectonic activity in the southern part of the Rhinegraben and its flanks than in the northern part (HÄGELE & WOHLENBERG 1970). (5) Generally speaking, the neotectonic (Cenozoic) displacements and deformations in Central Europe decrease with increasing distance to the Alps. (6) The main phases of crustal spreading along the Rhinegraben rift system are contemporaneous with the main phases of crust consumption in the Alpine fold belt. During the early Oligocene and the late Pliocene Rhinegraben taphrogenesis and Alpine orogenesis attained its tectonically most active stages. All these arguments favour the hypothesis that the observed regional stress field in Central Europe is more influenced by Alpine tectonism than by processes along the Mid-Atlantic edge of the concerned plate unit.

Contrary to the present state of stress, the paleo-strain vectors as deduced from fabric indicators in the Mesozoic cover of southern Germany (VOSSMERBÄUMER 1973, WAGNER 1964) reveal a tensile traction normal to the Rhinegraben axis. Tertiary dyke swarms and linear eruptions which appear at various places in southern Germany, the most conspicuous is the Heldburg group, are generally

NNE orientated, presuming a crustal dilatation normal to it (SCHRÖDER 1965). The same WNW—ESE tensile traction can be derived from the fabric analysis of horizontal stylolites in the area of the Hessen depression (BEIERSDORF 1969) as well as in southern France (WUNDERLICH 1973). Therefore, the state of stress during the early stages of the formation of the Rhinegraben rift system had been more adequate to the observed main faults of rifting, i. e. tensile traction normal to the major graben segments. The present stress field as calculated from fault-plane solutions of earthquakes (AHORNER et al. 1972) and from in situ stress measurements (GREINER, p. 120) reveals a direction of minimum stress oblique to the Rhinegraben, but normal to the (younger) graben of the Lower Rhine embayment. The present regional stress field with a compressive component normal to the curved Alpine fold belt had prevailed over the primary stress field with a nearly linear NNE-orientation of compressive stress. If interpreting the neotectonic activity and seismic competence of fault planes in the Alpine foreland, we must keep in mind a rotation of the principal axes of stress since the fault planes had been activated first. The active, sinistral strike-slip movements along the Rhinegraben agree with the present state of stress, but are not in congruence with the earlier stress field indicated by fabric analysis of the Mesozoic cover.

The Mesozoic to early Tertiary stress field with its prevalent NNE-striking axis of compressive stress seems related to the general front of plate collision between Eurasia and Africa in the Mediterranean normal to it (Fig. 13). The late Tertiary to Recent stress field had been modified by the local enhancement of plate collision in the Alpine fold belt which had caused a restricted strengthening of subduction and crust consumption. During the geosynclinal stage of Alpine orogenesis the downgoing slab of the foreland crust had been absorbed and shrunk together by the subsiding and straitening subduction zones. During Oligocene times the uplift of the Central Alps began and the uppermost, preponderant sedimentary fill of the narrowed trenches had been squeezed out, forming the Alpine nappe units. The creeping nappe movements had been impelled by gravity slide, prevalently in northwesterly directions. The rigid blocks of the foreland crust had been pushed forward in the same direction, normal to the front of the advancing nappe units. These movements had interacted the gravity slide of the same crustal blocks sideward away from the uplifting rift cushions. The resultant is the observed direction of lateral creep movements oblique to the Rhinegraben axis which is causing the internal and external tectonics of the active rift segment (Fig. 6, 9).

Fig. 12. When Alpine/peri-Alpine interrelations are under discussion, the Cenozoic volcanism in Central Europe should not be neglected. Obviously, most of the occurrences of Tertiary and Quaternary volcanics are related to the rift segments which cross or accompany the different volcanic provinces. The Chaîne des Puys along the western rim of the Limagne graben, the Kaiserstuhl in the southern Rhinegraben area (BARANYI, p. 226), the Vogelsberg in the Hessen depression (SCHENK, p. 294) the Siebengebirge at the southern end of the Lower Rhine embayment and the volcanic province in the rift zone of northern Bohemia (KOPECKÝ 1966) are the most conspicuous examples. On the other hand, some authors have mentioned a broad, linear band of Neogene volcanic events which crosses the major rift structures in east—westerly direction. It extends from the basaltic necks in Silesia to the Eifel volcanoes over a distance of nearly 750 km. Based on a gradual decrease of some prevalent absolute ages from east to west, DUNCAN et al. (1972) concluded an eastward drift of the European plate with a velocity of about 2.3 cm yr^{-1} over a stationary mantle plume. In opposition to that BURKE et al. (1973) postulated a southward movement of the European plate. They claim to have found evidence by the traces of mantle plumes from the Plomb du Cantal to the Chaîne des Puys (French Central Massif) and from the Kaiserstuhl volcano to the sub-Recent activity of the so-called Eifel hot spot. Many K/Ar age measurements of Cenozoic volcanic rocks, however, furnished by LIPPOLT et al. (p. 213 this book and earlier publications) have not been considered. Taking these results into account which demonstrate a broad scattering of age data for volcanic necks along both sides of the Rhinegraben, the requirements for a mantle plume beneath a north—south drifting European plate are not accomplished.

On the other hand, as discussed in the captions for the Fig. 4 and 9, a westward shifting of rift activity, from the Hessen depression to the Lower Rhine embayment and from the Bresse to the Limagne, is evident. The asymmetrical buckling of the subcrustal rift "cushion" underneath the western flank of the Rhinegraben and the Limagne could likewise be interpreted by a slight drifting of the crust over the mantle. The Central European volcanic belt from Silesia to the Eifel which has been explained by DUNCAN et al. as a continuous trace of a mantle plume, consists of two independent portions. The western part of the trace comprising the neovolcanoes of the Rhoen, the Vogelsberg, the Westerwald, the Siebengebirge, the Neuwied basin and the Eifel reveals a gradual decrease of ages from east to west measured on volcanic rocks which erupted during the climax of activity in the above mentioned areas. Likewise the hot plume of the Swabian volcano reveals a westward extension of the isotherms relative to the Miocene volcanic necks (CARLÉ, p. 208). From these observations we may conclude that a moderate ESE-ward drift of the European plate unit with a maximal drift rate of about 1 cm yr for the last 20 m.y. seems to be reliable. The Oligocene-Neogene dextral strike-slip motion of about 300 km along the Insubric fault zone, a shear element traversing the whole Alpine mountain range in east—westerly direction (see Fig. 10), as required by LAUBSCHER (1972), may be interpreted as transform element related to the ESE-ward plate movement in the same way as the Anatolian dextral strike-slip fault parallel to it (PAVONI 1969). Parallel to the deduced ESE-direction of plate motion a linear orientation of horizontal striae in the Middle Triassic (VOSSMERBÄUMER 1973) is observed, and normal to it a preferred orientation of horizontal stylolites in various stages of the Mesozoic cover (BEIERSDORF 1969, WUNDERLICH 1973), both in wide areas of Central and Western Europe.

The ESE-ward trend of perhaps the whole Eurasian plate may be combined with a slight southward shift of the Western European sub-plate unit west of the Rhinegraben rift system. The supposed large scale plate movements calculated from mantle plumes are of a higher order than the special sub-plate movements derived from the dilatation and strike-slip motion along the graben system. However, for a careful interpretation of the structure and development of the single grabens, we must keep in mind that the position of the crustal segments involved in graben tectonics are not fixed during the grabens' geological history relative to the underlain upper mantle, its hot spots and perhaps its rift cushions included. Nothing remained stable, neither the loci of volcanic activity nor the loci of tectonic displacements impelled by crust—mantle interaction.

Most of the Tertiary lava flows in Central Europe are predominantly derived from an alkali-olivine-basaltic to an olivine-nephelinitic magma which in accordance to experimental petrology studies must have originated from a depth of about 80—100 km. In most of the relevant studies the magmatic activity has been brought in connection with the described rift structures (WIMMENAUER 1972), in some cases with the mantle-derived "cushion" underneath the major graben segments (ILLIES 1972). Another model has been proposed by WEISKIRCHNER (1967), who explained the formation of the magmatic province of the Hegau by a mantle flow escaping the Alpine subduction zone. In the caption to Fig. 11 we have already conceded that mutual interrelations and subcrustal mass equalizations between the Alpine fold belt and its foreland seem likely. In this connection it may be striking that all major centers of volcanic activity appear approximately equidistant from the northern rim of the Alpine fold belt. The distance between the orogenic belt of the Alpides and the Neogene centers of volcanic activity in the western and northern foreland ranges between 200 and 350 km and the larger exposures of volcanics are arranged like a blurred wreath around the Alpine fold belt, but accentuated at the intersections with the rift system. Simultaneously, a general epeirogenic uptrend may be observed for the related areas during Neogene times. Most of the European "mittelgebirge", i. e. elevated blocks of the Hercynian mountain range, have been intensively uplifted during Plio-Pleistocene times, e. g. the Rhenish Schiefergebirge, the Harz, the Sudeten, the Bayrische Wald, the Black Forest, the Vosges, the Central Massif etc. Under this scope the Cenozoic rifting, block faulting and volcanic activity appear to be correlated to kinematic and rheologic interrelations between the Alpine orogenic belt and its unstable foreland.

The magmatic realm of the Cenozoic volcanism of the Rhinegraben rift system is derived from a depth of about 80 to 100 km. However, wide-ranging chemical differences between the specific locations and generations of individual lava flows reveal a magmatic differentiation of initially homogeneous magma masses in shallower magma chambers (WIMMENAUER 1972). A similar trend of fractional crystallization of a primary olivine-basaltic magma is postulated by KING & CHAPMAN (1972) to explain the compositional change within the younger suite of volcanic rocks in the Kenya rift valley. These processes predominantly occurred within the uppermost layer of the upper mantle. The observed mantle swell or "cushion" underneath the major graben segments apparently was predetermined to favour phase transitions and magmatic differentiations within the subcrustal mantle rise (ALTHAUS, p. 341). The sequences of volcanic rocks within the individual volcanic centers lead to the assumption that mantle-derived magma bodies in the rift "cushion" have been subjected to gravitational differentiation by subsidence of heavier phases and rising of lighter phases through the remaining magma. These processes apparently strengthened the buoyancy in the top area of the subcrustal rift "cushion" and additionally impelled the upsurging masses to rise and to penetrate the fissured basement of the graben segments (cf. BARANYI, p. 230). The described corresponding action of mantle diapirism and crustal thinning results in a reduction of the energy required for taphrogenesis.

Fig. 13. We have discussed continental rifting under the aspects of local conditions of the Rhinegraben and of regional conditions of the block faulting in Central Europe as related to the Alpine orogeny. But under a global consideration the Rhinegraben rift system is only an appendix of an intercontinental rift belt which comprises the East African rift valleys as well as the Red Sea—Dead Sea rift systems. The northern extension of the Rhinegraben system does not continue as a reactivated Paleozoic Mediterranean—Mjoesa ground-fissure through Holstein and to southern Norway as already mentioned by STILLE (1923). It continues as the central graben of the Netherlands and farther north as fault troughs in the central North Sea basin (RICHTER-BERNBURG, p. 28). The southern continuation of the rift system in France is indicated by the graben of Alès and further to the south by the El Panadés graben in northern Spain. On account of Neogene plate movements in the Western Mediterranean, especially the anti-clockwise rotation of the Corsica—Sardinia microplate (ALVAREZ 1972), the continuation throughout the Mediterranean appears somewhat blurred. Obvious graben structures have been observed in Sardinia (the Campidano) and in the strait of Sicily (ZARUDZKI 1972) where one of the border faults is exposed at the southern coast of Malta (ILLIES 1970). From the Malta shelf the system continues to the African continent and extends to the Hon graben in Libya. South of the Hon oasis the graben contours go astray. A line of Neogene basaltic flows, however, extends southward to the Tibesti massif (ILLIES 1970). From there, with a 1700 km shift to the east, the Red Sea rift system, with its northern annexes of the Suez graben and the Aqaba-Dead Sea rift, continues. The Red Sea graben terminates at the Afar triple rift junction where two other rift branches join, the Gulf of Aden ridge/rift system and the Ethiopian rift valley. The Sheba ridge of the Aden Gulf continues to the Carlsberg ridge/rift system. The Ethiopian rift continues to the Gregory rift valley in Kenya which splits up and goes astray in Northern Tanzania. However, this is not the southern end of the rift belt. North of the Lake Albert, about 500 km west of the Gregory rift, begins the western branch of the East African rift system. This one is even more spectacular because of the formation of a chain of long and partly abyssal graben lakes. Its main branch meets the coast of Mozambique and terminates near the town of Beira.

Without regard to the many local and regional details concerning the structure and development of the individual graben segments, its mutual interrelations in its superordinated constructional plan, its common symptoms of crust and mantle structures, its specific magmatic composition of the accompanying volcanism, its

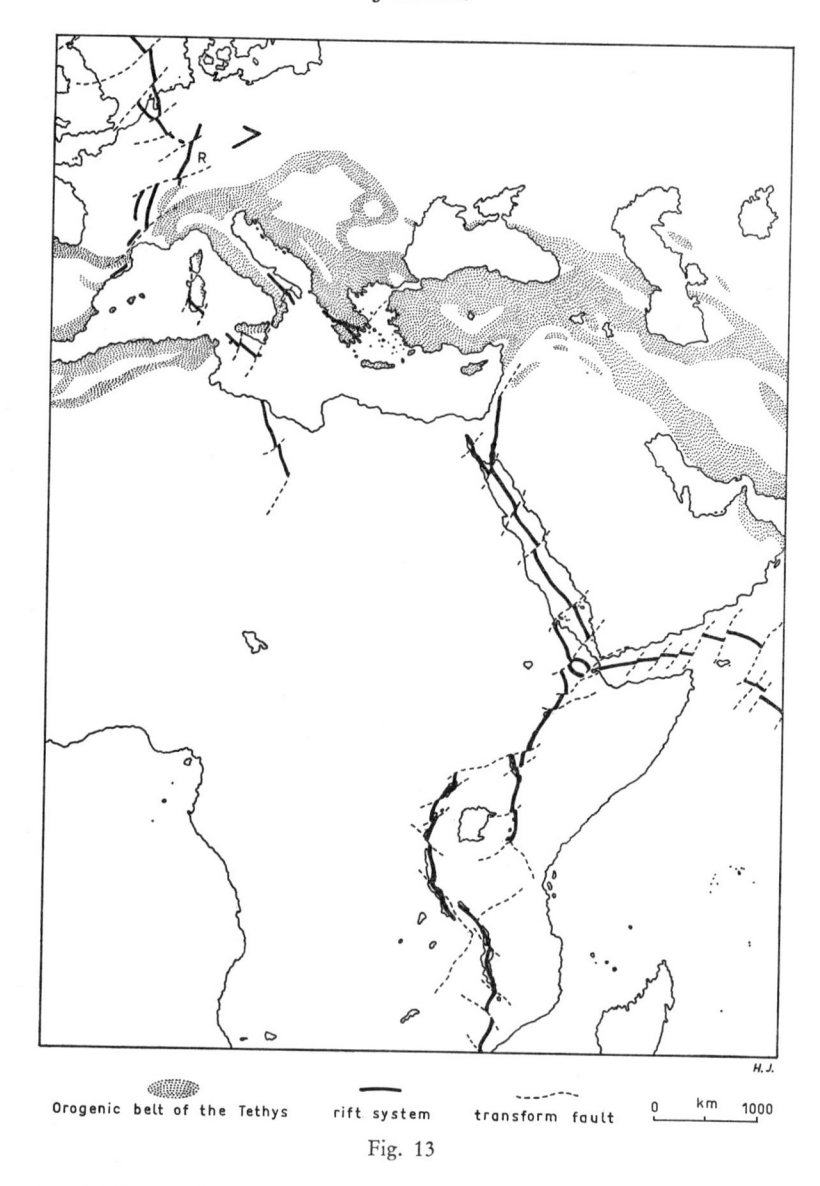

Orogenic belt of the Tethys   rift system   transform fault   0 — 1000 km

Fig. 13

anomalous high heat flow and its hot brines, its characteristic seismic activity and its consonant morphological features are typical and evident for taphrogenesis. The prevalent strike of tectonic elements remains the same, and the episodes of an accelerated subsidence and dip-slip movements indicate a phasic consonance or contemporaneity over wide distances of the rift belt. And, since especially indicated in the figure, the transverse features of lateral offset, i. e. the continental transform faults, show a directional arrangement similar to the pattern of oceanic transform faulting, even if most of them are somewhat deflected by the pre-existent basement framework (McConnell, p. 48, Illies 1972). All these observations manifest a global control of the intercontinental rift belt concerning its tectonic regime as well as its regional setting.

The intercontinental rift belt displays an intersection and interference with the 1500 km wide girdle of fold belts and subduction zones originating from the Tethys. Within this broad Tethys zone the front of specific plate collision between the Eurasian macro-plate in the north and the African and the Arabian plates in the south is split into a bundle of special subduction and fold segments which had advanced or retreated in interdependence on the fluctuating history of the individual tectogenes.

For the Rhinegraben and its neighbouring segments we have listed a series of mutual interrelations between its formation and development and, on the other hand, the formation and development of the orogenic belt of the Alps. The enumeration of such interrelations must be supplemented by the reference to coincidences of the phases of the Rhinegraben taphrogenesis with the phases of the Alpine orogenesis. Synchronously in early-Oligocene times when about 39 Ma ago Alpine folding, overthrusting, and nappe movements was in its tectonically most active stage (Trümpy 1972), the Rhinegraben taphrogeosyncline attained the first climax of its subsidence activity. Again when the Pliocene fold chains of the Swiss Jura mountains about 5 Ma ago invaded, like surfing waves, the southern mouth of the Rhinegraben, and Alpine folding and nappe tectonics had reached the subalpine part of the molasse trough, Rhinegraben subsidence and the elevation of its shoulders were reactivated to its second taphrogenic cycle (Illies 1972). In the same way, Kodyn (1960) has shown that the Cretaceous and Tertiary block tectonics in the Bohemian Massif took place in several phases that corresponded to the periods of increased orogenic activity in the Carpathians and in the Alps.

An interplay of forces and masses between a subduction zone and a rift zone is considered to control the tectonic evolution of the grabens south of the orogenic belt of the Tethys. The opening of the Red Sea—Gulf of Aden rift system seems strongly related to the folding of the Zagros range along the northern front of the Arabian plate. The formation of quasi oceanic crust in the Afar depression and the rising basaltic swell beneath the Gregory rift are compensated by the subduction of lithosphere induced by plate collisions in the Tethys. Continental rifting and alpinotype folding have acted antagonisticly, both controlled by the same interactional movements between the asthenosphere and the rigid shell above it, the lithosphere.

There is a remarkable coincidence of contemporaneous orogenic and taphrogenic activities affecting the Alpine fold belt as well as the graben system normal to it (Milanovsky 1972), whereas the inflluence of sea-floor spreading activity in the Mid-Atlantic and Carlsberg ridges is not of similar effect to the tectonic development of the continental graben and fold belts. The continental belts of the European—East African grabens, perhaps the Baikal graben system too, and the Alpine-Himalaya mountain ranges had acted as a reciprocating system compared with the ocean-floor system of the mid-oceanic rift/ridge belts and the circum-Pacific trench-subduction zones.

I am grateful to Dr. K. SCHÄFER for many stimulating discussions, criticism, and for reading and suggesting improvements in the manuscript. L. H. GRAY kindly revised the English text.

References

AHORNER, L. et al., 1972: Seismotektonische Traverse von der Nordsee bis zum Apennin. — Geol. Rdsch., **61**, 915—942.
ALVAREZ, W., 1972: Rotation of the Corsica — Sardinia Microplate. — Nature Phys. Sci., **235**, Febr. 7, 103—105.
ANDERLE, H.-J., 1970: Outlines of the Structural Development at the Northern End of the Upper Rhine Graben. — In: ILLIES, J. H. & MUELLER, ST. (eds.), Graben Problems, 97—102, Schweizerbart, Stuttgart.
ANSORGE, J. et al., 1972: Refraktionsseismische und reflexionsseismische Untersuchungen im Gebiet des Oberrheingrabens. — Paper presented 5th colloqu. Rhinegraben Research Group, 5 p., Karlsruhe.
BANKWITZ, P., 1971: Geologische Auswertung von geodätisch ermittelten rezenten Krustenbewegungen im Gebiet der DDR. — Petermanns Geogr. Mitt., **115**, 130—140.
BEIERSDORF, H., 1969: Druckspannungsindizien in Karbonatgesteinen Süd-Niedersachsens, Ost-Westfalens und Nord-Hessens. — Geol. Mitt., **8**, 217—262.
BOIGK, H., 1968: Gedanken zur Entwicklung des Niedersächsischen Tektogens. — Geol. Jb., **85**, 861—900.
BURKE, K. et al., 1973: Plumes and Concentric Plume Traces of the Eurasian Plate. — Nature Phys. Sci., **241**, Febr. 12, 128—129.
COGNÉ, J. et al., 1966: Les «Rifts» et les Failles de Décrochement en France. — Rev. Géogr. Phys. Géol. Dynam. (2), **8**, 123—131.
DUNCAN, R. A. et al., 1972: Mantle Plumes, Movement of the European Plate, and Polar Wandering. — Nature, **239**, Sept. 8, 82—86.
GOGUEL, J., 1957: Gravimétrie et Fossé Rhénan. — Verh. Koninkl. Nederl. Geol.-Mijnbouwk. Genootsch., Geol. Ser., **18**, 125—147.
GOSPODARIČ, R., 1969: Probleme der Bruchtektonik der NW-Dinariden. — Geol. Rdsch., **59**, 308—322.
GREINER, G., 1973: In-situ-Spannungsmessungen auf der Schwäbischen Alb. — In: MÜLLER, L. (ed.), Sonderforschungsbereich 77, Felsmechanik, 131—142, Karlsruhe.
HÄGELE, U. & WOHLENBERG, J., 1970: Recent Investigations on the Seismicity of the Rheingraben Rift System. — In: ILLIES, J. H. & MUELLER, ST. (eds.), Graben Problems, 167—170, Schweizerbart, Stuttgart.
ILLIES, J. H., 1962: Oberrheinisches Grundgebirge und Rheingraben. — Geol. Rdsch., **52**, 317—332.
— 1970: Graben Tectonics as Related to Crust-Mantle Interaction. — In: ILLIES, J. H. & MUELLER, ST. (eds.), Graben Problems, 4—27, Schweizerbart, Stuttgart.
— 1972: The Rhine Graben Rift System — Plate Tectonics and Transform Faulting. — Geophys. Surv., **1**, 27—60.
KING, B. C. & CHAPMAN, G. R., 1972: Volcanism of the Kenya rift valley. — Phil. Trans. R. Soc. Lond. A., **271**, 185—208.
KODYM, O. sen., 1960: Platform stage of the development of the Czech massif. — In: BUDAY, T. et al., Tectonic development of Czechoslovakia, 122—139, Praha.
KOPECKÝ, L., 1966: Tertiary volcanics. — In: SVOBODA, J. (ed.), Regional Geology of Czechoslovakia, pt. 1, 554—581, Prague.

Kubella, K., 1951: Zum tektonischen Werdegang des südlichen Taunus. — Abh. hess. Landesamt Bodenforsch., **3**, 1—81.

Landsberg, H., 1931: Der Erdbebenschwarm von Groß-Gerau 1869—1871. — Gerlands Beitr. Geophys., **34**, 367—392.

Laubscher, H. P., 1971: Das Alpen-Dinariden-Problem und die Palinspastik der südlichen Tethys. — Geol. Rdsch., **60**, 813—833.

Lefavrais-Raymond, A., 1962: Contribution à l'Étude Géologique de la Bresse d'après les Sondages Profonds. — Mém. Bur. Rech. Géol. Min., **16**, 1—170.

Mäussnest, O., 1969: Die Ergebnisse der magnetischen Bearbeitung des Schwäbischen Vulkans. — Jber. Mitt. oberrh. geol. Ver., **51**, 159—167.

Mégnien, C. et al., 1971: Structure tectonique des terrains sédimentaires au Nord-Ouest du Morvan. — Bull. B.R.G.M., (2), **1**, (3), 163—170.

Milanovsky, E. E., 1972: Contributions to the Problem of Spatial Interactions between Geosynclinal-Orogenic Belts and Rift Systems. — Vestn. Moscowsk. Univ., Ser. Geol., **4**, 3—18 (in Russian).

Paul, W., 1955: Zur Morphogenese des Schwarzwaldes. — Jh. geol. Landesamt Baden-Württ., **1**, 395—427.

Pavoni, N., 1969: Zonen lateraler horizontaler Verschiebung in der Erdkruste und daraus ableitbare Aussagen zur globalen Tektonik. — Geol. Rdsch., **59**, 56—77.

Plessmann, W., 1972: Horizontal-Stylolithen im französisch-schweizerischen Tafel- und Faltenjura und ihre Einpassung in den regionalen Rahmen. — Geol. Rdsch., **61**, 332—347.

Ritsema, A. R., 1969: Seismo-tectonic implications of a review of European Earthquake Mechanism. — Geol. Rdsch., **59**, 36—56.

Schmidt-Thomé, P., 1972: Tektonik. — In: Brinkmann, R. (ed.), Lehrbuch der Allgemeinen Geologie, t. 2, Enke, Stuttgart.

Schneider, G., 1971: Seismizität und Seismotektonik der Schwäbischen Alb. — 79 S., Enke, Stuttgart.

Schröder, B., 1965: Tektonik und Vulkanismus im oberpfälzer Bruchschollenland und im fränkischen Grabfeld. — Erlanger geol. Abh., **60**, 1—90.

Seibold, E., 1951: Das Schwäbische Lineament zwischen Fildergraben und Ries. — N. Jb. Geol. Paläont. Abh., **93**, 285—324.

Sonne, V., 1970: Das nördliche Mainzer Becken im Alttertiär. Betrachtungen zur Paläoorographie, Paläogeographie und Tektonik. — Oberrhein. geol. Abh., **19**, 1—28.

Straub, E. W., 1962: Die Erdöl- und Erdgaslagerstätten in Hessen und Rheinhessen. — Abh. geol. Landesamt Baden-Württ., **4**, 123—136.

Tollmann, A., 1969: Die Bruchtektonik in den Ostalpen. — Geol. Rdsch., **59**, 278—288.

Trümpy, R., 1972: Über die Geschwindigkeit der Krustenverkürzung in den Zentralalpen. — Geol. Rdsch., **61**, 961—964.

Vitard, M.-J., 1969: Études géophysiques en Bresse. — Bull. Soc. géol. France, (7), **11**, 330—337.

Voigt, E., 1954: Das Norddeutsch-Baltische Flachland im Rahmen des europäischen Schollenmosaiks. — Mitt. Geol. Staatsinst. Hamburg, **23**, 18—37.

Vossmerbäumer, H., 1973: Die „Lösungsrippeln" (Schmitt 1935) im Wellenkalk Frankens. — N. Jb. Geol. Paläont. Abh., **142**, 351—375.

Wagner, H. G., 1964: Kleintektonische Untersuchungen im Gebiet des Nördlinger Rieses. — Geol. Jb., **81**, 519—600.

Wagner, W., 1926: Goethe und der geologische Aufbau des Rochusbergs bei Bingen. — Notizbl. Ver. Erdkunde u. Hess. Geol. Landesanst., **8**, 224—231.

Weiskirchner, W., 1967: Bemerkungen zum Hegau-Vulkanismus. — Abh. geol. Landesamt Baden-Württ., **6**, 139—141.

Wiedemann, H. U., 1968: Eine neue Schichtlagerungs-Karte für die östliche Hälfte des „Schwäbischen Lineaments". — N. Jb. Geol. Paläont. Abh., **130**, 106—111.

WIMMENAUER, W., 1972: Gesteinsassoziationen des jungen Magmatismus in Mitteleuropa. — Tschermaks Min. Petr. Mitt., **18**, 56—63.
WUNDERLICH, H. G., 1973: Grundzüge der alpidischen Geodynamik Westeuropas. — N. Jb. Geol. Paläont. Mh., **1973**, 2, 99—112.
ZARUDZKI, E. F. K., 1972: The Strait of Sicily — a Geophysical Study. — Rev. Géogr. Phys. Géol. Dynam., (2), **14**, 11—28.